The study of the origin and evolution of cosmic elements attracts researchers from astrophysics, nuclear physics, meteoritics and many other disciplines. Articles from international specialists in all these fields are drawn together in this book, following a conference held in Paris in 1992.

Stellar and primordial nucleosynthesis, cosmic ray spallation and other processes leading to the formation of cosmic elements are discussed. Recent observational data are presented concerning cosmic abundances together with nuclear data and methods that evaluate the probability of the nuclear reactions involved. The evolution of stars and galaxies and the origin of our solar system and the universe are reviewed in the light of these developments. Articles summarise the current knowledge and address more topical issues making this text suitable for both graduate students and researchers.

Origin and Evolution of the Elements

Origin and Evolution of the Elements

Edited by

N. PRANTZOS and E. VANGIONI-FLAM
Institut d'Astrophysique de Paris

M. CASSÉ
Service d'Astrophysique, Centre d'Études de Saclay

CAMBRIDGE
UNIVERSITY PRESS

Published by the Press Syndicate of the University of Cambridge
The Pitt Building, Trumpington Street, Cambridge CB2 1RP
40 West 20th Street, New York, NY 10011-4211, USA
10 Stamford Road, Oakleigh, Melbourne 3166, Australia

© Cambridge University Press 1993

First published 1993

Printed in Great Britain at the University Press, Cambridge

British Library cataloguing in publication data available

Library of Congress cataloguing in publication data available

ISBN 0 521 434289 hardbaack

CONTENTS

GALACTIC EVOLUTION AND COSMOCHRONOLOGY

REMERCIEMENTS

Nous remercions chaleureusement
pour leur aide financière
les organismes suivants:

We sincerely thank
for their financial support
the following organisations:

C. E. A.
C. N. E. S.
C. N. R. S.
(pour le PICS N° 114: *Origine des Eléments Légers*)
E. S. A.
G. M. F. - F. N. A. C.
I. A. P.
Ministère des Affaires Etrangères
Observatoire de Paris

et pour leur participation
les organismes suivants:

and for their participation
the following organisations:

La Delégation Générale du Québec
Palais de Tokyo
FEMIS
Société Européenne de Combustion
THOMSON

PREFACE

Origin - Evolution - Elements: Few words can be found with a more profound meaning to an astrophysicist or a scientist. These three words were put together for the first time in the thirties, when it was realised that: i) most objects in the Universe have a more or less similar chemical composition (dominated by hydrogen and helium), a fact clearly pointing to a common *Origin* for the constituent elements and ii) the energy sources of the Sun and stars, thermonuclear reactions, transform light elements into heavier ones.

The role of stars, however, was obscured in the fourties by the possibility, advocated by G. Gamow, of a *primordial origin* for *all the elements*, a possibility leaving a marginal role for *Evolution*; but it was soon realised that the instability "gap" at nuclear mass numbers 5 and 8 was fatal to this idea. Thus the stellar origin of the elements, advocated by F. Hoyle, has increasingly won agreement in the early fifties, when E. Salpeter and F. Hoyle showed how the triple-alpha reaction in red giants could "bridge" the gap, and J. Morgan observed some very metal poor stars in the galactic halo.

In the mid-50ies the stage was already set (thanks to the classical work of M. and G. Burbidge, W. Fowler, F. Hoyle and A. Cameron), when Hubert Reeves entered the scene of Nuclear Astrophysics, starting with a Ph. D. Thesis in Cornell. His many contributions to the field in the past thirty five years (important nuclear reactions in advanced stellar evolution; implications of light element abundances for primordial nucleosynthesis and cosmology; production of the fragile lithium, beryllium and boron by cosmic rays; and many others) have had a profound impact on our understanding of stars and the Universe. This impact is well illustrated in many of the articles appearing in this volume, especially the articles by his former thesis advisor E. Salpeter, his former student J. Audouze, and H. Reeves himself.

Organizing a Symposium to celebrate H. Reeves' 60th birthday is then a fair tribute to his work; and it was natural to have it in Paris, in view of his contribution to the development of astrophysics and general scientific culture in France, his adoptive country. Still, celebration was not the only motivation for the Symposium. Indeed, several developments in recent years gave a new impetus to the field of Nuclear Astrophysics. Let us mention only a few of them:
- the explosion of SN1987A in the Large Magellanic Cloud, confirming early ideas on explosive nucleosynthesis (through the detection of the γ-ray lines from ^{56}Co decay) and on the core collapse of massive stars (through the epochal detection of the supernova neutrinos);
- the development of radioactive ion beam facilities (making possible for the first time the serious study of reactions of unstable nuclei), that will allow a better understanding of explosive nucleosynthesis;
- the detection of solar neutrinos by the GALLEX and SAGE collaborations, sheding new

light on an old problem, although no definite conclusions can be drawn yet;
- the development of new (ground and space) facilities in various spectral regions, with better and better signal/noise ratios that made possible the detection of tiny effects: e.g. the recent observations of Be (with the Anglo-Australian telescope) and B (with the Hubble Space Telescope) in very metal poor stars, or the detection of the 122 keV line from the ^{57}Co decay in SN1987A (with the Gamma-Ray Observatory).

These are only some of the recent observational and experimental developments in Nuclear Astropysics, that opened new windows on our understanding of the Origin and Evolution of the Elements. The celebration of H. Reeves' 60th birthday gave an excellent opportunity to bring together most of the world's specialists in order to discuss those new and exciting results. More than 120 people (and among them many of H. Reeves' collaborators, students and friends) have participated in the Symposium, that was held at the Institut d'Astrophysique de Paris (IAP), from June 22 to 25, 1992.

The succes of the meeting, which took place in an informal but stimulating atmosphere, is due to the contribution of many individuals and institutions. We wish to express our gratitude to: Dr. A. Omont, Director of the IAP, for his hospitality; Mrs. B. Raban, for her efficient work in the Secretariat of the Symposium; and Mrs. M.C. Pantalacci, M.C. Pelletan, and A. Placenti for their kind collaboration.

We are deeply indebted to Mrs. Yvette Oberto for her careful and devoted work during all the preparation of the Symposium and the editing of this book; it would have been difficult to accomplish all this without her invaluable contribution.

Dr. M. Combes, President of the Paris Observatory, is warmly aknowledged for his kind welcome in the majestic Salle Cassini, for the birthday cocktail. As a passionate lover of music, H. Reeves appreciated particularly the concert organized on this occasion in the Palais de Tokyo, thanks to the generous contribution of FEMIS, GMF and FNAC. We warmly thank Mr. J.C. Carrière and J. Gajos, President and Director, respectively, of FEMIS (Fondation Européenne des Métiers de l'Image et du Son); Mr. J.L. Pétriat and B. Allien, respectively President and Vice-President of GMF (Garantie Mutuelle des Fonctionnaires); and G. Huber, Communication Director in GMF-FNAC. Finally, we gratefully aknowledge the generous financial support of the organisations appearing in the list (next page), and the participation of Mr. J.M. Tulli of the Société Européenne de Combustion.

N. Prantzos
E. Vangioni-Flam
M. Cassé

Paris, October 1992

LIST OF PARTICIPANTS

ABIA Carlos	Universidad de Granada, Spain
AGUER Pierre	CSNSM, Orsay, France
ARNOULD Marcel	Inst. d'Astron. et d'Astrophys., Bruxelles, Belgium
AUDI Georges	CSNSM, Orsay, France
AUDOUZE Jean	Institut d'Astrophysique de Paris, France
BANIA Thomas	Boston University, Boston, USA
BARAFFE Isabelle	MPI für Astrophysik, Garching, Germany
BEAUDET Gilles	Université de Montréal, Canada
BECKER Stephen	Los Alamos National Laboratory, USA
BEGEMANN Friedrich	MPI für Chemie, Mainz, Germany
BERNAS Monique	IPN, Orsay, France
BOFFIN Henri	Inst. d'Astron. et d'Astrophys., Bruxelles, Belgium
BRAVO Eduardo	Universidad de Barcelona, Spain
BUSSO Maurizio	Osserv. Astron. di Torino, Pino Torinese, Italy
CALOI Vittoria	Istituto di Astrof. Spaziale, Frascati, Italy
CASSÉ Michel	CEA-CE Saclay, Gif-sur-Yvette, France
CAYREL Giusa	Observatoire de Meudon , France
CAYREL Roger	Observatoire de Paris, France
CESARSKY Catherine	CEA-CE Saclay, Gif-sur-Yvette, France
CHARBONNEL Corinne	Obs. Midi-Pyrénées, Toulouse, France
COC Alain	CSNSM Orsay, France
CUNHA Katia	University of Texas, Austin, USA
CZAJKOWSKI Serge	IPN, Orsay, France
DE LA REZA Ramiro	Obs. Nacional, Rio de Janeiro, Brazil
DELBOURGO-SALVADOR Pascale	CEA, Bruyères-le-Chatel, France
DUROUCHOUX Philippe	CEA-CE Saclay, Gif-sur-Yvette, France
DZITKO Hervé	CEA-CE Saclay, Gif-sur-Yvette, France
EL EID Mounib	Universitäts-Sternwarte Göttingen, Germany
EMERICH Claude	IAS, Verrières-le-Buisson, France
EPHERRE Marcelle	CERN, Geneva, Switzerland
FLIEGNER Jens	Universitäts-Sternwarte Göttingen, Germany
FORESTINI Manuel	Inst. d'Astron. et d'Astrophys., Bruxelles, Belgium
FRANCOIS Patrick	Observatoire de Paris, France
GAIGE Yves	Obs. Midi-Pyrénées, Toulouse, France
GALLINO Roberto	Istituto di Fisica Generale, Torino, Italy
GEISS Johannes	Physik. Inst., Univ. Bern, Switzerland
GILMORE Gerry	Institute of Astronomy, Cambridge, UK
GREVESSE Nicolas	Inst. d'Astrophys., Univ. de Liège, Belgium
GUILLEMAUD-MUELLER Dominique	IPN, Orsay, France
HAMMER Wolfgang	Inst. für Strahlenphysik, Univ. Stuttgart, Germany
HATTORI Makoto	Inst. of Phys. and Chem. Res., Saitama-Ken, Japan
HAYWOOD Misha	Observatoire de Besançon, France
HERZIG Klaus	Universitäts-Sternwarte Göttingen, Germany
HOWARD Michael	Inst. d'Astron. et d'Astrophys., Bruxelles, Belgium
JORISSEN Alain	European Southern Obs., Garching, Germany
KAJINO Taka	Tokyo Metropolitan University, Japan
KIENER Jurgen	CSNSM, Orsay, France
KLAPISCH Robert	CERN, Geneva, Switzerland
KRATZ Karl-Ludwig	Inst. für Kernchemie, Mainz, Germany
LAMBERT David	European Southern Obs., Garching, Germany

LATTANZIO John	Monash University, Clayton, Australia
LATTIMER James	Dept. of Earth and Space Sci., Stony Brook, USA
LEFEBVRE Anne	CSNSM, Orsay, France
LEMOINE Martin	Institut d'Astrophysique de Paris, France
LEQUEUX James	Observatoire de Meudon , France
MAGAIN Pierre	Inst. d'Astrophys., Univ. de Liège, Belgium
MARTIN Eduardo	Inst. de Astrophys. de Canarias, Tenerife, Spain
MASNOU Jean-Louis	Observatoire de Meudon, France
MENEGUZZI Maurice	CERFACS, Toulouse, France
MEYER Bradley	Clemson University, Clemson, USA
MEYER Jean-Paul	CEA-CE Saclay, Gif-sur-Yvette, France
MICHAUD Georges	Université de Montréal, Canada
MOLARO Paolo	Osserv. Astron. di Trieste, Trieste, Italy
MONTMERLE Thierry	CEA-CE Saclay, Gif-sur-Yvette, France
MUELLER Alex	IPN, Orsay, France
NEUFORGE Corinne	Inst. d'Astrophys., Univ. de Liège, Belgium
OTT Ulrich	MPI für Chemie, Mainz, Germany
PAGEL Bernard	Nordita, Copenhagen, Denmark
PEARSON Mike	Université de Montréal, Canada
PELLAS Paul	Museum d'Histoire Naturelle, Paris, France
PICHON Bernard	Observatoire de Meudon, France
PRANTZOS Nikos	Institut d'Astrophysique de Paris, France
RAISBECK Grant	CNSM, Orsay, France
RAITERI Claudia	Osserv. Astron. di Torino, Pino Torinese, Italy
RAMADURAI Souriraja	Tata Inst. of Fundamental Res., Bombay, India
REBOLO Raphael	Inst. de Astrophys. de Canarias, Tenerife, Spain
REEVES Hubert	CEA-CE Saclay, Gif-sur-Yvette, France
ROBIN Annie	Observatoire de Besançon, France
ROLFS Claus	Ruhr-Universität Bochum, Germany
RUIZ-LAPUENTE Pilar	Universidad de Barcelona, Spain
SALPETER Edwin	Cornell University, Ithaca, USA
SATO Katsuhiko	Faculty of Science, Univ. of Tokyo, Japan
SCHATZMAN Evry	Observatoire de Meudon, France
SCHRAMM David	University of Chicago , USA
SMITH Verne	University of Texas, Austin, USA
SORLIN Olivier	IPN, Orsay, France
SPITE Francois	Observatoire de Meudon, France
SPITE Monique	Observatoire de Meudon, France
STARRFIELD Sumner	Arizona State University, Tempe, USA
TERASAWA Nobuo	Inst. of Phys. and Chem. Res., Saitama-Ken, Japan
THIBAULT Catherine	CSNSM, Orsay, France
THIELEMANN Friedrich-Karl	Harvard University, Cambridge, USA
VANGIONI-FLAM Elisabeth	Institut d'Astrophysique de Paris, France
VAUCLAIR Sylvie	Obs. Midi-Pyrénées, Toulouse, France
VIDAL-MADJAR Alfred	Institut d'Astrophysique de Paris, France
VIGROUX Laurent	CEA-CE Saclay, Gif-sur-Yvette, France
WALTERS William	University of Maryland, College Park, USA
WASSERBURG Gerry	California Institute of Technology, Pasadena, USA
WISSHAK Klaus	Inst. für Kernphysik, Karlsruhe, Germany
YIOU Francoise	CSNSM, Orsay, France
ZANDA Brigitte	Museum d'Histoire Naturelle, Paris, France
ZHAO Gang	Inst. d'Astrophys., Univ. de Liège, Belgium

News from the Past?

J. AUDOUZE

Institut d'Astrophysique de Paris (France)

ABSTRACT

This introductory chapter is an attempt to highlight the important insights of Hubert Reeves in nuclear and particle astrophysics. His inspiring discoveries concern the origins : origin of the Universe, with his contributions to primordial nucleosynthesis – origin of cosmic rays and of the light elements lithium, beryllium and boron – origin of the solar system. The purpose of his research has always been to bring new ideas on the past. The basic argumentation is outlined here. It is also emphasized that most of his proposals still enlighten cosmology, nucleosynthesis and theoretical astrophysics in general.

1 INTRODUCTION

Cher Hubert,

Ayant été ton premier étudiant de recherche, je me sens à la fois flatté et heureux d'introduire ce colloque de recherche qui rassemble les meilleurs spécialistes du monde entier sur l'origine et l'évolution des éléments chimiques, et te souhaiter au nom de nous tous tes amis, un très joyeux anniversaire.

There is indeed much joy to be assembled around Hubert Reeves in this conference at the Paris Institute of Astrophysics, which I was most fortunate to head from 1978 to 1989, at a time when cosmology and particle astrophysics were introduced here.

The task of astrophysics is first to observe the Universe and its components and to use the laws of physics in an attempt to understand the origin and the evolution of the world as a whole and of all its components. I would like to show the relations which exist between i) the Universe and its structures, e.g. the astrophysical sites such as the galaxies and their clusters, the stars and their clusters, the planets and their systems..; ii) the physical laws like quantum physics (governing particle and nuclear physics), relativity and gravity, electromagnetism, thermodynamics..; and

iii) the evolution of the Universe.

Astronomers observe the skies and its many structures, while physicists show that the whole Universe is understandable and can be described. There is a class of scientists well represented in this conference (and Hubert Reeves is a prominent example) who are able to relate observation and theories and construct plausible models (or senarios) regarding the origin and the evolution of such structures and more generally of the whole content of the Universe.

By making a quick overview of the scientific achievements of Hubert, I would like to point out not only the success but also his "recipes" : a subtle mixture of american pragmatical scientific training with a recognized talent of lecturing and a european inclination for philosophy is at the origin of his contributions, which are all characterized by a constant inclination towards unification and simplicity. Among his work, performed during the golden age of nuclear and particle astrophysics, let me outline three most important results:

a) the recognition of the cosmological impact of the (D/H) ratio in the solar system (Geiss and Reeves 1972).

b) the demonstration that lithium, berryllium and boron (LiBeB) are synthetized by the interaction between the galactic cosmic rays (GCR) and the interstellar medium, and the relation between this nucleosynthesis and the GCR origin (Reeves et al. 1970, Meneguzzi et al. 1971).

c) the coincidence between the period of galactic rotation (10^8 years) and the time elapsed between the nucleosynthesis of short lived radioactive isotopes like ^{129}I or ^{244}Pu, and the isolation of the solar system material (Reeves and Johns 1976) followed by the "Bing Bang theory of solar system origin" (Reeves 1978a,b) according to which the solar system is born in a region affected by a set of almost simultaneous supernova explosions.

2 THERMONUCLEAR REACTIONS AND STELLAR EVOLUTION

The first domain in nuclear astrophysics studied by H. Reeves is the relation between thermonuclear reactions nucleosynthesis and stellar evolution which has been set up by Burbidge et al. (1957) and Cameron (1957), and has inspired the work of many astrophysicists during the 1955–1965 period, (F. Hoyle, W.A. Fowler, E.M. and G.R. Burbidge, A.G.W. Cameron, J.W. Truran, W.D. Arnett, D.D. Clayton and others).

This kind of work has been his main research activity during the period lasting from his thesis work (performed under the guidance of E. Salpeter and devoted to the nucleosynthetic phase of carbon burning; Reeves and Salpeter 1959) up to around 1965. Let me mention some of his contributions during that period, dealing with the triple-α reaction (Paquette and Reeves 1964), the computation of the reaction rates involving C, O and Ne (Reeves 1962) or the application of the optical model of the nuclear potential to the computation of the C+C and O+O reactions (Vogt, Michaud and Reeves 1965). Among the works published during this early (pre–french) period, I would like to point out two papers :

i) A comprehensive compendium of Gamow S factors which are the critical parameters for the computation of the thermonuclear reaction cross section σ

$$\sigma = \frac{S}{E} \, exp \left(- Z_1 Z_2 \, \frac{2Me^2}{hv} \right)$$

This paper entitled *Thermonuclear reaction rates* (Reeves 1966a) has been followed up by the series of papers by Fowler, Caughlan and Zimmermann (Caltech contributions). His tables of thermonuclear cross sections have been used in two subsequent works : the first dealing with the calculation of solar neutrino fluxes (Pochoda and Reeves 1964) and the second estimating the energy generation in stellar interiors taking into account not only the radiative losses but also the neutrino fluxes emitted by the stars.

ii) A pertinent analysis of the stellar neutron sources (Reeves 1966b). This paper discusses the relative importance of the $^{13}C(\alpha, n)^{16}O$ and the $^{22}Ne(\alpha, n)^{25}Mg$ reactions in the stellar neutron production necessary for the s and the r nucleosynthetic processes. The termination of the r process, i.e. the unlikely possibility of forming superheavy nuclei has been established later by Johns and Reeves (1973).

3 SPALLATION REACTIONS AND THE SAGA OF LiBeB

When H. Reeves arrived in France in 1965 he had the luck to collaborate closely with the "Orsay group" led at that time by René Bernas and including in particular E. Gradztajn, M. Epherre, F. Yiou, R. Klapish, etc. That group had a unique expertise in measuring the spallation cross sections, especially those concerning C, N and O i.e. the complex nuclei which are the most abundant in nature. Reeves, associated with this group and with E. Schatzman produced an important paper (Bernas et al. 1967, BGRS) providing a systematics for these reactions plus some ideas regarding the occurence of these high energy reactions in various sites and especially in the external zones of stars (the flares at the surface of the sun or the T Tauri stars) in connection with the still unsolved problem concerning the synthesis of the light elements lithium, beryllium and boron (LiBeB). It is indeed in this paper that the isospin rule was proposed to derive a rough estimate of the spallation cross sections σ. BGRS realized that these average cross sections are divided by a factor ~ 10 when the isospin (symetry) factor $T \simeq \dfrac{|N-Z|}{2}$ (where N and Z are the neutron and the proton numbers of the product) increases by 1. This simple rule has been improved in Audouze et al. (1967) who proposed

$$\sigma = \sigma_g \, e^{-P\Delta A} \, e^{-RT^2}$$

where σ_g is the geometrical cross section, ΔA the difference between the atomic number of the parent nucleus and the product. This formula, among other advantages, can be derived rather simply from the two–step model proposed by Serber (1947) to describe the spallation processes.

Regarding the synthesis of LiBeB, the theory was influenced at that time by B^{2}FH and Fowler et al. (1962) who believed that these light elements are produced by spallation reactions occuring at the surface of stars at various stages of their evolution but *mainly* during their early (T Tauri) phase. Ryter et al. (1970) showed at that time that arguments based on energetics were preventing a substantial LiBeB production by spallation reactions occuring at the surface of young stars. For the solution of this problem, H. Reeves brought an important new concept when he realized that in galactic cosmic rays $\dfrac{LiBeB}{CNO} \sim 0.20$ while $\dfrac{LiBeB}{CNO} \sim 10^{-5}$ in the standard (cosmic) abundances. This remark triggered the idea that most of the observed LiBeB is due to the interaction between GCR and the interstellar medium (ISM), occuring during their propagation in the Galaxy. This prescription has been concisely presented in Reeves et al. 1970 while the detailed model was developed in Meneguzzi et al. (1971; MAR) and Meneguzzi and Reeves (1975). The GCR sources are assumed to be distributed inside the galactic disk which itself is assumed to be a "leaky box". MAR managed to calculate not only the overall production of LiBeB (70% coming from the GCR H and He impinging on the ISM CNO, while the remaining 30% comes from the GCR CNO hitting the ISM H and He) but also were able to explain the GCR observed composition and the amount of ISM matter (6–8 g cm^{-2}) encountered by GCR whose transport age should be of the order of 10^7 years. The contribution of GCR to the ISM heating has also been estimated by MAR after some first estimates of Fowler et al. (1970).

Today, the GCR origin of all ^{6}Li,^{9}Be,^{10}B, ^{11}B and of about 10% of ^{7}Li seem to be universally accepted (Reeves 1974 and 1991). The origin of ^{7}Li is a problem in itself since for this exceptional case one should invoke several nucleosynthetic sites besides the GCR interaction, including primordial nucleosynthesis and thermonuclear processes occuring in red giants and/or in novae. There is still the puzzle of the $^{11}B/^{10}B$ ratio which is found to be equal to 4 in meteorites while all the available calculations lead to a value comprised between 2.2 and 2.8.

4 PRIMORDIAL NUCLEOSYNTHESIS

As stressed in the introductory remarks the main impact of H. Reeves on this problem is probably his work with Geiss (1972, followed up in 1981; see also Reeves 1972, Reeves and Bottinga 1972) where they convinced the astronomical community that in the solar system $D/H \sim 10^{-5}$ instead of the 10^{-4} value, obtained when one uses without care the early estimates of DH/H_2O made by Boato (1954).

The calculations of primordial nucleosynthesis are not complicated and were performed already by Peebles (1966) then by Wagoner et al. (1967). The difficulty of that topic which is discussed at length in this symposium comes mainly from the abundance determinations. Several astrophysicists deserve to be mentioned here, for their important work in this field : Vidal–Madjar, Laurent, Gry, Ferlet and York for D, Bania, Rood and Wilson for ^3He, Pagel, Kunth, Sargent, Lequeux, Peimbert, Terlevitch for ^4He, M. and F. Spite, Hobbs and Boesgaard for ^7Li.

The simplest model of primordial nucleosynthesis provides a rather satisfactory explanation for the origin of the primordial D, ^3He, ^4He, and ^7Li and also gives an estimate of the baryonic density $\Omega_b \sim 0.1$ and of the maximum number of lepton (neutrino) families ($N_\nu \sim 3$).

Two further contributions of H. Reeves should also be recalled: the one made with Y. David (David and Reeves 1980), where they investigated the possibility to reconcile large Ω_b values with some neutrino (especially the muon neutrino) degeneracy; and the second concerning the implications of inhomogeneities (possibly triggered by quark–hadron phase transitions) on primordial nucleosynthesis. With his talent to clarify the most complex physical situations by evoking beautiful images, H. Reeves proposed the quark–hadron "butterfly" which has to coincide with the ^7Li trench (Reeves et al. 1988, Reeves 1991a). These works confirm the fact that the ^7Li abundance constrains severely the hadronic inhomogeneities.

5 NUCLEOCHRONOLOGIES AND THE ORIGIN OF THE SOLAR SYSTEM

Time and evolution are constant themes in the research of H. Reeves. A quite recent paper (Reeves 1991b) expresses his views on nucleochronologies, some of which have been already expressed in Reeves and Johns (1976), or in Reeves and Meyer (1978). They can be summarized as follows :

i) The rate of star formation is almost constant throughout the life of the galaxy and consequently the rate of nucleosynthesis does not vary dramatically during this evolutionary period.

ii) The massive stars are mainly located in spiral arms; this explains the 10^8 years period between the synthesis of ^{129}I and ^{244}Pu and the formation of the solar system as mentioned above.

iii) The synthesis of ^{26}Al ($\tau \sim 10^6$ years) is related to the O–B association inside, (or at the vicinity of which) the solar nebula formed. The already quoted "Bing Bang" hypothesis comes from the remark that stars form in very restricted volumes of the galaxy. Inside these regions which represent about 10^{-7} of the total, star formations are not single events but are triggered by the rapid and dramatic evolution of several massive stars (Reeves 1978a,b).

6 A FEW LAST REMARKS

This presentation, in which I have not attempted to give justice to all the contributions of H. Reeves, is not an epilogue. Indeed, this symposium will convince you that the ideas and the insights of Hubert Reeves are still vivid. Particle and nuclear astrophysics will benefit for many years from his contributions and his influence, as I myself did in the past.

I would like to point on this occasion a few astrophysical problems on which my own perception is somewhat different from his. For instance, I think that the star formation and the nucleosynthetic rate are not as constant as claimed in Reeves (1991b), who analyzes galactic evolution in terms of simple models adopting the instantaneous recycling approximation. Concerning the deuterium evolution, H. Reeves favours models with little D destruction, assumption which could be found in conflict with the low values of ^{4}He primordial abundance; by contrast, in Vangioni–Flam and Audouze (1988), we advocated a large D destruction during the galactic evolution.

Science, and especially astrophysics, constitute a source of inspiration, enthusiasm and pleasure. It is because Hubert Reeves has convinced us that the contemplation and the study of the Universe is most exciting, and also because he showed to us how to make research and to build up simple, useful, and often robust models, that we are all happy and grateful to wish him again *"Bon anniversaire Hubert"*.

ACKNOWLEDGEMENTS: I gratefully acknowledge financial support of PICS No 114 for this Symposium.

REFERENCES

Audouze J, Epherre M, Reeves H, 1967 Nuclear Physics **A97** n°1

Bernas R, Gradsztajn E, Reeves H, Schatzman E, 1967 Ann. of Physics **44** 426

Boato G, 1954 Geochim. Cosmochim. Acta **6** 209

Burbidge E M, Burbidge G R, Fowler W A and Hoyle F, 1957 Rev. Mod. Phys. **29** 547

Cameron A G W, 1957, *Chalk River Report*, CRL-41

Caughlan G R and Fowler W A, 1988 *Atomic Data and Nuclear Data Tables* **40** 291

David and Reeves H, 1980 *Les Houches XXXII Physical Cosmology* Balian R, Audouze J, Schramm D (eds) p. 443

Fowler W A, Caughlan G R and Zimmermann B A, 1967 Ann. Rev. Astr. Ap. **5** 525

Fowler W A, Caughlan G R and Zimmermann B A, 1975 Ann. Rev. Astr. Ap. **13** 69

Fowler W A, Greenstein J E and Hoyle F, 1962 Geophys. J. Roy. Astron. Soc. **6** 148

Fowler W A, Reeves H and Silk J, 1970 Astrophys. J. **162** 49

Geiss J, Reeves H, 1972 Astron. Astrophys. **18** 126

Geiss J, Reeves H, 1981 Astron. Astrophys. **93** 189

Johns O, Reeves H, 1973 Astrophys. J. **186** 233

Meneguzzi M, Audouze J, Reeves H, 1971 Astron. Astrophys. **15** 337

Meneguzzi M, Reeves H, 1975 Astron. Astrophys. **40** 91

Paquette G, Reeves H, 1964 Astrophys. J. **140** 1319

Peebles P J E, 1966 Astrophys. J. **146** 542

Pochoda P, Reeves H, 1964 Planet. Space Sci. **12** 119

Reeves H, Salpeter E, 1959 Phys. Rev. **116** 1505

Reeves H, 1962 Astrophys. J. **135** 779

Reeves H, 1966a Stellar evolution, Plenum Press

Reeves H, 1966b, Astrophys. J. **146** 447

Reeves H and Bottinga, 1972, Nature **238** 326

Reeves H, Fowler W A, Hoyle F, 1970 Nature, **226** 727

Reeves H, 1972 Phys. Rev. **6** 3363

Reeves H, 1974 Ann. Rev., Astron. Astrophys. **12** 437

Reeves H, Johns O, 1976 Astrophys. J. **206** 958

Reeves H, 1978a *Protostars and planets*, Ed. T. Gehrels, Tucson Arizona, p. 399

Reeves H, 1978b *The origin of the solar system* J F Dermott (ed) University of Newcastle upon Tyne 24–3 to 9–4 1976, p.1

Reeves H., Meyer J P 1978, Astrophys. J., **266**, 613

Reeves H, Delbourgo–Salvador P, Salati P, Audouze J, 1988, Eur. J. Phys. **9** 179

Reeves H, 1991a Physics Reports **201** 335

Reeves H, 1991b, Astron. Astrophys. **244** 294

Ryter C, Reeves H, Gradsztajn E, Audouze J, 1970 Astrophys. Astron. **8** 389

Serber R, 1977, Phys. Rev. **72** 1114

Vangioni–Flam E and Audouze J, 1988, Astron. Astrophys. **193** 81

Vogt E, Michaud G, Reeves H, 1965 Phys. Lett. **19** 570

Wagoner R V, Fowler W A, Hoyle F, 1967, Astrophys. J. **148** 3

Two Theses on C, N, O Destruction

E.E. SALPETER

Cornell University, Ithaca, New York

I want to discuss two memorable Ph.D. theses, both from Cornell but 31 years apart. As you can guess, one was by our birthday celebrant, Hubert Reeves, in 1960 and the other by Lars Bildsten just last year. What the two theses have in common is not only high quality, but that they both deal with the destruction of the most abundant medium-light elements, C, N, and O. I will mostly give a historic comparison rather than a scientific discussion and I will not give references in general, but easily available accounts of at least the earlier parts of the thesis work will be found in:

H. Reeves and E.E. Salpeter 1959, Phys. Rev. 116, 1505;
Lars Bildsten, I. Wasserman and E.E. Salpeter (1990), Nucl. Phys. A 516, 77 and (1992), Ap. J. 384, 143.

The destruction mechanism and the nuclear outcome is very different in the two investigations, but so is the setting - late stages of ordinary stellar evolution for the earlier one, post-supernova neutron stars for the later one. The chronological order of the theses is in step with the order in stellar evolution, unlike an earlier anachronism:

When Hubert came to Cornell as a graduate student in 1955, the qualitative ideas on the exotic collapse leading to supernova explosions had already gotten surprisingly far, due in part to Fred Hoyle's insight and imagination. On the other hand, the theory of post-main-sequence evolution had developed surprisingly slowly, even though the basic principles are quite mundane. This holdup happened because numerical calculations were needed and electronic computers did not exist yet. While Hubert was working on his thesis, red giant evolution was pinned down fairly well, including the helium flash at the tip of the red giant branch. Little was known about the subsequent horizontal and asymptotic branches, although it was at least known that the evolutionary timescales after the helium flash were much shorter than before, but longer than a year. Fortunately, the latter fact was the main astronomical input Hubert needed for his calculations because the timescale is longer than dynamic times (so he could assume good thermal equilibrium) and longer than beta-decay times (so he could assume proximity to the beta-stability line). As a consequence he was able to leave the timescale as an adjustable parameter (anywhere from 1 year to 10^6 years, say) and present results which could subsequently be put into stellar evolution calculations. The thesis was thus fairly pure nuclear physics, not yet astronomy, but the distinction was beginning to blur. When I became a member of the I.A.U. (sometime before the 1958 Assembly in Moscow) Richard Wooley (one-time Astronomer Royal for Australia) expressed doubts to me about the mingling of physicists and astronomy - he was particularly dubious about youngsters like Hubert doing a thesis on the borderline between the two sciences!

Hubert's main task was to calculate the rates of various compound nucleus formations, initiated by $C^{12} + C^{12}$ collisions, etc., as well as the final outcome. As mentioned, opportunities for numerical calculations were limited before the advent of electronic computers. However, this gave him an opportunity to display his ingenuity in deciding which analytic approximation was justified in each part of the calculation. As a consequence, computational technique represented surprisingly little of a handicap, but the paucity of nuclear experimental results was a handicap. In fact, Hubert did not spend all of his time in Ithaca, but some of it back in his native Canada in Chalk River where there was a thriving nuclear physics group. This included, amongst others, Allen Bromley - not as a U.S. President's Science Advisor but as a young experimentalist. At any rate, the end result was a pretty creditable thermonuclear reaction rate as a function of temperature (from 5 to 9 $\times 10^8 K$) for the destruction of C^{12}. The predicted abundance ratios in the production of Na^{23}, Mg^{24}, Al^{27} were not quite so reliable on modern standards, but at least not bad. As a by-product, Hubert also calculated the production of neutrons through the sequence $C^{12}(p,\gamma)N^{13} \rightarrow C^{13}, C^{13}(\alpha,n)0^{16}$. The second thesis also produced neutrons as a by-product, but in a different manner.

The intervening 30 years have provided a different background for Lars Bildsten's thesis (carried out largely under the direction of Prof. Ira Wasserman). Nuclear astrophysics and stellar evolution up to a supernova event is thoroughly understood. Supernovae themselves are less well understood, but we certainly know lots about the formation of neutron stars. However, Lars' thesis concerns not the formation of neutron stars, nor their interior structure, but surface phenomena long after the neutron star was born. I am not referring to the outermost atmosphere of an isolated neutron star, the pulsar phenomena, but the aftermath of accretion from a companion or from the interstellar medium. X-ray astronomy has already shown that such accretion takes place today and produces surface X-rays and a large number of theoretical calculations have already been carried out. However, there are many unknown parameters and the microphysics is complicated, so we are still pretty ignorant. Most of the X-ray data so far is from continuum emission, but there was the hope that nuclear X-ray and γ-ray spectral lines would be observed (by GRO, ROSAT, AXAF, etc.) and give more quantitative information. The accreting material contains C, N and O at least with solar abundance and line emission from excited states of these nuclei is of interest. Lars' main aim was to calculate the production rate of such γ-ray lines, especially the 4.44 MeV line of C^{12}. It was hoped that calculating the rate of destruction of these nuclei would only be a minor by-product, but some unexpected effects make destruction most likely:

If magnetic fields are well below 5×10^{11} Gauss, protons and alpha particles in an accretion flow onto a neutron star surface are slowed down mainly by multiple, classical, Coulomb scattering. If the accretion luminosity is well below the "Eddington limit", the particles enter the atmosphere with almost free fall speeds (almost 100 MeV per nucleon) and can undergo nuclear reactions in collisions with ambient atmospheric protons. An appreciable fraction (but by no means all) of the incident alpha particles are broken up, the more interesting channel being $He^3 + n$ (see below). Lars Bildsten showed that incident C, N, and O nuclei have a rather different fate because of one detail in Coulomb scattering. The slowing down rate depends on Z^2/A (where Z and A are atomic charge and mass, respectively) which is equal for H and He but appreciably larger for C and heavier elements. For this reason, C-nuclei (and others) are stopped in layers of the upper atmosphere where protons (and alpha particles) still have almost their full incident velocity. The advection speed of the ambient atmosphere is quite low, so the residence time of the nuclei is quite long and the nuclei are almost certain to be hit often by incident p and α. These nuclear collisions produce some nuclear excitations and therefore some gamma ray line (especially the C^{12}-line at 4.438 MeV), but the final outcome is almost complete destruction of C^{12} and all heavier nuclei.

Unlike Hubert, Lars was able to perform numerical calculations on fast computers, but the real bottlenecks and challenges were not so very different. Once again, measurements of nuclear cross-sections (and especially angular distributions) were important and Lars also had to travel to visit nuclear experimentalists. Once again, knowing where to make what analytical approximation was more important than the actual computations, although here there was more emphasis on fluid dynamics. Once again the results were of interest to "real astronomers", but this time to X-ray and γ-ray astronomers for two different reasons.

X-ray bursts are thought be triggered by a nuclear-burning helium-flash a little below the surface of a neutron star which has hydrogen and helium-rich material flowing in from some form of steady accretion. The helium-flash itself does not involve heavier nuclei, but details of the onset of helium burning depend on just how the accreting hydrogen burns to form more helium. This hydrogen burning in turn depends on the abundance of the C, N, O elements and the absence of these elements makes for a more complex flash (the slower steady hydrogen burning leads to a mixed hydrogen-helium flash).

The other observational consequence is the production of nuclear γ-ray lines. As in Hubert's thesis, one by-product of the calculation is the production rate of neutrons. Since the accreting material is very hydrogen-rich, the important neutron reactions are not absorption by heavy elements but n-p recombination to form deuterons and the corresponding 2.2 MeV γ-ray line. Unfortunately this line is mostly produced fairly deep in the atmosphere and the spectral line energy is smeared out by Compton scattering. As a consequence, both this line and the nuclear excitation lines are not easy to observe and prospects for GRO are not very optimistic, at least for neutron stars accreting solar abundance material. Fortunately, there are cases where the accreting material is likely to be strongly enriched in helium and/or the C, N, O nuclei (the X-ray source U 1820-30 is one such example). Since all the nuclear γ-ray lines are dependent on these elements, such cases should be much more favorable. However, just how much more favorable is not at all clear at the moment, since there are intricate non-linear effects in the flow dynamics with multiple nuclear species. I am looking forward to further stimulating work on this topic, but I hope I will not have to wait another 31 years for the next thesis on C, N, O destruction.

STELLAR AND NUCLEAR DATA

Cosmic Abundances of the Elements

N. GREVESSE and A. NOELS

Institut d'Astrophysique, Université de Liège,
5, avenue de Cointe, B-4000 Liège (Belgium)

ABSTRACT

We update the table of abundances of Anders and Grevesse (1989) and we propose a new value of the Z/X ratio.

1 INTRODUCTION

Cosmic abundances is a somewhat misleading expression for solar system abundances. They have been the first to be determined in our Galaxy and they cover the whole range of elements and isotopes of the periodic table. Since the pioneering work of Suess and Urey (1956), Cameron (1989 and references therein) has devoted an extensive research to establishing a complete table of cosmic abundances, updating it continuously during at least two decades.

Two different sources have been used. On the one hand, the meteorites (carbonaceous CI chondrites) allow to measure the abundances of the elements as they were in the primeval solar nebula. The values are now known with great accuracy, better than about 10 % for most of the elements (Anders and Ebihara, 1982).

On the other hand, the Sun itself should have kept in its external layers the initial abundances for most of the elements. Solar abundances can be derived from spectroscopic analyses of the photosphere, sunspots, chromosphere, corona and from solar wind (SW) and solar energetic particle (SEP) measurements. Photospheric abundances are without any doubt the most accurate because of the very high quality of the solar photospheric spectra that extend now from the visible to the far infrared (Delbouille et al., 1973; Delbouille et al., 1981; Farmer and Norton, 1989). Furthermore, the physical conditions and physical processes are better known in these layers than anywhere else in the atmosphere of the Sun. The variability of the sunspots, corona, SW and SEP, as well as the departures from LTE in chromospheric and coronal layers make them less reliable indicators of the solar abundances although important progress has been made in recent years (Anders and Grevesse, 1989 and references therein).

Discrepancies between meteoritic and solar abundances have progressively gone away as the solar values have become more accurate due to improvements in the atomic data. The agreement is now remarkably good, except obviously for the depleted elements Li and Be. A detailed discussion is given in Anders and Grevesse (1989). Now that the iron problem is solved (see hereafter), it is our belief that meteoritic abundances really reflect the chemical composition in the external layers of the Sun.

It is now possible to give a quite accurate value for the ratio Z/X, which is essential for the calibration of the Sun. Moreover, the detailed distribution of the main contributors to Z is now known with a better accuracy. This distribution is used in the opacity computations which are so important in stellar structure studies.

We present here an updating of Anders and Grevesse (1989)'s results, reviewing successively the elements for which significant changes have occurred. These changes are to be found in the photospheric abundances since the meteoritic values have not been improved. A new value of the ratio Z/X is given in the conclusions.

2 HELIUM

Despite its high abundance, helium is unfortunately undetectable in the photospheric spectrum and in the meteorites. Solar wind and solar energetic particle measurements agree to a rather low value of N_{He}/N_H of about 4 % (Gloeckler and Geiss, 1989). The innermost giant planets, Jupiter and Saturn, show rather small He abundances, of the order of 6 % and 2 % respectively while the outermost Uranus and Neptune have a higher He abundance, of the order of 9 % ± 2 % (Conrath et al., 1989). Calibration of most recent standard solar models for Z = 0.02, computed with the new OPAL opacities (Iglesias et al., 1992), with the new photospheric abundance of iron (see hereafter) and the MHD equation of state (Däppen et al., 1988), leads to a helium mass fraction, Y (see Noels and Grevesse, 1992 for a review), of

$$Y = 0.28 \pm 0.01,$$

i.e.,

$$\frac{N_{He}}{N_H} = 10\% \pm 0.5\%.$$

3 LITHIUM - BERYLLIUM - BORON

These light elements are of particular interest since they are the only ones that can be partly destroyed during the early life of the Sun. Contrary to the other elements for which an agreement is generally met between photospheric and meteoritic abundances, differences in Li, Be and B between those two sources should be evaluated as precisely as possible because these elements can serve as strong constraints for the physical processes in the external layers of the Sun. These processes can increase the extent of the mixed region below the solar photosphere and burning of such light elements, especially Li, can take place, resulting in a depletion of the surface abundances. The recommended photospheric values are still the ones presented in Anders and Grevesse (1989). Starting from meteoritic, presumably initial values,

$$A_{Li,m} = 3.31 \pm 0.04,$$

$$A_{Be,m} = 1.42 \pm 0.04,$$

$$A_{B,m} = 2.88 \pm 0.04,$$

the present photospheric values are reduced to

$$A_{Li,\odot} = 1.16 \pm 0.10,$$

$$A_{Be,\odot} = 1.15 \pm 0.10,$$

$$A_{B,\odot} = 2.6 \pm 0.30,$$

with depletion rates of

$$R_{Li} = 140(+50, -40),$$

$$R_{Be} = 1.9(+0.7, -0.5).$$

The photospheric B is too uncertain to allow giving a depletion rate for B.

4 CARBON - NITROGEN - OXYGEN

In collaboration with A.J. Sauval, we have revisited the photospheric abundance of these crucial elements, partly lost in meteorites, using new data in the IR obtained by the ATMOS experiment (Farmer and Norton, 1989) and new transition probabilities computed by Hibbers et al. (1991) for NI and by Biémont et al. (1991) for OI. The solar data are to be found in Grevesse et al. (1990) for NI and Grevesse et al. (1991) for CI. OI lines have been remeasured for this work. The solar photospheric model of Holweger and Müller (1974) has been updated in gaseous and electronic pressures because of the new photospheric abundance of iron. It should be emphasized that Lambert (1978) was the first to insist on the interest of using all the indicators of the photospheric abundances of these elements, atoms as well as molecules.

4.1 Carbon

The abundance derived from the CI lines of Grevesse et al. (1991) is now

$$A_{C,\odot} = 8.55 \pm 0.05. \qquad 12\%$$

Preliminary results obtained with the same updated model for the other indicators of the C abundance, i.e. C_2 Swan and C_2 Phillips bands, CH(A-X) lines and CH variation-rotation lines (Grevesse et al., 1992) lead to a close agreement with the hereabove mentioned value, without increasing the dispersion.

4.2 Nitrogen

NI lines in Grevesse et al. (1990), updated with the new solar model, give an abundance of 8.00 ± 0.07. Adding the updated results from the NH vibration rotation lines in Grevesse et al. (1992), 7.95 ± 0.07, decreases slightly the N value to the recommended one :

$$A_{N,\odot} = 7.97 \pm 0.07.$$

4.3 Oxygen

Permitted OI lines are those given in Biémont et al. (1991). This set includes the triplet at 7770 Å. We have added two forbidden lines at 6300 and 6364 Å, for which we used the transition probabilities from Baluja and Zeippen (1988). We obtain

$$A_{O,\odot} = 8.87 \pm 0.07. \qquad 17\%$$

It has to be mentioned that there seems to be a hiatus between the value derived for permitted and forbidden lines. The triplet at 7770 Å leads to values of 8.85 and 8.81 depending on the enhancement factor of the damping parameter, 1 and 1.5 respectively, while the forbidden lines, insensitive to this factor, give a larger value of 8.95. These forbidden lines are very weak lines formed in LTE whereas the triplet lines, much stronger, are very sensitive to non LTE effects (Kiselman, 1991). However, these non LTE effects reducing the abundance derived from the triplet lines, should still increase the hiatus. Of course, we might not exclude unknown blends in the two forbidden lines although it seems unlikely to lower the abundance by the amount demanded to agree with the triplet lines. The mean OI value given here is in agreement with the value obtained from the OH vibration-rotation lines, 8.87 ± 0.05, found by Grevesse et al. (1992). The value recommended here gives a value of the C/O ratio of 0.5 instead of 0.4 as in Anders and Grevesse (1989). This ratio is of particular importance for the outer solar system chemistry for which a value of 0.5 is preferred (Prentice, 1991; Watson, 1991) to the lower value.

$$\left(\frac{C}{O}\right) = 0.479 \pm 0.1$$

5 NEON - ARGON

These noble gases are not seen in the solar photospheric spectrum nor in meteorites but their abundances can be derived from the coronal spectrum, SW and SEP. It is now well known that some physical processes affect the abundance pattern in the upper solar layers. A fractionation exists between low and high first ionization potential (FIP) elements, the transition occuring at about 10 eV. It was not obvious, at first, to decide which of these elements were affected by this fractionation, i.e. which of these elements kept their photospheric abundances. For this reason, Meyer (1989) suggested to use local galactic values derived from HII regions, HI gas and hot stars. The fractionation has been understood in two ways, either an enrichment of the low FIP elements (Geiss, 1982), or a depletion of the high FIP elements (Meyer, 1985; Stone, 1989). Recently, Feldman (1992), in agreement with Geiss (1982) and Meyer (1992), has argued that the high FIP elements, C, N, O, Ne, Ar..., are the ones that are not affected by the fractionation processes while the low FIP elements are enhanced by a factor of about 4.5 in the upper solar layers. This means that a scaling factor should be applied to the ordinate of the fractionation curve as shown in figure 3 of Anders and Grevesse (1989), the zero being decreased by 0.65 dex. Futhermore, a careful analysis of impulsive flare spectra reveals a photospheric pattern for low (Mg) and high (O, Ar, Ne) FIP elements except for very low FIP (Na, Ca) elements, which are enhanced by a factor of 2. Feldman (1992) therefore argues that photospheric abundances of Ne and Ar can be obtained directly from the analysis of impulsive flare spectra, IFS,

$$A_{Ne,IFS} = 8.14 \pm 0.10,$$

$$A_{Ar,IFS} = 6.76 \pm 0.20.$$

These values are in rather good agreement with the SEP measurements (Breneman and Stone, 1985),

$$A_{Ne,SEP} = 8.03 \pm 0.04,$$

$$A_{Ar,SEP} = 6.52 \pm 0.06,$$

as well as with the SW values (von Steiger and Geiss,1989). We suggest to adopt for the photospheric abundances of Ne and Ar, a weighted mean between these values, with a weight 2 for the SEP more accurate values and a weight 1 for the IFS values, which gives

$$A_{Ne,\odot} = 8.07 \pm 0.06,$$

$$A_{Ar,\odot} = 6.60 \pm 0.14.$$

6 IRON

The solar photospheric abundance of iron recommended in Anders and Grevesse (1989) was derived from low excitation FeI lines (Blackwell et al., 1984; Blackwell et al., 1986), using the very accurate Oxford transition probabilities (Blackwell et al., 1982). Its value, $A_{Fe,\odot} = 7.67 \pm 0.03$, is 40 % higher than the very accurate meteoritic value, $A_{Fe,m} = 7.51 \pm 0.01$. Since then, new accurate transition probabilities have been obtained for higher excitation lines in FeI as well as for FeII lines of solar interest. With these data, the photospheric abundance of iron has been revisited by different groups (Holweger et al., 1990; Pauls et al., 1990; Biémont et al., 1991; Holweger et al., 1991; Hannaford et al., 1992; Johansson et al., 1992). These analyses show that the photospheric abundance of iron now agrees with the meteoritic value,

$$A_{Fe,\odot} = A_{Fe,m} = 7.51 \pm 0.01.$$

The high value obtained from low excitation FeI lines still remains to be explained. A possible explanation of this discrepancy could come from a slight overestimation of the temperature distributon in the layers where those lines are formed, which would hardly affect higher excitation FeI lines and FeII lines.

7 LEAD

The photospheric abundance of Pb is essentially determined from four PbI lines, two of them being strongly blended. Grevesse and Meyer (1985) obtained a value, 1.85 ± 0.05, smaller than the meteoritic value, 2.05 ± 0.03. The photospheric value has been increased to 2.00 ± 0.05 by Youssef and Khalil (1989). We have rechecked this, using our updated model and found

$$A_{Pb,\odot} = 1.95 \pm 0.08,$$

still possibly slightly smaller than the meteoritic value

$$A_{Pb,m} = 2.05 \pm 0.03.$$

It has been stressed by Beer (1989) that the metoritic value was too high to explain a realistic age for the Galaxy, using the ^{235}U clock. Isotopes of lead are produced by the s- and r-processes in amounts that can be evaluated. The remaining fraction of ^{207}Pb, Pb_U, comes from radioactive decay of ^{235}U. The larger this Pb_U value, the larger the age of the Galaxy. Using s- and r-process models, Beer suggests an abundance

$$A_{Pb} = 1.97 \pm 0.03,$$

in agreement with our photospheric value but smaller than the meteoritic value. Apart from that argument, our result does not rule out the meteoritic value.

8 THORIUM

The ratio of these elements in G dwarfs of various ages has been used to constrain the age of the Galaxy. Th is a radioactive element, with a half life of 14 Gyr, while Nd is stable. Both elements are produced by nucleosynthesis through the r-process for Th while Nd is produced only partly by the r-process. Given a model for the chemical evolution of the Galaxy, the ratio Th/Nd as a function of age can be predicted. This relation can be confronted to the observed relation obtained from stars whose ages are derived by various ways. Butcher (1987) and more recently Morell et al. (1992) have found that the Galaxy cannot be older than 10-12 Gyr, which is in contradiction with the ages derived from stellar evolution results. Pagel (1989) is in favor of this constraint, using the ratio Th/Eu in G dwarfs.

The abundances of Th and Nd in these stars are generally derived from the analysis of a single line only. For Nd, the line is of good quality and unblended. In the Sun, the abundance of Nd derived from this line is in excellent agreement with the meteoritic results,

$$A_{Nd,\odot} = 1.50 \pm 0.06,$$

$$A_{Nd,m} = 1.47 \pm 0.01.$$

It is likely that the abundances of Nd in other G dwarfs are reliable. The situation is quite different for Th. The line observed is a weak line of ThII in the wing of a stronger FeI line. Its equivalent width is difficult to measure very accurately. Moreover, this line is known to be blended by a line of CoI whose contribution to the solar equivalent width of the observed ThII feature is about 30 % (Lawler et al., 1990). Evenso, the abundance of Th derived for the solar photosphere, using either the equivalent width or spectral synthesis, is definitely higher than the meteoritic value,

$$A_{Th,\odot} = 0.27 \pm 0.08$$

and

$$A_{Th,m} = 0.08 \pm 0.02,$$

respectively. This result is very hard to understand as there are no obvious reasons for the solar abundance of Th to be different from the meteoritic one. Therefore, we suspect another unknown blend in this line and we recommend a very careful analysis before any conclusion to be drawn.

Such a blend could possibly affect very differently stars of various effective temperatures, gravities and metallicities, as it is the case for the CoI blend. It is likely that the results obtained for Th abundances in G dwarfs without this blend are biased. A discussion of the age of the Galaxy should be postponed until this problem is solved.

9 MISCELLANEOUS

Two more photospheric values have been revised since the work of Anders and
Grevesse (1989, table 2), using new accurate transition probabilities. The first of
these is the solar abundance of Cd (Youssef et al., 1990), which is now

$$A_{Cd,\odot} = 1.77 \pm 0.11,$$

in excellent agreement with the meteoritic value,

$$A_{Cd,m} = 1.76 \pm 0.03.$$

The second one is the solar abundance of Sm (Biémont et al., 1989),

$$A_{Sm,\odot} = 1.01 \pm 0.06,$$

also in excellent agreement with the meteoritic result,

$$A_{Sm,m} = 0.97 \pm 0.01.$$

Some disagreements still seem to exist in the newly revised photospheric value of
Sc, using ScI and ScII lines, for which new accurate transition probabilities have
recently been measured. The value obtained by Neuforge (1992) is

$$A_{Sc,\odot} = \begin{array}{l} 3.14 \pm 0.12 \quad (\text{ScI}), \\ 3.20 \pm 0.07 \quad (\text{ScII}). \end{array}$$

The value obtained from the ScII lines, the best indicator of the Sc abundance, is
somewhat larger than the meteoritic value,

$$A_{Sc,m} = 3.09 \pm 0.04.$$

A similar situation prevails for Ti (Bizzarri et al., 1992),

$$A_{Ti,\odot} = \begin{array}{l} 5.03 \pm 0.03 \quad (\text{TiI}), \\ 5.04 \pm 0.04 \quad (\text{TiII}). \end{array}$$

Here, both values agree and are larger than the meteoritic abundance,

$$A_{Ti,m} = 4.93 \pm 0.02.$$

10 CONCLUSIONS

Most of the elements that are present in meteorites have photospheric abundances which agree perfectly well with the meteoritic values. The most striking case is the one of iron which resisted for such a long time before coming down to the meteoritic level. Some exceptions can still be found but we believe that most of them will be removed in a near future with the availability of still more accurate transition probabilities. As it has been suggested by many authors in the recent past, our feeling is that carbonaceous chondrites of type I exactly reflect the chemical composition of the solar photosphere.

The reductions in the abundances of important contributors to Z like O, C, N, Fe lead to a lowering in Z/X,

$$\frac{Z}{X} = 0.0245,$$

instead of $Z/X = 0.0275$ (Anders and Grevesse, 1989) used in most standard solar calibrations. Decreasing Z/X in the standard solar models should result in a slightly lower Y value, smaller than 0.28.

ACKNOWLEDGEMENTS

It is our pleasure to thank J. Sauval for continuous very helpful discussions. We thank the belgian Fonds National de la Recherche Scientifique for financial support.

REFERENCES

Anders, E., Ebihara, M. 1982, Geochim. Cosmochim. Acta **46**, 2363.

Anders, E., Grevesse, N. 1989, Geochim. Cosmochim. Acta **53**, 197.

Baluya, K.L., Zeippen, C.J. 1988, J. Phys. B., At. Mol. Phys. **21**, 1455.

Beer, H. 1989, in *Astrophysical Ages and Dating Methods*, 5th IAP Astrophysical Meeting, Eds. E. Vangioni-Flam, M. Cassé, J. Andouze, J. Than Thanh Van, Ed. Frontières, p. 349.

Biémont, E., Grevesse, N., Hannaford, P., Lowe, R.M. 1989, A & A **222**, 307.

Biémont, E., Baudoux, M., Kurucz, R.L., Ansbacher, W., Pinnington, E.H. 1991, A & A **249**, 539.

Biémont, E., Hibbert, A., Godefroid, M., Vaeck, N., Fawcett, B.C. 1991, Ap. J. **375**, 818.

Bizzarri, A., Huber, M.C.E., Noels, A., Grevesse, N., Bergeson,S.D., Tsekeris, P., Lawler, J.E. 1992, to be published.

Blackwell, D.E., Booth, A.J., Petford, A.D. 1984, A & A **132**, 236.

Blackwell, D.E., Petford, A.D., Simmons, G.J. 1982, MNRAS **201**, 595 and references therein.

Blackwell, D.E., Booth, A.J., Haddock, D.J., Petford, A.D., Leggett, S.K. 1986, MNRAS **220**, 549.

Breneman, H.H., Stone, E.C., 1985, Ap. J. **299**, L 57.

Butcher, H.R. 1987, Nature **328**, 127.

Cameron, A.G.W. 1989, in *Cosmic Abundances of Matter*, Ed. C.J. Waddington, AIP Conf. Proc. 183, Am. Inst. Physics, p. 349.

Conrath, B. et al. 1989, Science **246**, 1454.

Däppen, W., Mihalas, D., Hummer, D.G., Mihalas, B.W. 1988, Ap. J. **332**, 261.

Delbouille, L., Roland, G., Neven, L. 1973, *Photometric Atlas of the Solar Spectrum from λ 3000 Å to 10000 Å*, Institut d'Astrophysique, Université de Liège, Belgium.

Delbouille, L., Roland, G., Brault, J. W., Testerman, L. 1981, *Photometric Atlas of the Solar Spectrum from 1850 to 10000 cm^{-1}*, Kitt Peak National Observatory, Tucson.

Farmer, C.B., Norton, R.H. 1989, *A High-Resolution Atlas of the Infrared Spectrum of the Sun and the Earth Atmosphere from Space, Vol I : the Sun*, NASA Ref. Publ. 1224, Washington D.C.

Feldman, U. 1992, Physica Scripta (in press).

Geiss, J., 1982, Space Sci. Rev. **33**, 201.

Gloeckler, G., Geiss, J. 1989, in *Cosmic Abundances of Matter*, Ed. C.J. Waddington, AIP Conf. Proc. 183, Am. Inst. Physics, p. 49.

Grevesse, N., Meyer, J.P. 1985, Proc. 19th Int. Cosmic Ray Conf., La Jolla, August 11-23, OG Session, p. 5.

Grevesse, N., Sauval, A.J., Blomme, R. 1992, in *Infrared Solar Physics*, Eds. D. Rabin, J.T. Jefferies, IAU Symposium 154, Tucson, Kluwer, in press.

Grevesse, N., Lambert, D.L., Sauval, A.J., van Dishoeck, E.F., Farmer, C.B., Norton, R.H. 1990, A & A **232**, 225.

Grevesse, N., Lambert, D.L., Sauval, A.J., van Dishoeck, E.F., Farmer, C.B., Norton, R.H. 1991, A & A **242**, 488.

Hannaford, P., Lowe, R.M., Grevesse, N., Noels, A. 1992, A & A (in press).

Hibbert, A., Biémont, E., Godefroid, M., Vaeck, N. 1991, A & A Suppl. **28**, 505.

Holweger, H., Müller, E.A. 1974, Solar Physics **39**, 19.

Holweger, H., Heise, C., Kock, M. 1990, A & A **232**, 510.

Holweger, H., Bard, A., Kock, A., Kock, M. 1991, A & A **249**, 545.

Iglesias, C.A., Rogers, F.J., Wilson, B.G., 1992, preprint.

Johansson, S., Nave, G., Geller, M., Sauval, A.J., Grevesse, N. 1992, in *Infrared Solar Physics*, Eds. D. Rabin, J.T. Jefferies, IAU Symposium 154, Tucson, Kluwer, in press.

Kiselman, D. 1991, A & A **245**, L9.

Lambert, D.L. 1978, MNRAS **182**, 249.

Lawler, J.E., Whaling, W., Grevesse, N. 1990, Nature, **346**, 635.

Meyer, J.P. 1985, Ap. J. Suppl. **57**, 173.

Meyer, J.P. 1989, in *Cosmic Abundances of Matter*, Ed. C.J. Waddington, AIP Conf. Proc. 183, Am. Inst. Physics, p. 245 and references therein.

Meyer, J.P. 1992, this volume.

Morell, O., Källander, D., Butcher, H.R. 1992, A & A (in press).

Neuforge, C. 1992, this volume.

Noels, A., Grevesse, N. 1992, in *Inside the Stars*, IAU Symposium 145, Vienna, Eds. W. Weiss, A. Baglin, Publ. Astr. Soc. Pac. (in press).

Pagel, B.E.J. 1989, in *Evolutionary Phenomena in Galaxies*, Eds. J. Beckman, B.E.J. Pagel, Cambridge Univ. Press, p. 201.

Pauls, U., Grevesse, N., Huber, M.C.E. 1990, A & A **231**, 536.

Printice, A.J.R. 1991, private communication.

Stone, E.C., 1989, in *Cosmic Abundances of Matter*, Ed. C.J. Waddington, AIP conf. Proc. 183, Am. Inst. Physics, p. 72.

Suess, H.E., Urey, H.C. 1956, Rev. Mod. Phys. **28**, 53.

Wasson, J.T. 1991, private communication.

von Steiger, R., Geiss, J. 1989, A & A **225**, 222.

Youssef, N.H., Khalil, N.M. 1989, A & A **208**, 271.

Youssef, N.H., Dönszelmann, A., Grevesse, N. 1990, A & A **239**, 367.

Element Fractionation at Work in the Solar Atmosphere

Jean-Paul MEYER

Laboratory for High Energy Astrophysics, NASA/Goddard Space Flight Center, USA [1]
and Service d'Astrophysique, Centre d'Etudes de Saclay, 91191 Gif-sur-Yvette, France [2]

1   INTRODUCTORY THOUGHTS : GALACTIC AND SOLAR PHYSICS

The path that led to the realization that large scale element fractionation is taking place in the outer atmosphere of the Sun is not trivial. It shows, once more, the fruitfulness of cross fertilization between large scale Astrophysics (here, Galactic Cosmic Ray physics) and investigations in our own private laboratory, that Mother Nature has been kind enough to donate us : the Sun. Kind of a balanced back and forth movement not unlike Hubert's spirit. It also illustrates the fact, as will be shown below, that the wealth of information that can be gathered from the solar environment does not ease the understanding of problems, and forces us to face the fact that our simplified explanations for more distant phenomena may well be often far off describing reality.

The whole story starts with Galactic Cosmic Rays (GCR's). For decades, the dogma was that GCR's originate in supernovae, since GCR's are globally enriched in heavy elements, and since supernovae synthesize heavy elements, disperse them, provide ample energy for acceleration, and are observed to accelerate electrons. This view was certainly very appealing and dramatic : GCR's gave us a direct sample of the products of nucleosynthesis in supernovae ! It was not until the '70s that it was first noted that the *detailed* heavy element composition of GCR's seemed, strangely enough, to be ordered in terms of *atomic* physics parameters, and more precisely in terms of the First Ionization Potential (FIP) or of other related parameters, which control the tendency of an element to be neutral or ionized in a gas at ~ 5000 to 10000 K, and/or subjected to a radiation of comparable energy (Kristiansson 1971,1974; Havnès 1973; Cassé & Goret 1973,1978; Arnaud & Cassé 1985; Silberberg & Tsao 1990). This comparatively low temperature of the parent gas, together with the *lack of* a depletion, in GCR's, of refractory elements locked in grains in virtually all but the hottest interstellar medium (Cassé et al. 1975), first suggested that that the GCR nuclei originated in stellar surfaces (Meyer, Cassé & Reeves 1979). Meanwhile, it became progressively more and more clear that the *detailed* composition of GCR's was inconsistent with what could be reasonably expected from supernova nucleosynthesis (Arnould 1984; Meyer 1985b,1988; Prantzos et al. 1993) [3].

It were these findings regarding GCR's that led the same community of physicists, also engaged in the observation of other energetic particles, the Solar Energetic Particles (SEP's), to notice that the composition of these particles, too, could have something to do with FIP

---

[1]   NAS/NRC Senior Research Associate.

[2]   Permanent address.

[3]   There is, however, evidence for the presence in GCR's of *a component* originating in He-burning zones, possibly ejected by Wolf-Rayet stars, that accounts for the coupled excesses of ^{22}Ne and C in GCR's (Meyer 1981c,1985b; Cassé & Paul 1982; Prantzos at al. 1987).

(Hovestadt 1974; Webber 1975,1982; McGuire et al. 1979,1986; Cook et al. 1980,1984; Mewaldt 1980; Meyer 1981a,1985a,1991; Breneman & Stone 1985; Stone 1989; Garrard & Stone 1991; Cane et al. 1991; Reames et al. 1991; Reames 1993). The difficulty, here, was that the composition of SEP's is changing all the time. Meyer (1981a,1985a) and Breneman & Stone (1985) managed to separate out the *permanent* FIP-bias imprint on the data, clearly related to the composition of the heliospheric source material, from the rigidity-dependent variations of the composition resulting from variable conditions of particle acceleration. The SW and early spectroscopic analysis of the corona confirmed the ubiquity of the FIP-biased compositions in the solar outer atmosphere (Cook et al. 1980,1984; Meyer 1981b,1985b,1991; Veck & Parkinson 1981; Geiss & Bochsler 1985; Gloeckler & Geiss 1989).

These findings regarding the solar environment, the similarity between the GCR and the solar coronal composition, in turn, strongly supported the earlier arguments suggesting that the GCR nuclei had been first extracted from stellar atmospheres. And the parent gas of GCR's could now be specified more precisely : probably the coronae of F to M late-type stars which possess, like the Sun, a chromosphere at around ~ 7000 K (Meyer 1981c,1985b).

The ironic point in this story is that it is no longer so clear that it is indeed FIP that governs the GCR composition !!! The FIP of a particular element and the volatility of the chemical compounds it forms are, indeed, largely correlated. Therefore, an apparent correlation with FIP can be mimicked by an actual correlation with volatility, which would imply that GCR's are largely grain destruction products. One can, for instance, conceive a preferential acceleration of grain destruction products in supernova shock waves, in which the grains are being destroyed (Epstein 1980; Cesarsky & Bibring 1980; Bibring & Cesarsky 1981). The abundances of the few elements which are *exceptions* to this general correlation between FIP and volatility (Meyer 1981d) hint that volatility might, indeed, be the relevant parameter. But the question is still open (*e.g.*, Soutoul et al. 1991).

But, in the solar environment, we don't have to worry about this ambiguity ! It does not seem possible that the coronal composition anomalies be controlled by the vaporization of solid, meteoritic bodies (§ 4.4). Here, the relevant parameter is, definitely, FIP.

The present paper, devoted to the solar environment only, is an update of earlier summary given two years ago (Meyer 1991, hereafter Paper I). I refer to this earlier paper for a discussion of the basic data sets giving evidence for the main features of the coronal composition, and concentrate here on the, very rich, new results appeared since then.

2   BRIEF DESCRIPTION OF THE SOLAR ATMOSPHERE , TYPES OF COMPOSITION OBSERVATIONS , AND RELEVANT SITES

2.1   Scope : Observations Near and Far From Sun , and in Open- and Closed-Field Regions

Fig. 1 shows a simple-minded representation of the solar atmosphere, based on a spherically symmetric, locally plane parallel, approximation (Vernazza et al. 1981 [VAL]; Dryer 1982). It shows the upper part of the photosphere with its negative temperature gradient, up to the temperature minimum at ~ 4200 K, the chromosphere with its positive temperature gradient and in particular its "temperature plateau" around ~ 6500 K, the conventional "Transition Region" in which the temperature rises abruptly from ~ 30000 K to ~ 1.5 MK, and finally the corona with temperatures on this order (when quiet). The Solar Wind (SW) bulk velocity is also shown, and the effective SW expansion region, or "Interplanetary Medium", is conventionally defined as starting where this SW bulk velocity exceeds the SW proton thermal speed. This figure

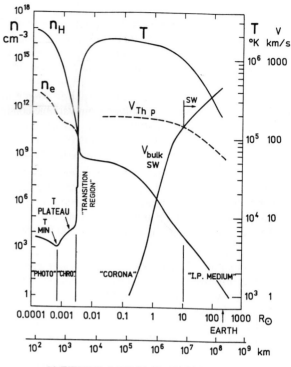

Fig. 1 : Idealized spherically symmetric (locally plane parallel) model of the solar atmosphere and heliosphere (after VAL and Dryer 1982). It displays the conventional definitions for the photosphere, T–minimum, chromosphere (T-plateau, see fig. 2), transition region, corona, and interplanetary medium (where the solar wind effectively expands, i.e. where its bulk velocity exceeds the thermal proton velocity). The n_H and n_e plots show that H is predominantly neutral in the photosphere and chromosphere, and entirely ionized above. Altitudes are above the $\tau_{5000 \, \AA} = 1$ level of the photosphere. Note the huge range of density variation across this figure, and the steep density gradient in the chromosphere (§ 2.1).

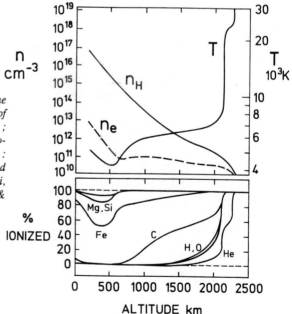

Fig. 2 : Top : enlargement of the photospheric-chromospheric part of fig. 1, with a linear altitude scale ; it shows, in particular, the chromospheric T-plateau. Bottom : associated calculated ionized fractions of H, He, C, O, Mg, Si, Fe (after VAL and Geiss & Bochsler 1985 ; § 2.1).

shows in particular the very steep density gradient in the chromosphere (n_H), and the predominantly neutral state of H in the chromosphere, as contrasted with its ionized state everywhere above (cf. n_e vs. n_H). An enlargement of the photospheric-chromospheric part of fig. 1 is shown in fig. 2.

Fig. 1 is only meant to very roughly fix ideas, especially since plane parallel models are totally unrealistic at low altitudes : the material is actually extremely clumpy (filling factor ~ 1 %), and essentially all the observed material in the chromosphere and lower corona lies within discrete loops, as depicted in fig. 4. In particular, most of the material observed at conventional "Transition Region" temperatures (~ 30000 K to ~ 1 MK) actually does *not* lie in an interface layer between chromospheric (~ 10000 K) and coronal (~ 1.5 MK) temperatures, but rather within individual, fairly isothermal loops (*e.g.*, Feldman 1983; Antiochos & Noci 1986). Spherical symmetry is not either realistic for the large scale description of the heliosphere (*e.g.*, Wang & Sheeley 1990; Wang et al. 1990; Sheeley 1992a; Withbroe 1988) ; but it is sufficient here to describe the relevant orders of magnitudes for the altitude.

Since in the chromosphere and lower corona the transport of gas is controlled by magnetic fields, the structure of these fields may be essential for the description and understanding of the composition fractionations that take place. Fig. 3 gives an artist's conception of the heliospheric magnetic fields, in which I will distinguish two types of regions : *(i) "closed-field regions"*, in which the low altitude field is dominated by closed loops ; they

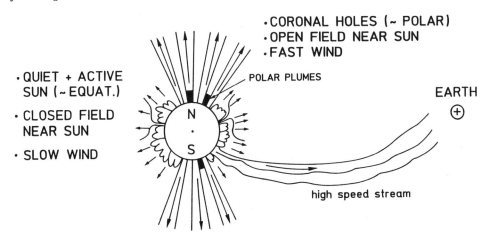

Fig. 3 : An artist's view of the heliosphere. Two types of regions are distinguished (§ 2.1). (i) Closed-field regions on the solar surface, i.e. the "quiet" and "active" Sun, predominant in the equatorial regions, which are associated with the slow SW regions in the distant interplanetary medium ; note that it is not clear whether the slow SW originates predominantly in evaporating large, old loops, or rather in narrow open-field channels within predominantly closed field regions, which fan out above the closed-loop altitude (§ 2.4). (ii) Open-field regions, i.e. coronal holes, predominant in the polar regions, which are associated with fast SW in the distant interplanetary medium ; fast SW streams are, however, sometimes found even near equator, where they can be observable from Earth. Observations on polar plumes (vertical structures within polar coronal holes, some 10 times denser than their environment) will be crucial in this study (§ 5.2.1).

correspond to the *"quiet"* and *"active"* [4] regions of the Sun, and are the source of the *slow Solar Wind (SW)* ; they are predominant in the solar equatorial regions ; *(ii) "open-field regions"*, in which the low altitude field opens directly into the interplanetary medium ; they correspond largely to larger *"coronal holes"*, which emit low density, *fast SW* ; they are largely confined to the polar regions of the Sun, although open-field, high-speed SW streams are also occasionally found at lower latitudes, from where they can easily reach the Earth ; the polar coronal holes contain denser vertical features, the "polar plumes", which will be important in the subsequent discussion. Let me insist that these denominations of "closed-" and "open-field regions" in the heliosphere refer to the structure of the field *at low altitude*, in the regions of the lower corona which are magnetically connected to the region of the heliosphere under consideration. Most closed loops do not extend beyond, say, 150000 km $\approx$ 0.20 R_0 above photosphere, and all fields are roughly radial around ~ 1 Mkm $\approx$ 1.50 R_0 (Schatten et al. 1969; Wang & Sheeley 1990) [5].

Of course, this description is extremely schematic. For instance, there are probably narrow open field channels within predominantly closed-field regions, and these may well be the actual source of the slow SW (fig. 3 ; § 2.4). Also, sunspots, while closely related with solar active regions, are widely open-field structures within predominantly closed-field regions (§ 5.1.2b).

Let me now review the various types of available composition observations, and the sites in the heliosphere which they are relevant to. These are illustrated in fig. 4. I will treat separately those observations which refer to the compositions at low altitude, in specific sites near the solar surface (~ 5000 to 40000 km, *i.e.* 0.01 to 0.06 R_0 above photosphere), and those relevant to large scale compositions of the outer corona and the interplanetary medium far away from Sun (~ 1 to 600 R_0).

2.2   Observations Relevant to Compositions *Near* Sun : Specific Sites on the Solar Surface

(a) Photosphere : The solar photospheric composition will serve us as a standard for normalization of all other compositions in the solar environment. It is essentially based on Anders & Grevesse's (1989) review, with the latest developments discussed by Grevesse and Noels (1993) in this volume. It is now very accurately known for most elements, the meteoritic and spectroscopic determinations agreeing remarkably well whenever available (including for Fe). There is also a perfect agreement (fig. 11) between the "photospheric" abundances of Ne (observable neither in the photospheric spectrum, nor in meteorites) derived *(i)* from the assumption that the Ne/O ratio in the solar photosphere is equal to that observed everywhere outside the solar system, in HII regions spanning a wide range of metallicities and in hot stars (Meyer 1989 [6]), and *(ii)* from EUV observations of Ne/O

---

[4]   See footnote # 8.

[5] R_0 = solar radius ; all altitudes are given above the photosphere ($\tau_{5000 Å} = 1$).

[6]   The recent infra-red Ne/O value in the Orion nebula now agrees with the visible determinations (Rubin et al. 1991). The constancy of the Ne/O ratio for all values of O/H is confirmed by additional observations of blue compact galaxies by Peña et al. (1991). See fig. 8 of Meyer (1989).

in the gases of an erupting prominence and of a compact flare, which have retained the composition of the original photospheric gas, as evidenced by the value of their O/Mg ratio (~ 24 as in the photosphere, rather than ~ 5 as in the corona ; Widing et al. 1986; Feldman & Widing 1990) [7] .

(b) γ-ray spectroscopy : γ-ray lines resulting from the nuclear interactions between energetic protons and heavier ions confined within a loop with the ambient material can inform us about the compositions of both target and beam (Murphy et al. 1991). The time scale of the emission can yield lower limits to the density of the bombarded medium, which seems often very high (~ 10^{15} cm^{-3}), typical of the lower chromosphere or even photosphere (fig. 1, 2, 4 ; Hua et al. 1989; Rieger 1991). The best studied event, that of 27 April 1981, had a long duration, so that a major fraction of the γ-ray emission could originate in the lower corona (fig. 4 ; Murphy et al. 1991; Rieger 1991; Hulot et al. 1992).

(c) EUV and X-ray spectroscopy : This field is currently thriving ! More and more *specific sites* are being observed at low altitudes in the corona (~ 5000 to 40000 km, *i.e.* 0.01 to 0.06 R_0) : *(i)* in closed-field regions, quiescent and erupting prominences, active regions, flaring loops [8], sunspots, and *(ii)* in open-field regions, polar plumes in coronal holes, and other parallel or diverging field regions. Average Mg/Ne ratios some ~ 5000 km ≈ 0.007 R_0 above the limb are also obtained both over quiescent + active closed-field regions and over coronal holes (fig. 4 ; Sylwester et al. 1984,1988; Lemen et al. 1986; Widing et al. 1986; Antonucci et al. 1987; Noci et al. 1988; Strong et al. 1988,1991; Sylwester 1990; Widing & Feldman 1989,1992a,b,c; Feldman & Widing 1990; Feldman et al. 1990; Doschek & Seeley 1990; Doschek & Bhatia 1990; Doschek et al. 1991; Philips & Feldman 1991; Fludra et al. 1991,1993; McKenzie & Feldman 1992; Sterling et al. 1992; Feldman 1992,1993; Widing 1993; Schmelz 1993; Schmelz & Fludra 1993; Saba & Strong 1992,1993a,b).

(d) Solar Energetic Particles (SEP's) from impulsive events (largely ^{3}He-rich events) : They sample the gas in or near a flaring loop (T ~ 4 to 30 MK), at typical altitudes of ~ 20000 km ≈ 0.03 R_0 in closed-field regions (fig. 4). In spite of large distortions of the composition in the acceleration process, these particles do give us some information regarding the composition of the gas from which they have been first extracted (*e.g.*, Reames 1990; Reames et al. 1990,1993).

---

[7] We stress that these determinations are *not* derived from the observation of FIP-biased material of coronal origin ; they, therefore, do *not* depend upon the *assumption* of equal depletions of O and Ne relative to Mg in the corona, as compared to photosphere (Breneman & Stone 1985; Meyer 1989).

[8] Quiet coronal loops have a typical temperature of 1.5 MK ; they are usually not bright enough to be observed (except when enhanced by limb brightening, § 5.1.2d and 5.2.2c). When magnetic energy is progressively released in a loop, the loop reaches typical temperatures of ~ 3 MK, and is called "active". A group of active loops is called an "active region". When magnetic energy is brutally released within an active loop, it can heat it to temperatures from ~ 5 up to ~ 30 MK, and the loop is called "flaring". A flaring loop within an active region often heats nearby active loops to some ~ 4 MK. Active and flaring loops are commonly observed in EUV's and X-rays.

2.3   Observations Relevant to Compositions *Far From* Sun : Large Scale Averages in the Interplanetary Medium

(a) Slow Solar Wind (SW) : It is by and large associated with quiet closed-field regions near Sun (fig. 4). However, as discussed in § 2.4, its material might well *not* originate in closed loops near solar surface. The earlier beam foil data and the ISEE-3 data refer to this slow wind (*e.g.*, Gloeckler & Geiss 1989).

(b) Dense Solar Wind : Coronal Mass Ejections (CME's) are large scale ejections of gas originating high in the corona, *not* caused by a flare on the Sun ; however, the same large scale perturbation of the coronal field often provokes both a high altitude CME and a flare on the solar surface (fig. 4 ; *e.g.*, Kahler 1992). Denser SW's are found associated with CME's : *both* the driver gas of the CME itself *and* ordinary, ambiant SW compressed by

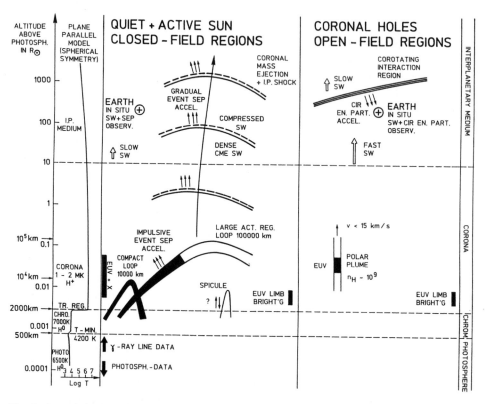

Fig. 4 : An artist's view of the various features in the solar atmosphere and interplanetary medium. The vertical scale is a logarithmic altitude scale above the photosphere ($\tau_{5000 \text{ Å}} = 1$). Left : plot of the temperature in the nominal, idealized spherically symmetric model of fig. 1. Center and right : relevant features in closed- and in open-field regions of the corona and associated interplanetary medium, shown at their approximate altitude. The blackened areas indicate the typical altitudes of the gases analyzed by spectroscopy. See text §§ 2.2 and 2.3.

the CME associated forward shock (fig. 4). They are observed by the AMPTE spacecraft within the Earth magnetosheath, when the magnetosphere is compressed by the high pressure of the dense SW (Gloeckler & Geiss 1989; von Steiger et al. 1992a).

(c) Fast Solar Wind : This coronal hole SW is associated with open-field structures near Sun (fig. 4). It has been observed by ISEE-3, AMPTE (within the Earth magnetosheath, when the magnetosphere is compressed by the high pressure of the fast SW), and Ulysses spacecrafts (Ipavich et al. 1986; Gloeckler et al. 1989; von Steiger et al. 1992a,b).

(d) Solar Energetic Particles (SEP's) from large, gradual events : Gradual event SEP's are accelerated out of the SW by CME's throughout the upper corona and interplanetary medium, up to beyond the orbit of Earth (fig. 4 ; *e.g.*, Kahler et al. 1978; Kahler et al. 1984; Cane et al. 1988; Kahler 1992). So, they sample the slow SW composition in the large- scale-corona and interplanetary medium associated with closed-field regions. They are therefore to be associated with low altitude material within closed loops *inasmuch as* the slow SW material itself is (§ 2.4). The data are very accurate and comprehensive (Meyer 1985a; Breneman & Stone 1985; McGuire et al. 1986; Stone 1989; Cane et al. 1991; Reames et al. 1991; Reames 1993). The derivation of the parent gas composition from the variable observed compositions resulting from rigidity (A/Q) dependent acceleration is well mastered (Meyer 1985a; Breneman & Stone 1985; Stone 1989; Paper I) [9]. In addition, Mazur (1991) has recently interpreted the energy dependence of the observed compositions in terms of Q/A-dependent acceleration, and shown that, at low energies, all SEP abundances approach those in their parent gas. This has, in particular, allowed him and Reames (1993) to derive, for the first time, the abundances of Fe (which earlier had to be assumed) and of H in the parent gas (footnote # 14 ; § 3.2).

(e) Energetic Particles from Corotating Interaction Regions (CIR) : In the corotating garden-hose spiral interplanetary field pattern, the overtaking of a low-speed SW stream by a coronal hole high-speed stream (CIR) generates a strong reverse shock, which accelerates particles out of the fast coronal hole SW (open-field region). This takes place some $\sim$ 3 a.u. $\approx$ 600 R_0 away from the Sun (fig. 4 ; *e.g.*, Reames et al. 1991).

2.4 A Genetic Relationship Between the Gases Observed at Low and at High Altitude ?

Let me now ask a key question : should the large scale compositions observed far from Sun be averages over those of the features observed in magnetically connected regions of the solar surface ? In other words : does the SW gas observed at large distance (either directly, or in the form of SEP or CIR accelerated particles) originate in the features observed on the solar surface ? For the open-field regions and their associated fast coronal hole SW, this is most likely the case. But for closed-field regions it might well *not* be the

---

[9] This derivation should be based on the meteoritic determination of the photospheric Fe abundance, which has now proven to be correct (*e.g.*, Grevesse 1992a). With this value, the Q/A-dependent correction to be applied to Breneman & Stone's average SEP abundances virtually disappears : so, we recommend to use their *uncorrected* SEP abundances as coronal abundances.

case. The low altitude observations primarily refer to active and flaring loops (§ 2.2). Is the slow SW essentially made of material from evaporating large, old, active (and quiet ?) loops ? This may be suggested by some recent Yohkoh observations. But a large contribution from active loop material at ~ 3 MK is *not* suggested by the slow SW freezing in temperatures of ~ 1.5 to 2 MK (Feldman et al. 1981). Though by and large associated with predominantly closed-field regions, the slow SW might actually well originate in *narrow* open-field structures structures squeezed between closed-loop systems at low altitude, which open up widely above the closed-loop level, some ~ 150000 km ≈ 0.20 R_0 above photosphere (Wang & Sheeley 1990; Wang et al. 1990; Sheeley 1992a [10] ; fig. 3).

3   BASIC CORONAL COMPOSITION PATTERN , AND THE PROBLEMS OF HYDROGEN AND HELIUM

3.1   Basic Heavy Element Composition Pattern

For the basic composition pattern of the corona as it was known some two years ago, I refer the reader to Paper I. As shown in fig. 5 taken from that paper, the bulk coronal heavy element composition differs from that of the photosphere by a bias controlled by the First Ionization Potential (FIP) of the element, the heavy elements with FIP < ~ 10 eV ("low-FIP elements") being all comparatively ~ 4 to 5 times as abundant as those with FIP > ~ 10 eV ("high-FIP elements"). The evidence for this bias comes mainly from accurate observations of gradual event SEP's, which are confirmed by slow SW data and by EUV and X-ray spectroscopy of coronal loop gas. It is therefore mainly relevant to *the interplanetary medium and probably the average large-scale upper-corona over closed-field regions*. It is *this* well-determined pattern that has been, more or less properly, referred to as *"coronal"* in the past. Here I will stick with this usual denomination, or call it *"large-scale-coronal"*. It is this pattern that appears schematically on fig. 6, 7 and 9. Let me recall that that same basic composition pattern is found in Galactic Cosmic Rays (Meyer 1985b).

3.2   Absolute Heavy Element Abundances Relative to Hydrogen

For a long time, however, it has been a difficult task to anchor this relative heavy element abundance pattern to H, since *(i)* in the corona fully stripped H has no lines and produces only a continuum, *(ii)* the SW data did not yield a clear cut behaviour for H, and *(iii)* it was not easy to interpret the highly variable H-to-heavy-element ratios observed in SEP's in terms of abundances in their source gas. The apparently clearest piece of evidence we had was a single X-ray line-to-continuum study by Veck & Parkinson (1981), yielding a "high" H abundance relative to heavies, indicating that low-FIP elements had photospheric abundances relative to H, while the high-FIP ones were depleted (fig. 6) (Meyer 1985b). But line-to-continuum studies are a delicate matter ! In the past few years, and especially

---

[10]   According to these authors, the SW speed is primarily controlled by the rate of divergence of the local field between the solar surface and ~ 1.5 R_0 above the surface : the *larger* the divergence, the *slower* the SW. In predominantly closed-field regions, there is no hindrance to a wide fanning out of narrow, squeezed open flux tubes above the closed coronal loops (say, above ~ 0.20 R_0). In coronal holes, by contrast, the flux tubes cannot expand much faster than $\propto r^2$.

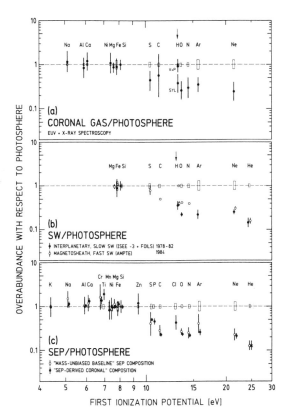

Fig. 5 : Overabundance of the elements in the corona relative to photosphere vs. First Ionization Potential (FIP), as it was known some two years ago (§ 3.1 ; from Paper I). Normalized to low-FIP Si. The boxes represent the uncertainties on the photospheric abundances. The data, from (a) coronal gas spectroscopy, (b) slow and magnetosheath Solar Wind, and (c) gradual event Solar Energetic Particle, are essentially relevant to the closed-field regions of the Sun (§§ 2.2c, 2.3a,d ; except for the magnetosheath SW data, § 2.3b,c). The general overabundance by a factor of ~ 4.5 of "low-FIP" elements with FIP > ~ 10 eV relative to the "high-FIP" heavy elements is conspicuous. Note that the SEP C/O/Ne ratios are equal to photospheric, and in particular that C [FIP = 11.3 eV] behaves strictly as a high-FIP element in SEP's, where its abundance is very well determined (§ 6.1.1a). Similarly, elements with very low FIP (K,Na,Al,Ca) do not show any trend for an overabundance relative to Mg,Fe,Si (footnote # 28).

very recently, several new, independent evidences have been obtained, showing that the above conclusion was erroneous and that the H abundance is "low" : *high-FIP elements have photospheric abundances relative to H (which simply behaves like heavier high-FIP elements), while low-FIP ones are enhanced.* This actually agrees with earlier findings from difficult visible forbidden lines studies. I now discuss these new (and old) evidences, summarized in fig. 6 :

(i) SW : a "low" H abundance, roughly consitent with a photospheric H-to-high-FIP-element ratio, is now found in both the slow SW and the high-pressure SW observed by AMPTE when the magnetosphere is compressed (Gloeckler & Geiss 1989) [11] . While the SW H-to-heavies ratio might differ somewhat from that in its source gas, it seems unlikely that fractionation during SW acceleration should lead to a *lowering* of this ratio, since it is H that drags along heavies in the SW (Geiss et al. 1970 ; Bürgi & Geiss 1986). So, the "low" H abundance found in the SW should be relevant to the coronal gas.

---

[11] Note that this high-pressure SW is a mixture of dense SW of CME driver gases, of slow SW compressed by the CME shocks, and of fast coronal-hole SW (in which almost no FIP-bias whatsoever is found, so that its H abundance relative to high- and low-FIP elements is irrelevant ; § 5.2.2b ; Gloeckler et al. 1989).

(ii) X-ray and EUV spectroscopy : in contrast to Veck & Parkinson (1981)'s earlier study, later line-to-continuum studies of Ca,Fe/H have yielded a "low" H abundance (Sylwester et al. 1984,1988; Lemen et al. 1986; Fludra et al. 1991,1993; Sterling et al. 1992). In addition, the very high luminosities of Na and Ca lines found in polar plumes (in which very large amplitudes of the FIP-bias are observed, § 5.2.1) qualitatively suggest that the abundances of these elements are strongly enhanced relative to H (Widing & Feldman 1992a,c; Feldman 1992).

(iii) H, C, O, Si lines near sunspots : studying the hot material around a sunspot, Feldman et al. (1990) have shown that the gas has a photospheric-type composition right above the sunspot (low Si/CNO), and a coronal-type composition in the neighbouring plages (high Si/CNO) (§ 5.1.2b). They managed to analyse also very low temperature lines, including the Ly α line of H°, and observed that the relative intensities of the H and C, O lines above the sunspot and in the plages behave alike, while the Si lines are comparatively much more enhanced in the plages. This seems to be clear evidence that in coronal-type compositions, Si is enhanced relative to H (and not C, O depleted).

(iv) : Gradual SEP events : in SEP's, it has always been a difficult task to relate the observed highly variable abundances of H (Q/A = 1) to those of heavier elements

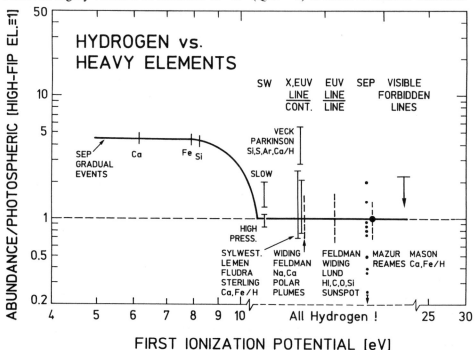

Fig. 6 : Anchoring the heavy element FIP-biased coronal composition pattern to Hydrogen. The basic gradual event SEP pattern of overabundances relative to photosphere is shown vs. FIP (fig. 5, 7), normalized to high-FIP Oxygen. But the abscissa scale has been interrupted, and all points refer to Hydrogen ! They are discussed in § 3.2. Most of the data imply that low-FIP heavy elements are overabundant (and not high-FIP ones underabundant) relative to Hydrogen, which behaves like heavier high-FIP elements.

(Q/A $\leq$ 0.5), due to the differences in their Q/A-dependent acceleration conditions. But recently, Mazur (1991) has shown that, at low energies, the SEP abundances approach those in their parent gas, whatever Q/A (§ 2.3d). The parent gas H/O ratios he inferred for 10 events are shown in fig. 6. While there is still a lot of scatter, these data clearly point to a "low" H abundance in the coronal gas. The low energy H/O ratio subsequently obtained by Reames (1993), averaging over a number of events, leads to the same conclusion.

(v) : Earlier visible forbidden line abundance studies : they are commonly considered quite unreliable, because of difficulties in calculating the excitation rates largely controlled by cascades and by radiative excitation (*e.g.*, Meyer 1985b). Note, however, that the one such study based on a thorough study of the atomic physics (Mason 1975a,b) did yield an enhancement of Fe/H and Ca/H by factors of ~ 2 to 3 (and possibly more) in the corona, *i.e.* a "low" H abundance in fig. 6 .

So, it is at last clear that, in absolute terms, *i.e.* relative to H, low-FIP heavy elements are enhanced in the corona (and *not* high-FIP elements depleted).

3.3   The Helium Problem

In the corona, the He abundance cannot be reliably determined from spectroscopy (*e.g.*, Feldman 1992). We therefore have information only from SW and energetic particle measurements.

Far from Sun, the He/H ratio in the various types of SW is typically 2 to 3 times lower than in the photosphere, and quite variable (Neugebauer 1981; Ogilvie et al 1989; Gloeckler & Geiss 1989; Gloeckler et al. 1989; ref. in Bürgi 1992a) ; the discussion of these variations goes far beyond the scope of this paper. Consistently, comparable deficiencies of He by a factor of ~ 2 relative to other high-FIP elements and H are found in the gradual event SEP data, which are accelerated out of the slow SW (Meyer 1985a; McGuire et al. 1986; Stone 1989; Mazur 1991; Cane et al. 1991; Reames et al. 1991; Reames 1993).

Regarding the open-field regions specifically, there is an apparent contradiction between the well defined, stable depletion of He/H by a factor of 2 found in the fast SW (above ref.) and the *absence* of a depletion relative to other high-FIP elements found in the CIR accelerated particles (fig. 9 ; Reames et al. 1991).

The key question is : to which point are these SW He deficiencies *relative to other high-FIP elements* due to fractionations between photosphere and corona, *and/or* in the SW formation region at a few R_0 ? In the SW formation region, He is actually expected to be less efficiently dragged along by H than all heavier elements in the SW (Geiss et al. 1970; Bürgi & Geiss 1986; Bürgi 1992a,b). One may therefore seriously consider the hypothesis that the low SW He abundance be due *only* to a fractionation at this high altitude level, and that He behaves like H and heavier high-FIP elements in the corona.

The *only* available piece of data relevant to low altitudes, below the SW formation region, is provided by impulsive event SEP's. It also yields ^{4}He/C,O ratios low by a factor of ~ 2 as compared to photosphere ! In all likelihood, these ratios should reflect the composition of the parent gas, in or near flaring loops (fig. 7 ; Reames et al. 1990,1993).

We conclude that, if our interpretation of the significance of the impulsive event SEP data is valid, the factor of ~ 2 deficiency of He relative to other high-FIP elements is primarily due to a fractionation between photosphere and corona, rather than in the SW formation region.

4   BASICS FOR MODELS

4.1   Main Points

I refer the reader to basic, general comments in § VI of Paper I. In brief, a FIP-biased composition implies an ion-neutral (i-n) fractionation. This can take place only in a medium in which neutrals and first ions *exist*, *i.e.* in a gas at T < ~ 10000 K, and not subjected to an intense UV or X-ray radiation [12] . Such gases are found in the chromosphere and in the photosphere (and possibly in transient jets originating in these media, such as possibly spicules [13]). Since the tenuous corona is, one way or another, fed, *either* from the much denser underlying chromosphere, *or* directly from the photosphere (via jets), the following consistent framework for models can be suggested : *(i)* the "low-FIP" elements found overabundant in the corona are those which were predominantly ionized in the chromosphere/photosphere, and the "high-FIP" elements, which are found not enhanced, those which were neutral ; this statement is roughly consistent with observation (fig. 2) ; and *(ii)* for some reason, the chromospheric/photospheric ions rise into the corona *on the average* ~ 4.5 times as efficiently as their neutral fellows, at least in closed-field regions.

Let me insist that the observations impose that the fractionation process be *independent of mass* [14]. The specific problem of the He deficiency must also be addressed.

4.2   The Thermostated Chromospheric Temperature-Plateau

Apart from the photosphere itself, the gas of the chromospheric T-plateau at ~ 6500 K constitutes an obvious material in which the i-n separation could take place (fig. 2).

Let me elaborate on the thermostating of this T-plateau (Athay 1981). The chromosphere receives mechanical energy. How does it radiate it away ? At ~ 6500 K, collisions cannot ionize H ; but they can excite it to the n = 2 level. From that level, visible photospheric

---

[12]   This condition probably allows to exclude i-n separation models based on thermal diffusion in the high T-gradients of the Transition Region, in which all elements are essentially ionized.

[13]   It is highly debated whether spicules represent, *either* a real movement of matter with a net mass of gas delivered to the corona, *or* a real movement of matter with *all* rising gas falling down again within the same channel, *or* merely a traveling temperature wave within a static gas.

[14]   Observations show the *lack of a mass fractionation*, both between H and other, heavy high-FIP elements (§ 3.2 ; fig. 6), and between Fe and Mg,Si among low-FIP elements (fig. 5, 7, 9). The latter statement is based on data on slow, dense and fast SW (Gloeckler & Geiss 1989), CIR energetic particles (Reames et al. 1991), γ-ray lines (Murphy et al. 1991) and X-ray lines (McKenzie & Feldman 1992; Schmelz 1993; see, however, Saba & Strong 1993b) (fig. 7, 9). In the rich gradual event SEP data, the Fe/Mg,Si ratio is variable, and its value in the parent gas had to be *assumed* in conventional analysis (Meyer 1985a; Breneman & Stone 1985; Paper I) ; based on Mazur's (1991) considerations (§ 2.3d), Reames (1993) has now been able to derive a Fe/Mg,Si ratio in the parent gas, which confirms the above statement.

photons above 3.4 eV can easily photoionize H. In brief, the H ionization is controlled by its collisional excitation to the n = 2 level. H can recombine, either radiatively, or collisionally ; in any case, the medium is optically thick for all emitted Ly-α and Ly-cont photons, which get reabsorbed on the spot. So, no energy can escape at these wavelengths, and H plays a negligible role in the radiative budget ! But the degree of H ionization controls n_e ! And we have just seen that it is controlled by the rate of collisional excitation to the n = 2 level, which is *extremely* sensitive to temperature : it increases by a factor of 1000 between 6000 and 8000 K ! Now, the main radiation loss in the chromosphere goes through optically thin Ca II and Mg II lines ~ 3 eV, whose excitation rates are $\propto n_e$. So, the thermostating cycle goes as follows :

$$T \uparrow \quad \to \quad n_{H^+} \approx n_e \uparrow\uparrow\uparrow \quad \to \quad Ca,Mg \ excit. \uparrow\uparrow\uparrow \quad \to \quad Ca,Mg \ emiss. \uparrow\uparrow\uparrow \quad \to \quad T \downarrow .$$

This stabilizing process works as long as an increase of the temperature can liberate more electrons, *i.e.* as long as $n_{H^+} \ll n_H$. When H becomes close to totally ionized, the number of free electrons can no longer increase, and the temperature diverges : this is the start of the transition region [15].

Two important remarks are in order regarding this thermostating of the chromosphere :

(i) Since the temperature in the chromospheric T-plateau is controlled by the H collisional excitation to n = 2, we expect the dividing line ionized and neutral elements in the chromosphere to lie by the Ly α = 10.2 eV energy. That is exactly where the step in the observed FIP pattern lies !!! If the i-n separation takes place in chromospheric T-plateau material, this coincidence is no chance !

(ii) This thermostating is due to very simple, basic physical phenomena, of very wide applicability. All that is required is the presence of predominantly neutral H, and the lack of very energetic radiations. It is therefore not surprising that main sequence stars which maintain neutral H on their surface, *i.e.* those over the broad range from F to M, are all observed to possess chromospheres very similar to that of the Sun, and with very similar T-plateau temperatures (*e.g.*, Linsky 1980). This brings us back to the GCR particles, and to the suggestion that they were first extracted from the outer atmosphere of ordinary later-type F to M stars (perhaps in the form of MeV energetic particles ; Meyer 1985b) : the step in their FIP pattern lies exactly at the same place as that in the solar coronal pattern (see, however, § 1) !

4.3   Ion-Neutral Fractionation in Chromospheric or Photospheric Material : Observational Clues ?

Unfortunately, the observed division between low- and high-FIP elements in the corona matches well that between predominantly ionized and predominantly neutral elements, *both* in the chromospheric T-plateau (VAL; Geiss & Bochsler 1985 ; fig. 2), *and* throughout the photosphere down to an altitude of − 60 km (Grevesse 1992).

---

[15] Although there concepts have been largely developed in the framework of unrealistic plane parallel models (e.g., VAL; Fontenla et al. 1990,1993), they have a more general applicability, *e.g.*, to the gas within the magnetic tubes of force at the footpoint of a loop.

One element might, however, give us a clue : Carbon. Carbon has a metastable level at 1.3 eV, with I.P. = 10.0 eV, which can be photoionized by Ly α photons (von Steiger & Geiss 1989). In the chromospheric T-plateau, the fraction of neutral C may therefore be very temperature dependent. According to the calculations of VAL (see also Geiss & Bochsler 1985), C is very roughly ~ 50 % ionized and ~ 50 % neutral in the T–plateau (fig. 2). More recent work however shows that the degree of ionization of C is very model dependent, and suggests that C may be much more neutral than previously estimated (Fontenla et al. 1990,1993; Fontenla 1992). In the photosphere, by contrast, where LTE prevails, C clearly remains essentially neutral down to an altitude of – 60 km (8000 K ; Grevesse 1992). But the degree of ionization of C is, of course, also very sensitive to any heating of photospheric/chromospheric gas during its ascent in any kind of upward jet (*e.g.*, von Steiger & Geiss 1989).

More generally, the position of the step in the FIP pattern is largely a reflection of the temperature of the material in which the i-n fractionation takes place. So, a closer look at the abundances, and possibly abundance *variations*, of elements with FIP close to the step (especially C, but also Si, S, and even perhaps O) may be of interest to coin down the medium in which the i–n fractionation takes place. I will address this point in § 6.

4.4   Possible Mechanisms for the Ion-Neutral Fractionation ?

Averaged over all sites in which material rises into corona within closed-field regions, chromospheric [16] ions rise into corona ~ 4.5 times as efficiently as chromospheric neutrals (discussion in Paper I). This process is independent of mass (§ 4.1). This is what any mechanism has to account for.

We do not know whether the mechanism is widespread, common, and always yielding locally an enhancement factor on the order of 4.5, or whether there are rarer, specific sites which inject material highly enriched in low-FIP ions (by factors >> 4.5). This problem of the genesis of the coronal composition is, of course, deeply related with the broader problem of the filling of the corona : which are the main channels by which matter rises from photosphere/chromosphere into corona, in particular in closed-field regions ? spicules, or similar, smaller jets ? chromospheric evaporation and loop dissipation ? steady, narrow open-field channels ? velocity filtering of chromospheric non-thermal tails (Scudder 1992) ?

The demand that the fractionation process be independent of mass imposes conditions on any model. The fractionation mechanism itself may be independent of mass. It may also depend on mass, but be efficient enough for the process to "saturate" : the gas to be supplied to the corona must then get, *either* emptied of *virtually all* its neutrals of all masses, *or* enriched with *virtually all* the ions of all masses from some neighbouring reservoir.

Three types of mechanisms have be considered, which have been recently reviewed by Hénoux & Somov (1993) :

---

[16]   Hereafter, I will often write "chromospheric" for brevity, while strictly speaking what is meant is : "chromospheric or photospheric".

(i) Simple diffusion of *neutrals* out of some gas, while ions are fixed by the magnetic field ; this gas, thus impoverished in neutral species, feeds the corona. Vauclair & Meyer (1985) considered gravitational settling (which leads to a "high" H, in contradiction with the recent evidences, § 3.2), and von Steiger & Geiss (1989) mainly jets. Simple diffusion seems clearly not enough efficient to do the job with realistic material densities (collisions !!), and within realistic time scales for solar conditions (discussion in Paper I).

(ii) Diffusion *driven by electromagnetic forces* : such models are currently being developed (Ip & Axford 1991; Steinitz 1992; Antiochos 1993; Hénoux & Somov 1993; Koutchmy & Lorrain 1992; Fontenla & Avrett 1993; Tagger et al. 1993). Let me insist that such models can work along two very different lines : *either* diffusion of *ions*, driven by an electric field, *into* a gas to be supplied to the corona (*e.g.*, Antiochos 1993) ; *or* diffusion of *neutrals,* e.g. driven by a pinching of the gas, *out of* such a gas (*e.g.*, Hénoux & Somov 1993).

(iii) The apparent excess of low-FIP elements has been interpreted as mimicked by an actual excess of *refractory* elements, originating in the evaporation of solid, meteoritic bodies at altitudes below ~ 7.5 R_0 (Lemaire 1990 ; cf. the "grain" hypothesis for GCR's in § 1). In view of the coronal density gradient (fig. 1), it seems very difficult to conceive that such an evaporation of meteoritic bodies could affect the gas composition of the various dynamic features observed as far down as ~ 5000 to 40000 km (0.01 to 0.06 R_0) above photosphere (§ 5). In addition, mirroring in the coronal fields will largely preclude a penetration of higher altitude ions (Geiss et al. 1992). The observed lack of low charge states in the SW also conflicts with the evaporation hypothesis (Geiss et al. 1992). In addition, fairly volatile Na, which is only partly condensed in most meteorites, (*e.g.*, Mason 1979; Meyer 1981d), is not less enhanced than other low-FIP elements in all available "coronal" composition data (fig. 5 ; Meyer 1985a; Breneman & Stone 1985; McGuire et al. 1986; Stone 1989; Paper I; Widing & Feldman 1992a,c; Reames et al. 1993). So, this type of model does not seem adequate.

So, clearly, the promising models are those of type *(ii)*. My purpose here is, however, not to discuss these models, but to (try to !) clarify the observational constraints they must meet, along two lines :

(i) try to identify the sites in which the i-n fractionation takes place, by investigating the variations of the amplitude of the FIP-bias in various sites (§ 5), and

(ii) look for variations of the temperature of the gas in which the i-n fractionation takes place, by investigating possible variations of the position of the step in the FIP-bias pattern, and in particular the behaviour of C (§ 4.3 and 6). More generally, variations of the *shape* if the FIP-bias pattern will be discussed in § 6.

5   VARIATIONS IN THE AMPLITUDE OF THE FIP BIAS

5.1   Closed Field Regions   ("Quiet" and "Active" Sun)

5.1.1 *Large distance averages (interplanetary medium and distant corona)*

We have direct observations of the slow SW, and very comprehensive data on gradual event SEP's, which are accelerated out of the upper coronal gas and the slow SW in the interplanetary medium (§ 2.3a,d). Both indicate a *permanent* large scale FIP-bias of the parent gas, by a factor of ~ 4.5 , which is quite stable. Its amplitude may fluctuate between factors of ~ 3.5 to ~ 6.5. Carbon always behaves strictly as a high-FIP element ! (fig. 5 ; Meyer 1985a; Breneman & Stone 1985; McGuire et al. 1986; Gloeckler & Geiss 1989; Stone 1989; Paper I; Garrard & Stone 1991; Cane et al. 1991; Reames et al. 1991; Reames 1993).

This well-determined basic pattern is shown schematically by the thick, solid line on fig. 7, and has been reproduced for reference in fig. 6 and 9. It is this pattern that has been, more or less properly, simply denoted *"coronal"* in the past. Though it is actually relevant only to *the interplanetary medium and the average large-scale upper-corona over closed-field regions,* I will stick here with this usual denomination, or denote it *"large-scale-coronal"*.

5.1.2 *Localized features near Sun   (lower corona)*

Here we have a wealth of recent data *(i)* from EUV and X-ray spectroscopy, on a number of elements such as O, Ne, Na, Mg, Si, Ar, Ca, Fe, and more particularly on the (low–FIP–Mg)/(high–FIP–Ne) ratio [17], and *(ii)* from impulsive event SEP's, which are believed to be accelerated in or close to the flare site.

All these data pertain to the gas composition at low altitudes, say, between 5000 and 40000 km (0.01 and 0.06 R_0) above the photosphere. I insist that *all* of them also pertain to gases at temperatures between ~ 30000 K and ~ 20 MK, *i.e. above* the region at ~ 6500 K where neutral and first ions can coexist, *above* the sites where the i-n fractionation can take place.

I will now analyse the data relevant to four different sites, in turn. The results, restricted to a few key elements characterizing the amplitude of the FIP-bias, are illustrated in fig. 7.

(a) Individual loop systems : active regions and flaring loops (EUV and X-rays) : individual loop systems can be observed only when they are active (~ 3 to ~ 5 MK) or flaring (~ 4 to ~ 20 MK).

Mg/Ne and Fe/Ne ratios covering the entire range from photospheric to ~ 4.5 times photospheric, *i.e.* the large-scale-coronal value (§ 5.1.1), are being observed. But higher

---

[17] From Mg VI/Ne VI, two ions whose emissivities peak around 400000 K and have very closely the same temperature dependence.

enhancements are *not* found [18] . Values close to photospheric are systematically found in small (h < ~ 10000 km ≈ 0.015 R_0), compact, loops, associated with a locally bipolar field [19], and coronal-like values in the observable footpoints of larger, dilute, loops, associated with a locally unipolar field (fig. 7 ; Widing & Feldman 1989,1992b; Feldman & Widing 1990; Strong et al. 1991; McKenzie & Feldman 1992; Feldman 1992; Sheeley 1992b; Saba & Strong 1993a,b).

The time evolution of one and the same active region has been observed over durations of a few days. As illustrated in fig. 8, each individual active region starts with small, compact loops with photospheric composition, which evolve into large, dilute loops with coronal composition : one sees the change in composition within one and the same active region, over time scales on the order of a day (Strong et al. 1991; McKenzie & Feldman 1992; Widing & Feldman 1992c; Saba & Strong 1992,1993a,b) [20] . It is even observed that, within one particular active region, newly emerging loops, at low altitude in the center of the active region have a photospheric composition, while larger loops at the outskirts of the same active region, often connecting to other regions of the solar surface, have a coronal composition (Sheeley 1992b).

This evolution of the composition of the loops forming an active region can be interpreted in two ways : *(i) either* the i-n fractionation takes place below the footpoint of every single loop, or at least every single *active*, loop (cf., *e.g.*, Antiochos 1993) ; it is then a very common phenomenon over the surface of the Sun ; this would be a very important conclusion indeed (hereafter *"local philosophy"*) ; *(ii) or* as loops age and grow, they reconnect with preexisting coronal loops, and their gases mix ; the emerging photospheric-type gas simply evaporates and is replaced by ambient coronal gas ; in that case, we have learned nothing regarding the sites of the i-n fractionation (!), and very specific sites *could* be responsible for the FIP-bias of the large scale corona (hereafter *"reconnection philosophy"*).

---

[18] This conclusion is based on all EUV studies of Mg/Ne (Widing & Feldman 1989,1992b; Feldman & Widing 1990; Feldman 1992), as well as on the X-ray studies of Fe/Ne by Strong et al. (1991) and McKenzie & Feldman (1992, hereafter MKF) based on Arnaud & Rothenflug's (1985, hereafter ARO) Fe ion fractions. In their current study of X-ray data from the FCS instrument, Saba & Strong (1993a,b, hereafter SS) compare abundances obtained with the Fe ion fractions of ARO and with those calculated recently by Arnaud & Raymond (1992, hereafter ARA) ; they also stress the dependence of the results on a possible resonance scattering affecting the Fe XVII line intensity, and on the poorly determined Fe XVIII emissivity. While with ARO, their range for Fe/Ne (~ 0.8 to 4 · photo) agrees with that of MKF, with the improved ARA estimates they obtain a much lower range (~ 0.4 to 2 · photo for most points, with an odd dependence of the mean abundance ratio with temperature). The two studies also differ on the Fe/Mg ratio, which is found constant and equal to photospheric (spread ≈ 30 %) by MKF, and variable between ~ 1 and 3 · photo for ARO and ~ 0.6 and 1.8 · photo for ARA in SS's study ; ironically, while MKF use ARO, SS's mean Fe/Mg agrees better with theirs for ARA than for ARO ! Some of the difference between the results of the two studies might be due to the narrower field of view of SS's instrument, which also samples preferentially the brightest, and most recently heated, sites within active regions. See footnotes # 29 and 31 for the discussion of the Ne/O ratio.

[19] It is in such a compact flaring loops that Feldman & Widing (1990) have directly determined the photospheric Ne and Ar abundances (§2.2a).

[20] Composition changes on a time scale of ~ 10 min. have even been found by Strong et al. (1991). But they disappear when the data are analyzed on the basis of Arnaud & Raymond's (1992) Fe ionization equilibria (Saba & Strong 1993a,b).

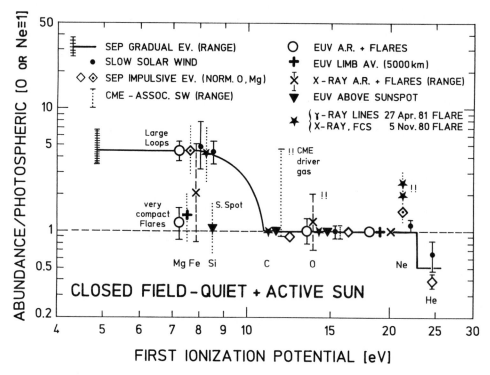

Fig. 7 : Overabundances relative to photosphere vs. FIP, for the principal elements observed in closed-field regions, associated with the quiet and active parts of the solar surface (§ 2.1). Normalized to high-FIP O or Ne. The basic gradual event SEP pattern (fig. 5) is schematically shown by the thick, solid line ; it can be seen that it agrees well with the slow SW points ; the amplitude of this "large-scale-coronal" FIP-bias fluctuates within a fairly narrow range, as indicated (§ 5.1.1). The various data are discussed in § 5.1 for the variation of the amplitude of the FIP-bias and § 6.1 for the variations of the shape of the pattern, especially of the position of the step (such variations are stressed by "!!" in the figure). This figure shows in particular : (i) that all amplitudes of the FIP-bias between no bias at all and the "large-scale-coronal" bias are observed on the solar surface, with no bias in very compact loops and large-scale-coronal bias in the footpoints of large loops (§ 5.1.2a) ; (ii) the associated lack of a FIP-bias in the average very low altitude limb (§ 5.1.2d) ; (iii) the lack of an excess of low-FIP Si above a sunspot (§ 5.1.2b) ; (iv) the suggestion from some X-ray data that O behaves as an intermediate-FIP element in active region and flare gases (low Ne/O ; § 6.1.2b ; fig. 11) ; (v) the presence of the FIP-bias in the impulsive event SEP's, the γ-ray line and FCS X-ray flare data, and their suggestion of an excess of Ne relative to C,O in more or less impulsive flares (§§ 5.1.2c, 6.1.2c ; in the impulsive flare data, C and He are normalized to O, and Ne is normalized to Mg ; these separate normalizations are required because selective acceleration mechanisms prevent us from relating the He,C,O and the Ne,Mg,Si abundances in the parent gas, Reames et al. 1993) ; (vi) the suggestion that C behaves as a low-FIP element in CME driver gas (§ 6.1.1b) ; (vii) the evidence from impulsive event SEP data that He is already deficient relative to other high-FIP elements in the coronal gas, below the SW formation region (§ 3.3).

Loop reconnection is observed to be an ubiquitous phenomenon, and the observed time evolution of the topography of active regions supports the its importance (Sheeley et al. 1975a,b). However, it is not clear whether reconnection can affect the larger part of the emerging bipolar flux, nor whether the dense photospheric-type gas can effectively evaporate and be replaced by more dilute ambiant coronal gas. The composition observations themselves may possibly suggest that the "reconnection philosophy" applies : if the fractionation takes place locally, it indeed sounds odd that all Mg/Ne ratios *between* photospheric and large-scale-coronal are observed, but that *higher* enhancements are *not*, while this is naturally understood if the photospheric-type gas is just being replaced by ambiant coronal gas.

(b) Sunspots (EUV) : in the gas situated right above the umbra of a sunspot, the Si/CNO ratio has been observed to be photospheric, while in the nearby plages (active regions) Si is enhanced as everywhere in the corona (fig. 7 ; Feldman et al. 1990; Doschek et al. 1991). Observations of the Mg VI/Ne VI line ratio indicate that his situation applies to some sunspots, but *not* to all sunspots as a rule (footnote # 17 ; Sheeley 1992b). We can immediately think of three possible explanations for this situation. In the framework of the "local philosophy" in which the composition above the sunspot is governed by what happens in the underlying chromosphere, *either (i)* no i-n fractionation takes place in the sunspot chromosphere because the field is too intense (Antiochos 1993), *or (ii)* the fractionation does take place, but Si and Mg are neutral in the cool sunspot chromosphere

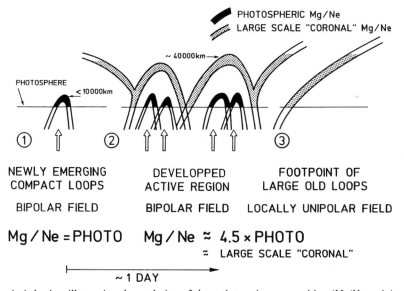

Fig. 8 : Artist's view illustrating the evolution of the active region composition (Mg/Ne ratio) over time scales of ~ a day (§ 5.1.2a). New compact, bipolar, loops emerge from the photosphere with photospheric Mg/Ne. Within ~ a day, they evolve into large, active regions with large-scale-coronal Mg/Ne, which are eventually observed as unipolar footpoints of large loops. It is even observed that low-lying compact loops with photospheric Mg/Ne continue to emerge from the photosphere below large scale active regions with large-scale-coronal Mg/Ne.

(Lites at al. 1987) and therefore behave as high-FIP elements [21]. In the framework of the "reconnection philosophy", *(iii)* the gas particles from reconnected ambiant coronal loops might be prevented from reaching the observed lower altitudes, due to mirroring in the very strong magnetic field gradient, which is indeed strongest vertically just above the sunspot [22].

(c) Impulsive SEP events (^{3}He-rich events) : they sample the gas composition at ~ 3 to ~ 20 MK in and/or near a flaring loop. In spite of very strong selective effects in the acceleration (depending on the Q/A ratios at 3 to 20 MK, not on FIP !), it seems clear that the material out of which the particles are accelerated has a coronal, not a photospheric composition (fig. 7 ; based on the comparative behaviors of Ne and MgSi ; Reames et al. 1990,1993).

(d)   Quiet and active Sun average composition at ~ 5000 km $\approx$ 0.007 R_0 above the limb (EUV) : limb brightening allows to observe an average Mg/Ne ratio for the "quiet Sun", probably pertaining to a mixture of quiescent and somewhat active loops (material clumpy ; filling factor ~ 1 %) at an altitude on the order of ~ 5000 km $\approx$ 0.007 R_0. The Mg/Ne ratio seems essentially photospheric (enhancement factor ~ 1.3; fig. 7 ; Widing & Feldman 1992c; Widing 1993 [23]) ! The most likely explanation for this lack of a FIP-bias is that, at the observed low altitudes, essentially newly emerging loops filled with photospheric gas are being observed (cf. *(a)* above ; fig. 4, 8) [24].

5.2   Open Field Regions
(Coronal Holes and Other High Speed Wind Streams)

5.2.1 Polar plumes and other diverging field regions

Polar plumes are denser, diverging field vertical structures within the polar coronal holes, whose composition is observed between altitudes of ~ 15000 and 40000 km (0.02 and ~ 0.06 R_0) (Ahmad & Withbroe 1977; Ahmad & Webb 1978; Widing & Feldman 1992a). Polar plumes, together with other diverging field regions, are the only places where *FIP-biases much higher than ~ 4.5*, the large scale average in closed-field regions, are being observed : Mg/Ne is typically enhanced by factors of ~ 8 to ~ 15, and possibly sometimes up to ~ 35, relative to photosphere (fig. 9 ; Widing & Feldman 1989,1992a,b; Feldman 1992) !

So we feel we have caught the site where the FIP-bias is being formed : open, diverging field structures !!! And fig. 10 does indeed suggest that the opening of the field is the keyparameter controlling the amplitude of the FIP bias (Widing & Feldman 1989,1992b;

---

[21] Abundance measurements of elements with much lower FIP, such as Na, Ca, Al could settle this issue.

[22] Anticipating § 5.2.2, we may also note that sunspots are widely open-field structures, and that the possible lack of a FIP-bias above some of them may be reminiscent of the low FIP-bias found in the large scale open-field coronal holes.

[23] Contrary to the earlier, very preliminary, reports by Feldman (1992) and Widing & Feldman (1992b).

[24] Other possible explanations : the i-n fractionation might proceed only in active, not in quiet, loops ("local philosophy") ; or : the quiescent loops might not reconnect as much as those associated with an active region ("reconnection philosophy").

Feldman 1992). Of course, we then expect the FIP-bias to be on the average larger in coronal holes, which are the main locus of open field structures. Let us check whether this is the case.

5.2.2 The average FIP-bias in coronal holes

We have three types of measurements relevant to the large scale FIP-bias in coronal holes :

(a) *Corotating interaction regions energetic particles (CIR)* : these are particles accelerated around 3 a.u. $\approx 600 R_0$ from the Sun, near the interface between a slow and a fast SW stream ; it is believed that the accelerated particles originate in the fast wind stream. In this population, Mg,Si,Fe/Ne,O,N ratios are enhanced by a factor of only ~ 2.5 relative to photosphere (fig. 9 ; Reames et al. 1991) [25].

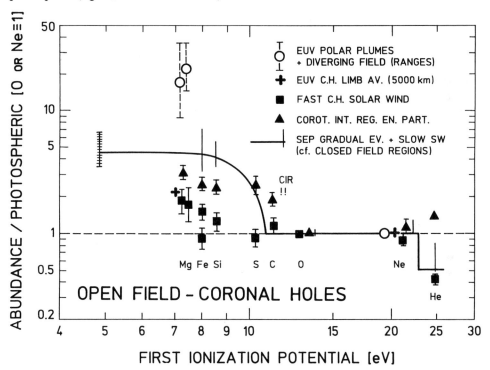

Fig. 9 : *Same as fig. 7, for open-field regions, i.e. coronal holes (§ 2.1). The basic gradual-event-SEP/slow-SW pattern in fig. 7 has been repeated for reference, although it refers to closed-field regions. The various data are discussed in § 5.2 and 6.2. This figure shows in particular : (i) the very strong FIP-biases observed in polar plumes and diverging field structures (§ 5.2.1) ; (ii) the fairly weak or even non-existent FIP-biases observed, by contrast, in the data relevant to the large scale coronal holes : very low altitude limb average, fast SW, and CIR energetic particles (§ 5.2.2) ; (iii) the evidence that C behaves as a low-FIP element, from CIR data (§ 6.2b) ; (iv) the apparent conflict between the fast SW data indicating that He is depleted like in closed-field regions (fig. 7) and the CIR data indicating that it is not (§ 3.3).*

---

[25] The behaviour of C is discussed in § 6.2b.

Fig. 10 : The Mg/Ne ratios, characterizing the amplitude of the FIP-bias, observed in a number of features on the solar surface, presented in order of increasing opening of their magnetic field (§§ 5.1.2a,d, 5.2.1, 5.2.2c ; adapted from Widing & Feldman 1992b). Solid points: detailed EUV emission measure analysis ; open points: estimates based on the Mg VI/Ne VI line ratio only. Only the closed error bars are significant ; they represent ranges of variation. The photospheric and "coronal" values are defined in §§ 2.2a and 3.1. This figure shows that the amplitude of the FIP-bias by and large seems to increase with the opening of the field structure. However, the low average coronal hole Mg/Ne ratio measured on the solar limb (as well as high altitude coronal hole data; § 5.2.2) shows that this relationship does not have a general applicability. Probably, part of the apparent correlation is due to the time evolution of the composition of active regions, which accompanies the growth of the scale size of their magnetic field (§§ 5.1.2a, 7.3.2; fig. 8).

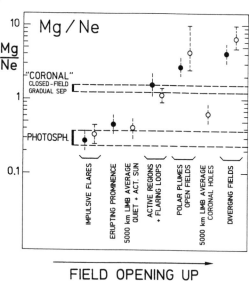

(b) Coronal hole SW : in the fast SW observed at 1 a.u. ≈ 200 R_0 , there is almost no FIP fractionation at all ! Mg,Si,Fe/Ne,O,C ratios are enhanced by factors between ~ 1 and ~ 2 relative to photosphere (fig. 9 ; Ipavich et al. 1986; Gloeckler et al. 1989; von Steiger et al. 1992a,b).

(c) Coronal hole average composition at ~ 5000 km ≈ 0.007 R_0 above the limb (EUV) : contrary to those on quiet Sun areas (§ 5.1.2d), coronal hole limb brightening data yield a significant average Mg/Ne enhancement, by factor of ~ 2.2 relative to photosphere (fig. 9 and 10 ; Widing & Feldman 1992c; Widing 1993). This enhancement, while lower than the large scale enhancement in closed-field regions, is quite comparable to that found in open-field regions at high altitudes (CIR energetic particles, fast SW).

5.2.3 Amplitude of the FIP-bias in open field regions - Conclusions

In polar plumes and other open, diverging field structures, we have Mg/Ne enhancements of ~ 8 up to ~ 35 relative to photosphere, much higher than the large scale enhancement of ~ 4.5 found above closed field regions. By contrast, all indicators of the *average* FIP-bias in coronal holes, both near and far from Sun, indicate low-FIP element enhancement factors between ~ 1 and ~ 2.5, distinctly smaller than those found in closed field regions (fig. 9). We are forced to conclude that polar plumes are not typical of the large scale coronal holes ; note that it is estimated that polar plumes provide some ~ 15 % of the mass of polar coronal holes (Ahmad & Withbroe 1977) [26] .

---

[26] Much higher figures are given by Ahmad & Webb (1978), based on an outflow velocity of > 100 km/s in the plume. Such high velocities are refuted by the lack of a departure from hydrostatic equilibrium found by Widing & Feldman (1992a), which implies outflow velocities of < 20 km/s.

6   VARIATIONS OF THE POSITION OF THE STEP IN THE FIP-BIAS PATTERN , AND OF ITS SHAPE

As discussed in § 4.3, the position of the step in the FIP-bias pattern, remarkably close to the Ly α energy [10.2 eV], is indicative of the temperature of the gas in which the i–n fractionation takes place (chromospheric T-plateau or photospheric material). In particular, we have seen that the ionized fraction of C may be as high as $\sim$ 50 % in the chromospheric T-plateau, and is certainly very temperature dependent in the neighbourhood of $\sim$ 7000 K, hence very sensitive to any heating in an upward jet of gas (§ 4.3). In order to help coining down the medium in which the i–n fractionation takes place, we now investigate possible *variations of the position of the step*, by looking more closely at the abundance variations of elements with FIP close to the step : Si [8.2 eV], especially C [11.3 eV], and even O [13.6 eV] [27].

Other variations of the *shape* of the pattern will also be investigated [28]. Recall, in particular, that the impulsive event SEP data indicate that the depletion of He [24.6 eV] by a factor of $\sim$ 2 to 3 relative to other high-FIP elements consistently observed in the SW and gradual event SEP data exists already in the coronal gas, *below* the SW formation region (§ 3.3).

6.1   Closed Field Regions ("Quiet" and "Active" Sun)

6.1.1 *Large distance averages (interplanetary medium and distant corona)*

(a) Slow SW and Gradual SEP events : in SEP's, the C/O/Ne ratios are remarkably constant, and equal to photospheric : C and O *always behave strictly* as high-FIP elements (fig. 5, 7 ; Meyer 1985a; Breneman & Stone 1985; McGuire et al. 1986; Stone 1989; Cane

---

[27] Note that the degrees of ionization of O and H are tightly related, due to the very large O – H charge exchange cross section of these two elements with virtually equal FIP's. Therefore, if even O behaves as an intermediate-FIP element, i.e. if a significant fraction of O is ionized in the medium in which the fractionation takes place, the same must be true for the Hydrogen gas itself.

[28] A few recent spectroscopic studies have suggested that K, Na and Ca, with FIP's as low as $\sim$ 4 to 6 eV, may be enhanced by an additional factor of $\sim$ 2 relative to Mg,Si,Fe with FIP $\sim$ 8 eV (Doschek et al. 1985; Feldman & Widing 1990; Philips & Feldman 1991; Widing & Feldman 1992a,b; Feldman 1992), an overenhancement which is *not* present the large scale average composition given by the SEP data (fig. 5; Breneman & Stone 1985; Paper I). However, none of these spectroscopic evidences is really compelling (Na determinations based on a single line ; Ca IX and X yielding conflicting Ca abundances ; Ca XIX possibly significantly emitting at lower temperatures than Fe XXIV ; dependance upon uncertain Fe ionization balance, see Fludra et al. 1991). Taken at face value, such high Na,Ca/Mg,Si,Fe ratios would imply an ion-neutral separation taking place near temperature-minimum ($\sim$ 4200 K), where Na,Ca are 100 % ionized while at least Fe is largely neutral (fig. 2 ; but how about Mg,Si ?). It is, however, extremely surprising that Ca/Fe ratios enhanced by the same factor of $\sim$ 2 are found in a polar plume, in which Mg/Ne is enhanced by a factor as large as $\sim$ 10 (Widing & Feldman 1992b), and in a compact flare, whose composition is otherwise *photospheric* (Feldman & Widing 1990) ! These data would imply, in particular, that the above low temperature fractionation affects even material completely unaffected by the main ion-neutral fractionation leading to the 10 eV step in the FIP pattern.

et al. 1991). The same is true for Ne/O in the slow SW, at least on the average (C/O is not observed) (Gloeckler and Geiss 1989). In the accurate SEP data, Si seems slightly under-abundant relative to Mg, *i.e.* it behaves slightly as an intermediate-FIP element, but the Si/Mg ratio is stable.

So, C behaves as if it were fully neutral in the selection gas associated with the slow SW (§ 2.4). This is consistent with a selection taking place in a gas originating in the photosphere above altitude – 60 km. In the chromospheric T-plateau, by contrast, the ionized fraction of C is not known with certainty ; *if* the earlier estimates by VAL are correct, C could be ~ 50 % ionized there (§ 4.3). C may also very easily get ionized in the slight heating of any upward jet of photospheric or chromospheric material. Now, consider any situation in which C is very significantly, though not entirely, ionized in the gas in which the fractionation takes place : say, ~ 50 % ionized. It means that each individual C atom spends ~ 50 % of its time as a neutral and ~ 50 % of its time as an ion. Then models in which *neutrals* are driven *out of* the gas to be supplied to the corona (*e.g.*, Hénoux & Somov 1993) may still be acceptable, because the C atoms are still subjected to this drift of neutrals 50 % of their time. But models in which *ions* are driven *into* this gas (*e.g.*, Antiochos 1993) then have a problem, because the C atoms are subjected to this drift of ions 50 % of their time, and should therefore be significantly enhanced relative to O and Ne, contrary to observation.

(b) CME driver gas : in the high-pressure SW observed near a CME shock front (CME driver gas itself and nearby shock-compressed ordinary SW), C/O varies by a factor on the order of ~ 4.5, roughly between the photospheric value and 4.5 times this value ! The higher values seem associated with higher SW freezing temperatures (~ 2 MK), and are thus presumably relevant to CME driver gases (fig. 7 ; von Steiger et al. 1992a). So, we have a hint that C may behave as a low-FIP element in CME driver gases. This would indicate a *slightly* higher temperature for the medium in which the i-n fractionation takes place. Do the gradual flares often associated with CME's slightly heat the chromospheric gas down below ? This result is all the more difficult to understand that it is generally believed that the gas ejected in CME's is just ambient coronal gas that has been in the corona long before the CME broke out (Kahler 1992).

6.1.2 Localized features near Sun   (lower corona)

(a) Impulsive SEP events (^{3}He-rich events) : these particles sample the gas in and/or near a flaring loop. They definitely show no enhancement of C/O. Ne/O in the parent gas cannot be easily determined, due to a strong preferential acceleration of Ne,Mg,Si. But the data do not suggest any deficiency of Ne in the gas relative to large-scale-coronal, as compared to O,Mg,Si (fig. 7 ; Reames et al. 1990,1993). See more discussion in § (c).

(b) X-ray and EUV spectroscopy of active regions and flares (O as an intermediate-FIP element ?) : the data sample the gas within active or flaring loops. We have no data on C/O. As for Ne/O, we have EUV and X-ray data on various sites with amplitudes of the

FIP-bias (Mg,Fe/Ne ratios) spanning the entire range between photospheric and large-scale-coronal compositions (fig. 11).

The EUV data on four sites suggest a constant Ne/O ratio, equal to the photospheric and gradual-SEP/SW values (Widing & Feldman 1989,1992b; Feldman & Widing 1990 ; fig. 11 and 7). A similar result is obtained from the X-ray emission measure analysis of the 25 Aug 1980 gradual flare with coronal Mg/Ne, observed by the FCS instrument in its "home" position (Schmelz 1993 ; fig. 11).

By contrast, comprehensive SOLEX X–ray data on a number of flares and active regions seem to indicate a negative correlation between the Ne/O and the Fe/Ne ratios (McKenzie & Feldman 1992 ; fig. 11). Normalizing to highest-FIP Ne, whose abundance relative to Fe varies by the usual factor of ~ 4.5, this means that the higher the amplitude of the FIP bias, the higher the O/Ne ratio : O seems to behave as an intermediate-FIP element (fig. 7). Very similar X-ray studies performed with the FCS instrument in the scanning mode also suggest a factor of ~ 3 variability of the Ne/O ratio (Strong et al. 1988; Saba & Strong 1992,1993a,b).

This result is difficult to understand *if* the slow SW has its origin in the material observed within active and flaring loops (§ 2.4). First, the determinations of Ne/O relevant to the photosphere (§ 2.2a) and to the large-scale-corona associated with the slow SW (§ 6.1.1a) do not suggest any difference in this ratio within errors, so that the *mere variation* of Ne/O with Fe/Ne is puzzling (fig. 11). Regarding absolute values of Ne/O [29], while the McKenzie and Feldman's X-ray ratios found for near-photospheric Fe/Ne are roughly consistent with the determinations of Ne/O relevant to the photosphere, those found for near-coronal Fe/Ne seem ~ 1.8 times lower than the large-scale-coronal values (slow SW and very accurate gradual event SEP determinations ; fig. 11 [30]). If these observations are confirmed, they therefore imply that the slow SW gas is, indeed, *not* genetically related with the active and flaring loop gas (§ 2.4) !! In the latter, the i-n fractionation has to take place at slightly higher temperatures than in the source regions of the slow SW.

(c) γ-ray line, X-ray and impulsive SEP event evidences for enhanced Ne abundances in flares : Ne/C,O ratios about 2.5 times *higher* than those found virtually everywhere else in the solar environment (both photosphere and corona) have been observed in gases associated with more or less impulsive flares with otherwise coronal-type composition (Mg,Si,Fe/C,O). Such high Ne/O ratios are systematically found in the various events observed with γ-ray lines : in the well studied 27 April 1981 long duration event, in which the observed material should largely lie in the moderate density lower corona (Murphy et

---

[29] These absolute values may not be final. Saba & Strong (1993a,b) find that the absolute values of the Ne/O ratio depend on still uncertain atomic physics. See footnote # 18. MKF's range for Ne/O is 0.08 to 0.20 (fig. 11) ; SS's Ne/O range is ~ 0.15 to 0.30 for ARO and ~ 0.10 to 0.23 for ARA ; ironically, while MKF use ARO, SS's range for Ne/O agrees with theirs better for ARA than for ARO (as for Fe/Mg) !

[30] The high pressure SW observed by the AMPTE instrument is a mixture of coronal hole (open-field regions) and CME associated wind (§ 2.3b,c ; Gloeckler and Geiss 1989; Gloeckler et al 1989). It is therefore much less relevant here than the slow SW.

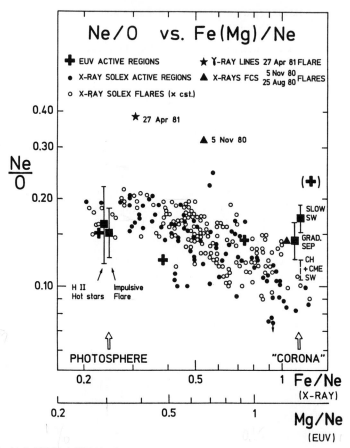

Fig. 11 : The high-FIP/high-FIP Ne/O ratios observed in active and flaring loops on the solar surface, vs. the Fe/Ne (or Mg/Ne) ratio characterizing the amplitude of the FIP-bias (fig. 7). For reference, I have plotted the points for the photosphere (HII regions + hot stars, and compact solar flare, § 2.2a) and for the large-scale-corona (gradual event SEP's and slow SW, § 6.1.1a; the "coronal hole + CME" point is much less relevant here, footnote # 30). They suggest equal Ne/O ratios in both media. While the EUV and one FCS X-ray points also suggest a constancy of Ne/O, the bulk of the SOLEX X-ray points suggest an anti-correlation between Ne/O and Fe/Ne, and a low Ne/O for coronal-type Fe/Ne values (§ 6.1.2b). This suggests that O behaves as an intermediate-FIP elements within active and flaring loops, while it definitely does not do so in the slow SW. The absolute values of the ratios (determined for active regions only) might, however, be not final (footnotes # 18 and 29). This figure also shows the γ-ray line and one FCS X-ray point showing exceptionally high Ne/O values in flares (§ 6.1.2c ; footnote # 31).

al. 1991; Rieger 1991; Hulot et al. 1992 ; fig. 11 and 7), as well as in a number of impulsive γ–ray events in which the observed gas should be very dense lower chromospheric or even photospheric gas (~ 10^{15} cm^{-3} ; Murphy 1992; Hua et al. 1989; Rieger 1991).

In the X-ray range, two flares have been observed by the FCS instrument in its "home" position, and have been subjected to an emission measure analysis. One of them was impulsive, the 5 Nov 1980 flare, and Schmelz & Fludra (1993) and Schmelz (1993) found in it a high Ne/O ratio similar to those found in γ-rays (fig. 11 and 7) [31].

Ne/C,O ratios enhanced by factors of ~ 3.5 are also observed among the particles accelerated in impulsive, ^{3}He-rich events (§ (a) ; Reames et al. 1988,1990,1993). But in these events the Ne enhancement clearly results, at least to first order, from the selective acceleration, rather than from the composition of the parent gas. This is strongly suggested by the similarity of the enhancements found for the Ne and Mg,Si abundances relative to C,O, as compared to coronal values (factors of ~ 3.5 and 2.4). As for the residual overabundance of Ne *relative to Mg,Si*, a factor of ~ 1.5, we do not know whether it results from the selective acceleration, or reflects the composition of the parent gas (fig. 7 ; Reames et al. 1993).

High Ne/O ratios certainly cannot be interpreted in terms of a preferential selection of ions out of a purely thermal gas ! In the framework of the "local philosophy", Shemi (1991) has suggested an i-n fractionation in the chromosphere bombarded by flare-associated soft X–rays, which can specifically photoionize Ne, and *not* O and C. The difficulty is that the specific photoionization of Ne takes place only if softer X-rays (< 0.85 keV) are first eliminated by absorption in overlaying layers (Yeh & Lindau 1985). This requires a high density (> $5 \cdot 10^{14}$ cm^{-3}) for the material in which the i-n separation has to take place. At such high densities, collisions make any separation very difficult, so that any deviation from the photospheric composition is difficult to imagine. Densities ~ 10^{15} cm^{-3} are, however, *indeed* indicated by the time scales of the γ-ray emission in most γ-ray flares, though not for the best studied 27 April 1981 γ-ray flare .

(d) Sunspots : I recall that one possible explanation for the photospheric value of the Mg,Si/CNO ratios above some sunspots (§ 5.1.2b ; fig. 7) is that Si and Mg are neutral in the cool sunspot photosphere/chromosphere (Lites et al. 1987). Observation of lower-FIP Na, Ca, Al abundances could settle this issue.

6.2   Open Field Regions   (Coronal Holes)

We have only two pieces of data, both relevant to the large scale, interplanetary composition :

(a) Coronal hole SW : C/O/Ne are found normal, equal to photospheric (fig. 9 ; Gloeckler et al. 1989; von Steiger et al. 1992a). But this is not very significant, since the coronal hole SW shows almost no FIP fractionation whatsoever (§ 5.2.2b).

(b) Corotating interaction regions energetic particles : in this population of particles accelerated out of coronal hole SW, Ne/O is normal, but C/O is enhanced by a factor of ~ 2, almost equal to the (weak) enhancement factor of ~ 2.5 of the Mg,Si,Fe/Ne,O,N ratios (§ 5.2.2a ; fig. 9 ; Reames et al. 1991, and ref. therein). So, *C (as well as S) seems*

---

[31]   Recall also that Strong et al.'s (1988) and Saba & Strong's (1993a,b) analysis of the FCS data in the scanning mode yields Ne/O ratios reaching values as high as 0.30, especially in the long duration events, when the ARO estimate of the Fe ion fractions is used (see footnotes # 18 and 29).

to behave almost as a low-FIP element ! Could this high C/O be due to a rigidity
dependent bias during particle acceleration in the CIR ? This seems very unlikely, since the
Fe/Mg,Si ratio is found very normal in the CIR energetic particles, while Fe and Mg,Si
have widely different Q/A ratios in the SW. So, this high C/O ratio might well indicate that,
at the base of coronal holes, the i-n fractionation takes place at a slightly higher temperature
than in the sources of the slow SW associated with predominantly closed-field regions
(§ 2.4).

7   SUMMARY   AND   DISCUSSION

I will first summarize separately the composition observations relevant to *(i)* the closed-
field regions of the solar surface ("quiet and active Sun") and their associated interplanetary
medium pervaded by slow SW, and *(ii)* the open-field regions (coronal holes) and their
associated fast SW streams (§ 2.1 ; fig. 3, 4). I will then conclude with some general
remarks about possible mechanisms.

7.1   Closed Field Regions Discussion   ("Quiet" and "Active" Sun)

Far from Sun, the composition of the slow SW is observed, both directly *and* through that
of gradual event SEP's, which are primarily accelerated out of this slow SW and should
reflect its composition (§ 2.3a,d ; fig. 4). So, both of these large distance compositions
reflect the composition of *the sources of the slow SW*, whatever these may be. While
some recent Yohkoh data suggest that the slow SW material might largely originate in
evaporating large, old, active (and quiet ?) loops, its freezing temperatures are much lower
than typical active region temperatures. Though associated with predominantly closed-field
regions, the slow SW might actually well originate in *narrow* open-field structures
squeezed between closed-loop systems at low altitude, which open up widely above the
closed-loop level (§ 2.4 ; fig. 3). If this is the case, the materials observed far from Sun
and on the solar surface (pertaining essentially to active and flaring loops ; § 2.2) are *not*
related, and their compositions have no reason to be the same, even on the average (§ 2.4).

At high altitude, the composition of the slow SW is FIP-biased relative to photospheric
composition. Heavy elements with FIP < ~ 10 eV are overabundant by a factor on order of
~ 4.5 relative to heavy elements with higher FIP *and to Hydrogen*, which behaves like
heavier high-FIP elements. But He is deficient by a factor of ~ 2 (§§ 3, 5.1.1 ; fig. 5, 7).
This bias is independent of mass (footnote # 14). It is *this* very well determined high
altitude composition above closed-field regions that is, more or less properly, denoted
"large-scale-coronal" or simply *"coronal"* (§ 5.1.1).

This certainly implies, at the sources of the slow SW, an i-n separation in a ~ 7000 K gas,
where neutral and first ions can exist, *i.e.* necessarily in photospheric or chromospheric
gas (or jets originating therein) (§ 4.1 ; fig. 1, 2) : on the average, ions must rise into
corona ~ 4.5 times as efficiently the neutrals, whatever their mass. These levels at
~ 7000 K lie *below* all sites observed above photosphere (T > 30000 K ; except maybe
for the γ-ray line data, § 2.2b) : we can't observe the regions where the fractionation
actually takes place (fig. 4) !

The factor of ~ 2 deficiency of He relative to H and heavier high-FIP elements probably exists in the coronal gas itself, and is thus *not* merely the result of an insufficient drag exerted on He in the SW formation region (impulsive event SEP data in fig. 7 ; § 3.3). It poses a specific challenge to models for fractionation between photosphere and corona.

On the surface of the Sun, observations of active and flaring loops show all intermediate compositions between photospheric and large-scale-coronal, but *no* low-FIP element enhancements larger than large-scale-coronal (§ 5.1.2 ; fig. 7) [32]. More precisely, active regions observations show the steady rise of material with photospheric composition within low, compact loops, and their gradual evolution into larger loop structures with coronal-type composition, over time scales of ~ a day (§ 5.1.2a ; fig. 8). It is not clear whether this evolution of the composition is due, *either (i)* to an i-n separation taking place very commonly at the footpoints of each and every (active ?) loop (*"local philosophy"*), *or (ii)* to reconnection with external, large coronal field structures allowing the evaporation of the photospheric-type gas and its replacement by ambient coronal gas (*"reconnection philosophy"* ; in that case we have learned nothing about the specific sites where the i-n separation of the material feeding the corona takes place !). The fact that low-FIP element enhancements larger than large-scale-coronal are *not* observed might possibly suggest that reconnection is predominant.

Remarkably enough, the only location (let alone compact loops just emerging from photosphere) in which a completely photospheric Mg,Si/CNO ratios are found is right above some sunspots. This might be due, *e.g.*, to the lower chromospheric temperature above sunspots, causing Si and Mg to be neutral there (observe very low-FIP Na,Ca,Al !), or to the high field preventing the fractionation ("local philosophy"), or to mirroring of downward coronal ions ("reconnection philosophy") (§§ 5.1.2b, 6.1.2d). It may also be reminiscent of the low FIP-bias observed in, likewise open-field, large scale coronal holes (§§ 5.2.2, 7.2).

We now summarize our knowledge of the exact shape of the FIP-bias pattern. There has been a search for a possible excess of *very* low-FIP elements K, Na, Ca, Al relative to Mg, Si, Fe in some media, but no solid evidence has been found (footnote # 28). Let me also recall at this point the He anomaly (§ 3.3).

The exact position of the step in the FIP-bias pattern is characteristic of the precise ionization pattern within the medium in which the i-n fractionation takes place. Roughly, the position of the step in the slow SW composition is consistent with both the photospheric and the chromospheric T-plateau ionization patterns (§ 4.3), the latter being controlled by the thermostating at the Ly α energy (§ 4.2). However, this may be not entirely true for Carbon, which has a metastable state subject to ionization by Ly α, and may thus play a crucial role in characterizing the fractionation medium. C is entirely neutral in the photosphere down to $-$ 60 km, but could be as much as ~ 50 % ionized in the chromospheric T-plateau, this estimate being however model dependent and not final (§ 4.3 ; fig. 2). In addition, the degree of ionization of C is very sensitive to any heating of the gas within an upward jet (§ 4.3). Now, two classes of models can be considered to

---

[32] There are still some questions with the atomic physics used in the analysis of the observations (footnote # 18).

enrich the corona with ionized elements : *either neutrals* are driven *out of* a gas to be supplied to the corona, *or ions* are driven *into* this gas (§ 4.4). In the slow SW, both O and C behave *strictly* as a high-FIP element (C/O/Ne = photospheric in the gradual event SEP observations ; § 6.1.1a ; fig. 5, 7). *If* C is indeed largely ionized in the gas in which the fractionation takes place, this poses problems to models in which *ions* are driven *into* a tube of rising gas at the chromospheric level : one would then expect an excess of C relative to O and Ne, contrary to the observations relevant to the slow SW (§ 6.1.1a).

There are indications of enhancements of C/O, *i.e.* for an i-n fractionation in a slightly warmer medium, in the sources of other, specific SW's : those of the CME driver gas (§ 6.1.1b ; fig. 7 ; not easy to explain !) and of the open-field coronal hole SW (§§ 6.2b, 7.2 ; fig. 9).

On the solar surface, some X-ray data yield low Ne/O ratios (*i.e.* enhanced O/Ne) within active and flaring loops with "coronal" Fe/Ne ratio. This may imply that even O behaves there as an intermediate-FIP element, hence a fractionation in a still warmer gas in the chromospheric part of the loop footpoints (§ 6.1.2b ; fig. 7, 11). These data might, however, not be final, and are not supported by other, related observations [33]. If confirmed, they imply that active and flaring loop material does not contribute significantly to the slow SW, in which Ne/O is definitely photospheric (§§ 6.1.1a, 2.4).

By contrast, γ-ray and X-ray spectroscopy (and *possibly* impulsive event SEP data) have also yielded *high* Ne/O ratios in more or less impulsive flare gases with otherwise coronal-type FIP-biased composition (*e.g.*, Mg/C,O) (§ 6.1.2c ; fig. 7, 11). This certainly cannot be understood merely in terms of certainly cannot be understood in terms of any i-n fractionation in a medium at or near LTE. In the framework of the "local philosophy", a specific X-ray photoionization of Ne in high density media has been considered (§ 6.1.2c).

7.2   Open Field Regions Discussion   (Coronal Holes)

The distant, interplanetary observations (as well as global low altitude data) indicate a lower, or even nonexistent FIP-bias as compared to photospheric composition (low-FIP element enhancement factors of ~ 1 to ~ 2.5) (§ 5.2.2 ; fig. 5, 9). So, the i-n fractionation seems to proceed less efficiently at the base of the coronal holes.

By contrast, specific features on the solar surface within polar coronal holes, the polar plumes, are the only place in which we definitely see FIP-biases much larger (factors of ~ 8 to ~ 35) than all high altitude averages !! (§ 5.2.1 ; fig. 9). On the one hand, this information is absolutely crucial, since it indicates us one specific site in which the i-n fractionation certainly takes place. On the other hand, we are forced to conclude that polar

---

[33] The atomic physics used in the X-ray analysis may be not final (footnote # 29). On the other hand, current EUV data, as well as one X-ray point based on a emission measure analysis, do not confirm the variability of Ne/O (§ 6.1.2b ; fig. 11). The *impulsive* event SEP composition, presumably also associated with flaring loops or their neighbourhood, definitely shows no excess of C over O, and do not suggest an excess of O relative to Ne (§ 6.1.2a). The same is true for the γ-ray line data (§ 6.1.2c ; fig. 7). Note that, if O indeed behaves as an intermediate-FIP elements in active loops, C should behave essentially as a low-FIP element. Unfortunately, we have no spectroscopic observation of C in active loops.

plumes are not typical of large scale coronal holes, or of open-field structures in general, as regards composition ; (they are estimated to contribute some ~ 15 % of the coronal hole SW flux) (§ 5.2.3).

In the open-field regions, by contrast to closed-field regions, observations suggest that C behaves rather as a low-FIP element (§ 6.2b ; fig. 9). This might indicate a slightly higher temperature for the fractionation medium than in the sources of the slow SW (like in the CME gas ; §§ 6.1.1b, 7.1).

The He/H deficiency by a factor of 2 is particularly stable in the coronal hole SW, but is *not* found in the CIR energetic particle data (§ 3.3).

7.3.   General Considerations

7.3.1 Composition and filling of the corona

The problem of the genesis of the FIP-biased large-scale-coronal composition is, of course, deeply related with the much more fundamental, as yet unsolved, problem of the filling of the corona : which are the main channels by which matter rises from photosphere/chromosphere into corona, in particular in closed-field regions ? spicules, or similar, smaller jets ? chromospheric evaporation and loop dissipation ? steady, narrow open-field channels ? velocity filtering of chromospheric non-thermal tails (Scudder 1992) ? The coronal composition probably cannot be understood before the main feeding channels have been traced down. But composition studies may, in turn, eventually turn out to help deciding among these channels. Note that the large-scale composition anomalies over closed-field regions are *large*, a factor of ~ 4.5 : the fractionation mechanism is likely to affect the main channels that supply the corona with Hydrogen. However, since we have an *excess* of low-FIP elements relative to H (rather than a *depletion* of high-FIP ones), this statement is *not* certainly true : the coronal heavy element composition *could* also be controlled by *minor* channels carrying essentially *only* low-FIP heavies, with virtually no H (and high-FIP heavies), hence very little mass : such channels could be distinct from major channels supplying the coronal mass (H, He).

7.3.2 Key parameters for the ion-neutral fractionation ?

Which can be the key parameters for the operation of the i-n fractionation ? We can consider several parameters :

(i) The density in the ~ 7000 K region ? Any i-n separation mechanism works against collisions. The lower the density in the ~ 7000 K region, the larger the chances of an efficient fractionation [34].

(ii) The density contrast between the ~ 7000 K region and the coronal region to be filled by new, uprising material ? If a coronal reservoir *with initially photospheric composition* is supplied with uprising FIP-biased gas, the lower the relative density of this reservoir, the lower the dilution of the uprising gas, and the higher the resulting coronal FIP-bias (Feldman 1992). The question is : does this situation apply ? It might apply in a closed

---

[34] It is, however, ironic that, in coronal holes, the FIP-bias is by far higher within polar plumes, which are ~ 10 times as dense as their neighbourhood, at least at the coronal level.

loop of field rising from the photosphere (§ 5.1.2a ; fig. 8). But, more generally, open coronal reservoirs with photospheric composition are just not observed (except above some sunspots, § 5.1.2b).

(iii) The field geometry, its degree of opening ? The data on very compact flares, larger active regions and polar plumes have first suggested that this be the key parameter : *the more open the field, the more efficient the i-n fractionation* (§§ 5.1.2a, 5.2.1 ; fig. 10). The observed *lack* of a large FIP-bias in large scale coronal holes, however, leads us to question this tantalizing view (§§ 5.2.2, 5.2.3 ; fig. 9, 10) [35]. Part of the apparent effect is certainly due to the time evolution of the composition of active regions, which accompanies the growth of the scale size of their magnetic field (§ 5.1.2a ; fig. 8). A comparison of the *large scale* FIP-biases above open- and closed-field regions is too ambiguous to be fruitful : *first*, both large scale compositions can result from averaging over various inputs of material into the corona, and, *second*, there is a distinct possibility that the SW above so-called "closed-field regions" actually originates in *more open* field structures that those of coronal holes (§ 2.4) !!

(iv) The field intensity ? This is one of the possible explanations for the lack of a FIP-bias observed right above some sunspots (*e.g.*, Antiochos' (1993) model ; § 5.1.2b).

(v) The fine structure of the field, and of the upward flow of material ? In any model in which the i-n fractionation results from a diffusion *across* a vertical flux tube, whose material rises into corona, the narrower the flux tube, the easier the separation. For instance, in a loop fractionation model à la Antiochos (1993), the well established very small scale of the energy releases within active loops (e.g., Holman 1985; Holman & Benka 1992) is certainly a favorable factor. It is, in addition, essential that the material *around* this particular flux tube, with a composition "complementary" to that of the gas within the tube, does *not* simultaneously rise into the corona ! Maybe this is the reason why the FIP-bias is found to be weaker in coronal holes, within which matter possibly rises too uniformly. There is, however, no evidence for a very pronounced fine structure in polar plumes : Widing & Feldman (1992a) estimate a filling factor of only ~ 3, at the coronal level.

7.3.3 Possible mechanisms

Mechanisms based on gravitational settling or *plain* diffusion of neutrals out of some flux tube, while ions are fixed by the magnetic field, are not enough efficient (and the first do not account for the behaviour of H). Clearly, the diffusion must, one way or another, be *driven* by electromagnetic forces : *either* electric fields driving the diffusion of *ions into* the gas to be supplied to the corona, *or* pinching of a flux tube driving the *neutrals out of* this gas. Several models are currently being developed along such lines, summarized by Hénoux & Somov (1993). The demand to account for the mass independence of the FIP-bias is a strong requirement for all models. The He anomaly must also be accounted for. The hypothesis that the coronal composition anomalies be due to the evaporation of meteoritic material does not seem adequate (§ 4.4).

---

[35] Recall also that there may be *no* FIP-bias above some sunspots, with widely open field. But there are several possible explanations of this observation (§§ 5.1.2b , 7.1)

ACKNOWLEDGMENTS

This paper would not exist without a constant, close interaction with Uri Feldman, Don Reames, Julia Saba, Niel Sheeley, and Ken Widing. It also owes much of its substance to fruitful discussions with Spiro Antiochos, Johannes Geiss, George Gloeckler, Nicolas Grevesse, Jean-Claude Hénoux, Gordon Holman, Fred Ipavich, Serge Koutchmy, Helen Mason, Joe Mazur, Art Poland, Reuven Ramaty, Joan Schmelz, Keith Strong, Rudi von Steiger, and Steve White. I also acknowledge receipt of a National Research Council - GSFC Senior Research Associateship.

REFERENCES

Ahmad, I.A., & Withbroe, G.L. 1977, Solar Phys. **53**, 397.
Ahmad, I.A., & Webb, D.F. 1978, Solar Phys. **58**, 323.
Anders, E., & Grevesse, N. 1989, Geochim. Cosmochim. Acta **53**, 197.
Antiochos, S.K. 1993, Adv. Space Res., in press.
Antiochos, S.K., & Noci, G. 1986, ApJ **301**, 440.
Antonucci, E., Marocchi, D., Gabriel, A.H., & Doschek, G.A. 1987, A&A **188**, 159.
Arnaud, M., & Cassé, M. 1985, A&A **144**, 64.
Arnaud, M., & Raymond, J. 1992, ApJ, **398**, 394.
Arnaud, M., & Rothenflug, R. 1985, A&A Suppl. **60**, 425.
Arnould, M. 1984, Adv. Space Res. **4**, N°2-3, 45.
Athay, R.G. 1981, ApJ **250**, 709.
Bibring, J.P., & Cesarsky, C.J. 1981, 17st Intern. Cosmic Ray Conf., Paris, **2**, 289.
Bürgi, A. 1992a, J. Geophys. Res. **97**, 3137.
Bürgi, A. 1992b, in : Solar Wind 7, E. Marsch & R. Schwenn eds., (Pergamon Press), in press.
Bürgi, A., & Geiss, J. 1986, Solar Phys. **103**, 347.
Breneman, H.H., & Stone, E.C. 1985, ApJ Letters **299**, L57.
Cane, H.V., Reames, D.V., & von Rosenvinge, T.T. 1988, J. Geophys. Res. **93**, 9555.
Cane, H.V., Reames, D.V., & von Rosenvinge, T.T. 1991, ApJ **373**, 675.
Cassé, M., & Goret, P. 1973, 13th Intern. Cosmic Ray Conf., Denver, **1**, 584.
Cassé, M., & Goret, P. 1978, ApJ **221**, 703.
Cassé, M., Goret, P., & Cesarsky, C.J. 1975, 14th Intern. Cosmic Ray Conf., Munich, **2**, 646.
Cassé, M., & Paul, J.A. 1982, ApJ **258**, 860.
Cesarsky, C.J., & Bibring, J.P. 1980, in : IAU Symp. N° 94,
        G. Setti, G. Spada, & A.W. Wolfendale eds., (Kluwer), p. 361.
Cook, W.R., Stone, E.C., & Vogt, R.E. 1980, ApJ Letters **238**, L97.
Cook, W.R., Stone, E.C., & Vogt, R.E. 1984, ApJ **279**, 827.
Doschek, G.A., & Bhatia A.K. 1990, ApJ **358**, 338.
Doschek, G.A., Dere, K.P., & Lund, P.A. 1991, ApJ **381**, 583.
Doschek, G.A., Feldman, U., & Seeley, J.F. 1985, MNRAS **217**, 317.
Doschek, G.A., & Seeley, J.F. 1990, ApJ **348**, 341.
Dryer, M. 1982, Space Sci. Rev. **33**, 233.
Epstein, R.I. 1980, MNRAS **193**, 723.
Feldman, U. 1983, ApJ **275**, 367.
Feldman, U. 1992, Physica Scripta **46**, 202.
Feldman, U. 1993, Adv. Space Res., in press.
Feldman, U., & Widing, K.G. 1990, ApJ **363**, 292.
Feldman, U., Widing, K.G., & Lund, P.A. 1990, ApJ Letters **364**, L21.

Feldman, W.C., Asbridge, J.R., Bame, S.J., Fenimore, E.E., & Gosling, J.T. 1981,
    J. Geophys. Res. **86**, 5408.
Fludra, A., Bentley, R.D., Culhane, J.L., Doschek, G.A., Hiei, E., Watenabe, T.,
    & Philips, K.J.H 1993, Adv. Space Res., in press.
Fludra, A., Bentley, R.D., Culhane, J.L., Lemen, J.R., & Sylwester, J. 1991,
    Adv. Space Res. **11**, N° 1, 155.
Fontenla, J.M. 1992, private communication.
Fontenla, J.M., & Avrett, E.H. 1993, in : SOHO Workshop on Coronal Streamers,
    Coronal Loops, and Coronal and Solar Wind Composition,
    Annapolis, Aug. 1992, V. Domingo ed., (ESA Publ.), in press.
Fontenla, J.M., Avrett, E.H., & Loeser, R. 1990, ApJ **355**, 700.
Fontenla, J.M., Avrett, E.H., & Loeser, R. 1993, ApJ, in press.
Garrard, T.L., & Stone, E.C. 1991, 22nd Intern. Cosmic Ray Conf., Dublin, **3**, 331.
Geiss, J., & Bochsler, P. 1985, in : Isotopic Ratios in the Solar System,
    ed. by CNES (Paris : Capaduès Editions), p. 213.
Geiss, J., Hirt, P., & Leutwyler, H. 1970, Solar Phys. **12**, 458.
Geiss, J., Ogilvie, K.W., von Steiger, R., Mall, U., Gloeckler, G., Galvin, A.B.,
    Ipavich, F., Wilken, B., & Gliem, F. 1992, in : Solar Wind 7,
    E. Marsch & R. Schwenn eds., (Pergamon Press), in press.
Gloeckler, G., & Geiss, J. 1989, in : Cosmic Abundances of Matter,
    C.J. Waddington ed., AIP Conf. Proc. 183, (The Amer. Inst. of Physics), p. 49.
Gloeckler, G., Ipavich, F.M., Hamilton, D.C., Wilken, B.,& Kremser, G. 1989, Eos **70**, 424.
Grevesse, N. 1992, private communication.
Grevesse, N., & Noels, A. 1993, this volume.
Havnès, O. 1973, A&A **24**, 435.
Hénoux, J.C., & Somov, B.V. 1993, in : SOHO Workshop on Coronal Streamers,
    Coronal Loops, and Coronal and Solar Wind Composition,
    Annapolis, Aug. 1992, V. Domingo ed., (ESA Publ.), in press.
Holman, G.D. 1985, ApJ **293**, 584.
Holman, G.D., & Benka, S. 1992, ApJ Letters, in press.
Hovestadt, D. 1974, in : Solar Wind III, C.T. Russell ed., (Univ. of Calif. Los Angeles), p. 2.
Hua, X.M., Ramaty, R., & Lingenfelter, R.E. 1989, ApJ **341**, 516.
Hulot, E., Vilmer, N., Chupp, E.L., Dennis, B.R., & Kane, S.R. 1992, A&A **256**, 273.
Ip, W.H., & Axford, W.I. 1991, Adv. Space Res. **11**, N° 1, 247.
Ipavich, F.M., Galvin, A.B., Gloeckler, G., Hovestadt, D., Bame, S.J., Klecker, B.,
    Scholer, M., Fisk, L.A., & Fan, C.Y. 1986, J.G.R. **91**, 4133.
Kahler, S.W. 1992, Ann. Rev. Astr. Ap., **30**, 113.
Kahler, S.W., Hildner, E., & Van Hollebeke, M.A.I. 1978, Solar Phys. **57**, 429.
Kahler, S.W., Sheeley, N.R., Howard, R.A., Koomen, M.J., Michels, D.J., McGuire, R.E.,
    von Rosenvinge, T.T. & Reames, D.V. 1984, J. Geophys. Res. **89**, 9683.
Koutchmy, S., & Lorrain, P. 1992, private communication.
Kristiansson, K. 1971, Ap. Space Sci. **14**, 485.
Kristiansson, K. 1974, Ap. Space Sci. **30**, 417.
Lemaire, J. 1990, ApJ **360**, 288.
Lemen, J.R., Sylwester, J., & Bentley, R.D. 1986, Adv. Space Res. **6**, No. 6, 245.
Linsky, J.L. 1980, Ann. Rev. Astr. Ap. **18**, 439.
Lites, B.W., Skumanich, A., Rees, D.E., Murphy, G.A., & Carlsson, M. 1987, ApJ **318**, 930.
Mason, B. 1979, Data of Geochemistry / Cosmochemistry / Meteorites,
    Geological Survey Professional Paper # 440-B-1 (U.S. Govt. Printing Office)
Mason, H.E. 1975a, MNRAS **170**, 651.
Mason, H.E. 1975b, MNRAS **171**, 119.

Mazur, J.E. 1991, PhD Thesis, U. of Maryland.
McGuire, R.E., von Rosenvinge, T.T., & McDonald, F.B. 1979,
    16th Intern. Cosmic Ray Conf., Kyoto, 5, 61.
McGuire, R.E., von Rosenvinge, T.T., & McDonald, F.B. 1986, ApJ 301, 938.
McKenzie, D.L., & Feldman, U. 1992, ApJ 389, 764.
Mewaldt, R.A. 1980, in : Proc. Conf. Ancient Sun,
    R.O. Pepin, J.A. Eddy, and R.B. Merrill eds., (Pergamon), p. 81.
Meyer, J.P. 1981a, 17st Intern. Cosmic Ray Conf., Paris, 3, 145.
Meyer, J.P. 1981b, 17st Intern. Cosmic Ray Conf., Paris, 3, 149.
Meyer, J.P. 1981c, 17st Intern. Cosmic Ray Conf., Paris, 2, 265.
Meyer, J.P. 1981d, 17st Intern. Cosmic Ray Conf., Paris, 2, 281.
Meyer, J.P. 1985a, ApJ Suppl. 57, 151.
Meyer, J.P. 1985b, ApJ Suppl. 57, 173.
Meyer, J.P. 1988, in : Origin & Distribution of the Elements,
    G.J. Mathews ed., (World Scientific), p. 310.
Meyer, J.P. 1989, in : Cosmic Abundances of Matter,
    C.J. Waddington ed., AIP Conf. Proc. 183, (The Amer. Inst. of Physics), p. 245.
Meyer, J.P. 1991, Adv. Space Res. 11, N° 1, 269 [Paper I].
Meyer, J.P., Cassé, M., & Reeves, H. 1979, 16th Intern. Cosmic Ray Conf., Kyoto, 12, 108.
Murphy, R.J. 1992, private communication.
Murphy, R.J., Ramaty, R., Kozlovsky, B., & Reames, D.V. 1991, ApJ 371, 793.
Neugebauer, M. 1981, Fund. Cosmic Physics 7, 131.
Noci, G., Spadaro, D., Zappalà, R.A., & Zuccarello, F. 1988, A&A 198, 311.
Ogilvie, K.W., Coplan, M.A., Bochsler, P., & Geiss, J. 1989, Solar Phys. 124, 167.
Peña, M., Ruiz, M.sT., & Maza, J. 1991, A&A 251, 417.
Phillips, K.J.H., & Feldman, U. 1991, ApJ 379, 401.
Prantzos, N., Arnould, M., & Arcoragi, J.P. 1987, ApJ 315, 209.
Prantzos, N., Cassé, M., & Vangioni-Flam, E. 1993, this volume.
Reames, D.V. 1990, ApJ Suppl. 73, 235.
Reames, D.V. 1993, in : SOHO Workshop on Coronal Streamers,
    Coronal Loops, and Coronal and Solar Wind Composition,
    Annapolis, Aug. 1992, V. Domingo ed., (ESA Publ.), in press.
Reames, D.V., Cane, H.V., & von Rosenvinge, T.T. 1990, ApJ 357, 259.
Reames, D.V., Meyer, J.P., & von Rosenvinge, T.T. 1993, in preparation.
Reames, D.V., Ramaty, R., & von Rosenvinge, T.T. 1988, ApJ Letters 332, L87.
Reames, D.V., Richardson, I.G., & Barbier, L. 1991, ApJ Letters 382, L43.
Rieger, E. 1991, in : Gamma-Ray Line Astrophysics, P. Durouchoux and
    N. Prantzos eds., AIP Conf. Proc. 232, (The Amer. Inst. of Physics), p. 421.
Rubin, R.H., Simpson, J.P., Haas, M.R., & Erickson, E.F. 1991, ApJ 374, 564.
Saba, J.L.R., & Strong, K. 1992, private communication.
Saba, J.L.R., & Strong, K. 1993a, in : SOHO Workshop on Coronal Streamers,
    Coronal Loops, and Coronal and Solar Wind Composition,
    Annapolis, Aug. 1992, V. Domingo ed., (ESA Publ.), in press.
Saba, J.L.R., & Strong, K. 1993b, Adv. Space Res., in press.
Schatten, K.H., Wilcox, J.M., & Ness, N.F. 1969, Solar Phys. 6, 442.
Schmeltz, J.T. 1993, submitted to ApJ.
Schmeltz, J.T., & Fludra, A. 1993, Adv. Space Res., in press.
Scudder, J.D. 1992, ApJ, 398, 319.
Sheeley, N.R. 1992a, in : Solar Wind 7,
    E. Marsch & R. Schwenn eds., (Pergamon Press), in press.
Sheeley, N.R. 1992b, private communication.

Sheeley, N.R., Bohlin, J.D., Brueckner, G.E., Purcell, J.D., Scherrer, V.E.,
    & Tousey, R. 1975a, ApJ Letters **196**, L129.
Sheeley, N.R., Bohlin, J.D., Brueckner, G.E., Purcell, J.D., Scherrer, V.E.,
    & Tousey, R. 1975b, Solar Phys. **40**, 103.
Shemi, A. 1991, MNRAS **251**, 221.
Silberberg, R., & Tsao, C.H. 1990, ApJ Letters **352**, L49.
Soutoul, A., Cesarsky, C.J., Ferrando, P., & Webber W.R. 1991,
    22nd Intern. Cosmic Ray Conf., Dublin, **2**, 408.
Steinitz, R. 1992, in : Solar Wind 7, E. Marsch & R. Schwenn eds., (Pergamon Press), in press.
Sterling, A.C., Doschek, G.A. & Feldman, U. 1992, submitted to ApJ.
Stone, E.C. 1989, in : Cosmic Abundances of Matter,
    C.J. Waddington ed., AIP Conf. Proc. 183, (The Amer. Inst. of Physics), p. 72.
Strong, K., Claflin, E.S., Lemen, J.R., & Linford, G.A. 1988, Adv. Space Res. **8**, N° 11, 167.
Strong, K., Lemen, J.R., & Linford, G.A. 1991, Adv. Space Res. **11**, N° 1, 151.
Sylwester, J. 1990, in : The Dynamic Sun, EPS Solar Meeting,
    (Publ. Debrecen Heliophysical. Observ.), p. 212.
Sylwester, J., Lemen, J.R., & Mewe, R. 1984, Nature **310**, 665.
Sylwester, J., Zolcinski-Couet, M.C., Bentley, R.D., & Lemen, J.R. 1988,
    J. de Physique **49**, Suppl. au N° 3, C1-189.
Tagger, M., Shukurov, A.M., & Falgarone, E. 1993, submitted to A&A.
Vauclair, S., & Meyer, J.P. 1985, 19th Intern. Cosmic Ray Conf., La Jolla, **4**, 233.
Veck, N.J., & Parkinson, J.H. 1981, MNRAS **197**, 41.
Vernazza, J.E., Avrett, E.H., & Loeser, R. 1981, ApJ Suppl. **45**, 635 [VAL] .
von Steiger, R., Christon, S.P., Gloeckler, G., & Ipavitch, F.M. 1992a, ApJ **389**, 791.
von Steiger, R., & Geiss, J. 1989, A&A **225**, 222.
von Steiger, R., Geiss, J., Gloeckler, G., Balsiger, H., Galvin, A.B., Mall, U.,
    & Wliken, B. 1992b, in : Solar Wind 7,
    E. Marsch & R. Schwenn eds., (Pergamon Press), in press.
Wang, Y.M., & Sheeley, N.R. 1990, ApJ **355**, 726.
Wang, Y.M., Sheeley, N.R., & Nash, A.G. 1990, Nature **347**, 439.
Webber, W.R. 1975, 14th Intern. Cosmic Ray Conf., Munich, **5**, 1597.
Webber, W.R. 1982, ApJ **255**, 329.
Widing, K.G. 1993, in : SOHO Workshop on Coronal Streamers,
    Coronal Loops, and Coronal and Solar Wind Composition,
    Annapolis, Aug. 1992, V. Domingo ed., (ESA Publ.), in press.
Widing, K.G., & Feldman, U. 1989, ApJ **344**, 1046.
Widing, K.G., & Feldman, U. 1992a, ApJ **392**, 715.
Widing, K.G., & Feldman, U. 1992b, in : Solar Wind 7,
    E. Marsch & R. Schwenn eds., (Pergamon Press), in press.
Widing, K.G., & Feldman, U. 1992c, private communication.
Widing, K.G., Feldman, U., & Bhatia, A.K. 1986, ApJ **308**, 982.
Withbroe, G.L. 1988, ApJ **325**, 442.
Yeh, J.J., & Lindau, I. 1985, Atom. Data & Nucl. Data Tables **32**, 1.

A Revision of the Solar Abundance of Scandium

C. NEUFORGE

Aspirant of the belgian FNRS
Institut d'Astrophysique, Université de Liège,
5, avenue de Cointe, B-4000 Liège (Belgium)

1 INTRODUCTION

Although Sc is not a key element, its solar abundance has not been updated since quite a long time. Its value and uncertainty depend strongly on the oscillator strengths available. The theoretical as well as experimental gf-values used in earlier works (see e.g. Biémont, 1974) are without any doubt of lower accuracy than the ones that can be derived today. High accuracy gf-values have recently been measured for a large number of ScI and ScII lines present in the solar spectrum (Lawler and Dakin, 1989). We apply these new data to revise the abundance of Sc.

2 RESULTS

The different lines of ScI and ScII in the solar photospheric spectrum (Moore et al., 1966) have been remeasured from the solar atlas of Delbouille et al. (1973). The shapes of the lines have been carefully analyzed in order to avoid blends and to take care of the effects of hyperfine structure. We had to reject many ScI lines identified by Moore et al. (1966) because they lead to abundances so large that ScI is probably only a minor contributor to these lines. Furthermore, for some of them, the difference in the branching fractions between Lawler and Dakin (1989) and Kurucz (1988) were much too large. The oscillator strengths we use are from Lawler and Dakin (1989). They obtained these values from lifetimes measured by Marsden et al. (1988) together with branching fractions for a very large number of ScI and ScII lines. The uncertainties of these experimental data are small i.e. 10 % for most of the lines up to 20 % for some very faint lines (see Tables 1 and 2). We compared these values to the new theoretical gf-values calculated by Kurucz (1988). The new experimental data are generally 5 to 15 % smaller than the theoretical values but the spread in the differences increases quite a lot when the gf-values of the lines decrease. The abundances have been derived using the

Holweger and Müller (1974) photospheric model with a microturbulent velocity of 1 km s^{-1}. Note that the electronic and gaseous pressures have been recalculated with the new solar photospheric abundance of iron (Holweger et al., 1990, 1991; Biémont et al., 1991; Hannaford et al., 1992). If the hyperfine structure (HFS) broadening is not important for ScI where all the lines are very faint, we took it into account in ScII where some of the lines are much stronger. Nevertheless, the effects of HFS on the abundance are very small (10 % at most for the stronger ScII lines). Results are given in Tables 1 (ScI) and 2 (ScII) where we give successively the wavelength, excitation potential of the lower level, gf-value and uncertainty, equivalent width and solar abundance of Sc. As already mentioned, HFS has no influence on the abundances derived from ScI lines and only decreases the results derived from the stronger ScII lines by at most 10 % (the results reported take HFS into account). As usually, for the stronger lines, the damping constant has been increased by a factor 1.5. But neither the uncertainties in this enhancement factor nor in the microturbulent velocity have significant effects on the abundances. Mean values derived from ScI and ScII are respectively $A_{Sc} = 3.138 \pm 0.119$ and $A_{Sc} = 3.203 \pm 0.074$ (in the usual scale where log $N_H = 12.00$), while the meteoritic value is $A_{Sc} = 3.10 \pm 0.04$ (Anders and Grevesse, 1989).

3 CONCLUSIONS

Our values are derived from the most accurate transition probabilities available (Lawler and Dakin, 1989). Although taken with their uncertainties, they do not exclude an agreement with the meteoritic abundance, two problems show up. Firstly, the large spread in the results from the very faint ScI lines (0.119), much larger than the mean uncertainty of the log gf values (0.026). Whereas the spread in the ScII results (0.074) is hardly larger than the mean uncertainty in the log gf (0.058). Secondly, the best indicator of the Sc abundance in the sun is without any doubt ScII, by far the most important contributor to the scandium abundance and also insensitive to the solar model used. However, it leads to the highest A_{Sc}, about 25 % larger than the meteoritic value. We are reluctant to admit this difference as real as many recent studies have shown that the use of accurate transition probabilities always lead to solar values in agreement with meteorites.

REFERENCES

Anders, E., Grevesse, N. 1989, Geochim. Cosmochim. Acta **53**, 197.
Biémont, E., 1974, Solar Physics **39**, 305.
Biémont, E., 1978, MNRAS **184**, 683.
Biémont, E., Baudoux, M., Kurucz, R.L., Ansbacher, W., Pinnington, E.H. 1991, A & A **539**, 545.

Delbouille, L., Roland, G., Neven, L. 1973, Photometric Atlas of the Solar Spectrum from λ 3000 to 10000. Institut d'Astrophysique, Université de Liège, Belgium.

Hannaford, P., Lowe, R.M., Grevesse, N., Noels, A., 1992, A & A (in press).

Holweger, H., Müller, E.A. 1974, Solar Physics **39**, 19.

Holweger, H., Heise, C., Kock, M. 1990, A & A **232**, 510.

Holweger, H., Bard, A., Kock, A., Kock, M. 1991, A & A **249**, 545.

Kurucz, R.L. 1988, Transactions of the International Astronomical Union, ed. M. McNally, Vol. XXVIII, Kluwer, Dordrecht, p. 168.

Lawler, J.E., Dakin, J.T. 1989, J. Opt. Soc. Am. B **6**, 1457.

Marsden, G.C., den Hartog, E.A., Lawler, J.E., Dakin, J.T., Roberts, V.D. 1988, J. Opt. Soc. Am. B **5**, 606.

Moore C.E., Minnaert, M.G.J., Houtgast, J., 1966, The Solar Spectrum 2935 Å to 8770 Å, NBS Monograph 61.

Table 1. Data and results for ScI and ScII

	λ	χ_{exc}	log gf	W_λ	A_{Sc}
ScI	4743.821	1.448	0.442 ± 0.022	7.70	3.135
	4753.172	0.000	− 1.659 ± 0.020	1.00	2.910
	5081.572	1.448	0.469 ± 0.022	9.20	3.146
	5083.716	1.440	0.284 ± 0.021	6.15	3.132
	5356.087	1.865	0.168 ± 0.043	1.83	3.087
	5484.618	1.851	0.148 ± 0.040	3.15	3.324
	5671.809	1.448	0.495 ± 0.023	12.90	3.241
	5686.833	1.440	0.376 ± 0.017	8.20	3.130
ScII	4246.835	0.315	0.242 ± 0.020	152.00	3.097
	4400.396	0.605	− 0.537 ± 0.043	92.30	3.170
	4420.667	0.618	− 2.273 ± 0.076	14.70	3.152
	4431.360	0.605	− 1.969 ± 0.050	29.70	3.231
	4670.409	1.357	− 0.576 ± 0.046	60.20	3.142
	5239.821	1.455	− 0.765 ± 0.042	48.70	3.166
	5334.218	1.497	− 2.203 ± 0.088	3.70	3.190
	5357.185	1.502	− 2.111 ± 0.079	3.30	3.054
	5526.817	1.768	0.024 ± 0.050	79.20	3.234
	5640.984	1.500	− 1.131 ± 0.053	36.70	3.274
	5658.340	1.497	− 1.208 ± 0.039	32.50	3.297
	5667.147	1.500	− 1.309 ± 0.048	28.50	3.282
	5669.034	1.500	− 1.200 ± 0.047	33.50	3.310
	5684.194	1.507	− 1.074 ± 0.043	36.00	3.208
	5604.593	1.357	− 1.309 ± 0.146	35.80	3.240

Experimental Nuclear Astrophysics

C. ROLFS

Ruhr-Universität Bochum, Experimentalphysik III, Bochum, Germany

1. INTRODUCTION

Investigations during the last 50 years have shown that we are connected to distant space and time through a common cosmic heritage: the chemical elements. These elements were created in the hot interiors of remote and long-vanished stars as well as in the very early universe. The detailed understanding of this heritage combines astrophysics and nuclear physics, leading to the field of nuclear astrophysics. Impressive progress has been made in this field over the last decades, but there remain puzzles and problems which challenge the basic ideas underlying nucleosynthesis in stars and elsewhere[1]. Thus the ultimate goal of the field has not been attained and much work is needed on all its aspects: experiment, theory, and observation.

An understanding of most of the critical features of nuclear burning in stars (and elsewhere), such as time scales, energy production, and nucleosynthesis of the elements, hinges directly on the magnitude of the reaction rate for a given stellar temperature (thermonuclear energy region), which in turn depend on the energy dependence and absolute value of the cross section of the reaction of interest. The desired cross sections are among the smallest measured in the nuclear laboratory, often requiring long data collection times with painstaking attention to background[2-4].

Charged-particle-induced reactions occur over a wide range of temperatures (T_9 = 0.01 to 10; T_9 in units of 10^9 K). The ratio of the corresponding center-of-mass energy of a particular reaction to the height of the associated Coulomb barrier, E/E_C, ranges from about 0.01 to 1. This sub-Coulomb-barrier range is thus the range most commonly of interest to nuclear astrophysics. The steepest energy dependence in $\sigma(E)$ is contained in the penetration factor for the Coulomb and angular momentum barriers. For low incident energies and for s-waves, this factor is approximately proportional to $\exp(-2\pi\eta)$, where

η is the Sommerfeld parameter $(2\pi\eta = 2\pi Z_1 Z_2/\hbar v)$. In this expression, Z_1 and Z_2 are the integral nuclear charges of the two interacting particles in the entrance channel, and v is their relative velocity. It is convenient to factor out this energy dependence, as well as an additional factor of $1/E$ that arises from the de Broglie wavelength squared, and write

$$\sigma(E) = S(E)\ E^{-1}\ \exp(-2\pi\eta). \qquad (1)$$

The function S(E) defined by this equation contains all of the strictly nuclear effects and is therefore referred to as the nuclear or astrophysical S-factor. For nonresonant reactions the S-factor is a smooth function of energy that varies much less rapidly with beam energy than $\sigma(E)$. The stellar reaction rate, $\langle\sigma v\rangle$, is obtained by weighting $\sigma(E)$ by a Maxwell-Boltzmann velocity distribution. For nonresonant reactions, one finds[2-4] $\langle\sigma v\rangle \approx S(E) \approx S(0)$; for an isolated narrow resonance, $\langle\sigma v\rangle \approx (\omega\gamma)\exp(-E_R/kT)$; here $\omega\gamma$ is the strength and E_R the energy of the resonance. With improved experimental techniques, direct measurements of $\sigma(E)$ can be extended toward lower energies, but one is rarely able to reach the relevant energy regions for quiescent (hydrostatic) stellar burning. The observed energy dependence of $\sigma(E)$, or equivalently of S(E), must therefore be extrapolated into the stellar energy region (essentially to zero energy). Of course, the basis for extrapolation is improved if very low energy data with high accuracy are available. However, these extrapolated data represent only lower limits of the stellar reaction rate. If there are bound or unbound resonances near the particle threshold, they can completely dominate the reaction rate for low stellar temperatures. The experimental investigation of such resonances near the particle threshold represents a major challenge to the experimenter.

Due to the absence of a Coulomb barrier in neutron-induced reactions, the relevant cross section, and thus the stellar reaction rate, can be measured directly at the relevant stellar energies. It was found that the Maxwellian-averaged cross sections are relatively independent of temperature for most nuclides between 10 and 100 keV. From the viewpoint of data compilation, it is then sufficient to choose the most convenient and common energy, which is found to be 30 keV. Improved neutron fluxes, either with a Maxwell–Boltzmann velocity distribution (Karlsruhe) or a known distribution over a wide range of energies (Los Alamos), combined with efficient detectors (such as the 4π BaF_2 crystal ball at Karlsruhe) have provided precise data for many relevant reactions (for details, see refs. 5–7 and references therein).

In the following the discussion will be restricted to the case of charged-particle-induced reactions. First, improvements of experimental techniques will be discussed and, second, two illustrative examples of recent measurements will be presented.

2. IMPROVED AND NEW EXPERIMENTAL TECHNIQUES

Because the cross section $\sigma(E)$ drops by many orders of magnitude for energies between the Coulomb barrier and the point at which E/E_C is about 0.01, experimental investigations of $\sigma(E)$ require the availability of high ion-beam currents, stable and pure targets, and efficient detectors (see refs. 2–4 and references therein).

The need for high currents establishes the requirements for ion sources, accelerators, beam transport systems, and beam integration. For example, a 100 kV accelerator has been built at the Ruhr–Universität Bochum with beam currents (H^+ and He^+ ions) of the order of 1 mA on target. This accelerator allowed direct measurements of several reactions to energies as low as $E/E_C \simeq 0.01$ (sect. 3). The construction of smaller accelerators (30 and 50 kV) with similar beam power down to about 5 kV are presently implemented and tested at Bochum. The 30 kV accelerator has already been used[7] in a test run at the underground laboratory Gran Sasso.

The requirement of stable and pure targets can be fulfilled by windowless gas targets or specially prepared solid targets. In recent years the implantation technique has proven as an excellent means for producing isotopically enriched targets capable of withstanding high beam powers. Through proper preparation and choice of the implantation host material (the backing), such targets can be made free of contaminants. The implantation technique also appears to be a useful tool for the production of radioactive targets, e.g. ^{22}Na (ref. 8).

A perfect detector would have 100% detection efficiency, 4π geometry, unique identification of the reaction products, insensitivity to background radiation, high energy and time resolution, and reasonable costs. Obviously, such perfection is unattainable; for each specific experiment, these requirements conflict, and compromises must be made in the choice and design of the detectors.

In the case of neutron detection with high efficiency, the target is surrounded in 4π geometry with a material (e.g. polyethylene, graphite or heavy water), which moderates the emitted high-energy neutrons. These neutrons are then detected with ^{3}He-filled proportional counters imbedded in the material, with a total efficiency of about 30-40% (Caltech, Bochum, and Stuttgart). In the Stuttgart-design[9], 8 proportional counters are each placed at an inner and outer circle around the target; since high-energy (or low-energy) neutrons require a long (or short) distance for their moderation in the material, they should be detected predominantly in the outer (or inner) counters. This feature was verified experimentally; this detector allows thus for a "rough" spectroscopy of the neutrons. This feature was one of the criteria to identify a previously observed "resonance" in ^{22}Ne$(\alpha,n)^{25}$Mg as arising from the contaminant reaction ^{11}B$(\alpha,n)^{14}$N. These detectors have been used in many improved measurements of (α,n) reactions to very low energies. The sensitivity of these detectors is however limited by the yield contributions from cosmic rays: typically about 0.1 counts per second. A reduction of the cosmic rays by nearly 3 orders of magnitude can be achieved by carrying out such experiments deep underground, such as in the Gran Sasso Laboratory.

When the ejectiles are charged particles, they are usually observed with an array of surface barrier detectors around the target, leading to an efficiency of nearly 100%. For low counting rates new detection techniques (such as ΔE-E "hybrid" detectors, in combination with pulse shape inspection) are needed, which are presently exploited in several laboratories. The limit of yield measurements is determined also here by the contributions from cosmic rays. Active shieldings can provide an improvement against the myonic component (by about a factor 5) but not against the electromagnetic component of the cosmic ray showers; again, the best solution is to go underground.

Finally, the detection of γ-rays in capture reactions can be carried out with crystal balls made of NaI, BGO or BaF_2 crystals leading to nearly 100% efficiency for MeV γ-rays. A cheaper (but similar) solution is to surround the target by a thick scintillator (e.g. a NaI crystal 30 cm thick, with a central cylindrical hole to install the target and beam pipe), where the energies of the emitted γ-rays (e.g. $\gamma\gamma$ cascades) are summed leading to a peak in the spectrum at an energy E_γ = Q + E (Q = reaction Q-value, E = c.m. energy). Such a summing crystal has been implemented[10] recently at Bochum: if the events above room background ($\gtrsim$ 3 MeV) are added up to the summing peak, the total detector efficieny of about 80% is independent from the γ-ray decay scheme of the reaction. The detector appears thus very useful for a reinvestigation of many important capture reactions. The limit of the detector is again determined by cosmic rays; low cross section measurements require thus to go underground.

In the case of radioactive targets (or ion beams) the detection of capture γ-rays with standard detectors is hampered by the intense radiation from the target (or beam: note that a 1 nA particle beam of ^{13}N corresponds to an activity of about 1 Ci). An ideal detector for such studies should have a threshold property, i.e. zero efficiency for the main γ-ray flux of the target or beam: mainly 0.51 MeV) and a high and nearly uniform efficiency for (say) E_γ = 3-9 MeV capture γ-rays. A detector developed[11] is based on the photodisintegration of deuterium (2.225 MeV threshold) with subsequent counting of the moderated neutrons. The detector measures angle-integrated γ-ray fluxes with, however, a rather low efficiency ($\simeq$ 0.04%). Alternatively, radiative capture γ-rays might be observed with a detector based on the e^+e^- pair production, provided that the quantum energies of all γ-rays from the radioactivity were below the 1.022 MeV threshold.

Since all the recoiling nuclei B produced in the capture reaction A(x,γ)B travel in essentially the same direction as the beam, their observation with detectors having 100% efficiency would greatly improve the experimental sensitivity. In this approach, thin targets (with no backings) or windowless gas targets are needed. However, there are some obvious problems if a detector is placed in the beam direction: one observes not only the capture products B, but also the incident beam (including any beam contaminants), small angle elastic scattering products, and background events (e.g. from multiple scattering processes). These problems can be solved by various techniques, such as combining a Wien filter and magnetic and electrostatic analyzers with particle identification in detector telescopes. One such system has been built at Caltech: work on $^1H(^{12}C,\gamma)^{13}N$ has shown[12] its usefulness to purify the beam even though the mass difference between the beam and residual nuclei is only one mass unit. With a beam purification of the order 10^{10} one could also detect the residual nuclei B via their characteristic radioactivity signatures (total efficiency of residual nuclei $\approx$ 25%).

Sometimes it is easier to determine the cross section of a reaction via measurement of the inverse reaction, where both cross sections are related by the principle of detailed balance. A recent example of this approach is the reaction $^8Li(\alpha,n)^{11}B$, which is difficult to measure directly due to the instable ^8Li ($T_{1/2} \approx$ 1 s). The inverse reaction $^{11}B(n,\alpha)^8Li$ was thus studied recently[7,13]. However, due to the high Q-value of 6 MeV several excited states in ^{11}B can be fed via the direct reaction, while the inverse reaction always starts with ^{11}B in its ground state. Thus, such studies provide only partial information on the cross section, i.e. a lower limit.

Another example are radiative capture reactions, $X(a,\gamma)Y$, where the inverse reaction, $Y(\gamma,a)X$, is also called photodisintegration. Due to phase space factors, $\sigma_{\gamma a}$ is much larger than $\sigma_{a\gamma}$. This might be taken as indicating that, instead of directly studying the capture reaction, it would be easier to investigate the photodisintegration reaction, at a given E. However, photodisintegration reactions are not easy to carry out from a techniacl point of view. Another approach was suggested in recent years[14]. When a charged nuclear projectile Y moves with moderately high kinetic energy through the Coulomb field of a high-Z nucleus such as lead, strong electromagnetic fields are present for a short time. The effect of the time-varying electromagnetic field is equivalent to a high-flux spectrum of virtual photons. These photons can lead to Coulomb dissociation of the projectiles, $Y + Z \rightarrow X + a + Z$. The copious source of virtual photons offers a promising way to study the photodisintegration process. Several cases have been studied recently, including reactions involving instable nuclei[7]. However, it is not yet clear how far this method can be pushed in practice. Of course, the method provides again only partial information on the capture cross section, namely σ_{cap} to the ground state of Y only. Fortunately, for light nuclides this capture process is often the dominant process.

In the hot and explosive burning phases of stars, where the thermal energy approaches the Coulomb barrier ($E/E_C \simeq 1$), nuclear burning times can be measured in seconds. If the lifetime of a radioactive nucleus is longer than or of the same order as the burning time, that nucleus will be involved in the nuclear burning processes. A quantitative understanding of the observed nuclear ashes from these stars requires a knowledge of $\langle \sigma v \rangle$, and this in turn requires measuring (p,γ), (α,γ), and other reactions involving these nuclides. If the half-life of the nuclides is longer than a day or so, they may be made into a radioactive target. However, in a great majority of interesting cases, the half-life is shorter (e.g. $T_{1/2} = 10$ min for ^{13}N). In this case, the only direct method is to produce the radioactive nuclides in an accelerator, separate them, accelerate them in a second accelerator, and finally allow the radioactive ion beam (RIB) to interact with a H_2 or He target. All of this must be achieved in a time shorter than the decay lifetime of the radioactive nuclides. To develop RIB several proposals (e.g. TRIUMF and CERN) have been made. A successful project has already been carried at Louvain-la-Neuve[7]: study of the reaction ^{13}N$(p,\gamma)^{14}$O via ^{1}H$(^{13}$N$,\gamma)^{14}$O. It is clear that additional research and technical development are needed on all aspects of the probelm, before reactions involving short-lived nuclides can be routinely studied in the laboratory.

3. ELECTRON SCREENING

In equation 1 it is assumed that the Coulomb potential of the target nucleus and projectile is that resulting from bare nuclei, and thus the potential would extend to infinity. However, for nuclear reactions studied in the laboratory, the target nuclei are usually in the form of neutral atoms or molecules. The atomic (or molecular) electron cloud surrounding the target nucleus acts as a screening potential[15]: an incoming charged projectile experiences no repulsive Coulomb force until it penetrates the electron cloud; thus, the projectile effectively sees a reduced Coulomb barrier. This in turn leads to a higher cross section, $\sigma_s(E)$, than would be the case for bare nuclei, $\sigma_b(E)$, with an enhancement factor $f(E) = \sigma_s(E)/\sigma_b(E) \simeq \exp(\pi\eta U_e/E)$, where U_e is the electron screening potential (e.g. $U_e \simeq Z_1 Z_2 e^2/R_a$, with R_a an atomic radius). Note that $f(E)$ increases exponentially with decreasing incident energy. For energy ratios $E/U_e \gtrsim 1000$, shielding effects are negligible and laboratory experiments can be regarded as essentially measuring $\sigma_b(E)$. However, for $E/U_e \lesssim 100$, shielding effects cannot be neglected and become important for understanding low-energy data. Relatively small enhancements from electron screening at energy ratios $E/U_e \simeq 100$ can cause significant errors in the extrapolation of cross sections to lower energies, if the curve of the cross section is forced to follow the trend of the enhanced cross sections, without correction for the screening. Notice that for astrophysical and other applications (stellar and fusion plasmas) the value of $\sigma_b(E)$ must be known because the screening in these applications is quite different from that in laboratory nuclear reaction studies, and $\sigma_b E)$ must be explicitly included for each situation. Low-energy studies of ^{3}He(d,p)^{4}He clearly showed[16] for the first time such screening effects, as well as their dependence on the aggregate state of the target. Similar studies have been carried out recently for the ^{6}Li(p,α)^{3}He, ^{6}Li(d,α)^{4}He and ^{7}Li(p,α)^{4}He reactions[17], where the effects of electron screening have also been clearly observed. The available data provide values for the screening potential U_e, which are however significantly higher than those expected from the united atom test. Clearly, a thorough understanding

of screening effects is not possible at this time requiring additional
efforts in theory as well as in experiment. With a low-energy
accelerator available at Gran Sasso in the near future (sect. 2),
improved data for the above and other reactions appear possible.

4. ATOMIC EFFECTS ON α-α SCATTERING TO ^8Be GROUND STATE

The resonant α-α scattering to the ^{8}Be ground state represents
the first step in the triple-α process[1-4]. The parameters of this
s-wave resonane (Γ_R = 13.6±3.4 eV, E_R = 184.24±0.10 keV, both in the
laboratory system) were determined[18] in a difficult experiment at
Zürich in 1968. The experiment introduced the crossing beams
technique in nuclear physics; the collision of ^{4}He$^+$-particles with a
supersomic ^{4}He jet led to a total experimental energy resolution of ΔE
= 95 eV. The elastic scattering yield was inferred from kinematic
coincidences between pairs of detectors placed at the scattering
angles θ_{lab} = 45º/45º. This experiment also provided the first
observation of the influence of atomic electrons on nuclear resonance
reactions: the E_R = 184 keV resonance was not characterized by a
single structure (a single "dip" for θ_{lab} = 45º) but was split into
three apparent dips (Fig. 1). The observed multiple structure was
attributed to atomic effects: the experiment was not performed with
bare nuclides, but with ions and atoms, which lead to possible
rearrangement of the 3 electrons of the entrance channel into various
low-lying levels of the compound atom Be$^+$. Assuming the lowest
3-electron configurations, the above resonance parameters Γ_R and E_R
for bare nuclides were extracted from the data. However, subsequent
analyses by Merzbacher et al.[19] based on atomic physics models could
not reproduce the data (e.g. dotted curve in Fig. 1 with a "best" fit
for Γ_R = 5.0 eV) and questioned the validity of the values quoted for
the resonance parameters. It was suggested that improved α-α elastic
scattering data were highly desirable. Such an attempt was carried
out recently[20]. The basic experimental approach was identical to that
of the Zürich group: a ^{4}He$^+$ ion beam collided perpendicularly with a
neutral atomic ^{4}He beam to minimize the effects of Doppler broadening.
However, the total energy resolution was reduced significantly (ΔE =

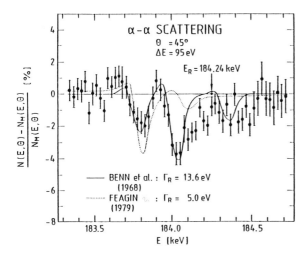

FIGURE 1

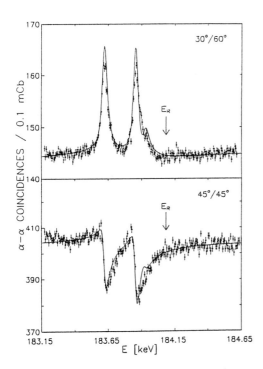

FIGURE 2

26 eV) by improvements in ion beam spread and Doppler broadening. In addition, the ^{4}He$^+$ beam current on target was 1000 times as large as that in the previous work; this resulted in a significant improvement in counting statistics (better than 0.5% per 3.0 eV energy step). In addition to measurements at 45°/45°, data were also obtained at 30°/60°, where the resonance effect was expected to be larger and of opposite sign (a "peak"). The results (Fig. 2) show that the resonance is split into 2 structures, not the 3 structures suggested previously. The structures arise predominantly from 2-electron configurations with the 3. electron ionized. The deduced width, Γ_R = 11.14±0.50 eV, is significantly smaller, where the uncertainty is reduced by nearly a factor of 7. The E_R = 184.07±0.10 keV parameter is in fair agreement with previous work.

REFERENCES

1. W.A. Fowler, Rev.Mod.Phys. 56(1984)149
2. C. Rolfs and W.S. Rodney, Cauldrons in the cosmos, University of Chicago Press, 1988
3. C.A. Barnes, Nuclear Astrophysics, ed. Lozano et al., Springer-Verlag (1989) p.197
4. C. Rolfs and C.A. Barnes, Ann.Rev.Nucl.Part.Sci. 40(1990)45
5. G.J. Mathews and R.A. Ward, Rep.Progr.Phys. 48(1985)1371
6. F. Käppeler et al., Rep.Progr.Phys. 52(1989)945
7. Nuclei in the Cosmos, Proc. Int. Conf. (Karlsruhe, July 1992), ed. Käppeler et al.
8. H. Ravn et al., Nucl.Instr.Meth. B58(1991)174
9. H.W. Drotleff, Thesis, Ruhr-Universität Bochum (1992)
10. M. Junker, Diplomarbeit, Ruhr-Universität Bochum (1992)
11. S. Seuthe et al., Nucl.Instr.Meth. A272(1988)814
12. M. Smith et al., Nucl.Instr.Meth. A306(1991)233
13. T. Paradellis et al., Zeitsch.Phys. A337(1990)211
14. G. Baur et al., Nucl.Phys. A458(1986)188
15. H.J. Assenbaum et al., Zeitsch.Phys. A327(1987)461
16. S. Engstler et al., Phys.Lett.B202(1988)179
17. S. Engstler et al., Zeitsch.Phys. A342(1992)471 and Phys.Lett. B279(1992)20
18. J. Benn et al., Nucl. Phys. A106(1968)296
19. J.M. Feagin, Thesis, University of North Carolina (1979)
20. S. Wüstenbecker et al., Zeitsch.Phys. (in press) and Thesis, Ruhr-Universität Bochum (1991)

On the Possibility of Making Radioactive Beams

D. GUILLEMAUD-MUELLER

Institut de Physique Nucléaire 91406 Orsay Cedex

1 INTRODUCTION

The production and use of unstable nuclei as projectiles to induce nuclear reactions has been a field of increasing interest during the last years. For astrophysical purposes, two domains related to unstable nuclei are interesting. The first concerns cross-section measurements and study of direct reactions at low energy (or at high energy by using inverse reactions). A second topic is the study of properties of nuclei far from stability which contribute to nucleosynthesis in explosive processes in order to understand isotopic abundance of the elements. The present contribution is essentially aimed at the latter aspect.

2 PRODUCTION OF RADIOACTIVE EXOTIC BEAMS

In order to produce beams of exotic nuclei two methods are available. With the first technique, ISOL, the nuclei are produced through a proton-induced primary reaction at rest in the target, diffused out of the target and separated in mass by an isotope separator (a well known example is ISOLDE at CERN). Post-acceleration of ISOL beams is now, for the first time, available at the Louvain la Neuve facility 'Decrock (1991)'.

The alternative technique at GANIL uses fragmentation of heavy-ion beams at some tens of MeV/n. A large number of fragments with various A and Z are formed with velocities close to that of the projectile. Preparing secondary beams from the fragmentation reaction of the primary GANIL beams called for the construction of a dedicated spectrometer with the following properties: 1) a large collection angle around 0° and very low background, in particular a complete elimination of the primary beam, in order to detect small cross section events; 2) good particle identification properties in order to discriminate without ambiguity the great number of produced nuclei; 3) a sufficient number of degrees of freedom in the ion optics in order to provide secondary beams of reasonable optical quality. Because conventional magnetic spectrometers only partly meet these requirements, the following solution was chosen for LISE: a magnetic spectrometer (see fig.1) made up of two identical dipoles (D1 and D2) in a double achromatic arrangement. An atomic Z-dependent selection criterion can be introduced by inserting an energy degrader in the intermediate focal plane between D1 and D2.

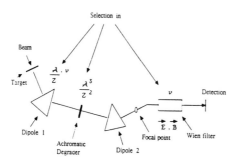

The second magnet is then set to a magnetic rigidity which corresponds to the energy-loss that the "wanted" isotope has experienced in the degrader. Thus, the transmission of this nucleus is optimized at the first achromatic focal point. A detailed description of LISE and its operation is given in 'Anne (1987)'. A much enhanced purification is achieved by an extension which has been installed recently 'Mueller (1991)'.

Fig.1: The LISE spectrometer

A Wien filter, mounted behind an achromatic deviation introduces a velocity selection. At the exit of the LISE particle identification is made in semiconductor detectors through the measurement of energy-loss ΔE, total energy E and time of flight TOF through the whole spectrometer. ΔE-TOF identification plots show in fig.2 how the different selection steps act together taking as example the separation of ^{20}Mg among the fragments from a primary ^{24}Mg beam at 90 MeV/n. Fig.2a shows the A/Z selection : stable isotopes produced with rates many orders of magnitude higher than the proton drip-line nucleus ^{20}Mg, are strongly suppressed. Insertion of an energy degrader mainly selects isotones of ^{20}Mg (fig.2b).

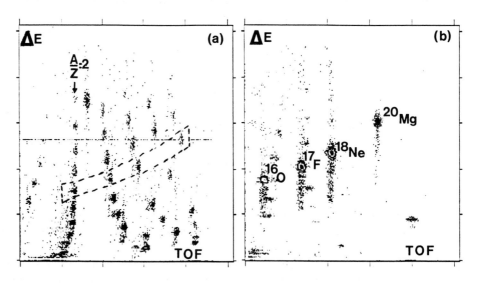

Fig.2: a)the A/Z selection of LISE b) the selection with a degrader (see text)

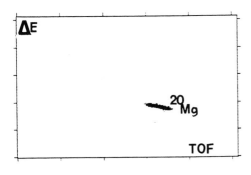

Fig 3: Selection with the velocity filter

The subsequent velocity selection by the Wien filter allows to obtain almost a pure ^{20}Mg beam (fig.3). The example is taken from a study of the decay properties of ^{20}Mg (observing β, proton, γ) recently made with LISE. Indeed, the β-decay of ^{20}Mg populates states in ^{20}Na important for the Hot CNO break out reaction ^{19}Ne(p,γ)^{20}Na 'Kubono (1991)'. This experiment is presently under analysis and results are expected to be available very soon.

3 AS AN EXAMPLE:THE ^{48}CA/^{46}CA ISOTOPIC RATIO

Another astrophysical problem which has been recently studied at LISE concerns the heavy calcium isotope abundance anomalies, ^{48}Ca/^{46}Ca = 56 in the solar system, up to 250 in the inclusions of the Allende meteorite. Indeed, the abundancy ratios are difficult to explain in rapid neutron capture scenarios using predicted nuclear decay properties ($T_{1/2}$, P_n) of the progenitor isotopes. Already in 1986, Hillebrandt et al. 'Hillebrand (1986)' had requested to measure the missing β-decay properties of the "key"-isotopes influencing the low production of ^{46}Ca as compared to ^{48}Ca in their β-delayed neutron model, i.e. ^{44}S and $^{45-47}$Cl. These very neutron-rich isotopes of astrophysical interest can be produced at LISE using fragmentation of a ^{48}Ca beam. The β-decay properties $T_{1/2}$, P_n of these isotopes could be studied through the observation of β-delayed neutrons 'Sorlin (1991)'. With the measured $T_{1/2}$ of ^{44}S and ^{45}Cl being considerably shorter than the predictions 'Klapdor (1984)' used in 'Hillebrandt (1986)', together with the higher, respectively lower P_n values of ^{46}Cl and ^{47}Cl, the conclusions of Hillebrandt et al. concerning the low production of ^{46}Ca in astrophysical scenarios are in principle strenghtened. As can be seen from fig.4 the experimental values may even turn out to be more favourable for a β delayed neutron process than assumed.

Successive neutron capture in the S and Cl chains is halted at the N=28 turning point nuclei where β-decay starts to dominate $T_{1/2}(\beta) < T_{1/2}(n)$. Of particular importance is the result that due to the short $T_{1/2}(^{45}$Cl) feeding of ^{46,47}Cl is reduced. These isotopes were found to be the main progenitors of ^{46}Ca in the 1986 approach. Moreover, even if the neutron capture in the Cl chain would proceed up to ^{46,47}Cl, their P_n values would again result in a low ^{46}Ca abundance (see fig.4). Hence with the present data, the problem of producing "very little" ^{46}Ca seems to be solved. But, on the other hand, may moderate the rather strict constraint on the neutron-exposure time requested by Hillebrandt et al.

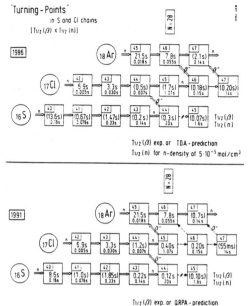

Fig. 4: Neutron-capture path in the S to Ar chains for a stellar temperature of T=8.10^8 K and a neutron density of 5.10^{-5} mol.cm^{-3}; upper part: status in 1986, lower part: present status. With our new experimental data 'Sorlin (1991)', at both N=28 "turning point" isotopes, ^{44}S and ^{45}Cl, β-decay back to stability starts to dominate over further neutron capture. Hence, the possible A=46,47 progenitors of ^{46}Ca will be produced in small amounts only. With this, large ^{48}Ca/^{46}Ca ratios can be obtained, as required to explain the observed abundances in meteorites.

4 CONCLUSION

The two preceeding examples may demonstrate that heavy-ion projectile fragmentation is a powerful tool to reach nuclei of importance in astrophysical processes which were hiterto unattainable. Note that radioactive beams from projectile fragmentation are also used for reactions which occur in astrophysical scenarios by studying Coulomb break-up which is just the inverse to the (p,γ) process 'Kiener (1991)'.

5 REFERENCES

R. Anne et al., Nucl. Inst. Meth. A257 (1987) 215
P. Decrock et al., 2nd Int. Conf. on Radioactive Beams, Th. Delbar ed, IOP (1991) 121
F.-K. Hillebrandt et al., Int. Conf. on Weak and Electromagnetic Interactions in Nuclei, H.V. Klapdor ed., Springer Verlag (1986) 987
J. Kiener et al., 2nd Int. Conf. on Radioactive Beams, Th. Delbar ed, IOP (1991) 311
H.V. Klapdor et al., At.Data Nucl. Data Tables 31 (1984) 81
S. Kubono et al., 2nd Int. Conf. on Radioactive Beams, Th. Delbar ed, IOP (1991) 317
A.C. Mueller et al., Nucl. Inst. Meth. B56/57 (1990) 559
O. Sorlin, Thesis IPN Orsay, IPNO-T 91.04 and to be published

Investigation of the $^{16}\mathrm{O}(\alpha,\gamma)^{20}\mathrm{Ne}$ rate at low energies [*]

H. KNEE, A. DENKER, H. DROTLEFF, M. GROßE, A. MAYER, G. WOLF and J.W. HAMMER

Institut für Strahlenphysik der Universität Stuttgart,
Allmandring 3, D-7000 Stuttgart-80 (Germany)

The reaction $^{16}\mathrm{O}(\alpha,\gamma)^{20}\mathrm{Ne}$ proceeds in stars very weakly and its reaction rate at astrophysical relevant temperatures could only be determined theoretically by nuclear systematics (Fowler et al., 1975). There is a strong demand for new experimental data at lower energies to reduce the discrepancies between semiempirical extrapolations and theoretical model calculations (Descouvemont and Baye, 1983; Langanke, 1984; Hahn et al., 1987; Baye and Descouvemont, 1988).

The aim of this work was the experimental determination of the excitation function and the reaction rate in the energy range $E_{\alpha,\mathrm{lab}} = 1\,000 - 3\,500$ keV to obtain reaction rates by direct observation. The cross section would be of interest even at much lower energies, but at 1 MeV, outside the resonances, a level of about 50 pikobarn is reached for the cross section measurements, which is at present the limit of sensitivity. The parameters of all resonances in this range had to be determined as well as the nonresonant direct capture contribution (DC). At least upper limits had to be given. This reaction is also of interest as reverse reaction in hot scenarios and for the testing of theoretical models.

This experiment is characterized by the application of high He^+-currents in normal kinematics, a windowless gastarget with enriched $^{16}\mathrm{O}$, because solid state oxygen targets suffer from poor stability with respect to high currents. Special precautions have to be made working with oxygen in a gas target. The experiment is further characterized by detectors of high efficiency (about 90 .. 120 % with respect to a 3"x3" NaI) and high resolution using intrinsic germanium to obtain nearly complete spectroscopic informations from the reaction and to distinguish it from background events.

In two experiments all resonances have been measured occuring at $E_{\alpha,\mathrm{lab}} = 1\,116$,

[*]Supported by the DFG, Bonn

1 317, 2 490, 3 036, 3 074 and 3 363 keV corresponding to states in ^{20}Ne at $E_x = 5\,624$ (3^-), 5 785 (1^-), 6 722 (0^+), 7 156 (3^-), 7 189 (0^+) and 7 421 (2^+) keV (see Fig. 1). Two states above $(E_x = 4.97$ MeV, $2^-)$ and below the threshold $(E_x = 4.25$ MeV, $4^+)$ are not contributing to the reaction rate, the first having unnatural parity and the second having high spin and too much distance from the threshold. A third experiment was dedicated to the search for the nonresonant direct capture.

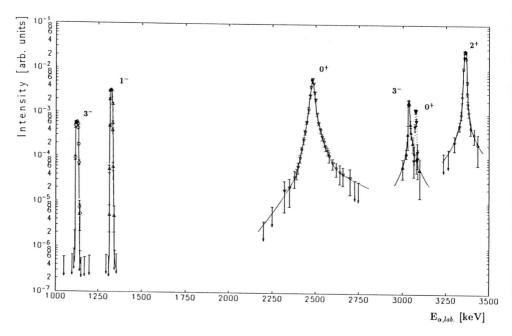

Fig. 1 : Yield curve of the ^{16}O$(\alpha, \gamma)^{20}$Ne–reaction in the energy range $E_{\alpha,\mathrm{lab}} = 1.0$ .... 3.5 MeV. For the yield between the resonances upper limits were determined, but are not shown in this diagram. They are of the order of magnitude of $5 \cdot 10^{-8}$ .... $8 \cdot 10^{-7}$ in the arbitrary units of this diagram.

The experiments have been performed at the 21°–beamline of the Stuttgart DYNA-MITRON accelerator using the windowless gastarget system RHINOCEROS, which had been operated in the static mode with 4 pumping stages and a pressure reduction of about 6 orders of magnitude in this case. Isotopic enriched (99.98 %) oxygen–16 was recirculated. The reaction chamber was a small stainless steel cell of 60 mm diameter with watercooled apertures at the entrance and the outlet. This setup gives a low gas pressure at the beam dump, concentrating the reaction zone to the target chamber and the apertures : the effective length of the target zone was about 90 mm, which fits well to the diameter of the detectors. The target pressure was 0.5 mbar, the pressure reduction before and behind the target chamber being 16 : 1

with 5 mm diameter apertures, which were not hit by the beam, because the beam defining aperture is located two stages before this chamber (about 30 cm). The He^+ beam currents were about 0.25 mA with narrow profile at the reaction chamber to perform 100 % transmission. The ionized oxygen, produced by the stopping of the ion beam, is chemically very active. Therefore the walls of the target cell and the apertures were plated with 20 μm of pure gold to prevent oxydation of these parts and also to avoid background reactions with the carbon (^{13}C) content of the stainless steel. Otherwise one gets an oxid layer at all surfaces, which leads to spurious yield contributions and large errors in the yield determination.

The germanium detectors were set up in very close geometry under 90°. A heavy shielding with pure iron was applied. For beam monitoring purposes two silicon detectors were used at 30° and 60° to the beam direction. In the relevant energy region all resonances have been measured with high accuracy as it is shown in Fig. 1. Monte Carlo simulations of the extended source geometry in the experimental chamber are performed, to improve the yield calculations and to obtain information on the γ-branchings, because discrepancies to the values from literature have been found.

A very serious problem was found to be the background reaction $^{17}O(\alpha, n)^{20}Ne$, even though the remaining very low ^{17}O concentration in the target gas used, because its yield is higher by some orders of magnitude than that from $^{16}O(\alpha, \gamma)^{20}Ne$. The neutrons produced by $^{17}O(\alpha, n)^{20}Ne$ lead to neutron induced gammas and are damaging the γ-detectors. Furthermore the $^{17}O(\alpha, n)^{20}Ne$-reaction leads to the same final nucleus and in part to the same gamma-transitions, making it absolutely necessary to observe and evaluate all γ-transitions, to obtain all branchings and to pick out those transitions which are clear fingerprints of the $^{16}O(\alpha, \gamma)^{20}Ne$-reaction.

In all measured spectra no indication for the direct capture (DC) contribution could be observed, but it is possible to deduce from our data new and considerably lower lying upper limits.

Besides the γ-measurements also the elastic scattering of α-particles at ^{16}O was investigated, because the cross section deviates from the Rutherford-law. The elastic scattering is always used in gas target experiments for the normalization of data. The result of these experimental runs is shown in Fig. 2. For comparison a small amount of argon was added to the target gas. Furthermore the consumption of oxygen by chemical reactions in the target system was investigated. We obtained good agreement with previous mass spectrometric measurements. The pressure in the reaction chamber was stabilized by an automatic control.

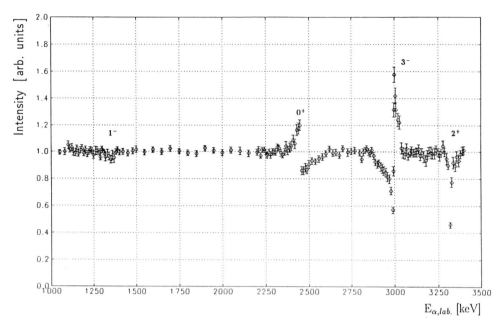

Fig. 2: Yield curve for the elastic scattering of α–particles on ^{16}O normalized to pure Rutherford scattering (on argon).

The ^{16}O$(\alpha,\gamma)^{20}$Ne capture reaction is very important for the testing of theoretical models of α–capture by light nuclei. Both nuclei in the entrance channel have closed shells and therefore the basic assumptions of the models are simple and reliable. If theory and experiment disagree, it would have far reaching consequences for the theory. The reaction ^{16}O$(\alpha,\gamma)^{20}$Ne was already analyzed using a phenomenological R–matrix theory (Hahn et al.; 1987) and also microscopic calculations (generator coordinate method) (Descouvemont and Baye, 1983; Baye and Descouvemont, 1988) and semimicroscopic calculations (Langanke, 1984). In the phenomenological approach it is not possible to the nonresonant part of the cross section because interference with the resonant amplitude can occur. The microscopic calculations cannot reproduce the energetic behaviour, of the astrophysical S–factor (Baye and Descouvemont, 1988), especially at some resonances. Therefore an optical model, based on a direct reaction mechanism, is applied to this problem, which reduces the many body problem of the interacting nuclei to the potential interaction of two particles (H. Oberhummer and G.Staudt, 1991; H.Abele and G. Staudt, 1992). Complicating is the fact, that two states have to be interpreted as $(^{12}$C $+ \, ^{8}$Be) states.

First calculations with this model using a double folded ^{16}O–α potential show good agreement with the experimental data of Hahn et al. (1987), however these data consist only of a few points. A more conclusive test should be possible by using our new experimental data. These new calculations are in progress.

The S–factor of $^{16}O(\alpha, \gamma)^{20}$Ne at about 300–450 keV ranges from 0.4 MeV·b (Fowler et al., 1975), 0.7 MeV·b to 1.6 MeV·b (Langanke, 1984) and 2.54 MeV·b (Descouvemont and Baye, 1983). From the upper limits for the DC γ–transitions, obtained from this measurement an upper limit for the DC contribution to the S–factor is determined to 0.9 MeV·b, which is in disagreement to the theoretical values of Langanke and Descouvemont but agrees with the result of Hahn et al., where only the S_0–factor was determined directly (0.26 ± 0.07 MeV·b) and the total S–factor (0.7 ± 0.3 MeV·b, upper levels 2.3 ... 2.8 MeV·b) was obtained by projection.

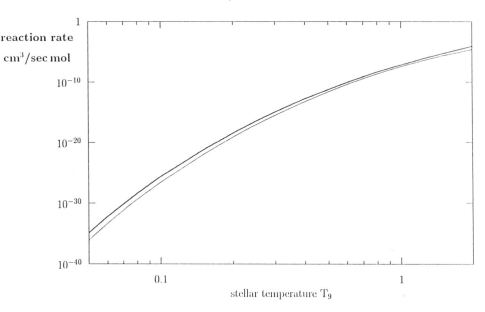

Fig. 3: Reaction rate of the Direct Capture contribution in $^{16}O(\alpha, \gamma)^{20}$Ne. The upper solid curve was deduced from the upper limits of this measurement, the lower curve was obtained from the FCZ evaluation without the rate of the resonant part.

REFERENCES

Abele, H., Staudt, G., 1992, subm. to Phys. Rev. C

Baye, D., Descouvemont, P., 1988, Phys. Rev. **C 38**, 2463

Descouvemont, P., Baye, D., 1983, Phys. Lett. **127 B**, 286

Fowler, W.A., Caughlan, C.R., Zimmerman, B.A., 1975, Ann. Rev. Astron. Astrophys. **13**, 69 (1975)

Hahn, K.H., Chang, T.R., Donughue, B.W., Filippone, B.W., 1987, Phys. Rev. **C 36**, 892

Langanke, K., 1984, Z. Phys. **A 317**, 325

Oberhummer, H., Staudt, G., 1991, Nuclei in the Cosmos, Ed. H. Oberhummer, Springer-Verlag, Berlin 1991

LIGHT ELEMENTS: THE BIG BANG AND COSMIC RAYS

Primordial Abundances of Hydrogen and Helium Isotopes

JOHANNES GEISS

Physikalisches Institut, University of Bern, and Max–Planck–Institut für Kernphysik, Heidelberg

ABSTRACT

The primordial abundances of the four lightest stable nuclei and lithium-7 reflect the physical conditions in the early universe, allowing to test cosmological concepts and to determine universal parameters such as expansion rate during the radiation dominated epoch, universal baryon density, or the possible existence of inhomogeneities at the time of nucleosynthesis. The primoridal ^{4}He/H ratio, derived from helium observations in metal–poor HII regions, is in excellent agreement with the prediction of the Standard Big–Bang Nucleosynthesis (SBBN) theory for three neutrino flavours. A new assessment of protosolar abundances gives $[(D+^3He)/H]_{\mathrm{PROTOSOLAR}}$ $= (4.1\pm1.0)10^{-5}$ and $[D/H]_{\mathrm{PROTOSOLAR}} = (2.6\pm1.0)10^{-5}$. Extrapolating these values to the primordial abundance ratios leads to universal baryonic densities far below the critical density of the universe. Observations and experiments presently under way or being prepared will further reduce uncertainties in the protosolar and present–day abundances of D, ^{3}He and D+^{3}He and also improve our understanding of the galactic evolution of the isotopes of hydrogen and helium.

1. INTRODUCTION

When in 1957 Burbidge, Burbidge, Fowler and Hoyle demonstrated that the elements and their isotopes, as we find them in the universe, could have been produced by a set of well defined nuclear processes operating in the interior of stars, there remained a small number of minor species, for which a stellar origin could not readily be postulated. These species, called the x-component, included the isotopes deuterium and helium–3, as well as the elements lithium, beryllium and boron. Since the steady state theory was a prominent cosmological view at that time, and little was known about the universal abundances of these x–nuclei, no satisfactory explanation for their origin could be found. The hypothesis of a local solar system production (Fowler, Greenstein and Hoyle, 1962) did not prevail in the light of new observations and energy considerations (Ryter et al 1970). Thus these x–nuclei of minor abundance became a cause for major concern to those who studied the origin of elements.

Today, this puzzle has been resolved (cf. Reeves 1974). The x–component actually comprises nuclei which stem from two totally differents settings:

i) Thermonuclear reactions in the proton–neutron gas cooling down from its early, high temperature state (Peebles 1966; Wagoner, Fowler and Hoyle 1967), producing significant amounts of ^{1}H, ^{2}H, ^{3}He, ^{4}He, and ^{7}Li; and

ii) non–thermal spallation reactions in interstellar or near–stellar space (Reeves, Fowler and Hoyle 1970; Meneguzzi, Audouze and Reeves 1971), providing the source for virtually all ^{6}Li, ^{9}Be, and ^{10}Be, as well as a significant fraction of ^{7}Li and ^{11}B.

In addition there is

iii) the set of hydrostatic and explosive thermonuclear reaction processes in stars set for by Burbidge, Burbidge, Fowler and Hoyle (1957), Cameron (1957), Reeves and Salpeter (1959), and further developed by many authors, producing all other nuclear species including the remainder of ^{7}Li and ^{11}B. It is well to remember on the occasion of his 60th birthday, that Hubert Reeves has contributed many original ideas and important theoretical studies to all three of these nucleosynthetic mechanisms.

In this article we discuss the observations on the occurence of the four lightest nuclei, the isotopes of hydrogen and helium. In order to establish their primordial abundances, it is necessary to understand their evolution with time, and thus it becomes crucial to make observations in different objects or reservoirs. These include: Solar system objects which were formed from the protosolar nebula, representing a sample of the interstellar galactic medium as it existed 4.6 Gyrs ago; the local galactic medium; the interstellar gas and the surfaces of stars in our galaxy and in other stellar systems.

We shall compare the primordial abundances derived from observations against the predictions of the Standard Model of Big–Bang Nucleosynthesis (SBBN), which assumes that during the epochs of neutron–proton freeze–out and nucleosynthesis, the universe was homogeneous, and its expansion was determined by photons, electrons and relativistic, non–degenerate neutrinos. The predicted SBBN abundances, calculated by Walker et al (1991) for N_ν=3 and with the latest decay parameters and reaction rates are shown in Figure 1. A general discussion of BBN theory is given in this volume by D.N. Schramm, K. Sato and others.

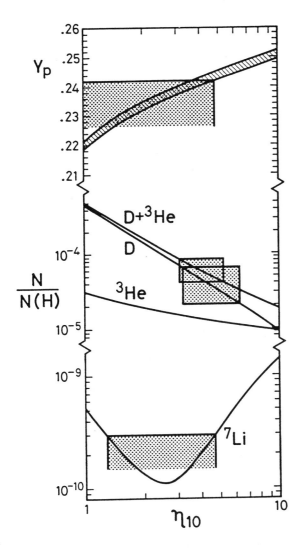

Figure 1: Predicted abundances of the five primordial nuclei, calculated for 3 neutrino flavours ($N_\nu = 3$) and a neutron mean-life of 882 to 896 seconds (after Walker et al. 1991) are compared with the primordial abundances derived from observation ($\eta_{10}=10^{10}n_B/n_\gamma$). The upper limit on Y_P, the primordial mass fraction of helium is from Pagel et al. (1992). The limits on the primordial (D+^{3}He)/H and D/H ratios are based on the protosolar values given in tables 2–4. For the upper limit on ^{7}Li, cf. section 6.

TABLE 1

Mass Fraction of Primordial Helium Y_P Derived from Observation

Y_P	Reference
.230 ± .004	Lequeux et al. 1979
.234 ± .008	Kunth & Sargent 1983; without II Zw 40
.232 ± .004	Peimbert 1985
.237 ± .005	Pagel et al. 1986
.230 ± .006	Torres-Peimbert et al. 1989
.229 ± .006	Pagel & Simonson 1989
.230 ± .015	Steigman, Gallagher & Schramm 1989
.229 ± .004	Walker et al. 1991; O/H → 0
.231 ± .003	Walker et al. 1991; N/H → 0
.230 ± .007	Walker et al. 1991; C/H → 0
.225 ± .005	Pagel 1990; only low metallicity
.220 ± .010	Fuller et al. 1991; only low metallicity
.228 ± .005	Pagel et al. 1992

TABLE 2

Derivation of the Protosolar $D+^3He$ Abundance

Description	Ratio	Value	
Solar Wind			
APOLLO	$^3He/^4He$	$(4.25\pm.21)10^{-4}$	a
ISEE-3, Average Flux Ratio	$^3He/^4He$	$(4.88\pm.48)10^{-4}$	b
ISEE-3, Average of Ratios	$^3He/^4He$	$4.37\ 10^{-4}$	b
Adopted average	$^3He/^4He$	$(4.50\pm.40)10^{-4}$	c
Protosolar	$(D+^3He)/^4He$	$(4.09\pm.87)10^{-4}$	c
Protosolar	$^4He/H$	$0.10\pm.01$	d
Protosolar	$(D+^3He)/H$	$(4.1\pm1.0)10^{-5}$	c

a Geiss et al. (1970; 1972)
b Coplan et al. (1984); Bochsler (1984)
c see text
d Turck-Chièze et al. (1988)

2. HELIUM

The theory of ^{4}He production in the early universe is straightforward (Hoyle and Tayler 1964; Peebles 1966). For SBBN conditions, the resulting abundance depends on weak interaction theory and the number of neutrino flavours $N_\nu=3$ (ALEPH 1989; DELPHI 1989; L3 1989; OPAL 1989). Assuming that the masses of these neutrinos are low enough and their chemical potential negligible during the BBN epoch, and adopting a neutron halflife of 889.8 s (Paul et al 1989; cf. Olive et al. 1990) the SBBN model predicts a primordial helium mass fraction of

$$Y_{p,\text{Theory}} = 0.228 + 0.023 \log \eta_{10} \tag{1}$$

where η_{10} is 10^{10} times the baryon/photon ratio (cf. Pagel 1991, Walker et al 1991). For a present–day baryon density of $\simeq 5 \ 10^{-31} \text{g/cm}^3$, we have $\eta_{10} \simeq 3$, and

$$Y_{p,\text{Theory}} = 0.239 \tag{2}$$

The data most relevant for deriving Y_p, the primordial helium abundance, are from HII regions. Y_p is obtained by extrapolating to zero "Metallicity". Such extrapolations to vanishing O, C or N abundance have been carried out for many years, leading to consistent results (cf. Table 1). However, the procedure is not without problems, because the relative production of He, O, C and N is variable, it depends on stellar composition and mass (Chiosi and Meader 1986). Moreover, excessive local production could affect the correlation between metal and He abundances. Pagel et al (1992) have studied these problems in great depth: Because winds from Wolf–Rayet stars and their progenitors locally add products of H–burning (He and N), they omitted H II regions with strong WR emissions and obtained consistent and superior regression lines (Figure 2), leading to $Y_p = 0.228 \pm .005$ (s.e.). For the purpose of our discussion we adopt their upper limit (95 % confidence)

$$Y_p < 0.242, \tag{3}$$

which allows for systematic errors. The match between theoretical prediction (equ. 1 and 2) and observation (Table 1; equ. 3) of Y_p is a most remarkable success of astronomical observation, particle physics and cosmological theory, demonstrating that the basic concepts of primordial nucleosynthesis are sound.

3. PROTOSOLAR DEUTERIUM AND HELIUM–3

The sum D+^{3}He is of direct cosmological significance. Both species are products of thermonuclear reactions in the early universe, but much of the primordial D has since been converted to ^{3}He. Thus the sum D+^{3}He is much more invariant against stellar

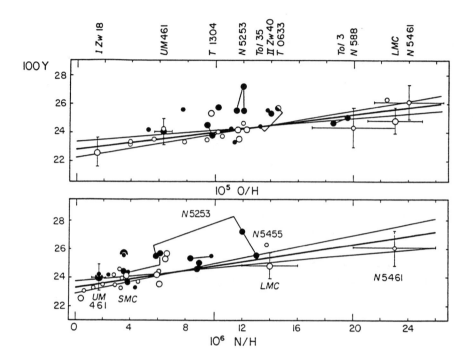

Figure 2: Regressions of helium against O/H and N/H for objects with definite detections of broad WR features (filled circles) and without definite detections (open circles). Regression lines (with ±1σ limits) are shown for the latter category. A few typical error bars are given. From Pagel et al. (1992).

processing than D alone.

Until the late 1960s it was generally assumed that the proportion of heavy water in the oceans ($1.6 \ 10^{-4}$) represents the cosmic abundance of deuterium. This view had been strengthened by Boato's (1954) results on the D/H ratio in carbonaceous chondrites. With a cosmic D/H ratio of $1.6 \ 10^{-4}$, deuterium would have been more abundant than Si or Fe. Yet among the stable nuclei D has by far the lowest binding energy per nucleon; it is much easier to find ways to destroy it than to produce it. In fact, deuterium posed the most difficult problem among the x–nuclei, and for a while it was even considered to be of a local origin, a product of slow neutron capture in the early history of the solar system (Fowler, Greenstein and Hoyle 1962).

3.1 Protosolar Deuterium plus Helium–3

In the 1960s it became clear that the noble gases in meteorites include a "solar" and a "planetary" component (Signer and Suess 1963). Detailed analyses showed that much of the solar component was located just below the surfaces of the meteoritic grains (Hintenberger, Vilcsek and Wänke 1965; Eberhardt, Geiss and Grögler 1965; Wänke 1965), indicating that it represents captured solar wind particles. There was a problem, however, with the measured $^4He/^3He$ ratios of a few times 10^{-4}, because it should have been increased by deuterium burning in the early Sun to a value above 10^{-3} (Geiss, Eberhardt and Signer 1965). This puzzle was a motive for conducting the Apollo Solar Wind Composition experiments by which solar wind particles were captured in foils exposed on the moon under controlled conditions and, after return to Earth, analysed in the laboratory. The $^3He/^4He$ ratios resulting from five Apollo experiments are given in figure 3, the weighed average of $4.25 \ 10^{-4}$ is given in Table 2. Based on the observed abundance of beryllium in the solar photosphere (Grevesse 1968), Geiss and Reeves (1972) showed that ^3He in solar surface material could not have been further processed at any time during solar history and concluded, that the $^3He/^4He$ ratio in the solar wind represents the $(D+^3He)/^4He$ ratio in the protosolar gas. After correcting for original solar ^3He, they went on to derive a D/H ratio of $(2.5\pm1.0) \ 10^{-5}$ for the protosolar gas. Geiss and Reeves (1972) compared this new abundance value with the predictions of big–bang production calculations by Wagoner, Fowler and Hoyle (1967), and showed that the universal (baryonic) density was of the order of $3 \ 10^{-31}$ g/cm. Independently, Black (1972) came to similar conclusions from an interpretation of helium isotope data in meteorites. The protosolar value thus derived is almost an order of magnitude lower than this ratio in sea water. An enrichment by such a large factor could only be due to chemical reactions at low temperature (Geiss and Reeves 1972). This view was confirmed when extremely high D/H ratios were found in some special meteorites (Robert, Merlivat and Javoy 1979). We know now, that high D/H ratios are common in hydrogen–

poor solar system bodies (cf. Eberhardt et al 1987; Owen, Lutz and de Bergh 1986), and this is attributed to D–enrichment in condensable molecules by ion–molecule reactions that went on in the cold molecular cloud out of which the solar system formed (Kolodny et al 1980; Geiss and Reeves 1981).

The method of deriving protosolar D and ^{3}He abundances from solar wind data has endured (cf. Anders and Grevesse 1989; Walker et al 1991), and the D+^{3}He abundance in the interstellar gas 4.6 billion years ago has come to be accepted as an important cosmological parameter. In view of more recent evidence and improved understanding of solar processes we present here a new quantitative assessment.

3.2 New Assessment of Protosolar Deuterium plus Helium–3.

With the ISEE–3 Ion Composition Instrument, Coplan et al (1984) determined ^{3}He/^{4}He in the solar wind. Figure 3 shows a contour plot with all the data obtained during a 4 year period, averages are given in Table 2. The large fluctuations indicated in Figure 3 are mainly due to counting statistics, but both the Apollo and ISEE–3 results clearly show genuine variations of ^{3}He/^{4}He in the solar wind, an indication that the average solar wind and solar surface ratios may not be exactly the same.

In fact we have to assess the effects of three mechanisms before we can accurately deduce the protosolar D+^{3}He abundance from solar wind measurements. (1) Models of corona expansion indicate that ^{3}He may be somewhat favoured over ^{4}He in the solar wind acceleration process (Bürgi and Geiss 1986). (2) The chromospheric process that leads to strong enhancements of elements with low first ionisation potential (the FIP effect), do not seem to include a significant mass fractionation (Meyer 1981; von Steiger and Geiss 1989), since the elements H to Ar with FIPs between 13 eV and 22 eV occur in solar proportion (Geiss 1973). (3) ^{3}He in material of the solar outer convective zone could not have been depleted (Geiss and Reeves 1972), but it may have been augmented by secular mixing (Schatzmann 1970, Bochsler and Geiss 1973; Lebreton and Maeder 1987; Bochsler, Geiss and Maeder 1990): In the course of solar history a bulge of ^{3}He was built up by incomplete hydrogen burning at intermediate solar depth, and even a modest admixture to the outer convective zone of material from below could increase the ^{3}He/^{4}He ratio at the solar surface, although the recently established absence of significant differential rotation in the middle and upper layers of the Sun has made such an increase less likely. In any case, the addition of surplus ^{3}He would mainly occur late in solar history, i.e. during and after the epoch when some of the lunar samples received their solar wind irradiation (Geiss 1973). The difference in the ^{3}He/^{4}He ratio between old and younger samples is small (Benkert et al 1988), indicating again that the secular increase of ^{3}He at the solar surface, if

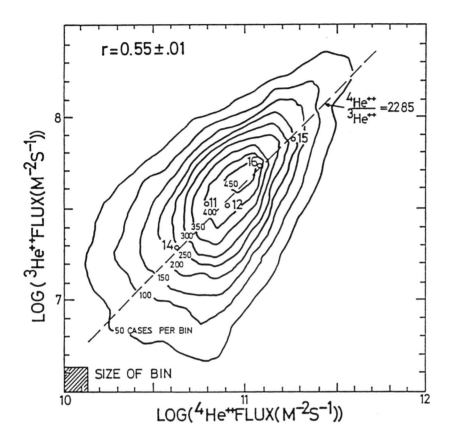

Figure 3: The ^{3}He/^{4}He abundance ratio in the solar wind as measured with the Apollo SWC experiments (Geiss et al. 1972) and with the ICI instrument on board the ISEE-3 satellite (Coplan et al. 1984; Bochsler 1984). The results from the Apollo 11 – 16 experiments are shown as circles. The ISEE-3 data (representing many thousand spectra taken in the 4-year period August 1978 to July 1982) are presented in a contour plot. The averages derived from the two data sets are in agreement. The larger variability of the ISEE-3 data is partly due to statistical fluctuations and partly due to real short-time variations.

TABLE 3

Deuterium and Helium-3 Abundances

Description	Ratio	Value	Ref.
Protosolar	$(D+{}^3He)/{}^4He$	$(4.09\pm.87)10^{-4}$	a
Protosolar	$(D+{}^3He/H$	$(4.1\pm1.0)10^{-5}$	a
Non-solar ("Planetary")			
Component in Meteorites	${}^3He/{}^4He$	$(1.5\pm.03)10^{-4}$	b
Protosolar, adopted	${}^3He/{}^4He$	$(1.5\pm.30)10^{-4}$	c
Protosolar	$D/{}^4He$	$(2.59\pm.89)10^{-4}$	c
Protosolar	D/H	$(2.6\pm1.0)10^{-5}$	c
Interstellar, IUE, Lyman	D/H	$(.5\ to\ 2)10^{-5}$	d
Interstellar, HST, Lyman	D/H	$(1.5^{+.07}_{-.18})10^{-5}$	e
Interstellar, 92 cm	D/H	$\leq 3\ 10^{-5}$	f

a see Table 1
b Eberhardt (1974); Frick and Moniot (1977)
c see text
d estimated averages; cf. Vidal-Madjar (1991), Boesgaard and Steigman (1985)
e towards Capella, Linsky et al. (1992)
f Blitz and Heiles (1987)

TABLE 4

Cosmological Implications of Protosolar $D+{}^3He$ and D

$D+{}^3He$		
Protosolar $(D+{}^3He)/H$	$(4.1 \pm 1.0)10^{-5}$	
Primordial $(D+{}^3He)/H$	$(4.1 - 8.4)10^{-5}$	
$n_B/n_\gamma = \eta$	$(3.1 - 5.25)10^{-10}$	
Baryonic density ρ_B	$(2.2 - 3.7)10^{-31}$ g/cm^3	a
$\Omega_B = \rho_B/\rho_c$	$0.047 - 0.079$	b
	$0.012 - 0.020$	c
D		
Protosolar D/H	$(2.6 \pm 1.0)10^{-5}$	
Primordial D/H	$(2.1 - 6.6)10^{-5}$	
$n_B/n_\gamma = \eta$	$(3.2 - 6.35)10^{-10}$	
Baryonic density ρ_B	$(2.3 - 4.5)10^{-31}$ g/cm^3	a
$\Omega_B = \rho_B/\rho_c$	$0.048 - 0.095$	b
	$0.012 - 0.024$	c

a with $T_0 = 2.75$ K
b with $T_0 = 2.75$ K and $H_0 = 50$ km s^{-1} Mpc^{-1}
c with $T_0 = 2.75$ K and $H_0 = 100$ km s^{-1} Mpc^{-1}

it occured at all, is not larger than ten percent. The lunar data also indicate that helium isotope fractionation in the corona is limited (cf. Geiss and Bochsler 1991).

We conclude, that the effects of mechanisms (1) to (3) are limited and adopt a factor of $\alpha = 1.1 \pm 0.2$ for the ratio between ^{3}He/^{4}He in the Sun directly after deuterium burning. The errors given for α are conservative. They could be further reduced by refined calculations of turbulent mixing (cf. S. Vauclair, this volume) and diffusive separation (cf. G. Michaud, this volume) in the solar surface layers, comparing ^{3}He with ^{4}He, Li and Be. Also, forthcoming solar wind studies should help us further to reduce the uncertainty of α. With the Ulysses–SWICS instrument (Gloeckler et al 1992) the ^{3}He/^{4}He ratio above the polar coronal holes will be determined, which is important for better understanding possible differences in the dynamics of the two helium isotopes in the chromosphere or the corona. The SOHO–CELIAS instrument (Hovestadt et al 1989) will separate isotopes of several elements, allowing to recognize mass discrimination in the solar wind source region.

With He/H $= 0.10 \pm 0.01$ as the original solar value (Turck–Chièze et al, 1988), we obtain the protosolar value (cf. table 2)

$$\left\{\frac{D + ^3He}{H}\right\}_{PROTOSOLAR} = (4.1 \pm 1.0)10^{-5} \qquad (4)$$

The well established solar system abundance systematics of elements indicates that our system formed from ordinary interstellar material 4.6 Gyrs ago. Thus we may take the protosolar value of $(D+^3He)/H = (4.1\pm1.0)10^{-5}$ as being typical for the interstellar gas of that epoch. As mentioned above, we expect the change of $(D+^3He)/H$ due to stellar processing to be modest, at least until the time of the birth of the Sun. Using the results of the most recent work on the galactic evolution of light nuclei (Steigman and Tosi, 1992), we have estimated the primordial $(D+^3He)/H$ given in Table 4. The constraints placed on the baryonic density are discussed in section 6.

4. DEUTERIUM

4.1 The D/H Ratio in the Protosolar Gas
The protosolar deuterium abundance can be derived from the protosolar D+^{3}He abundance by correcting for ^{3}He. The "planetary" noble gas component in meteorites provides us with a helium sample that is unaffected by solar deuterium burning, its ^{3}He/^{4}He ratio of $(1.5\pm.03)10^{-4}$ (Eberhardt 1974; Frick and Moniot 1977) is well established by the stepwise heating technique. There remains the question, though, how representative this sample really is for the protosolar gas? Element abundances in the "planetary" component of meteoritic noble gases are highly fractionated. If this

is due to a mass fractionation (in the gas phase, at the time of fixation in the solid, or afterwards by gas loss), ^{3}He/^{4}He would also have been affected. If, on the other hand, the observed element fractionation is related to atomic radius or ionisation potential (Göbel, Ott and Begemann 1978) the effects on isotopic ratios would be small. In order to account for these uncertainties, an error larger than the square root of $\Delta m/m$ is adopted in Table 3. Further studies on the planetary component in meteorites, or a determination of the isotopic composition of Jovian helium, e.g. by Galileo (cf. Nieman et al 1992) could confirm that we deal with the correct protosolar ^{3}He/^{4}He ratio.

This leads us to a general observation: Trapped gases in meteorites and in lunar material carry valuable information about the history of the Sun, meteorites, moon, and the solar system as a whole, but some questions as to the origin and nature of the gases remain not just in the case of the "planetary component". For instance, the nature of the two solar–type noble gas components (Benkert et al 1988; Kerridge et al 1991; Wieler, Baur and Signer 1992) in lunar material and in meteorites is still unclear. Without direct measurements in the contemporary solar wind (Geiss et al 1970; 1972), we might not even be sure which of the trapped components represents a fairly unaltered solar wind sample. A case in point is the nitrogen in lunar surface material: There is no agreement on which of the two (or more) isotopically different components is the present–day solar wind (Kerridge 1975; Geiss and Boschler 1991; Kerridge et al 1991), an uncertainty that is unlikely to be removed before ^{15}N/^{14}N is really measured in the solar wind or – perhaps – in Jupiter's atmosphere.

Fortunately, the uncertainty is the protosolar ^{3}He/^{4}He ratio has only a limited influence on the resulting protosolar D/H, because ^{3}He is the smaller component in the D+^{3}He sum. By substracting protosolar ^{3}He from D+^{3}He we obtain (cf Table 3)

$$(D/H)_{\text{PROTOSOLAR}} = (2.6 \pm 1.0)10^{-5} \qquad (5)$$

Since all errors in Tables 2 and 3 are conservative, we consider this a reliable result. The value derived here is very close to the $(2.5\pm1.0)10^{-5}$ given by Geiss and Reeves (1972) twenty years ago.

The D/H ratio in the Jovian atmosphere has been investigated by different methods. The most recent value of $(2\pm1)10^{-5}$ is given by Gautier and Owen (1989). Since mass fractionations in Jupiter's atmosphere are probably not very pronounced, this result supports the protosolar D/H ratio derived from the solar wind data.

Today, it is widely accepted that the protosolar D/H ratio of a few time 10^{-5} places a firm, cosmologically significant limit on the baryonic density of the universe. The decrease in primordial deuterium production with increasing density is so steep that it is difficult to obtain $\Omega_B=1$, even with drastic deviations from the standard model (cf. Sato and Terasawa, 1991), and this perception fuels the search for dark, non–baryonic matter.

4.2 The Interstellar D/H Abundance Ratio

Most of the data on the D/H ratio in the interstellar medium have been derived from observations of Lyman absorption lines in the spectra of stars (Rogerson and York, 1973; Vidal–Madjar and Gry 1984). The results were recently reviewed by Vidal–Madjar (1991). The large variability of the D/H ratios obtained in the direction of different stars came as a surprise, and its causes have been investigated by several authors (cf. Gry et al 1984; Vidal–Madjar 1991) who concluded, that at least some of the variability represents genuine local differences in the D/H ratio. Table 3 includes a range for the average D/H ratio in the interstellar gas as derived by several authors.

Recently, Linsky et al (1992) investigated the Lyman absorption lines in the direction of Capella with the HST Goddard High Resolution Spectrograph. A high signal/noise ratio allowed the authors to determine both the gas temperature and the turbulent velocity along the line of site and to analyze and compare line profiles in detail. Although there is as yet only a single such observation, we include the result in Table 3. Using the HST for measuring D/H along several lines of sight may well lead to a much better determination of the average D/H in the interstellar gas.

Blitz and Heiles (1987) have compared the 21 cm and 92 cm radio emissions of H and D in the antisolar direction and derived the upper limit for the interstellar D/H ratio included in Table 3. With improved sensitivity the radio technique would add an independent method for interstellar deuterium studies.

With the data presented in Table 3 it is not yet possible to determine a difference between the protosolar and the present–day interstellar D/H ratio, which would help us in choosing the correct model of galactic evolution. In any case, the effect of stellar processing is less severe and less uncertain in the protosolar gas (cf. Steigman and Tosi 1992). Moreover, we have very detailed information on the contributions from the various stellar processes to this galactic sample. Thus we have used the protosolar D/H ratio and the results of Steigman and Tosi (1922) on deuterium destruction in the solar ring of the galaxy, for deriving the bounds for the primordial D/H ratio given in Table 4.

5. HELIUM–3.

The primordial production of both D and He^3 decreases with increasing baryonic density (cf. Figure 1). Their galactic evolution, however, is different: Unlike deuterium, He^3 is not only destroyed by stars but also efficiently produced, as is demonstrated by the high ^{3}He/H ratio ($> 10^{-3}$) recently observed in a planetary nebula (Rood, Bania and Wilson, 1992). Thus it is possible that the average ^{3}He/H ratio in the present day interstellar gas is higher than the primordial ratio (Dearborn, Schramm and Steigman, 1986; Bania, Rood and Wilson 1987, Steigman and Tosi 1992). The ^{3}He abundance in the interstellar gas is derived from observations of the 3.46 cm hyperfine line of ^{3}He$^+$. The ^{3}He/H ratio is found to be strongly variable, with a tendency of rising values with increasing galactocentric distance (Bania, Rood and Wilson, 1987). As these authors point out, the lowest observed ^{3}He abundances can be used to establish an upper limit for the primordial abundance (and a lower limit for the baryon density), providing it is certain that the ^{3}He/H ratio in the studied galactic sample has not been decreased since the time of BBN. The method of collecting particles for laboratory analysis, successfully applied to solar wind isotope studies, can also be used for investigating helium isotopes in the neutral interstellar gas permeating the solar system (Geiss 1973). A first attempt has been made by Lind et al (1991), using the passive LDEF satellite. It is not yet certain whether, from this first experiment, an interstellar ^{3}He/^{4}He ratio can be derived with uncertainties low enough to compete with existing estimates. However, the experiment demonstrated that with some refinement interstellar helium can be unambiguously separated from particles of atmospheric or solar origin.

Presently, the best estimate of protosolar He^3/He^4 is derived from the planetary component of the gases in meteorites. Adopting the meteoritic data (Eberhardt 1974; Frick and Moniot 1977) and the error bars estimated in section 4, we have

$$(^3\text{He}/^4\text{He})_{\text{PROTOSOLAR}} = (1.5 \pm 0.3)10^{-5} \qquad (6)$$

It would be desirable to confirm this value by an independent method, such as in situ mass spectrometry in the Jovian atmosphere (Nieman et al 1992; see sect. 4).

Improved data on the protosolar and present–day ^{3}He abundances and the production of the ^{3}He (Rood, Bania and Wilson 1992) could provide a much needed control on models of the post–BBN evolution of the light nuclei.

6. CONCLUSION AND OUTLOOK.

Table 4 shows the bounds on primordial (D+^{3}He)/H and D/H, the cosmic baryon/photon ratio η, the baryonic density ρ_B and Ω_B, the ratio of baryonic density to critical density, as derived from the protosolar values given in this paper and SBBN theory. The lower bounds on the abundances (or upper limits on $n_B/n\gamma$) are more rigorous than the upper abundance bounds (or lower $n_B/n\gamma$ limits), because they depend less critically on details of galactic evolution models (cf. the data in Table 4 and results and discussions by Clayton, 1985; Steigman and Tosi, 1992, and others).

The observational bounds are compared in Figure 1 with the predicted abundances. Included is the upper limit for Y_p (equ. 3) adopted from Pagel et al (1992). To complete Figure 1, we have included a conservative upper limit for primordial ^{7}Li. The newest results on ^{7}Li$_p$ are given by F. Spite, G. Michaud, H. Reeves and S. Vauclair in this volume. There is a range in η, for which observed abundances and SBBN predictions agree for all five priomordial species, as has been noted for several years. Thus the observations are consistent with homogeneous primordial nucleosynthesis, but do not exclude some heterogeneity.

All four primordial abundance ratios shown in Figure 1 give upper limits for the baryonic density that are far below the critical density, a result of fundamental importance that has been known for quite some time. This does not mean, however, that the study of primordial abundances has lost its interest. On the contrary, progress from new methods and experiments may be expected in many areas, as has been described in the preceeding sections.

The upper limit on the universal baryon density derived from deuterium and the other light nuclei has spawned a worldwide search for non-baryonic, dark matter. However, aside from baryons, electrons and photons, not a single primordial particle has yet been detected. What if non-baryonic matter in quantities sufficient to close the universe or even only to account for the dark matter in clusters of galaxies is not found? Or if other observations demand a drastic change in the cosmological model? We need to ascertain the primordial abundances of the lightest nuclei as rigorously and precisely as we can, so that future refinements or even fundamental changes in cosmological theory can be reliably tested against them.

ACKNOWLEDGEMENTS
For many years, the author has enjoyed and benefitted from discussions with Hubert Reeves about astrophysics, cosmology and even more esoteric subjects. He also acknowledges discussions with P. Bochsler, P. Eberhardt, A. Maeder, P. Signer, R. von Steiger, H. Voelk, and T.L. Wilson, and he thanks B.E. Pagel, G. Steigman, D.N. Schramm, A. Vidal–Madjar and T.L. Wilson for providing him with new and partly unpublished results and material. This work was in part supported by the Swiss National Science Foundation.

REFERENCES

ALEPH Collaboration (1989) *Phys. Letters*, **B 231**, 519

Anders E and Grevesse N (1989) *Geochim. Cosmochim. Acta*, **53**, 197

Bania T M, Rood R T and Wilson T L (1987) *Astrophys. J.*, **323**, 30

Benkert J-P, Baur H, Pedroni A, Wieler R and Signer P (1988) *Lunar Planet. Sci. Conf.*, **XIX**, 59

Black D C (1972) *Geochim. Cosmochim. Acta*, **36**, 347

Blitz L and Heiles C (1987) *Astrophys. J.*, **313**, L95

Boato G (1954) *Geochim. Cosmochim. Acta*, **6**, 209

Bochsler P (1984) Helium and Oxygen in the Solar Wind, *University of Bern Habilitation Thesis*

Bochsler P and Geiss J (1973) *Solar Phys.*, **32**, 3

Bochsler P, Geiss J and Maeder A (1990) *Solar Phys.*, **128**, 203

Boesgaard A M and Steigman G (1985) *Ann. Rev. Astron. Astrophys.* , **23**, 319

Burbidge E M, Burbidge G R, Fowler W A and Hoyle F (1957) *Rev. Mod. Phys.*, **29**, 547

Bürgi A and Geiss J (1986) *Solar Phys.*, **103**, 347

Cameron A G W (1957) Stellar Evolution, Nuclear Astrophysics and Nucleogenesis, *Chalk River Report CRL-41*

Choisi C and Maeder A (1986) *Ann. Rev. Astron. Astrophys.*, **24**, 329

Clayton D D (1985) *Astrophys. J.*, **290**, 428

Coplan M A, Ogilvie K W, Bochsler P and Geiss J (1984) *Solar Phys.* , **93**, 415

Dearborn D S P, Schramm D N and Steigman G (1986) *Astrophys. J.* , **302**, 35

DELPHI Collaboration (1989) *Phys. Letters*, **B 231**, 539

Eberhardt P (1974) *Earth Planet. Sci. Lett.*, **24**, 182

Eberhardt P, Geiss J and Grögler N (1966) *Earth Planet. Sci. Lett.*, **1**, 7

Eberhardt P, Dolder U, Schulte W, Krankowsky D, Lämmerzahl P, Hoffman J H, Hodges R R, Berthelier J J and Illiano J M (1987) *Astron. Astrophys.*, **187**, 435

Fowler W A, Greenstein J L and Hoyle F (1962) *Geophys. J. R.A.S.*, **6**, 148

Frick U and Moniot R K (1977) *Proc. 8th Lunar Planet. Sci. Conf.*, 229

Fuller G M, Boyd R N and Kale J D (1991) *Astrophys. J.*, **371**, L11

Gautier D and Owen T (1989) in *Origin and Evolution of Planetary and Satellite Atmospheres*, S K Atreya et al (Eds), University of Arizona Press, Tucson, 487

Geiss J (1973) *Proc. 13th Int. Cosmic Ray Conf.*, Conf. Papers **5**, 3375

Geiss J and Bochsler P (1991) *The Sun in Time*, C P Sonett et al (Eds), University of Arizona Press, Tucson, 98

Geiss J, Eberhardt P and Signer P (1966) Experimental determination of the solar wind composition, Proposal to NASA for *Apollo experiment S-080*

Geiss J, Eberhardt P, Bühler F, Meister J and Signer P (1970) *J. Geophys. Res.*, **75**,

5972

Geiss J, Bühler F, Cerutti H, Eberhardt P and Filleux Ch (1972) *Apollo Preliminary Science Report, NASA SP-315*, Section 14

Geiss J and Reeves H (1972) *Astron. Astrophys.*, **18**, 126

Geiss J and Reeves H (1981) *Astron. AStrophys.*, **93**, 189

Gloeckler G et al (1992) *Astron. Astrophys. Suppl. Ser.*, **92**, 267

Göbel R, Ott U and Begemann F (1978) *J. Geophys. Res.*, **83**, 855

Grevesse N (1968) *Solar Phys.*, **5**, 159

Gry C, Lamers H and Vidal-Madjar A (1984) *Astron. Astrophys.* , **137**, 29

Hintenberger H, Vilcsek E and Wänke H (1965) *Z. Naturforschung*, **20a**, 939

Hovestadt D et al (1989) *ESA SP-1104*, 69

Hoyle F and Tayler R J (1964) *Nature*, **203**, 1108

Kerridge J F (1975) *Science*, **188**, 162

Kerridge J F, Signer P, Wieler R, Pepin R and Becker R H (1991) *The Sun in Time*, C P Sonett et al (Eds), University of Arizona Press, Tucson, 389

Kolodny Y, Kerridge J F and Kaplan I R (1980) *Earth Planet. Sci. Lett.*, **46**, 149

Kunth D and Sargent W L (1983) *Astrophys. J.*, **273**, 81

L3 Collaboration (1989) *Phys. Letters*, **B 231**, 509

Lebreton Y and Maeder A (1987) *Astron. Astrophys.*, **93**, 189

Lequeux J, Peimbert M, Rayo J F and Torres-Peimbert S (1979) *Astron. Astrophys.*, **80**, 155

Lind D, Geiss J, Bühler F and Eugster O (1991) *NASA Conf. Publication 3134*, 585

Linsky J L, Brown A, Gayley K, Diplas A, Savage B D, Ayres T R, Landsman W, Shore S N and Heap S R (1992) *Abstract*

Meneguzzi M, Audouze J and Reeves H (1971) *Astron. Astrophys.* , **15**, 337

Meyer J P (1981) *17th Int. Cosm. Ray Conf.*, **3**, 145

Nieman H B, Harpold D N, Atreya S K, Carignan G R, Hunten D M and Owen T (1992) *Space Sci. Rev.*, **60**, 111

Olive K A, Schramm D N, Steigman G and Walker T P (1990) *Phys. Lett.* , **B 236**, 454

OPAL Collaboration (1989) *Phys. Letters*, **231**, 530

Owen T, Lutz B L and de Bergh C (1986) *Nature*, **320**, 244

Pagel B E J (1990) in *Baryonic Dark Matter*, D Lynden-Bell and G Gilmore (Eds), Kluwer, 237

Pagel B E J (1991) *Second IAC Winter School*, Tenerife

Pagel B E J, Terlevich R J and Melnick J (1986) *Pub. Astron. Soc. Pacific*, **98**, 1005

Pagel B E J and Simonson E A (1989) *Rev. Mex. Astron. Astrofis.* , **18**, 153

Pagel B E J, Simonson E A, Terlevich R J and Edmunds M G (1992) *Mon. Not. R. Astr. Soc.*, **255**, 325

Paul W, Anton F, Paul L, Paul S and Mampe W (1989) *Zs. Phys.*, **A 503**, 473

Peebles P J E (1966) *Astrophys. J.*, **146**, 542

Peimbert M (1985) in *Star Forming Dwarf Galaxies*, D Kunth et al (Eds), Ed. Frontières, 403

Reeves H (1974) *Ann. Rev. Astron. Astrophys.*, **12**, 437

Reeves H and Salpeter E E (1959) *Phys. Rev.*, **116**, 1505

Reeves H, Fowler W A and Hoyle F (1970) *Nature*, **226**, 727

Robert F, Merlivat L and Javoy M (1979) *Nature*, **282**, 785

Rogerson J B and York D G (1973) *Astrophys. J. Letters*, **186**, L95

Rood R T, Bania T M and Wilson T L (1992) *Nature*, **355**, Feb 13

Ryter C, Reeves H, Gradsztajn E and Audouze J (1970) *Astron. Astrophys.*, **8**, 389

Sato K and Terasawa N (1991), *Phys. Scripta*, **T36**, 60

Schatzman E (1970) *Physics and Astrophys.* Lecture Notes, CERN 70–31

Signer P and Suess H E (1963) in *Earth Science and Meteoritics*, J Geiss and E D Goldberg (Eds), North Holland, Amsterdam, 241

Steigman G, Gallagher J S and Schramm D N (1989) *Comments Astrophys.*, **14**, 97

Steigman G and Tosi M (1992), preprint

Turck-Chièze, Cahen S, Cassé M and Doom C (1988) *Astrophys. J.*, **335**, 415

Vidal-Madjar A (1991) *Adv. Space Res.*, **11**, 97

Vidal-Madjar A and Gry C (1984) *Astron. Astrophys.*, **138**, 285

von Steiger R and Geiss J (1989) *Astron. Astrophys.*, **225**, 222

Wänke H (1965) *Z. Naturforschung*, **20a**, 946

Wagoner R V, Fowler W A and Hoyle F (1967) *Astrophys. J.*, **148**, 3

Walker T P, Steigman G, Schramm D N, Olive K A and Kang H-S (1991) *Astrophys. J.*, **376**, 51

Wieler R, Baur H and Signer P (1992) *23rd Lun. Planet. Sci. Conf.*

The 3-Helium Abundance in H II Regions and Planetary Nebulae

T.M. BANIA, R.T. ROOD, AND T.L. WILSON

Boston University, University of Virginia, and MPI für Radioastronomie

1 INTRODUCTION: 3-HELIUM IN THE INTERSTELLAR MEDIUM

Helium-3 is created by cosmological nucleosynthesis with an abundance of $^3\text{He}/\text{H} \approx 2 \times 10^{-5}$ by number (*cf.*, Reeves 1974), but then augmented by stellar nucleosynthesis (Audouze & Tinsley 1976; Rood, Steigman & Tinsley 1976). Stars comparable in mass to the Sun should contribute a large fraction of the present ^3He abundance in interstellar material: winds from these stars during their first and second ascent of the red giant branch, as well as planetary nebulae created at the end of the stars' lives, are expected to have $^3\text{He}/\text{H}$ about 100 times the cosmic value. These stars are also thought to be the principal source of new material to the interstellar medium (ISM), and measurement of the present ^3He abundance should therefore be an important diagnostic of chemical evolution in the Galaxy, as well as an essential prelude to determining the primordial cosmic abundance. In fact ^3He has several advantages over other tracers used to study ISM chemical evolution. For example, ^3He abundances: (1) can be determined for sources anywhere in the galaxy whereas deuterium (D) abundances are known accurately only for the local ISM; (2) are the only direct probes of the mass influx to the ISM from stars of approximately one solar mass; and (3) are less ambiguous to interpret than ^{13}C abundances since helium is chemically inert.

Over a decade ago we began a program to measure the galactic ^3He abundance using observations of the 8.665 GHz (3.46 cm) hyperfine line of $^3\text{He}^+$ (*cf.*, Rood, Bania & Wilson 1984; Bania, Rood & Wilson 1987 [BRW] and references therein). Until recently it had been possible to detect ^3He only in giant H II regions, where it is already well mixed into the interstellar medium. We report here an improved abundance for the first-detected ^3He source, the planetary nebula NGC 3242. We measure $^3\text{He}/\text{H} \gtrsim 10^{-3}$, consistent with stellar models. We also report improved abundances for a sample of galactic H II regions based on new $^3\text{He}^+$ measurements. The derived $^3\text{He}/\text{H}$ abundance ratios for H II regions span the range from 1.1 to 4.5×10^{-5} by number. These abundances are thus between ~ 0.5 and ~ 2.3 times the cosmic value.

2 ABUNDANCE OF 3-HELIUM IN NGC 3242

Although we have been measuring the ^{3}He/H abundance ratio in galactic H II regions for over a decade, the direct detection of ^{3}He at a source of its presumed stellar production was a goal that was realized only recently with the discovery of ^{3}He$^+$ emission in the planetary nebula NGC 3242 (Rood, Bania & Wilson 1992 [RBW]). Figure 1 shows the ^{3}He$^+$ spectrum measured for NGC 3242 . Since this spectrum results from about twice the integration time reported by RBW, we have redetermined the ^{3}He abundance. As in RBW we assume NGC 3242 to be a homogeneous and isothermal spherical source having an angular size, Θ, of 25″ and a distance from the Sun, d, of 1 kpc. Using least-squares gaussian fits for the observed line parameters, the derived abundance of ^{3}He is

$$^3\text{He/H} \approx 1.4 \times 10^{-3} \left(\frac{\Theta}{25''}\right)^{-1.5} \left(\frac{d}{1\,\text{kpc}}\right)^{-0.5}$$

Figure 1 The ^{3}He$^+$ spectrum of the planetary nebula NGC 3242. The ^{3}He$^+$ emission line (near 8666 GHz) and the H171η recombination line are prominent. A baseline was removed from the spectrum which is the result of a 50.5 hour integration with the 100 m telescope of the MPIfR. The spectrum also has been smoothed to a velocity resolution of 8 km sec^{-1}. Shown also are least-squares gaussian fits to the H171η and ^{3}He$^+$ emission lines.

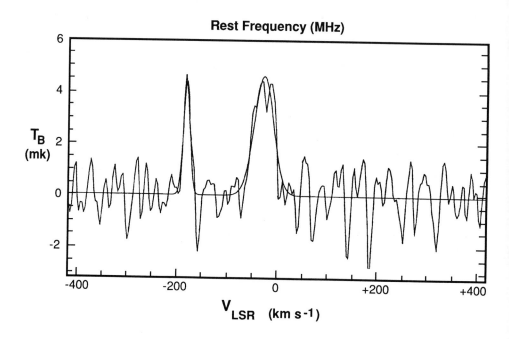

Given the uncertainties, we consider our result to be consistent with theoretical expectations for the production of ^{3}He by stellar nucleosynthesis. We now have direct evidence that the ^{3}He produced in the outer parts of stars with mass $\sim$1 $M_\odot$ during main-sequence hydrogen burning, and mixed to the surface during the red-giant phase, survives until the ejection of a planetary nebula.

3 ABUNDANCE OF 3-HELIUM IN GALACTIC H II REGIONS

With the 43 m National Radio Astronomy Observatory telescope in Green Bank, we have detected or obtained significant limits for the ^{3}He abundance in a sample of galactic H II regions. We have accumulated substantially more integration time for many of these sources since the BRW analysis. Based on these improved measurements, Balser *et al.* (1992) have derived new ^{3}He abundances using spherical homogeneous isothermal LTE models that are constrained by the continuum and recombination line observations of the sources. The results of this modelling are summarized below in a table. Listed for each source is the measured ^{3}He$^+$ line intensity (or limit), the model electron temperature and density, the observed value of y^+, the derived ^{3}He$^+$/H$^+$ and ^{3}He/H abundance ratios by number, and the source galactocentric distance taken from BRW. The ^{4}He$^+$/H$^+$ abundance ratios by number (y^+) were determined in the manner of Peimbert *et al.* (1992) using observations of the 91α, 114β, and 130γ transitions of hydrogen and helium recombination lines.

The ^{3}He$^+$/H$^+$ and ^{3}He/H abundance ratios were calculated as in BRW. Note that these abundances should still be viewed with some caution since no correction has been made for source structure. BRW show that for our source sample such corrections can *increase* the ^{3}He/H ratio by a factor of $\lesssim 3$. (We intend to use the Very Large Array to obtain the high resolution data we need to make better source structure corrections.)

Figure 2 shows our derived ^{3}He/H abundance ratios plotted as a function of the source Galactic radius. Although there are substantial source-to-source variations, there is apparently no large galactic-scale ^{3}He abundance gradient. There also is apparently no dependence on excitation conditions since there are no significant correlations with ionization parameters of H II regions. The derived ^{3}He/H abundance ratios for detected sources span the range from 1.1 to 4.5×10^{-5} by number and have an average value of 2.7×10^{-5}. These abundances are thus between $\sim$0.5 and $\sim$2.3 times the nominal cosmic value with an average ^{3}He enrichment of $\sim$35%.

Abundance of ^{3}He in Galactic H II Regions

Source	$T_A(^3He^+)$ (mK)	T_E (K)	N_E (cm^{-3})	y^+ ($^4He^+/H^+$)	$^3He^+/H^+$ (10^{-5})	$^3He/H$ (10^{-5})	R_G (kpc)
W3	2.8	8000	1660	0.074	3.7	4.3	10.1
G133.8+1.4	1.9	8000	410	0.072	2.4	2.8	10.1
S206	3.6	8000	120	0.096	2.5	2.5	11.5
S209	1.8	11000	110	0.099	1.1	1.1	18.9
Ori A	< 7.3	9000	3300	0.094	< 4.2	< 4.2	8.9
S311	2.8	9000	110	0.092	4.5	4.5	11.0
Sgr B2	3.2	8000	305	0.072	2.1	2.5	0.1
M17S	6.9	8000	860	0.090	2.0	2.0	6.4
G24.8+0.1	< 0.5	5500	100	0.060	< 0.6	< 0.9	3.6
W43	2.8	5600	305	0.071	1.6	1.9	4.6
W49	< 2.0	9000	410	0.090	< 0.6	< 0.6	8.2
W51	4.9	7500	710	0.083	2.1	2.2	6.6
NGC 7538	2.7	8600	435	0.097	3.6	3.6	9.9
S162	< 0.2	9000	89	0.083	< 0.3	< 0.3	10.3

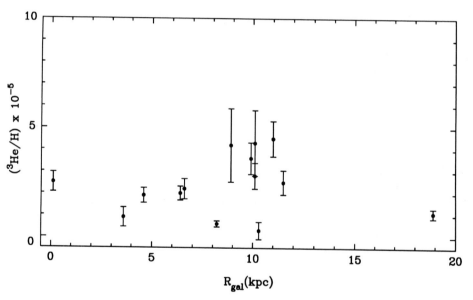

Figure 2 The ^{3}He/H abundance ratio derived for a sample of galactic H II regions is plotted as a function of galactocentric radius. The $\pm 1\ \sigma$ error bars reflect only measurement errors; neither source structure effects nor possible systematic errors are taken into account.

Assuming that stellar nucleosynthesis returns a net amount of ^{3}He to the ISM (Dearborn, *et al.* 1986), one should use the lowest measured ^{3}He/H ratio to obtain the primordial abundance. Currently S209 has the lowest measured abundance of 1.1×10^{-5}. If upper limits are used, then the ratio becomes $\lesssim 0.6 \times 10^{-5}$. (This is the average ratio of the upper limits listed in the table, excluding Ori A.) In the standard Big Bang model, the primordial ^{3}He and D abundances together can be used to constrain η, the baryon-to-photon ratio (*cf.*, Steigman & Tosi 1992 for a recent analysis).

We thank Dana Balser for calculating the H II region models used in the ^{3}He abundance derivations and for help with the 100 m observations. Portions of this research was supported by grants from NATO and the NSF (AST–9121169).

4 REFERENCES

Audouze, J. & Tinsley, B.M. *Ann. Rev. Astr. Ap.*, **14**, 43–79 (1976).

Balser, D.S., Bania, T.M., Rood, R.T. & Wilson, T.L. *Ap. J.*, in preparation (1992).

Bania, T.M., Rood, R.T. & Wilson, T.L. *Ap. J.*, **323**, 30–43 (1987).

Dearborn, D.S.P., Schramm, D.N. & Steigman, G. *Ap. J.*, **302**, 35–38 (198

Peimbert, M., Rodriguez, L.F., Bania, T.M., Rood, R.T., and Wilson, T.L. *Ap. J.*, in the press (1992).

Reeves, H. *Ann. Rev. Astr. Ap.*, **12**, 437–469 (1974).

Rood, R.T., Bania, T.M. & Wilson, T.L. *Ap. J.*, **280**, 629–647 (1984).

Rood, R.T., Bania, T.M. & Wilson, T.L. *Nature*, **355**, 618–620 (1992).

Rood, R.T., Steigman, G. & Tinsley, B.M. *Ap. J. (Letters)*, **207**, L57–L60 (1976).

Steigman, G. & Tosi, M. preprint (1992).

Primordial Nucleosynthesis

D.N. SCHRAMM

The University of Chicago,
and
NASA/Fermilab Astrophysics Center,
Fermi National Accelerator Laboratory

ABSTRACT

Hubert Reeves has made critical contributions to primordial nucleosynthesis. This paper highlights some of those contributions while reviewing the current state of the subject. Primordial nucleosynthesis provides (with the microwave backround radiation) one of the two quantitative experimental tests of the big bang cosmological model. The standard homogeneous-isotropic calculation fits the light element abundances ranging from ^{1}H at 76% and ^{4}He at 24% by mass through ^{2}H and ^{3}He at parts in 10^5 down to ^{7}Li at parts in 10^{10}. (Reeves' work in elucidating the origin of some light elements via cosmic ray processes was critical to the establishment of the cosmological significance of ^{2}H and ^{7}Li and their use as cosmological density probes.) This paper also notes how the recent LEP (and SLC) results on the number of neutrinos are a positive laboratory test of the standard scenario. The possible alternate scenario of quark-hadron induced inhomogeneities is discussed in depth. It is shown that when these scenarios are made to fit the observed abundances accurately, the resulting conclusions on the baryonic density relative to the critical density, Ω_b, remain approximately the same as in the standard homogeneous case, thus adding to the robustness of the conclusion that $\Omega_b \sim 0.06$. This latter point is the driving force behind the need for non-baryonic dark matter (assuming $\Omega_{total} = 1$) and the need for dark baryonic matter, since $\Omega_{visible} < \Omega_b$. The recent Pop II boron and beryllium results are also discussed and shown to be a consequence of Reeves' cosmic ray spallation processes rather than primordial nucleosynthesis.

1. INTRODUCTION

Perhaps Hubert Reeves' most significant scientific contributions have been in the area of light element nucleosynthesis. In particular, the work of Hubert, his students and collaborators in eliminating[1] the prevailing light element synthesis model of the 1960s and eventually showing that ^{6}Li, ^{9}Be, ^{10}B and ^{11}B were made via galactic cosmic ray spallation[2] was critical in establishing the cosmolgical origin for ^{2}H (and some ^{7}Li.) This cosmological origin necessarily required the baryon density of the universe to be well below the critical value. These arguments were summarized in the early 1970s in a paper[3] I was pleased to co-author with Hubert, Jean Audouze and Willy Fowler.

This paper will attempt to put that work into current perspective by reviewing the present status of primordial nucleosynthesis. After first reviewing the history, this paper will make special emphasis of the remarkable agreement of the observed light element abundances with the calculations. It should be remembered that this agreeement is one of the two prime tests of the Big Bang itself (the other being the microwave background). This agreement works only if the baryon density is well below the cosmological critical value. The review will also mention the nucleosynthesis prediction of the number of neutrino families and its subsequent verification by LEP (and SLC).

Also discussed is the possibility that a first order quark-hadron phase transition could have produced variations from the standard homogeneous model. It will be shown that contrary to initial indications, first order quark-hadron inspired results are consistent with the homogeneous model results.

Finally, a discussion of the recent boron and beryllium results for Population II stars will be made. It will be shown that all observations, to date, are best explained by adopting the Reeves, Fowler and Hoyle[2] model for galactic cosmic ray production to Pop II environments and not by any cosmological process.

This report will draw on two recent reviews (references [4] and [5]).

2. HISTORY OF BIG BANG NUCLEOSYNTHESIS

It should be noted that there is a symbiotic connection between primordial nucleosynthesis (hereafter referred to as Big Bang Nucleosynthesis or BBN) and the $3K$ background dating back to Gamow and his associates, Alpher and Herman. The initial BBN calculations of Gamow's group[6] assumed pure neutrons as an initial condition and thus were not particularly accurate, but their inaccuracies had little effect on the group's predictions for a background radiation.

Once Hayashi[7] recognized the role of neutron-proton equilibration, the framework for BBN calculations themselves has not varied significantly. The work of Alpher, Follin and Herman[8] and Taylor and Hoyle[9], preceeding the discovery of the 3K background, and Peebles[10] and Wagoner, Fowler and Hoyle,[11] immediately following the discovery, and the more recent work of our group of collaborators[12,13,14,15] all do essentially the same basic calculation, the results of which are shown in Figure 1. As far as the calculation itself goes, solving the reaction network is relatively simple by the standards of explosive nucleosynthesis calculations in supernovae, with the changes over the last 25 years being mainly in terms of more recent nuclear reaction rates as input, not as any great calculational insight (although the current Kawano code[15] is somewhat streamlined relative to the earlier Wagoner code[11].) With the possible exception of ^{7}Li yields, the reaction rate changes over the past 25 years have not had any major affect. The one key improved input is a better neutron lifetime determination.[16]

BIG BANG NUCLEOSYNTHESIS

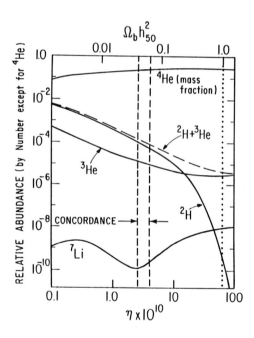

Figure 1. Big Bang Nucleosynthesis abundance yields (mass fraction) versus baryon density for a homogeneous universe.

With the exception of the effects of elementary particle assumptions to which we will return, the real excitement for BBN over the last 25 years has not really been in re-doing the basic calculation. Instead, the true action is focused on understanding the evolution of the light element abundances and using that information to make power-ful conclusions. And it is in this arena that Hubert Reeves' contributions have been so important. In the 1960's, the main focus was on ^{4}He which is very insensitive to the baryon density. The agreement between BBN predictions and observations helped support the basic Big Bang model but gave no significant information, at that time, with regard to density. In fact, in the mid-1960's, the other light iso-topes (which are, in principle, capable of giving density information) were generally assumed to have been made during the T-Tauri phase of stellar evolution,[17] and so, were not then taken to have cosmological significance. It was during the 1970's that BBN fully developed as a tool for probing the Universe. This possibility was in part stimulated by Ryter, Reeves, Gradstajn and Audouze[1] who showed that the T-Tauri mechanism for light element synthesis failed. Furthermore, ^{2}H abundance determinations improved significantly with solar wind measurements[18] and the in-terstellar work from the Copernicus satellite.[19] Hubert's recognition of the potential significance of a deuterium abundance determination from the Apollo-11 ^{3}He solar wind measurement and his collaboration with Johannes Geiss on the experimental results was particularly insightful. Reeves, Audouze, Fowler and Schramm[3] argued for cosmological ^{2}H and were able to place a constraint on the baryon density exclud-ing a universe closed with baryons. Subsequently, the ^{2}H arguments were cemented when Epstein, Lattimer and Schramm[20] proved that no realistic astrophysical pro-cess other than the Big Bang could produce significant ^{2}H. It was also interesting that the baryon density implied by BBN was in good agreement with the density implied by the dark galactic halos.[21]

By the late 1970's, a complimentary argument to ^{2}H had also developed using ^{3}He. In particular, it was argued[22] that, unlike ^{2}H, ^{3}He was made in stars; thus, its abundance would increase with time. Since ^{3}He like ^{2}H monotonically decreased with cosmological baryon density, this argument could be used to place a lower limit on the baryon density[23] using ^{3}He measurements from solar wind[18] or interstel-lar determinations.[24] Since the bulk of the ^{2}H was converted in stars to ^{3}He, the constraint was shown to be quite restrictive.[13]

It was interesting that the lower boundary from ^{3}He and the upper boundary from ^{2}H yielded the requirement that ^{7}Li be near its minimum of ^{7}Li/H $\sim 10^{-10}$, which was verified by the Pop II Li measurements of Spite and Spite,[25] hence, yielding the situation emphasized by Yang et al.[13] that the light element abundances are consistent over nine orders of magnitude with BBN, but only if the cosmological baryon density is constrained to be around 5% of the critical value.

The other development of the 70's for BBN was the explicit calculation of Steigman, Schramm and Gunn,[26] showing that the number of neutrino generations, N_ν, had to be small to avoid overproduction of ^{4}He. (Earlier work[27,9] had commented about a dependence on the energy density of exotic particles but had not done an explicit calculations probing N_ν.) This will subsequently be referred to as the SSG limit. To put this in perspective, one should remember that the mid-1970's also saw the discovery of charm, bottom and tau, so that it almost seemed as if each new detector produced new particle discoveries, and yet, cosmology was arguing against this "conventional" wisdom. Over the years, the SSG limit on N_ν improved with ^{4}He abundance measurements, neutron lifetime measurements and with limits on the lower bound to the baryon density, hovering at $N_\nu \lesssim 4$ for most of the 1980's and dropping to slightly lower than 4 just before LEP and SLC turned on.[14,28,29] This cosmological prediction is confirmed by the recent LEP and SLC results[30] where $N_\nu = 2.99 \pm 0.05$.

The power of homogeneous BBN comes from the fact that essentially all of the physics input is well determined in the terrestrial laboratory. The appropriate temperature regimes, 0.1 to $1 MeV$, are well explored in nuclear physics labs. Thus, what nuclei do under such conditions is not a matter of guesswork, but is precisely known. In fact, it is known for these temperatures far better than it is for the centers of stars like our sun. The center of the sun is only a little over 1 keV, thus, below the energy where nuclear reaction rates yield significant results in laboratory experiments, and only the long times and higher densities available in stars enable anything to take place.

To calculate what happens in the Big Bang, all one has to do is follow what a gas of baryons with density ρ_b does as the universe expands and cools. As far as nuclear reactions are concerned, the only relevant region is from a little above 1 MeV ($\sim 10^{10}$K) down to a little below 100 keV ($\sim 10^9$ K). At higher temperatures, no complex nuclei other than free single neutrons and protons can exist, and the ratio of neutrons to protons, n/p, is just determined by $n/p = e^{-Q/T}$, where $Q = (m_n - m_p)c^2 \sim 1.3$ MeV. Equilibrium applies because the weak interaction rates are much faster than the expansion of the universe at temperatures much above 10^{10}K. At temperatures much below 10^9K, the electrostatic repulsion of nuclei prevents nuclear reactions from proceeding as fast as the cosmological expansion separates the particles.

After the weak interaction drops out of equilibrium, a little above 10^{10}K, the ratio of neutrons to protons changes more slowly due to free neutrons decaying to protons, and similar transformations of neutrons to protons via interactions with the ambient leptons. By the time the universe reaches 10^9K (0.1 MeV), the ratio is slightly below 1/7. For temperatures above 10^9K, no significant abundance of complex nuclei can exist due to the continued existence of gammas with energies greater than MeV. Note that the high photon to baryon ratio in the universe ($\sim 10^{10}$) enables significant population of the MeV high energy Boltzman tail until $T \lesssim 0.1$MeV.

Once the temperature drops to about 10^9K, sufficient abundances of nuclei can exist in statistical equilibrium through reactions such as $n + p \leftrightarrow\ ^2$H $+\gamma$ and ^{2}H $+p \leftrightarrow\ ^3$He $+\gamma$ and ^{2}H $+n \leftrightarrow\ ^3$H $+\gamma$, which in turn react to yield ^{4}He. Since ^{4}He is the most tightly bound nucleus in the region, the flow of reactions converts almost all the neutrons that exist at 10^9K into ^{4}He. The flow essentially stops there because there are no stable nuclei at either mass-5 or mass-8. Since the baryon density at Big Bang Nucleosynthesis is relatively low (about 1% the density of terrestrial air) and the time-scale short ($t \lesssim 10^2 sec$), only reactions involving two-particle collisions occur. It can be seen that combining the most abundant nuclei, protons and ^{4}He via two body interactions always leads to unstable mass-5. Even when one combines ^{4}He with rarer nuclei like ^{3}H or ^{3}He, we still get only to mass-7, which, when hit by a proton, the most abundant nucleus around, yields mass-8. (As we will discuss, a loophole around the mass-8 gap can be found if $n/p > 1$, so that excess neutrons exist, but for the standard case $n/p < 1$). Eventually, ^{3}H decays radioactively to ^{3}He, and any mass-7 made radioactively decays to ^{7}Li. Thus, Big Bang Nucleosynthesis makes ^{4}He with traces of ^{2}H, ^{3}He, and ^{7}Li. (Also, all the protons left over that did not capture neutrons remain as hydrogen.) For standard homogeneous BBN, all other chemical elements are made later in stars and in related processes. (Stars jump the mass-5 and -8 instability by having gravity compress the matter to sufficient densities and have much longer times available so that three-body collisions can occur.)

3. INHOMOGENEOUS BIG BANG NUCLEOSYNTHESIS
As noted above, BBN yields all agree with observations using only one freely adjustable parameter, η, or equivalently, ρ_b. Thus, BBN can make strong statements regarding ρ_b if the observed light element abundances cannot be fit with any alternative theory. The most significant alternative that has been discussed involves quark-hadron transition inspired inhomogeneities.[31] While inhomogeneity models had been looked at previously (c.f. ref. [13]) and were found to make little difference, the quark-hadron inspired models had the added ingredient of variations in n/p ratios. Cosmologists are well aware that current trends in lattice gauge calculations imply that the transition is probably second order or not a phase transition at all.

Nevertheless, it has been important to explore the maximal cosmolgical impact that can occur. This maximal impact requires a first-order phase transition.

The initial claim by Applegate *et al.*, followed by a similar argument from Alcock *et al.*, that $\Omega_b \sim 1$ might be possible, created tremendous interest. Their argument was that if the quark-hadron transition was a first-order phase transition, then it was possible that large inhomogeneities could develop at $T \gtrsim 100$ MeV. The preferential diffusion of neutrons versus protons out of the high density regions could lead to Big Bang Nucleosynthesis occurring under conditions with both density inhomogeneities and variable neutron/proton ratios. In the first round of calculations, it was claimed that such conditions might allow $\Omega_b \sim 1$, while fitting the observed primordial abundances of ^{4}He, ^{2}H, ^{3}He with an overproduction of ^{7}Li. Since ^{7}Li is the most recent of the cosmological abundance constraints and has a different observed abundance in Pop I stars versus the traditionally more primitive Pop II stars,[25] some argued that perhaps some special depletion process might be going on to reduce the excess ^{7}Li.

At first it appeared that if the lithium constraint could be surmounted, then the constraints of standard Big Bang Nucleosynthesis might disintegrate. (Although several of us including Hubert Reeves emphasized that the number of parameters needed to fit the light elements was somewhat larger for these non-standard models, nonetheless, a non-trivial loophole appeared to be forming.) To further stimulate the flow through the loophole, Malaney and Fowler showed that, in addition to looking at the diffusion of neutrons out of high density regions, one must also look at the subsequent effect of excess neutrons diffusing back into the high density regions as the nucleosynthesis goes to completion in the low density regions. (The initial calculations treated the two regions separately.) Malaney and Fowler argued that for certain phase transition parameter values (e.g. nucleation site separations $\sim 10m$ at the time of the transition), this back diffusion could destroy much of the excess lithium.

However, Kurki-Suonio, Matzner, Olive and Schramm,[32] the Tokyo group[33] (which at times has included Hubert as a collaborator), and the Livermore group[34] have eventually argued that in their detailed diffusion models, the back diffusion not only affects ^{7}Li, but also the other light nuclei as well. They find that for $\Omega_b \sim 1$, ^{4}He is also overproduced (although it does go to a minimum for similar parameter values as does the lithium). One can understand why these models might tend to overproduce ^{4}He and ^{7}Li by remembering that in standard homogeneous Big Bang Nucleosynthesis, high baryon densities lead to excesses in these nuclei. As back diffusion evens out the effects of the initial fluctuation, the averaged result should approach the homogeneous value. Furthermore, it can be argued that any narrow range of parameters, such as those which yield relatively low lithium and helium, are unrealistic since in most realisitic phase transitions there are distributions of parameter values (distribution of nucleation sites, separations, density fluctuations, etc.). Therefore, narrow minima are washed out which would bring the ^{7}Li and ^{4}He values back up to their excessive levels for all parameter values with $\Omega \sim 1$. Furthermore, Freese and Adams[35] and Baym[36] have argued that the boundary between the two phases may be fractal-like rather than smooth. The large surface area of a fractal-like boundary would allow more interaction between the regions and minimize exotic effects.

Figure 2 shows the updated results of Kurki-Suonio et al.[32] for varying spacing l with the constraints from the different light element abundances. Notice that the Li, ^{2}D, and even the ^{4}He constraint do not allow $\Omega_b \sim 1$. Note also that with the Pop II ^{7}Li constraint, the results for Ω_b are quite similar to the standard model with a slight excess in Ω_b possible if l is tuned to ~ 10. Thus, even an optimally tuned first order quark-hadron transition is not able to alter the basic conclusions of homogeneous BBN regarding Ω_b. (It also cannot significantly change the N_ν argument.) In fact, the main role that a quark-hadron option has played for BBN is to show how robust the standard model results are.

"QUARK–HADRON" NUCLEOSYNTHESIS

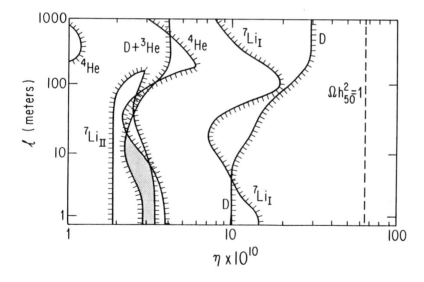

Figure 2. The updated Kurki-Suonio, Matzner, Olive and Schramm[32] results showing that even allowing for a 1st order quark-hadron transition with variable spacing l between nucleation sites, the light element abundances constrain the baryon to photon ratio, η, and thus Ω_b to essentially the same values as those obtained in the homogeneous case with only slightly large Ω_b's possible with $l \sim 10$.

4. BORON, BERYLLIUM AND THE SPALLATION PROCESS

While quark-hadron inspired variations have not been able to alter the basic conclusions of BBN, an important question remains, namely, is there an observable signature that could differentiate quark-hadron inspired variations from the homogeneous model? On the theoretical side, this point has been debatable. Several authors[37,38,39] have argued that because of the high n/p region in the inhomogeneous models, leakage beyond the mass 5 and 8 instability gaps can occur and traces of ^{9}Be, ^{10}B, ^{11}B and maybe even r-process elements can be produced. Thus, detection of nuclei beyond ^{7}Li in primitive objects may be a signature. However, Tarasawa and Sato[33] have argued that such leakage is negligible. Because of this debate as well as the recent experimental results, we have started theoretically examining this question ourselves. However, before discussing our results, let us first comment on some recent observations of Be and B in primitive Pop II stars.

In particular, there has been much recent attention given to reports[40,41] of beryllium lines being observed in extreme Pop II stars. For one very metal-poor Pop II star, HD 140283, boron was also observed.[42] The observations yielded

$$\frac{Be}{H} \sim 10^{-12.9\pm0.3} \tag{1}$$

which represents a combination of the two Be/H measurements with Gilmore et al.[40] obtaining a factor of ~ 3 higher Be/H than Ryan et al.[41] The boron was measured using the Hubble Space Telescope where a value was obtained[42] of

$$\frac{B}{H} \sim 10^{-12.1\pm0.1}. \tag{2}$$

The resulting boron to beryllium ratio is

$$\frac{B}{Be} \sim 5^{+4}_{-2}. \tag{3}$$

This particular star has its iron abundance depleted relative to the standard Pop I (present galactic disk) iron abundance by a factor of about $10^{-2.6}$, and its oxygen is depleted relative to Pop I by about $10^{-2.1}$. The high oxygen to iron in extreme Pop II stars is well understood[43] as due to heavy element production in massive Type II supernova producing high oxygen to iron, whereas later Pop I abundances also get a significant admixture of low-mass, slow-to-explode, Type I supernova ejecta where iron is dominant over oxygen. Because oxygen is chiefly made in Type II supernova, whereas iron has at least two significant sources, we feel it is mandatory to use oxygen as a measure of the Type II supernova contribution to such stars. In this regard, it is important to note that the Be/O for these stars is, within experimental errors, the same as Be/O for those high surface temperature Pop I stars whose convective zones are not deep enough to destroy their original Be. Thus, contrary to some initial claims, the Be/H observation does not require cosmological origin, only a scaling with oxygen of the same process that produced Be in the Pop I stars.

The presumed process that produced Be and B in Pop I stars (as well as the ^{6}Li), is thought to be Reeves' cosmic ray spallation.[3] For Be and B, such spallation comes from the breakup of heavy nuclei such as CNO and Ne, Mg, Si, S, Ca and Fe by protons and alphas. As noted by Epstein *et al.*,[44] for lithium one must also include alpha plus alpha fusion processes as well. This latter point was well noted by Steigman and Walker[45] who emphasized that Be and B spallation production on Pop II abundances would imply a significant enhancement of lithium from alpha-alpha relative to the reduced production of Be and B from depleted heavy nuclei. While the ^{6}Li so produced would be destroyed at the base of the convective zones in the stars observed,[46,47] the ^{7}Li would survive and might result in observable variations in the Spite[25] Pop II lithium plateau.

Perhaps most critical to any spallation origin is the resultant B/Be ratio. It is also known, from actual measurements, that the cosmic rays themselves[48] show $\frac{B}{Be} \sim 14$ (and B and Be are pure spallation products in the cosmic rays) with a carbon to oxygen ratio exceeding unity (Pop I has $C/O \lesssim 0.5$). Since spallation off carbon favors B relative to Be (mass 11 requires only a single nucleon ejected from mass 12), whereas oxygen being farther from either shows less favoritism, the cosmic ray observations are actually an upper limit on what B/Be ratio one might expect in Pop I cosmic rays. However, of more concern here is the lower limit on B/Be achievable by a spallation process.[49] We note that cosmic ray spectra that are flatter than $E^{-2.6}$ will be less favorable towards boron production. This is because the ^{11}B production threshold is below that for ^{9}Be. Thus, steeper spectra favor B relative to Be, whereas flatter specta remove the role of the threshold effects and yield relatively higher Be. Furthermore, Pop II composition has a lower C/O ratio than does Pop I. Like iron, carbon is not a pure Type II supernova product. Thus, spallation on pure Type II ejecta would have targets of oxygen, neon, magnesium, silicon, etc., but less carbon and nitrogen than Pop I. Recent GRO/EGRET gamma ray results show extragalactic high energy spectra with $\sim E^{-2.0}$. Thus, flat spectra may be quite reasonable.

We have carried out spallation calculations for flat spectra on Pop II material. The cross sections we used for the spallation calculations are a combination of all measured cross section data[50] and our semi-empirical estimates.[5,49] For comparison, we also used the semi-empirical cross sections of Silberberg and Tsao.[51] The resultant ratio is

$$\frac{B}{Be} \simeq 7.6 \qquad (4)$$

(from our semi-empirical cross sections), and $\frac{B}{Be} \simeq 8.6$ (Silberberg and Tsao). In other words, optimizing Be yields can still not get a B/Be ratio below 7.6. Since this is within one sigma of the observations on HD 140283, it is obvious that the present observations are quite consistent with spallation.

It is important to note that if spallation processes do indeed produce the observed Be and B in Pop II stars, then the cosmic ray flux must be stronger than it is in the present Galaxy. Remember that the present Pop I abundance of Be and B and ^{6}Li can be explained by the present cosmic ray flux hitting the Pop I CNO abundances[1,3] integrated over the lifetime of the Galactic disk prior to the formation of the observed stars. However, for these Pop II stars, the CNO and heavier element abundances are down and the stars presumably formed relatively early, before the disk formed. While some Galactic evolution models[52] expect this pre-disk formation epoch to be several Gyr long, it is nonetheless shorter than the age of the disk. (If the pre-disk time is merely the massive star stellar evolution time scale, then it can be very short.) The shorter time scale thus requires a consumately higher flux if the ratios to oxygen observed in Pop I are to be retained in the Pop II objects. Of course, many galactic evolution models[52] predict higher early supernova rates which produce just such a higher cosmic ray flux, so consistent models do exist.

From the above, we, at present, see no cause to invoke anything other than spallation; however, if the uncertainties in the B/Be ratio are decreased and the ratio remains below 7.6, then spallation would fail. Furthermore, if, as O/H decreases, Be/O and B/O are found to exceed significantly the ratio observed for higher oxygen abundances, then we would have to conclude that there is primordial cosmological production of Be and B.

Figures 3A and 3B show the trace element yields in a standard homogeneous BBN calculation with Figure 3A showing ^{2}H, ^{3}He, ^{6}Li and ^{7}Li, and Figure 3B showing the ^{9}Be, ^{10}B and ^{11}B yields. This work is part of an extensive study of $A \geq 6$ BBN by Thomas et al.,[53] using a more extensive reaction network than previously used. Note, in particular, that ^{9}Be/H yields are always less than 10^{-14} regardless of $\eta = n_b/n_\gamma$. (Also note that for the standard model, B/Be $\gg 10$ unless $\eta \lesssim 3 \times 10^{-10}$.) In other words, homogeneous BBN cannot yield Be/H consistent with the Pop II stellar observations. To explore preliminarily the alternative of inhomogeneous models, we have taken our extensive network and looked at high n/p ratios. For regions with $n/p > 3$, we can obtain Be/H $\sim 10^{-14}$ but no more for parameter values that still fit the $A \leq 7$ abundances. However, any realistic model will have a significant dilution of this material with low n/p regions. Thus, we tentatively view the achievement of such values as somewhat problematic, as do Tarasawa and Sato.[33] We will continue to explore a full inhomogeneous model which includes regions of extremely high n/p to see how robust any leakage to $A > 7$ truly is. Such an exploration is just beginning.

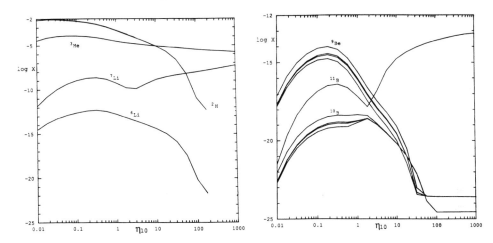

Figure 3A. The standard homogeneous BBN yields showing ^{2}H, ^{3}He, ^{6}Li and ^{7}Li for 6 orders of magnitude in η_b/η_γ. Note that ^{6}Li is always negligible relative to ^{7}Li.

Figure 3B. The standard homogeneous BBN yields for ^{9}Be, ^{10}B and ^{11}B. The various curves for ^{9}Be and ^{10}B represent different cross section assumptions. The ^{11}B yield is double humped due to production both directly as ^{11}B and also as ^{11}C which beta decays to ^{11}B.

If some Be and B can be shown to be cosmological, it will have great implications for Big Bang Nucleosynthesis. If simple inhomogeneities are unable to produce it, then more exotic ones will be required. The source of such inhomogeneities is either the quark-hadron transition or some other activity around that same cosmological epoch (no earlier than the electro-weak transition) so that density variations are retained. Of course, whatever these variations might be, they must not alter the spectacular agreement for $A \leq 7$ abundances for N_ν.

5. LIMITS ON Ω_b AND DARK MATTER REQUIREMENTS

The success and robustness of BBN in the face of the Be and B results as well as the quark-hadron variations gives renewed confidence to the limits on the baryon density constraints. Let us convert this density regime into units of the critical cosmological density for the allowed range of Hubble expansion rates. For the Big

Bang nucleosynthesis constraints,[14,32] the dimensionless baryon density, Ω_b, that fraction of the critical density that is in baryons, is less than 0.11 and greater than 0.02 for $0.04 \lesssim h_0 \lesssim 0.7$, where h_0 is the Hubble constant in units of 100km/sec/Mpc. The lower bound on h_0 comes from direct observational limits and the upper bound from age of the universe constraints.[54] The constraint on Ω_b still means that the universe *cannot be closed with baryonic matter.* (This point was made 20 years ago[3] and has proven to be remarkably strong.) If the universe is truly at its critical density, then nonbaryonic matter is required. This argument has led to one of the major areas of research at the particle-cosmology interface, namely, the search for non-baryonic dark matter.

Another important conclusion regarding the allowed range in baryon density is that it is in very good agreement with the density implied from the dynamics of galaxies, *including their dark halos.* An early version of this argument, using only deuterium, was described over fifteen years ago.[21] As time has gone on, the argument has strengthened, and the fact remains that galaxy dynamics and nucleosynthesis agree at about 5% of the critical density. Thus, if the universe is indeed at its critical density, as many of us believe, it requires most matter not to be associated with galaxies and their halos, as well as to be nonbaryonic. Let us put the nucleosynthetic arguments in context.

The arguments requiring some sort of dark matter fall into separate and quite distinct areas. These arguments are summarized in Figure 4. First are the arguments using Newtonian mechanics applied to various astronomical systems that show that there is more matter present than the amount that is shining. It should be noted that these arguments reliably demonstrate that galactic halos seem to have a mass ~ 10 times the visible mass.

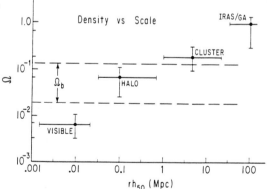

Figure 4. Implied densities versus the scale of the measurements.

Note however that Big Bang nucleosynthesis requires that the bulk of the baryons in the universe be dark since $\Omega_{vis} << \Omega_b$. Thus, the dark halos could in principle be baryonic.[21] Recently arguments on very large scales[55] (bigger than cluster of galaxies) hint that Ω on those scales is indeed greater than Ω_b, thus forcing us to need non-baryonic matter.

An Ω of unity is, of course, preferred on theoretical grounds since that is the only long-lived natural value for Ω, and inflation,[56] or something like it, provided the early universe with the mechanism to achieve that value and thereby solve the flatness and smoothness problems. Note that our need for exotica is not dependent on the existence of dark galatic halos. This point is frequently forgotten, not only by some members of the popular press but occasionally by active workers in the field.

Some baryonic dark matter must exist since we know that the lower bound from Big Bang nucleosynthesis is greater than the upper limits on the amount of visible matter in the universe. However, we do not know what form this baryonic dark matter is in. It could be either in condensed objects in the halo, such as brown dwarfs and jupiters (objects with $\lesssim 0.08 M_\odot$ so they are not bright shining stars), or in black holes (which at the time of nucleosynthesis would have been baryons). Or, if the baryonic dark matter is not in the halo, it could be in hot intergalactic gas, hot enough not to show absorption lines in the Gunn-Peterson test, but not so hot as to be seen in the x-rays. Evidence for some hot gas is found in clusters of galaxies. However, the amount of gas in clusters would not be enough to make up the entire missing baryonic matter. Another possible hiding place for the dark baryons would be failed galaxies, large clumps of baryons that condense gravitationally but did not produce stars. Such clumps are predicted in galaxy formation scenarios that include large amounts of biasing where only some fraction of the clumps shine.

Hegyi and Olive[57] have argued that dark baryonic halos are unlikely. However, they do allow for the loopholes mentioned above of low mass objects or of massive black holes. It is worth noting that these loopholes are not that unlikely. If we look at the initial mass function for stars forming with Pop I composition, we know that the mass function falls off roughly like a power law for standard size stars as was shown by Salpeter. Or, even if we apply the Miller-Scalo mass function, the fall off is only a little steeper. In both cases there is some sort of lower cut-off near $0.1 M_\odot$. However, we do not know the origin of this mass function and its shape. No true star formation model based on fundamental physics predicts it.

We do believe that whatever is the origin of this mass function, it is probably related to the metalicity of the materials, since metalicity affects cooling rates, etc. It is not unreasonable to expect the initial mass function that was present in the primordial material which had no heavy elements (only the products of big bang nucleosynthesis) would be peaked either much higher than the present mass function or much lower— higher if the lower cooling from low metals resulted in larger clumps, or lower if some sort of rapid cooling processes ("cooling flows") were set up during the initial star formation epoch, as seems to be the case in some primative galaxies. In either case, moving either higher or lower produces the bulk of the stellar population in either brown dwarfs and jupiters or in massive black holes. Thus, the most likely scenarios are that a first generation of condensed objects would be in a form of dark baryonic matter that could make up the halos, and could explain why there is an interesting coincidence between the implied mass in halos and the implied amount of baryonic material. However, it should also be remembered that a consequence of this scenario is to have the condensation of the objects occur prior to the formation of the disk. If the first large objects to form are less than galactic mass, as many scenarios imply, then mergers are necessary for eventual galaxy-size objects. Mergers stimulate star formation while putting early objects into halos rather than disks. Mathews et al.[58] have recently developed a galactic evolution model which does just that and gives a reasonable scenario for chemical evolution. Thus, while making halos out of exotic material may be more exciting, it is certainly not impossible for the halos to be in the form of dark baryons. The new microlensing projects by groups in France, the U.S., Australia and Poland should eventually test this possibility.

Non-baryonic matter can be divided following Bond and Szalay[59] into two major categories for cosmological purposes: hot dark matter (HDM) and cold dark matter (CDM). Hot dark matter is matter that is relativistic until just before the epoch of galaxy formation, the best example being low mass neutrinos with $m_\nu \sim 20 eV$. (Remember $\Omega_\nu \sim \frac{m_\nu(ev)}{100 h_0^2}$). Cold dark matter is matter that is moving slowly at the epoch of galaxy formation. Because it is moving slowly, it can clump on very small scales, whereas HDM tends to have more difficulty in being confined on small scales. Examples of CDM could be massive neutrino-like particles with masses, M_x, greater than several GeV or the lightest super-symmetric particle which is presumed to be stable and might also have masses of several GeVc. Following Michael Turner, all such weakly interacting massive particles are called "WIMPS." Axions, while very light, would also be moving very slowly[60] and, thus, would clump on small scales. Or, one could also go to non-elementary particle candidates, such as planetary mass blackholes or quark nuggets of strange quark matter, possibly produced at the quark-hadron transition.[61] Another possibility would be any sort of massive toplogical remnant left over from some early phase transition. Note that CDM would clump in halos, thus requiring the dark baryonic matter to be out between galaxies, whereas

HDM would allow baryonic halos.

When thinking about dark matter candidates, one should remember the basic work by Zeldovich,[62] by Lee and Weinberg,[63] and by others,[64] which showed that for a weakly interacting particle, one can obtain closure densities, either if the particle is very light, $\sim$20eV, or if the particle is very massive, $\sim$3GeV. This occurs because, if the particle is much lighter than the decoupling temperature, then its number density is the number density of photons (to within spin factors and small corrections), and so the mass density is in direct proportion to the particle mass, since the number density is fixed. However, if the mass of the particle is much greater than the decoupling temperature, then annihilations will deplete the particle number. Thus, as the temperature of the expanding universe drops below the rest mass of the particle, the number is depleted via annihilations. For normal weakly interacting particles, decoupling occurs at a temperature of $\sim$1 MeV, so higher mass particles are depleted. It should also be noted that the curve of density versus particle mass turns over again (see Figure 5) once the mass of the WIMP exceeds the mass of the coupling boson[65,66] so that the annihilation cross section varies as $\frac{1}{M_x^2}$, independent of the mass of the coupling boson. In this latter case, $\Omega = 1$ can be obtained for $M_x \sim 1TeV \sim (3K \times M_{Planck})^{1/2}$, where $3K$ and M_{Planck} are the only energy scales left in the calculation (see Figure 5). The curve turns over again as multi-channel effects occur.

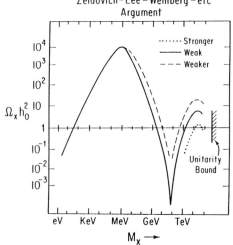

Figure 5. $\Omega_x h_o^2$ versus M_x for weakly interacting particles showing three or more crossings of $\Omega h_o^2 = 1$. Note also how the curve shifts at high M_x for interactions weaker or stronger than normal weak interaction (where normal weak is that of neutrino coupling through Z^o). Extreme strong couplings reach a unitarity limit at $M_x \sim 340$ TeV.

A few years ago the preferred candidate particle was probably a few GeV mass WIMP. However, LEP's lack of discovery of any new particle coupling to the Z° with $M_x \lesssim 45GeV$, coupled with underground ionization experiments[67] constrains that candidate. Other constraints for particles not fully coupled to the Z^0 were discussed by Ellis et al.[68] Although candidates can be discussed with M $\sim$ 3 to 20 GeV that avoid the current experimental constraints, it is probably fair to say that the current favorites tend to be at least as massive as 20 GeV and interact more weakly than a neutrino. Of course, an HDM ν_τ with $m_{\nu_\tau} \sim 20 \pm 10eV$ is still a fine candidate as long as galaxy formation proceeds by some mechanism other than adiabatic gaussian matter fluctuations.[69] This latter candidate becomes particularly attractive if recent hints from the gallium experiments require the solution to the solar neutrino problem to have neutrino mixing with $\nu_e - \nu_\mu$ mass scales of $\sim 10^{-3}$eV, making eV mass scales for ν_τ quite plausible in these see-saw type models where $m_{\nu_\tau} \sim m_{\nu_\mu}(m_{top}/m_{charm})^2$.

6. CONCLUSION

Primordial nucleosynthesis has become one of the cornerstones of modern cosmology. Hubert Reeves has played a significant role in the laying of this cornerstone.

7. ACKNOWLEDGEMENTS

In addition to thanking Hubert Reeves for the twenty years I have had the pleasure of knowing him, I would also like to thank my recent collaborators, Gary Steigman, Keith Olive, Michael Turner, Rocky Kolb, Grant Mathews, George Fuller and Terry Walker for many useful discussions. I would further like to thank Doug Duncan, Lew Hobbs and Don York for valuable discussion regarding the astronomical observations and John Simpson, Mark Wiedenbeck and John Wefel regarding spallation cross sections and cosmic ray observations. This work was supported in part by NSF Grant AST 90-22629, by NASA Grant NAGW-1321 and by DoE Grant DE-FG02-91-ER40606 at the University of Chicago, and by the DoE and NASA Grant NAGW-1340 at the NASA/Fermilab Astrophysics Center.

8. REFERENCES

1. C. Ryter, H. Reeves, E. Gradstajn, and J. Audouze (1970) Astron. and Astrophys., 8, 389.

2. A. Reeves, W. Fowler, and F. Hoyle (1970) Nature, 226, 727.

3. H. Reeves, J. Audouze, W.A. Fowler, and D.N. Schramm (1973) ApJ, 179, 909.

4. D.N. Schramm (1991) in Nobel Symposium No. 79: The Birth and Early Evolution of Our Universe, Gräftåvalen, Sweden, June 1990, Physica Scripta T36, 22.

5. D.N. Schramm, B. Fields, and D. Thomas (1992) in Proc. of Quark Matter 91, Gatlinburg, Tennessee, November 1991, Nucl. Phys., in press.

6. R.A. Alpher, H. Bethe, and G. Gamow (1948) Phys. Rev., 73, 803.

7. C. Hayashi (1950) Prog. Theor. Phys, 5, 224.

8. R.A. Alpher, J.W. Follin, and R.C. Herman (1953) Phys. Rev., 92, 1347.

9. R. Taylor and F. Hoyle (1964) Nature, 203, 1108.

10. P.J.E. Peebles (1966) Phys. Rev. Lett., 16, 410.

11. P. Wagoner, W.A. Fowler, and F. Hoyle (1967) ApJ, 148, 3.

12. D.N. Schramm and R.V. Wagoner (1977) Ann. Rev. of Nuc. Sci., 27, 37.
K. Olive, D.N. Schramm, G. Steigman, M. Turner, and J. Yang (1981) ApJ, 246, 557;
A. Boesgaard and G. Steigman (1985) Ann. Rev. of Astron. and Astrophys., 23, 319.

13. J. Yang, M. Turner, G. Steigman, D.N. Schramm and K. Olive (1984) ApJ, 281, 493.

14. K. Olive, D.N. Schramm, G. Steigman, and T. Walker (1990) Phys. Lett B, 236, 454.
T. Walker, G. Steigman, D.N. Schramm, K. Olive, and H.-S. Kang (1991) ApJ, 376, 51.

15. L. Kawano, D.N. Schramm, and G. Steigman (1988) ApJ, 327, 750 (1988).

16. W. Mampe, P. Ageron, C. Bates, J.M. Pendlebury, and A. Steyerl (1989) Phys. Rev. Lett., 63, 593.

17. W. Fowler, J. Greenstein, and F. Hoyle (1962) Geophys. J.R.A.S., 6, 6.

18. J. Geiss and H. Reeves (1971) Astron. and Astrophys., 18, 126;
D. Black (1971) Nature, 234, 148.

19. J. Rogerson and D. York (1973) ApJ, 186, L95.

20. R. Epstein, J. Lattimer, and D.N. Schramm (1976) Nature, 263, 198.

21. J. R. Gott, III, J. Gunn, D.N. Schramm, and B.M. Tinsley (1974) ApJ, 194, 543.

22. R.T. Rood, G. Steigman, and B.M. Tinsley (1976) ApJ, 207, L57.

23. J. Yang, D.N. Schramm, G. Steigman, and R.T. Rood (1979) ApJ, 227, 697.

24. T. Wilson, R.T. Rood, and T. Bania (1983) in Proc. of the ESO Workshop on Primordial Healing, ed. P. Shaver and D. Knuth (Garching: European Southern Observatory).

25. M. Spite and F. Spite (1982) Astron. and Astrophys., 115, 357;
R. Rebolo, P. Molaro, and J. Beckman (1988) Astron. and Astrophys., 192, 192;
L. Hobbs and C. Pilachowski (1988) ApJ, 326, L23.

26. G. Steigman, D.N. Schramm, and J. Gunn (1977) Phys. Lett., 66B, 202.

27. V.F. Schvartzman (1969) JETP Letters, 9, 184;
    P.J.E. Peebles (1971) Physical Cosmology (Princeton University Press).
28. D.N. Schramm and L. Kawano (1989) Nuc. Inst. and Methods A, 284, 84.
29. B. Pagel (1989) in Proc. of 1989 Rencontres de Moriond.
30. ALEPH, L3, OPAL, DELPHI results, 1991 Lepton-Photon meeting at CERN.
31. R. Scherrer, J. Applegate, and C. Hogan (1987) Phys. Rev. D, 35, 1151;
    C. Alcock, G. Fyuller, and G. Mathews (1987) ApJ, 320, 439;
    W.A. Fowler and R. Malaney (1988) ApJ, 333, 14.
32. H. Kurki-Suonio, R. Matzner, K. Olive, and D.N. Schramm (1990) ApJ, 353, 406.
33. K. Sato and N. Tarasawa (1991) in The Birth and Early Evolution of Our Universe: Proceedings of Nobel Symposium 79, J. S. Nilsson, B. Gustafson, and B.S. Skagerstam, eds., (Singapore, World Scientific) 60.
34. G.J. Mathews, B.S. Meyer, C.R. Alcock, and G.M. Fuller (1990) ApJ, 358, 36.
35. K. Freese and F. Adams (1990) Phys. Rev. D., 41, 2449.
36. G. Baym (1991) private communication.
37. L. Kawano, W. Fowler, R. Malaney (1990) Caltech-Kellogg preprint.
38. R. Boyd and T. Kajino (1990) ApJ, 359, 267.
39. J. Applegate, C. Hogan, and R. Scherrer (1988) ApJ, 329, 572.
40. G. Gilmore, B. Edvardson, and P. Nissen (1991) ApJ, 378, 17.
41. S. Ryan, M. Bessel, R. Sytherland, and J. Norris (1990) ApJ, 348, L57.
    S.G. Ryan, J.E. Norris, M.S. Bessell, and C.P. Deliyannis (1992) ApJ, 388, 184.
42. D. Duncan, D. Lambert, and D. Lemke (1991) ApJ, submitted.
43. J.C. Wheeler, C. Sneden, and J. Truran (1989) Ann. Rev. Astro. Astro., 27, 279.
    H. Reeves, W.A. Fowler, and F. Hoyle (1970) Nature, 226, 727.
44. R. Epstein, W.D. Arnett, and D.N. Schramm (1974) ApJ, 190, L13.
    R. Epstein, W.D. Arnett, and D.N. Schramm (1976) ApJ Supp., 31, 111.
45. G. Steigman and T. Walker (1992) ApJ, 385, L13.
46. L. Brown and D.N. Schramm (1988) ApJ, 329, L103.
47. M. Pinsonnealt, C. Deliyannis, and P. Demarque (1991) ApJ Supplement, in press.
48. J. Simpson (1983) Ann. Rev. Nuc. and Part. Sci., 33, 323.
49. T. Walker, G. Steigman, D.N. Schramm, K. Olive, and B. Fields (1992) ApJ, submitted.
50. S. Reed and V. Viola (1984) At. Data Nuc. Data Tables, 31, 359.
    Gupta and Webber (1989) ApJ, 340, 1124.

51. R. Silberberg and T. Tsao (1973) ApJ Suppl, 25, 325.
    G. Rudstam, Z.F. Naturforschung (1966) 21A, 1027.
52. G. Mathews and D.N. Schramm (1991) Fermilab preprint, ApJ, submitted.
    A. Burkert, J. Truran, and G. Hensler (1992) ApJ, 391, 651.
53. D. Thomas, D.N. Schramm, K. Olive, and B. Fields, in preparation (1992).
54. K. Freese and D.N. Schramm (1984) Nucl. Phys., B233, 167.
55. C. Fisher (1992) Ph.D. Thesis, University of California at Berkeley, and references therein.
56. A. Guth (1981) Phys. Rev. D, 23, 347.
    A. Linde (1990) Particle Physics and Inflationary Cosmology, (New York, Harwood Academic Publishers.
57. D. Hegyi and K. Olive (1986) ApJ, 303, 56.
58. G. Mathews, G. Bazan, J.J. Cowan, and D.N. Schramm (1992) Physics Reports, in press.
59. R. Bond and A. Szalay (1982) in Proc. Texas Relativistic Astrophysical Symposium, Austin, Texas.
60. M. Turner, F. Wilczek, and A. Zee (1983) Phys. Lett. B, 125, 35 (1983); 125, 519.
61. M. Crawford and D.N. Schramm (1982) Nature, 298, 538.
    See C. Alcock and A. Olinto (1988) Ann. Rev. Nuc. Part. Phys., 38, 161.
62. Ya. Zeldovich (1965) Adv. Astron. and Astrophys., 3, 241;
63. B. Lee and S. Weinberg (1977) Phys. Rev. Lett., 39, 165.
64. H-Y. Chiu (1966) Phys. Rev. Lett., 17, 712.
    C.P. Hut (1977) Phys. Lett. B, 69, 85;
    K. Sato and H. Koyayashi (1977) Prog. Theor. Phys., 58, 1775.
65. D. Brahm and L. Hall (1990) Phys. Rev. D, 41, 1067.
66. K. Griest, M. Kamionkowski, and M. Turner (1990) Phys. Rev. D., 41, 3565.
    K. Olive and M. Srednicki (1989) Phys. Lett. B, 230, 78.
67. D. Caldwell et al. (1990) in Proc. La Thuile, ed. M. Greco.
68. J. Ellis, D. Nanopoulos, L. Roskowski and D.N. Schramm (1990) Phys. Lett. B, 245, 251.
69. D. N. Schramm (1988) in Proc. Berkeley Workshop on Particle Astrophysics.

Primordial Abundances of Be and B from Standard Big Bang Nucleosynthesis

P. DELBOURGO–SALVADOR[1,2], E. VANGIONI–FLAM[1]

1. Institut d'Astrophysique de Paris, CNRS
2. Centre d'Etudes de Bruyères-le-Châtel, France

ABSTRACT

We evaluate the primordial abundances of beryllium and boron isotopes in the light of updated nuclear reaction rates, and determine their sensitivity to the present baryon density in the framework of the standard Big Bang model.

1. INTRODUCTION

Before the recent and numerous studies of Non-uniform Density Models (NDM) of primordial nucleosynthesis (Alcock et al. 1987; Malaney and Fowler 1988; Boyd and Kajino 1989), ^{9}Be, ^{10}B and ^{11}B were generally considered to be produced by spallation reactions induced by the galactic cosmic rays (GCR) on the interstellar medium (e.g. Meneguzzi, Audouze and Reeves 1971); their primordial abundances were considered to be negligible. But, NDM calculations of primordial nucleosynthesis showed that in the neutron rich regions resulting from the quark–hadron phase transition, ^{9}Be, ^{7}Li and heavier elements could be significantly produced. The primordial abundance of ^{9}Be could then be a test for the NDM models (Kajino and Boyd, 1990).

Moreover, the first measurements of the abundances of Be and B in old halo stars (with [Fe/H]$\leq$–2) have been recently made (Ryan et al. 1990, Gilmore et al. 1991, Ryan et al. 1992, Duncan et al. 1991), indicating that these abundances are larger than what GCR can produce, at least in standard models of galactic chemical evolution. Though potentially interesting, NDM scenarios meet with difficulties and several studies have been devoted recently to alternative models of chemical evolution which could explain the observations, for example a revision of the cosmic ray model at the early phases of the galaxy (Vangioni–Flam et al. 1990, Ryan et al. 1992, Prantzos et al. 1992).

The aim of this paper is to reconsider the primordial abundances of ^{9}Be, ^{10}B, ^{11}B in the framework of the standard model of Big Bang nucleosynthesis, with updated values for nuclear reactions rates. Indeed, some of the nuclear reactions leading to the production of these elements have been studied again theoretically and experimentally. The obtained values will serve to measure the departure from NDM primordial nucleosynthesis and as initial conditions to galactic chemical evolution models.

2. STANDARD MODEL OF PRIMORDIAL NUCLEOSYNTHESIS

The calculations of primordial nucleosynthesis have been made with a code built at the Institut d'Astrophysique de Paris (Delbourgo–Salvador et al. 1984). 27 nuclei are included with all reaction rates updated according to Caughlan and Fowler (1988) except for few reactions which will be discussed below. The adopted number of light neutrino families is three as is widely believed now ($N\nu$=2.99±0.05, after the LEP results, ALEPH and DELPHI coll. 1989, 1991). The most recent evaluation for the neutron half life time is adopted : $\tau_{\frac{1}{2}}$=10.3 min (W. Mampe et al. 1989; Hikasa et al. 1992).

Primordial abundances are calculated as a function of only one parameter Ω_b, the ratio of the current baryonic density of the Universe to the critical density. Figure 1 shows the final abundances of D, ^{3}He, ^{4}He, ^{7}Li, ^{9}Be, ^{10}B and ^{11}B as a function of η, (the baryon to photon ratio) related to the value of the Hubble constant H and the temperature of the cosmological background (T_o) by :

$$\Omega_b \, h^2 \; = \; 0.015 \; \eta_{10} \, (\frac{T_o}{2.7})^3$$

where η_{10} is $\eta/10^{10}$ and H $= 50 \, h$ km/s/Mpc.

The limits on η are set by the conservative limits on the abundance of primordial ^{4}He: $0.22 < Y_p < 0.24$ (e.g. Pagel et al., 1992), which give the most reliable interval: $1 \leq \eta_{10} \leq 3$. We do not consider here the delicate question of the primordial values of D and ^{7}Li, since they are affected by the stellar evolution and the chemical evolution of the galaxy. In the following sections, we discuss the nuclear reactions leading to the production and destruction of Be and B and give an estimate of their primordial abundances.

3. PRIMORDIAL NUCLEOSYNTHESIS OF ^{9}BE, ^{10}B AND ^{11}B

3.1 ^{9}Be

In standard Big Bang nucleosynthesis the main reaction of production of ^{9}Be is ^{7}Li(T,n)^{9}Be, which has been recently measured independently by two groups (Barhoumi et al. 1992; Brune et al. 1991). The two measurements are in good agreement (within 20%). These values differ from the latest theoretical evaluations (Boyd and Kajino 1990) by a factor 3–7. Other reactions leading to ^{9}Be via ^{8}Li are important in neutron rich zones of NDM models, but not in the standard model. As for the destruction of ^{9}Be, new measurements of the ^{9}Be(p,γ)^{10}B rate should be made to confirm the results of Sierk and Tombrello (1973). (see below).

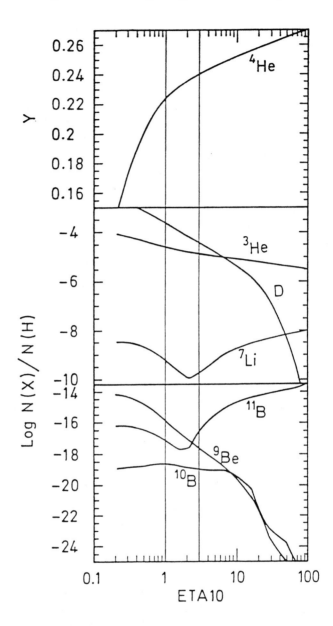

FIGURE 1: Final abundances of the light isotopes (by number, except for ⁴He, for which the mass fraction Y_P is given) as function of η, in the Standard Big Bang model. The limits on η are obtained by considering only Y_P, as derived from observations.

3.2 ^{10}B

^{10}B is produced via two reactions :
– ^{9}Be(p,γ)^{10}B, which contributes to about 90% of the final abundance; and
– ^{6}Li(α,n)^{10}B for the 10% left.

Table 1 shows the results of calculations with or without each one of these reactions. The uncertainty on the final abundance of ^{10}B is linearly related to the one on the ^{9}Be(p,γ)^{10} B rate. The final abundance of ^{9}Be is not affected by these variations. Very recently the Bochum–Munster group (Angulo–Perez 1992) and Youn et al. (1991) have measured the reaction destroying ^{10}B : ^{10}B(p,α)^{7}Be. They found a low energy resonance which enhances the rate by about a factor of ten compared to the Caughlan and Fowler (1988) rate, but the results are still under study (Angulo–Perez 1992). Table 1 shows the effect of an enhancement of this rate and of the ^{10}B(p,γ) rate by a factor of 10 on the final abundance of ^{10}B.

3.3 ^{11}B

The ^{11}B abundance curve shows the same feature (a valley around $\eta \sim 2$) as the ^{7}Li abundance curve (see Figure 1). This is due to the fact that ^{11}B is produced via two processes :
– ^{7}Li(α,γ)^{11}B at low density
– ^{7}Be(α,γ)^{11}C(β^+)^{11}B at high density.

The participation of the two radioactive elements (^{7}Be $\rightarrow$ ^{7}Li, ^{11}C $\rightarrow$ ^{11}B) delays the role of the ^{11}B(p,α)2α reaction in the destruction of ^{11}B. One can see on Table 2 that for any value of η these two processes are dominant. An evaluation of these two resonant alpha captures on ^{7}Li and ^{7}Be have been made by Hardie et al. (1984). New measurements of ^{11}B(p,α)2α by the Bochum–Munster group seems in agreement with the Caughlan and Fowler (1988) rate.

The uncertainty of the ^{11}B final abundance is linearly related to the one of the alpha capture rate. The uncertainties on the other reaction rates have an effect of a few percent only (see Table 2).

4. DISCUSSION AND CONCLUSION

This study leads to the following intervals for the primordial abundances (by number) of ^{9}Be, ^{10}B, ^{11}B, in the framework of the standard Big Bang model :

$$2.5\ 10^{-18} < (^{9}Be/H)_P < 1.8\ 10^{-16}$$
$$1.0\ 10^{-19} < (^{10}B/H)_P < 2.1\ 10^{-19}$$
$$2.5\ 10^{-18} < (^{11}B/H)_P < 3.2\ 10^{-17}$$

for: $1.0 < \eta_{10} < 3.0$.

Table 1

^{10}B production and destruction for $\eta=0.67$
(plateau in the ^{10}B abundance curve).
^{9}Be is unchanged: $X(^{9}\text{Be})=4.5\ 10^{-15}$.

	Final abundance $X(^{10}\text{B})$ by mass
Full network calculations	1.6 10^{-18}
Without the reaction: $^{9}\text{Be}(p,\gamma)^{10}\text{B}$	1.1 10^{-19}
Without the reaction: $^{6}\text{Li}(\alpha,\gamma)^{10}\text{B}$	1.5 10^{-18}
Rate of destruction $^{10}\text{B}(p,\gamma)\star\ 10$ and $^{10}\text{B}(p,\alpha)\star\ 10$	9.75 10^{-19}

Table 2

Production and destruction of ^{11}B
for different values of η

η	X(^{11}B) final abundance by mass						
	Full network	without $\eta(\alpha,\gamma)$	without $^{7}\text{Be}(\alpha,\gamma)$	without $^{9}\text{Be}(\tau,n)$	without $^{8}\text{Li}(\alpha,n)$	rate of $^{8}\text{Li}(\alpha,n)\star 10$	$^{7}\text{Li}(\alpha,\gamma)\star 2$ $^{7}\text{Be}(\alpha,\gamma)\star 2$
0.67	1.75 10^{-16}	1.89 10^{-19}	1.75 10^{-16}	1.75 10^{-16}	1. 75 10^{-16}	1.76 10^{-16}	3.49 10^{-16}
2.11	2.47 10^{-17}	1.78 10^{-17}	6.86 10^{-18}	2.47 10^{-17}	2.47 10^{-17}	2.47 10^{-17}	4.94 10^{-17}
6.67	7.43 10^{-15}	7.43 10^{-15}	4.12 10^{-19}	7.43 10^{-15}	7.42 10^{-15}	7.42 10^{-15}	1.48 10^{-14}
21.1	4.92 10^{-14}	4.92 10^{-14}	3.55 10^{-21}	4.92 10^{-14}	4.92 10^{-14}	4.93 10^{-14}	9.86 10^{-14}
66.7	1.37 10^{-13}	1.37 10^{-13}	2.15 10^{-22}	1.37 10^{-13}	1.37 10^{-13}		

These results are in agreement with the recent ones of the Chicago group (Schramm, this volume). They complete the most recent studies on this subject which had not included the nucleosynthesis of these isotopes (Walker et al. 1991). We can note a difference with the results of Kajino and Boyd (1990) concerning ^{10}B and ^{9}Be, due to our use of the recently measured reaction rate of ^{7}Li(T, n)^{9}Be (see Sec 3.1).

These abundances are much lower than the Be and B abundances recently measured in old halo stars [log (Be/H)$\simeq$-13, log(B/H)$\simeq$-12.1, for [Fe/H]$\simeq$-2.7, according to Gilmore et al. 1991, Ryan et al. 1992, Duncan et al. 1992]. NDM models could be an alternative, since they synthesize significant amounts of Be and B under certain conditions (Kajino and Boyd 1990). But this scenario meets with difficulties, because of the large obtained abundances of ^{7}Li and even ^{4}He (Kurki–Suonio et al. 1990).

More recent models reduce somewhat the inhomogeneities during primordial nucleosynthesis (due to neutron back diffusion, late time mixing), even if they were large just after the phase transition. Consequently, the results of these NDM models should be quite similar to the standard model results, the range of η being extended by a factor of two higher than in the standard model.

Finally, Be and B abundances in old stars cannot currently help to decide if some inhomogeneous primordial nucleosynthesis took place or not; observations of a plateau at very low metallicities ([Fe/H]$\simeq$-4) would be a clear indication for that. Recent work on the evolution of these elements in the galaxy (Vangioni–Flam et al. 1990, Prantzos et al. 1992, Ryan et al. 1992, Feltzing and Gustafsson 1992) have revealed that inhomogeneous models are not mandatory to explain the synthesis of Be and B. Indeed, various hypotheses on the behaviour of cosmic rays in the early galaxy lead to a reasonable fit of the most recent observations. In this context, it is worth mentionning that in the standard model Be and B are produced in similar amounts, in the relevant range of η (Fig. 1). In inhomogeneous models Be is more generously produced than B; this is opposite to the observed B/Be ratio ($7 \leq$B/Be$\leq$14) in halo stars, which is naturally explained in terms of cosmic ray spallation.

ACKNOWLEDGEMENTS
This work is supported in part by the CNRS "Programme International de Collaboration Scientifique" (PICS n°114).

REFERENCES
Alcock, C.R., Fuller, G.M. and Mathews, G,J, 1987, ApJ. **300**, 439
ALEPH Collaboration : 1989, Phys. Lett. B., **231**, 519
Angulo–Perez, C., 1992, private communication
Barhoumi, S., Bogaert, G., Coc, A., Aguer, P. Kiener, J., Lefebvre, A., Thibaud,

J.P., Baumann, F., Friedeben, H., Rolfs, C. and Delbourgo–Salvador, P., 1991, Nucl. Phys., **A 535**, p. 107

Boyd, R.N. and Kajino, T., 1989, Ap.J., **336**, L55

Brune, C.R., Kavanagh, R.W., Kellog, S.E. and Wang, T.R., 1991, Phys. Rev., **43**, 875

Caughlan, G. and Fowler, W.A., 1988, Atom. Dat., Nucl. Dat. Tab. **40**, 291

Delbourgo–Salvador, P., Vangioni–Flam, E., Malinie, G. and Audouze, J., 1984, in : *The Big Bang and Georges Lemaitre* ed. Berger A. (Reidel P .C.), p. 113

DELPHI Collaboration : 1989, Phys. Lett. B., **231**,539

Duncan, D., Lambert, D. and Lemke, D., 1992, Ap. J., submitted

Feltzing, S., Gustafsson, B., 1992, Ap. J., submitted

Gilmore, G., Edvardsson, B. and Nissen, P.E., 1991, Ap. J., **378**, 216

Hardie, G., Filippone, B.W., Elwyn, A.J., Wiescher, M., Segel, R.E., 1984, Phys. Rev., C **29**,1199

Hikasa, K. et al. 1992, Phys. Rev. **D45**, 51

Kajino, T. and Boyd, R.N., 1990, Ap.J., **359**,267

Kurki–Suonio, H., Matzner, R.A., Olive, K.A., Schramm, D.N., 1990, Ap. J., **353**, 406

Malaney, R.A. and Fowler, W.A., 1988, Ap.J., **333**, 14

Mampe, W., Ageron, P., Bates, C., Pendlebury, J.M. and Steyerl, A., 1989, Phys. Rev. Lett., **63**, 593

Meneguzzi, M., Audouze, J., Reeves, H., 1971, Astron. Astrophys., **15**, 337

Pagel, B.E.J., Simonson, E.A., Terlevich, R.J., Edmunds, M.G., 1992, M.N.R.A.S., **255**, 325

Prantzos, N., Cassé, M., 1992, these proceedings

Prantzos, N., Cassé, M., Vangioni–Flam, E., Ap. J., 1993, in press

Reeves, H., Meyer, J.P., 1978, Ap. J., **266**, 613

Ryan, S.G., Bessel, M.S., Sutherland, R.S. and Norris, J.E., 1990, Ap. J., **348**,L57

Ryan, S.G., Norris,J.E., Bessel, M.S. and Deliyannis, C.P., 1992, Ap. J., **388**,184

Siesk, A.J., Tombrello, T.A., 1973, Nucl. Phys., **210**.341

Vangioni–Flam, E., Cassé, M., Audouze, J., Oberto, Y., 1990, Ap. J., **364**,568

Walker, T., Steigman, G., Schramm, D.N., Olive, K., Kang, H.S., 1991, Ap. J., **376**, 51

Youn, M., Chung, H.T., Kim, J.C., Bhang, H.C. and Chung, K.H., 1991, Nucl. Phys., **A 533**, 321

B/Be Ratio in Inhomogeneous Universe

N. TERASAWA

The Institute of Physical and Chemical Research(RIKEN),
Hirosawa 2-1, Wako, Saitama 351-01, Japan

ABSTRACT

The B/Be ratio observed in HD140283 by the Hubble Telescope is apparently about 10 and smaller than the predicted value by conventional cosmic-ray spallation calculations. Though the B/Be ratio in the inhomogeneous universe can be as small as 1 in limited parameter ranges, it is constrained to be as large as the standard homogeneous value if we demand that the abundances of light elements such as D, ^{3}He, ^{4}He, and ^{7}Liare consistent with observations. The crucial point is, furthermore, the predicted abundances of B and Be are orders of magnitude smaller than the observed values. Hence the observed B/Be ratio is suggested to be purely a problem of galactic chemical evolution.

1. INTRODUCTION

Recently boron has been detected in HD140283 by the Hubble Telescope (Lambert 1991). The observed B/Be ratio is about 10 though conventional cosmic-ray spallation calculation predict a B/Be ratio of 10-30. Since the B/Be ratio in inhomogeneous universe is expected to be as small as 1, this fact may implicate the contribution of beryllium synthesized by the inhomogeneous primordial nucleosynthesis. However, Terasawa and Sato 1990, have shown that the ^{9}Be abundance can not be so large in the parameter ranges where the predicted abundances of light elements such as D, ^{3}He, ^{4}He, and ^{7}Li are consistent with observations. This is because the predicted abundances of these elements do not decrease so dramatically by inhomogeneity if the neutron diffusion during the nucleosynthesis is correctly taken into account.

In the present work, we show the abundances of ^{9}Be and ^{11}B in the inhomogeneous universe calculated with extended network including elements up to ^{14}N in the multi-zone scheme. Some neutron-rich nuclei are added to the network and some related reaction rates are updated according to the recent experimental results.

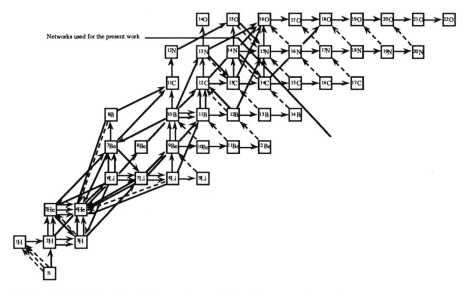

Fig.1 – The Networks of elements used for the present work.

2. ASSUMPTIONS

Spherical symmetry is assumed for the space which is divide into 100 shells and the stretching function method is adopted. The diffusion equation for neutrons,

$$\frac{dn}{dt} = \nabla(D_n \nabla n) \tag{1}$$

coupled with nucleosynthesis is solved in a completely implicit manner. The simple version of a flux-limiter is applied for the diffusion coefficient to avoid the overestimate of the neutron diffusion.

For the initial ($T = 10^{11}$K) density profile, we assume a top-hat profile; the density of inner shells is R times higher than that of surrounding low-density shells. The parameter f_v refers to the volume fraction of the initial high-density shells in the whole region. Another parameter $d(T = 1\mathrm{MeV})$ refers to the radius of the site in problem normalized by the size at T=1MeV.

Some neutron-rich elements such as ^{9}Li are added to the network which includes the elements up to ^{14}N. Reaction rates are updated according to Caughlan and Fowler 1988 and those relating to ^{9}Be are added and updated by the table given in Malaney et al. 1991. The reaction rate for ^{8}Li$(\alpha, n)^{11}$B is, furthermore, updated by Boyd et al.

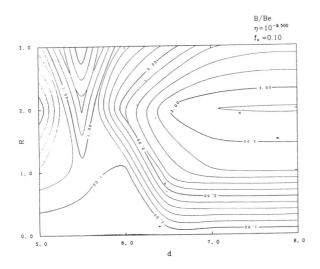

Fig.2 – The B/Be ratio displayed on a $d - R$ plane. Contours are labeled with $\log(\text{B/Be})$.

1992, which is confirmed to cause negligible differences to the abundances of related elements. The lifetime of neutrons is taken to be 887.6 sec (Mampe et al. 1989) and the number of neutrino species is 3.

As it is shown in the figure, the B/Be ratio turn out to be smaller than 1 in the limited parameter range ($d(\text{T} = 1\text{MeV}) \sim 10^{5.5}\text{cm}, R \sim 1000, f_v = 0.11$). However, the abundances of (D+ ^{3}He) and ^{7}Li are too large to be consistent with observational constraints in this range. Furthermore, the predicted abundances of B and Be themselves are orders of magnitude smaller than observed values.

3. CONCLUSIONS

The B/Be ratio can be as small as 1 in the highly inhomogeneous universe if the scale of the density irregularity is in appropriate range. The B/Be ratio is, however, constrained to be as small as the value predicted in the standard homogeneous universe if the consistency of the light element abundances, such as ^{4}He, D, ^{3}He, and ^{7}Li is demanded. The predicted values of ^{9}Be/H and ^{11}B/H are orders of magnitude smaller than the observed ones. Hence the problem of the B/Be ratio observed in HD140283 can not be resolved by inhomogeneous nucleosynthesis. The updated reaction rates for ^{8}Li$(\alpha, n)^{11}$B causes negligible differences to the abundances of related elements.

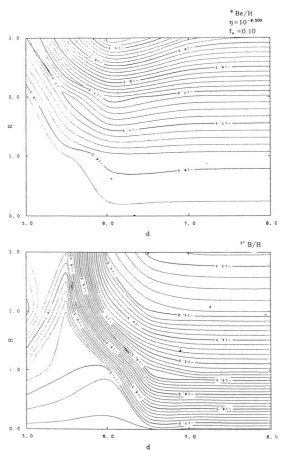

Fig.3 – ^{9}Be abundance (top) and ^{11}B abundance (bottom). Contours are labeled with log(A/H).

REFERENCES

Boyd, R. N. et al. 1992, Phys. Rev. Lett. **68**, 1283.

Lambert, D. L. 1991, at *21st Int. Astr. Union Gen. Assembly*, Buenos Aires.

Malaney, R. A. et al. 1991, ApJ, **372**, 1.

Mampe, W. et al.1989, *Phys. Rev. Lett.*, **63**, 593.

Terasawa, N. and Sato, K., 1990, ApJ, **362**,L47.

BERYLLIUM IN METAL POOR STARS

GERARD GILMORE

Institute of Astronomy, Madingley Rd., Cambridge CB3 0HA, England

INTRODUCTION

The relationship between the abundance of beryllium and that of heavier elements, especially CNO and Fe, is a direct probe of the history of cosmic rays, magnetic fields, supernovae and the mixing of the interstellar medium in the Galaxy. For the most metal poor stars in the Galaxy, this relationship is not only uniquely powerful as a probe of the physical conditions in the proto Galaxy, but also provides a test of the predictions of Big Bang nucleosynthesis models when extrapolated back to zero heavy element abundance. Recent echelle spectrograph observations have extended the range of stars for which beryllium abundances are available to metallicities approaching [Fe/H]= -2.8 (Gilmore etal 1991,1992; Ryan etal, 1992), a factor of thirty lower than was available a year ago. In this paper we consider the reliability of the present data, and the general conclusions concerning the evolution of the Galaxy and the primordial abundances of the light elements which these results require.

THE PRECISION OF AVAILABLE BERYLLIUM ABUNDANCES

The precision with which the abundances of beryllium and heavier elements have been determined is difficult to define. Factors entering the calculation include not only the obvious observational parameters of spectral resolution and signal-to-noise ratio, but also the stellar modelling and the atomic and molecular physics uncertainties. For present purposes, it is important to distinguish carefully between *random* and *systematic* errors.

Beryllium is visible as a weak doublet with much stronger lines of VII, TiII, and OH within ~ 0.2Å. Hence, there is a threshold in resolution below which the beryllium lines are visible only as wings on strong and complex features. The consequence of this is most clearly seen by comparing figure 3 of Rebolo etal (1988) with figure 1 of Gilmore etal (1991). The observations of Rebolo etal were obtained at a resolution of ~ 0.28Å, so that neither of the beryllium lines is resolved from other stronger features. The derived beryllium abundances are necessarily of lower precision than the more recent observations, which have a resolution $\lesssim 0.1$Å. Since the Rebolo etal data were the only abundance measurements of beryllium for stars more metal poor than [Fe/H]$\lesssim$ −0.6 prior to the recent work, the most conservative attitude to available data is to use only the most recent abundance data for stars more metal poor than the old disk, or [Fe/H]$\lesssim$ −0.6. Conversely, it will also clearly be of interest to study the abundance of beryllium in the more metal rich stars of the thin disk with high spectral resolution.

When the data for metal poor stars are restricted in this way, we have available data for a total of 8 stars with [Fe/H] $\lesssim$ −0.6 from Gilmore etal (1992), and for 3 stars from Ryan etal (1992). There is one star in common to these studies, HD140283, in addition to consistent upper limits for an additional star, HD84937. For HD140283 Gilmore etal derive [Be/H]= −12.97, while Ryan etal derive [Be/H]= −13.25. Of this difference of 0.28dex, ∼ 0.2dex is caused by a systematic difference in the adopted stellar gravity. The stars observed twice by Gilmore etal have derived beryllium abundances which agree to ∼ 0.18dex between observations. The *random* errors in the beryllium abundances of very metal poor stars are clearly surprisingly small, based on this simple comparison, with internal random errors in the beryllium abundance being $\lesssim$ 0.15dex.

Iron abundances for the same stars observed by Gilmore etal were derived from equivalent widths measured in visual spectra. We included only spectral lines for which the oscillator strength is known to an accuracy better than ±0.10 dex, and which are weaker than about 80 mÅ in the spectra of the metal poor stars studied here, and assumed the LTE approximation.

The iron abundance for the 8 halo stars derived from FeII lines is larger than the abundance derived from FeI lines, the average value of [FeII/FeI] being 0.15 dex. This may suggest that we have systematically overestimated the log g values by about 0.4 dex, or it could be a non-LTE effect. It is, however, interesting that the rms deviation of [FeII/FeI] is 0.08 dex only, suggesting that the *differential* log g values have been determined to an accuracy better than ± 0.20 dex. Similarly, the absolute [Fe/H] values may be underestimated by about 0.15 dex, but the differential [Fe/H] values have errors less than 0.08 dex. This is important, because only the *differential* values of log g and [Fe/H] are relevant for the value of the slope of the log$(N_{\mathrm{Be}}/N_{\mathrm{H}})$ − [Fe/H] relation for halo stars.

A similar analysis of Mg, Ca, Ti and Cr for our programme stars shows that the rms deviation in [X/Fe] for X = Mg, Ca, Ti, Cr, is as small as 0.06 dex, confirming that very accurate differential metal abundances have indeed been obtained. Similarly, those stars observed twice have derived oxygen abundances which are consistent to

$\lesssim 0.05$dex from the two spectra. Conservatively, we adopt an internal error of 0.1dex in the oxygen abundances.

Note that although there is no reason to suspect their existence, the possibility of *systematic* zero point offsets in the beryllium abundance scale cannot yet be ruled out. This makes it difficult as yet to determine the reliability of an extrapolation of the relationship observed between the abundances of beryllium and heavier elements, especially oxygen, back to zero abundance, and hence to deduce the primordial beryllium abundance. This extrapolation is of course further uncertain due to our poor understanding of the physics of cosmic ray generation in the very early Galaxy. For present purposes we restrict discussion to the implications of present data for Galactic cosmic ray production models of beryllium and lithium.

The level of internal consistency apparent in the data allows a quite precise determination of the relationship between the abundances of beryllium and oxygen in halo stars. Additionally, the statistical fits to this relationship provide strong confirmation of the internal consistency of the data derived above. The beryllium abundances for the ten different metal poor stars with good abundances published by Gilmore etal (1992) and by Ryan etal (1992) are shown in Figure 1. Also shown is a linear least squares fit to the halo stars. Within a formal uncertainty of $\sim$ 25%, the relationship between the logarithmic beryllium abundance and the logarithmic oxygen abundance has slope unity, so that $N_{Be} \propto N_O^{1\pm0.25}$. This result is derived by both non-linear weighted least squares and by maximum likelihood techniques, and in both cases the statistical significance of the fit supports the adopted internal error estimates.

IMPLICATIONS OF THE OBSERVED BERYLLIUM *VS* OXYGEN RELATIONSHIP

The evolution in the early Galaxy of the light elements created by spallation has been well reviewed recently by Reeves (1991). The essential physics is described by

$$\frac{d(N_X/N_H)}{dt} \propto \zeta(t) \times \int \sigma(E) \times \Phi(E,t)\,dE - \text{loss terms,} \qquad (1)$$

where N_X is the abundance of the element being created through spallation of CNO nuclei and N_H the hydrogen abundance, $\zeta(t)$ is the relative abundance of the target heavy-elements which are spalled, $\sigma(E)$ is the reaction cross section which is dependent on the energy E of the cosmic rays, and $\Phi(E,t)$ is the time-dependent cosmic ray flux spectrum. $\sigma(E)$ is measured in a laboratory and for present purposes can be assumed known. For the spallogenic production of lithium the low energy production is dominated by ^{4}He + ^{4}He reactions. Since ^{4}He is an abundant primordial element, $\zeta(t)$ for Li production is effectively constant, and the Li production rate depends only on $\Phi(E,t)$.

The heavy elements of interest as targets, particularly oxygen, are created in supernovae from high-mass stars. Thus $\zeta(t)$ is a measure of the integrated number of supernovae in the relevant part of the proto-Galaxy. Cosmic rays require kinetic energy for their acceleration, and this energy input is also probably dominated by winds and supernovae from high-mass stars. Thus, the cosmic-ray flux prior to any losses is also dependant on the total number of supernovae in the relevant part of the proto-Galaxy, and may also be parameterised by the heavy element abundance, $\zeta(t)$. Thus one expects, in the simplest case, a *quadratic* dependence of the beryllium abundance on the oxygen abundance, in contradiction with the results of figure 1.

The two limiting cases of spallation production are easily seen from equation 1 – the production may be limited by either the target abundance or the cosmic ray abundance, or by both. Observations require one of these limiting cases to be appropriate for the halo phase of galactic formation. Preliminary models of cosmic ray limited production have been presented by Feltzing & Gustafsson (1992), though no detailed models of target limited spallation which reproduce the data are yet available.

The production of lithium is potentially of interest. The targets for production of lithium are ^{4}He nuclei, α particles, which are always abundant, since the primordial helium abundance is so high. Allowing for this dependence, Steigman & Walker (1992) show that the lithium production will follow $\log(N_{\mathrm{Li}}/N_{\mathrm{Be}}) = -[\mathrm{Fe/H}]$, if the cosmic ray energy spectrum at early times was like that observed now. With that energy spectrum, spallation provides a production ratio $(N_{\mathrm{Li}}/N_{\mathrm{Be}}) \approx 200$ at $[\mathrm{Fe/H}]=-2.3$. The normalisation of this ratio is sensitive to the cosmic ray energy spectrum, and will be lower by a multiplicative factor of a few for an energy spectrum which also reproduces current measurements of the Be/B ratio.

An intriguing feature of the linear dependance of the spallation production of lithium on (logarithmic) heavy element abundance, is that the beryllium abundance itself has just the same linear dependance, but with opposite slope. These compensating dependencies on the metallicity of the ISM cancel out. Thus the predicted spallation production of lithium consequent upon the spallation production of beryllium is independent of [Fe/H], at an (uncertain) level of approximately $\log(N_{\mathrm{Li}}) = -10.4$. This spallation production, being independent of [Fe/H], would not be visible in the 'Population II plateau', but acts as a zero point offset to the primordial lithium abundance. The resulting primordial abundance, in the absence of later depletion, is consequently uncertain, but is near $\log(N_{\mathrm{Li}}) = -10.1$.

Figure 1: Beryllium abundance *vs* logarithmic oxygen abundance relative to the solar value for the metal poor stars observed by Gilmore etal and by Ryan etal. The line is a least squares fit to the data for the stars with [O/H] < −1.

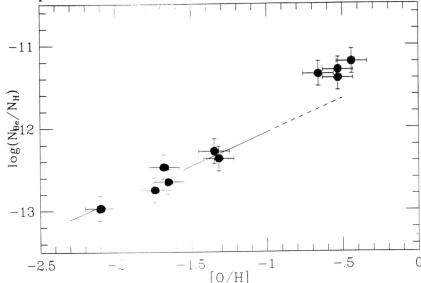

Figure 2: Observed abundances of Li, Be and B in metal poor stars. The solid dots are lithium mean values for all stars in the metallicity intervals [Fe/H] ≤ −2.5, −2.5 < [Fe/H] ≤ −2.0, −2.0 < [Fe/H] ≤ −1.5 and −1.5 < [Fe/H] ≤ −1.3, with mean errors indicated. The dashed line is the spallative production of lithium predicted to occur with the spallative production of Be and B if the early Galactic cosmic ray energy spectrum were like that now.

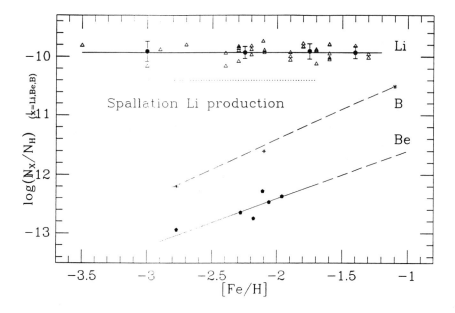

Nearly one-half of the lithium produced by early spallation is however ^{6}Li. This element is extremely fragile, and could have been completely destroyed even if initially very abundant in the stars of interest here, if rotational mixing is the predominant depletion mechanism. If however diffusive settling is the dominant destruction mechanism, then the ^{6}Li will be depleted in roughly the same proportion as is the ^{7}Li, and one might still see some cosmic ray production. Particularly interesting in this regard is the measurement of a ratio of ^{6}Li/^{7}Li of ~ 0.05 in the halo star HD84937, one of the stars for which Gilmore etal have measured beryllium. Since standard Big Bang models predict insignificant ^{6}Li production, while spallation with the present day energy spectrum predicts roughly equal amounts of ^{6}Li and ^{7}Li, this observation both precludes efficient deep mixing, and requires at least 10% of the Li in this Pop II star to be made by spallation production. The simplest prediction from the observed beryllium abundance is then that some $\sim 30\%$ of the Population II lithium 'plateau' abundance is created by cosmic ray spallation. This is illustrated in Figure 2.

CONCLUSIONS
Recent high resolution observations of very metal poor stars allow the derivation of abundances of beryllium, iron and oxygen with internal random errors of $\lesssim 0.15$dex, $\lesssim 0.10$dex and $\lesssim 0.10$dex respectively. The resulting relationship between the beryllium and oxygen abundances in stars with [Fe/H]$\lesssim -1.5$ is $N_{Be} \propto N_O^{1.\pm 0.25}$. This result is inconsistent with published models of cosmic ray spallation production, and requires that the cosmic ray production be either target limited or cosmic ray limited. Spallative lithium production with this beryllium production is predicted to be independent of [Fe/H]. This lithium can provide a significant contribution to the Pop II Li plateau abundance, though in an abundance which is sensitively dependant on the cosmic ray energy spectrum in the early Galaxy.

References
Duncan, D., Lambert, D., & Lemke, M. *Astrophys. J.* submitted (1992)
Feltzing, S., & Gustafsson, B., *Astrophys. J.* submitted (1992)
Gilmore, G., Edvardsson, B., Nissen, P.E. *Astrophys. J.* **378**, 17 (1991)
Gilmore, G., Gustafsson, B., Edvardsson, B., Nissen, P.E. *Nature* **357**, 379 (1992)
Rebolo, R., Molaro, P., Abia, C., Beckman, J.E., *Astron. Astrophys.* **193** 193 (1988)
Reeves, H. *Physics Reports* **201**, 336 (1991)
Ryan, S.G., Norris, J.E., Bessell, M.S. Deliyannis, C. *Astrophys. J.*, **388** 184 (1992)
Smith, V., Lambert, D., & Nissen, P. 1992 in preparation
Steigman, G., Walker, T.P. *Astrophys. J.* **385**, L13 (1992)

Beryllium abundances in unevolved metal deficient stars

R. REBOLO[1], R. J. GARCIA LOPEZ[1], E. L. MARTIN[1],
J. E. BECKMAN[1], C. D. MCKEITH[2], J. K. WEBB[3], AND
Y. V. PAVLENKO[4]

1. Instituto de Astrofísica de Canarias

2. Dept. of Pure and Applied Physics, The Queen's University of Belfast

3. University of New South Wales

4. The Principal Observatory of Ukraine

1 ABSTRACT

We present beryllium abundances for three unevolved metal poor stars with [Fe/H] in the range -0.4 to -1.6. A spectral synthesis analysis has been conducted to derive the abundances, and the main results can be summarized as follows: (i) We have detected the Be II doublet in the three stars. Their Be abundances suggest a slight increase of the galactic abundance through the metallicity range studied. (ii) These abundances are considerably larger than those of stars in the literature with [Fe/H]≤ -1.7, indicating that a strong Be enrichment process took place in the very early evolution of the Galaxy.

2 INTRODUCTION

Measurement of beryllium abundances in metal poor stars is needed to explain the origin of this element. More than twenty years ago Menneguzzi, Audouze and Reeves (1971) suggested that the beryllium found in the Galaxy could be produced in spallation reactions between Cosmic Rays and C,N,O nuclei in the Interstellar Medium. Observational studies aimed to define the galactic evolution of beryllium began in the eighties (Molaro and Beckman 1984; Rebolo et al. 1988). In this last work beryllium was detected for first time in stars with [Fe/H]~ -1, with an abundance considerably lower than that in the Sun. This implies that there has indeed been galactic enrichment, which can be explained using the spallation mechanism (e.g.

Vangioni-Flam et al. 1990). More recently, certain models of inhomogeneous Big Bang nucleosynthesis have suggested a primordial production of beryllium with an abundance in a range detectable with present-day techniques (e.g. Malaney and Fowler 1989). In contrast, the calculations by Terasawa and Sato (1990) indicate that such primordial production is unlikely. In any case only additional observational efforts could resolve this controversy. A few new beryllium measurements have been obtained very recently (Gilmore et al. 1991, Ryan et al. 1992). In this paper we present beryllium abundances for three additional stars with metallicities in the range $-1.6 < [\text{Fe/H}] < -0.4$.

3 OBSERVATIONS. STELLAR PARAMETERS. ANALYSIS

The observations of two of the stars (HD132475 and HD38510) were carried out during July of 1990 and February of 1991. We used the UCL echelle spectrograph with an IPCS detector at the Anglo-Australian Telescope. This configuration provided a dispersion of 0.024 Å/pixel at λ 3130 Å. The third star (HD 165908) was observed in August of 1991 with the IACUB (Instituto de Astrofísica de Canarias-University of Belfast) echelle spectrograph, attached to the Cassegrain focus of the 2.5 m Nordic Optical Telescope at the Observatorio del Roque de los Muchachos (La Palma). The dispersion obtained in this case was 0.017 Å/pixel with a Thompson CCD camera. These observations belong to a large programme designed to measure light element abundances to test nucleosynthesis and mixing processes. We carried out a standard data reduction using the IRAF package.

The stars HD132475 and HD165908 were selected from a review of lithium studies in metal poor stars (Deliyannis, Demarque and Kawaler 1990), and HD38510 was taken from the work of Norris, Bessell and Pickles (1985). Their metallicities are [Fe/H]=-1.6, -0.4 and -0.9, respectively. The effective temperatures are in the range $5500 - 6000$ K, and their surface gravities are close to $\log g \sim 4.0$.

To derive the beryllium abundance for each star we carried out a spectral synthesis analysis in order to fit the profile of the weaker line of the Be II doublet, which is clearly resolved in our spectra. The synthetic spectra were computed using the code ABEL6 (see for details Pavlenko 1992). This program allows to compute LTE atomic and molecular synthetic profiles for a given model atmosphere. The model atmospheres employed are by Kurucz (1992, private communication). Our error bar estimate for a beryllium determination is 0.3 dex, which includes errors in the synthesis fit as well as uncertainties in the adopted stellar parameters. Figure 1 shows the fit obtained in the case of HD132475.

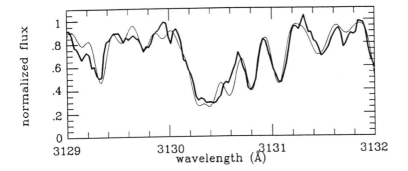

Figure 1: Fit of the synthetic spectrum (thin line) to the observed spectrum (thick line) in the region of the Be II λ 3131.06 Å line, for the star HD132475 ([Fe/H]=−1.6).

4 BERYLLIUM ABUNDANCES

We detected the Be II doublet in the 3 observed stars. The Be abundances obtained

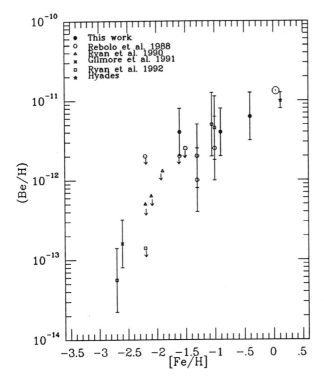

Figure 2: Be abundance versus metallicity for our stars and stars taken from the literature. Note the strong increase in abundance in the range −3 < [Fe/H] < −1.5.

are in the range log N(Be)=0.6–0.9 (where abundances are in the customary scale log N(H)=12) which is considerably larger than the detection found in HD140283 ([Fe/H]=-2.7; Gilmore et al. 1991) and the upper limits established in other very metal poor stars (Rebolo et al. 1988; Ryan et al. 1992). In Figure 2 we plot Be abundances versus metallicity for our stars and others available in the literature. Our aim is to delineate the galactic evolution of beryllium. The Sun and the Hyades are included as representative of Population I stars. It is apparent from the figure that a strong Be enrichment process must have taken place during the very early evolution of the Galaxy, while metallicity was increasing from −3 to −1.5. This behaviour together with the well known lithium plateau in the same metallicity range (Rebolo, Molaro and Beckman 1988) puts strong constraints on the early history of ISM spallation reactions.

6 REFERENCES

Deliyannis, C. P., Demarque, P., and Kawaler, S. D. 1990, *Ap. J. Suppl.*, **73**, 21.

Gilmore, G., Edvardsson, B., and Nissen, P. E. 1991, *Ap. J.*, **378**, 17.

Hobbs, L. M., and Duncan, D. K. 1987, *Ap. J.*, **317**, 796.

Malaney, R. A., and Fowler, W. A. 1989, *Ap. J.*, **345**, L5.

Menneguzzi, M., Audouze, J., and Reeves, H. 1971, *Astr. Ap.*, **50**, 337

Molaro P., Beckman J.E. 1984, *Astr. Ap.*, **139**, 394

Pavlenko, Y. V. 1992, *Soviet Ast.*, **35**, 212.

Rebolo, R., Molaro, P., Abia, C., and Beckman, J. E. 1988, *Astr. Ap.*, **193**, 193.

Rebolo, R., Molaro, P., and Beckman, J. E. 1988, *Astr. Ap.*, **192**, 192.

Ryan, S., Bessell, M., Sutherland, R., and Norris, J. 1990, *Ap. J.*, **348**, L57.

Ryan, S. G., Norris, J. E., Bessell, M. S., and Deliyannis, C. P. 1992, *Ap. J.*, **388**, 184.

Terasawa, N., and Sato, K. 1990, *Ap. J.*, **362**, L47.

Vangioni-Flam, E., Casse, M., Audouze, J., and Oberto, Y. 1990, *Ap. J.*, **364**, 568.

Beryllium abundance in HD 140283

P.MOLARO[1], F.CASTELLI[2], L. PASQUINI[3]

[1] Astronomical Observatory of Trieste
[2] CNR-Gruppo Nazionale Astronomia, Unità di Ricerca di Trieste
[3] European Southern Observatory

1 ABSTRACT

High resolution (FWHM $\approx$ 8 km s^{-1}) CASPEC blue spectra of the V=7.2 mag halo star HD 140283 are analyzed in the region of the Be II $\lambda\lambda$ 3130Å lines. The fainter of the Be II doublet, which is less blended and normally used for beryllium abundances, is undetected in our spectrum. By means of a synthetic spectrum analysis an upper limit of log (Be/H) = -12.90 is placed for the beryllium abundance. When combined with the boron determination in HD 140283 by Duncan et al. (1992) the B/Be ratio becomes greater than 10 and can be entirely explained by cosmic rays spallation.

2 INTRODUCTION

Beryllium has a special place in the nucleosynthesis processes. It is the lightest nuclide not synthesized in the standard Big Bang and it is not produced in stars where, on the contrary, is destroyed. The currently accepted explanation for its existence is via spallation of heavy nuclei by energetic cosmic rays in the interstellar medium. Recently it has been pointed out that alternative, non-standard Big Bang models produce appreciable quantities of Be , i.e. log Be/H$\approx$=-13, nearly three orders of magnitude over the standard theory and about two orders of magnitude below the present solar abundance (Malaney and Fowler, 1989; Kajino and Boyd, 1990). These models fail to produce boron and the B/Be ratio has been proposed as a diagnostic for the presence of inhomogeneities at the epoch of primordial nucleosynthesis. Cosmic rays cannot account for B/Be ratios smaller than $\approx$ 10 and lower ratios require an additional source for Be.

HD 140283 is the only halo star for which abundances for both B and Be are presently available. Ryan et al. (1992) (RNBD) and Gilmore et al. (1991, 1992) (GEN, GGEN) claimed the detection of the faint Be II lines, and boron has been measured by Duncan et al. (1992) (DLL). These determinations give a B/Be ratio of 10^{+5}_{-4}, which is at the minimum value allowed by cosmic ray spallation synthesis. The new observations presented here suggest that the beryllium abundance in HD 140283 might be substantially lower than previously estimated.

3 OBSERVATIONS

The observations were carried out La Silla (Chile) in March 1992 using the 3.6m ESO telescope and CASPEC mounted with the long camera and the 31.6 lines/mm echelle. The detector was a blue sensitive 22 μm square pixel GEC CCD (ESO N. 14). The entrance slit was of 1 arcsec giving a nominal resolving power of R$\approx$ 41000. The actual resolution, as measured from the profile of comparison lamp lines and from unblended lines in the stellar spectra is R $\approx$ 36000. For each exposure the Be II lines are simultaneously recorded in two adjacent orders. The average of two orders for a 2 hour exposure is shown in Figure 1. The S/N ratio is $\approx$ 60, which is about a factor of two better than that achieved by RNBD, GEN and GGEN.

4 RESULTS

The atmospheric parameters for HD 140283 are rather uncertain and an exhaustive discussion is given by RNBD and DLL. We adopted a Kurucz new model computed for T_{eff}=5750 K, log g = 3.5, [M/H]=-2.5 and ξ = 1.5 km s^{-1}, because the energy distribution yielded by it well fits the observed one (Oke and Gunn 1983) dereddened for E(B-V)=0.02 (Ryan et al., 1991). To match the observed spectrum we modified the C, O, and Fe abundances. We derived [Fe/H]=-2.73 from the atomic iron lines, and [C/H]=-2.3 and [O/H]=-1.9 from the OH and CH molecular lines present in the 3110-3150 Å region. Solar abundances are those of Anders and Grevesse (1989).

In Fig 1 the synthetic spectra computed with log(Be/H)=-13.35 (top panel) and log(Be/H)=-12.90 (bottom panel) are compared with the observations. The highest beryllium abundance is about the value derived by RBND, GEN and GGEN, when the differences in the various atmospheric parameters are considered. On the top of the figure the label lists the lines which contribute to the synthetic spectrum according to the notation introduced by Kurucz and Avrett (1981). In particular, Be II is indicated by 4.01, CH by 106 and OH by 108. Inspection reveals that the stronger of the Be II doublet at $\lambda\lambda$ 3130.420 Å is rather well reproduced by the highest abundance, but the fainter line at $\lambda\lambda$ 3131.065 Å is not seen. The absence of the fainter line is also notable in the individual spectra used in the average, which is a guarantee against possible cosmic rays events in correspondence of the line. Fig. 1 shows that the stronger line is blended with OH at $\lambda\lambda$ 3130.433 Å and CH at $\lambda\lambda$ 3130.370 Å and $\lambda\lambda$ 3130.290 Å. The fainter is free from contaminants and is more reliable for the abundance determination.

From this analysis a safe upper limit to the beryllium abundance in HD 140283 can be placed at log(Be/H)=-12.90. With this upper limit the B/Be ratio becomes greater than 10 leaving little room for extra nucleosynthetic sources in addition to the cosmic rays spallation.

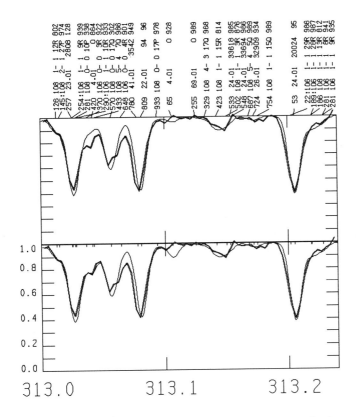

Figure 1: The Be λλ 3130Å region of HD 140283. The synthetic spectrum (thin line) is for log(Be/H)=-13.35 (top panel) and log(Be/H)=-12.90 (bottom panel).

4 REFERENCES

Anders, E. Grevesse, N. 1989 *Geochim. Cosmochim. Acta* **53**, 197.

Duncan D. K., Lambert D. L., Lemke M.: 1992 Ap. J. in press

Gilmore, G., Edvardsson, B., Nissen, P.E. 1991, ApJ **378**, 17.

Gilmore, G., Gustafsson, B., Edvardsson, B., Nissen, P.E. 1992, Nature **357**, 379.

Kajino T. and Boyd R.N. 1990 Ap.J. **359**, 267

Kurucz, R. L., Avrett E. H., 1981 SAO Sp. Rep. **362**

Malaney R. A. and Fowler W. A. 1989, Ap. J. 345, 15.

Oke, J.B. and Gunn, J.E. 1983, Ap. J. **266**, 713.

Ryan S.G.,Norris J.E., Bessell M.S. 1991 Astron. J. **102**, 303.

Ryan S.G., Norris J.E., Bessell M.S., Deliyannis C. P. 1992 Ap. J. **388** 184.

Cosmic Rays and LiBeB Isotopes in the Galaxy

N. PRANTZOS[1,2], M. CASSE[2], E. VANGIONI-FLAM[1]

1. Institut d' Astrophysique de Paris-CNRS
2. Centre d' Etudes de Saclay-CEA, FRANCE

1 INTRODUCTION

Cosmological nucleosynthesis in the framework of the standard big bang model has been very successful in explaining the origin of the lightest isotopes $H, ^2H, ^3He, ^4He$, and of a part of ^7Li (e.g. Schramm, this volume). The abundance of this latter isotope resulting from standard big bang nucleosynthesis (SBBN) calculations is compatible with Li observations in low metallicity stars, which show a "plateau" at $\log(\text{Li}/\text{H}) \sim -10$ (e.g. Spite and Spite, 1982 and this volume) for $[\text{Fe}/\text{H}] < -1$ (the usual notation $[\text{A}/\text{B}] = \log(\text{A}/\text{B}) - \log(\text{A}/\text{B})_\odot$ is adopted). The other light isotopes (^6Li, ^9Be, ^{10}B and ^{11}B) are thought to be produced by spallation reactions between protons, ^4He and CNO nuclei, during the propagation of galactic cosmic rays (GCR) in the interstellar medium (ISM), as suggested by Reeves, Fowler and Hoyle (1970) and Meneguzzi, Audouze and Reeves (1971); see also Reeves, this volume.

The recent observations of Be (Gilmore et al. 1991, 1992; Ryan et al. 1992) and B (Duncan et al. 1992) in low metallicity stars show that those light elements were present in the early galaxy in amounts larger than "conventional" GCR nucleosynthesis can account for (e.g. Vangioni-Flam et al. 1990). The absence of a Be (or B) "plateau" at low metallicity (Fig. 1), however, "cooled" the initial excitement about a possible cosmological origin for those elements. Still, the observed linearity between the abundances of BeB and Fe (the galactic "clock") is at variance with the expectations of the standard GCR scenario, which predicts a "secondary" behaviour for those light elements, i.e. a quadratic rather than a linear relationship (see Sec. 2). The observed B/Be ratio ($\sim$10 at all metallicities and 10$\pm$4 in the star HD140283) is, however, quite compatible with a GCR origin!

In the following we discuss briefly the problems encountered by the standard scenario of LiBeB production by GCR (Sec. 2), we present a new model for the confinement of CR in the early Galaxy (Sec. 3) that satisfies the observational constraints (developed in detail in Prantzos, Cassé and Vangioni-Flam 1992; hereafter PCF) and we discuss some recent relevant ideas concerning, in particular, the origin of GCR (Sec. 4).

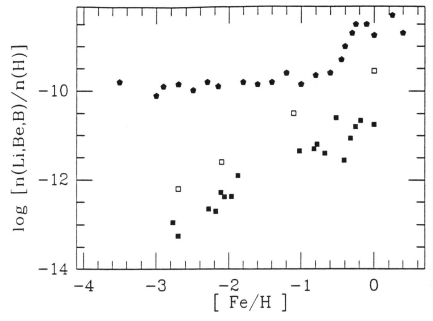

FIGURE 1: Li (pentagons), Be(black squares) and B(open squares) abundances as a function of metallicity. Only the upper envelope of the Li abundances (Spite 1991) is given. Solar system values are uncertain by a factor of ~1.5-2.

2 THE STANDARD MODEL OF LIBEB PRODUCTION BY GCR

LiBeB isotopes are produced by GCR protons and alphas impinging on the CNO nuclei of the ISM (A-component in the following), and by GCR CNO nuclei impinging on ISM protons and alphas (B-component in the following). The contribution of the latter component to the LiBeB production is evaluated today to ~20 % of the former one (Meneguzzi and Reeves 1975), and this must also have been the case in the past *if the composition of cosmic rays evolves like the one of the ISM.* This assumption is supported by current ideas on the origin of GCR (Cassé and Goret 1978; Meyer 1985), suggesting that the bulk of GCR particles originate mainly in the chromospheres of solar-type stars. The production rate of the light (L) nuclei is given by:

$$\frac{dY_L}{dt} = <\sigma F_{p\alpha}^{GCR}> Y_{CNO}^{ISM} + <\sigma F_{CNO}^{GCR}> Y_{p\alpha}^{ISM}\zeta + (<\sigma F_{\alpha}^{GCR}> Y_{\alpha}^{ISM}) \quad (1)$$

where Y are the abundances (by number) of the various species, σ and F are the relevant cross-sections and GCR fluxes respectively (folded with the proper energy spectrum), ζ a deceleration term to account for the LiBeB produced by the CNO in flight (B-component), and the last term (C-component, in parenthesis) describes the production of ^{6,7}Li nuclei by fusion reactions between GCR alpha particles and ISM ones; in the early Galaxy, when CNO nuclei were 10^2-10^3 times less abundant than today, this mechanism turns out to be of crucial importance, as already noticed by Steigman and Walker (1992; see also PCF).

Eq. (1) can be combined with the simple model of galactic chemical evolution, i.e. a closed box with instantaneous recycling approximation and star formation rate (SFR) proportional to the gas fraction σ_g: SFR=$\nu\sigma_g$). A well known result of that model is the relationship between σ_g and the metallicity Z ($Z \sim Y_{CNO}^{ISM}$ in Eq. 1):

$$Z \sim y\ ln(\frac{1}{\sigma_g}) \sim y\ \nu\ t \tag{2}$$

where y is the *yield* of a stellar generation (e.g. Tinsley 1980). The last term in Eq. (2) indicates a linear increase of Z with time and can be used to eliminate t from Eq. (1). Assuming that the GCR flux F^{GCR} does not vary by much (see below) and that B-comp ~ 0.2 A-comp *always*, one obtains from Eq. (1) the evolution of Be and B abundances:

$$Y_L \propto \frac{Z^2}{2\ y\ \nu} \tag{3}$$

i.e. the well known quadratic relationship between a secondary and a primary element. As emphasized in the Intrtoduction, that relationship is not satisfied by the recent observations of Be and B and leads to theoretical abundances $\sim$30 times lower than the observed ones at [Fe/H]$\sim$-3. Notice that this is not valid for ^{6,7}Li, since their production is dominated by the third term in Eq. (1), yielding a linear relationship between their abundances and time (or Z).

Trying to overcome this difficulty Ryan et al. (1992) proposed *outflow* during the halo phase as a possible solution. Indeed, since Hartwick (1976) it is known that the observed metallicity distribution of the halo stars can best be reproduced if substantial outflow is assumed. This is equivalent to a reduced *effective yield* : $y_{eff} \sim y/10$. Eq. (3) is then transformed to:

$$Y_L^{Out} \propto \frac{Z^2}{2\ y_{eff}\ \nu} \sim 10\ Y_L \tag{4}$$

i.e. much larger Be and B abundances in the halo can be obtained for the same metallicity, but the Be vs. Fe relationship is still a quadratic one. This is confirmed by detailed numerical models in PCF who showed, however, that Li is overproduced in that case, due to the role of $\alpha+\alpha$ reactions.

Notice that those results are obtained (analytically or numerically) with a GCR flux proportional to the rate of supernovae (SN), which are currently thought to be the accelerating agents of the GCR. Since the SN rate is usually assumed to be proportional to the gas fraction, the GCR flux is found to vary by less than a factor of ~20 between the early halo and today; this is a much smaller variation than the one of Z, justifying the assumption leading from Eq. (1) to Eq. (3). One might think that a very large SN rate in the early Galaxy (e.g. assuming an exceptional star formation rate, or a bi-modal star formation favouring massive stars) could solve the problem. The only effect, however, is to obtain the *same* Be vs. Fe relationship *earlier* in time, since large quantities of Fe are rapidly produced in that case (unless those stars are assumed not to produce Fe; but then, unacceptably high [O/Fe] ratios are obtained in the halo).

3 COSMIC RAY CONFINEMENT IN THE EARLY GALAXY

It is currently thought that GCR are accelerated by SN shock waves and propagate in the Galaxy by diffusing on the irregularities of the galactic magnetic field and suffering losses from ionisation, nuclear reactions and leakage, according to the "leaky box" model for the galactic disk (e.g. Cesarsky and Soutoul, this volume). The interstellar CR flux spectrum at equilibrium is given by:

$$F(E) = \frac{1}{\omega(E)} \int_E^\infty q(E') \; exp[-\frac{R(E') - R(E)}{\Lambda}] \; dE' \tag{5}$$

where $\omega(E)$ (in MeV gr^{-1} cm^2) represents ionisation losses, $R(E) = \int_0^E \omega^{-1}(E')dE'$ (in gr cm^{-2}) represents the *ionisation range*, Λ (also in gr cm^{-2}) is the *escape length* from the galactic disk, and $q(E)$ the *injection (source) spectrum*. From the observed LiBeB/CNO abundance ratio in GCR the escape length is evaluated today to $\Lambda_o \sim$ 10 gr cm^{-2} up to $E \sim$3-4 GeV/n. Notice that GCR particles are escaping from the galactic plane, i.e. the box is considered to be "leaky" only in the direction vertical to the plane and "closed" along the other two directions. The corresponding geometrical length, i.e. the thickness of the disk, is $H_o \sim 1$ kpc.

In the early Galaxy the escape length might have been quite different, because of the different geometry (larger vertical dimension H) and the larger gas fraction σ_g. The difficulties to evaluate it are underlined in PCF, where it is shown that under some plausible assumptions (in particular, assuming that the diffusion coefficient of GCR remains constant during galactic evolution) one is lead to $\Lambda \propto \sigma_g H$. This means that Λ might have been much larger in the halo phase, by factors as high as ~150: indeed, σ_g (halo/today) ~ 10, and H (halo/today) ~ 15, since we can assume that

GCR were confined during the halo phase up to a height of 15 kpc (i.e. the current disk radius).

A larger Λ means that GCR particles spend more time in the ISM before escaping from the Galaxy. In PCF it is shown that, since more and more particles are kept inside the galactic "box" (which becomes less "leaky"), *the total flux increases* approximately as $\sqrt{\Lambda}$, but the low-energy (10-100 MeV/n) flux is almost independent of Λ; moreover, since GCR particles are more subject to ionisation losses before leaving the Galaxy, *the equilibrium spectra become flatter up to higher energies*. These features appear clearly in Figs. 2a,b. The larger GCR flux favours greatly the production of the LiBeB isotopes in the early Galaxy (and particularly the production of the Be and B isotopes). The flatter GCR spectra help to keep relatively low the Li/Be production ratio at that epoch, when Li production was extremely favoured by the $\alpha + \alpha$ reactions. This helps to avoid an overproduction of Li whenever Be is significantly produced at low metallicities.

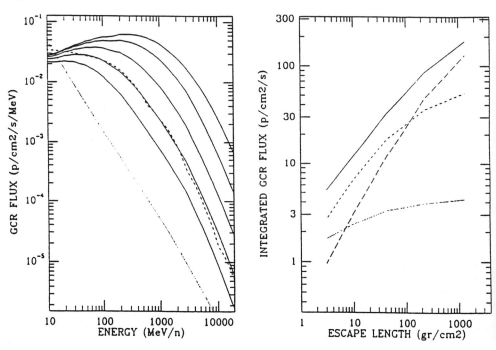

FIGURE 2: (a) Interstellar GCR spectra as a function of the escape length Λ (solid lines); larger fluxes correspond to larger Λ (3, 10, 40, 200 and 1200 g/cm², respectively). The dashed line represents a "realistic" current interstellar GCR spectrum and the dotted line a realistic source spectrum (see PCFa); (b) Dependence of GCR flux on Λ. *Solid line:* total flux; *long dashes:* flux in the 1000-10000 MeV/n range; *short dashes:* flux in the 100-1000 MeV/n range; *dotted line:* flux in the 10-100 MeV/n range.

An application of those prescriptions to a detailed model of galactic chemical evolution that reproduces well most of the observational features of the solar neighborhood (see PCF), leads to the results presented in Figs. 3a,b,c,d. Now the agreement between theory and observations is satisfactory for all the LiBeB isotopes: Be and B are produced to their observed amounts at low metallicity (within error bars), without any associated overproduction of Li; all isotopes, except ^{7}Li for which no stellar source is introduced in the model, reach their cosmic abundances by the time of the solar system formation, if normalisation to the solar Be abundance is adopted (this is not quite true for ^{11}B; all "standard" GCR scenarii give ^{11}B/^{10}B $\sim$2.5, instead of the observed value of $\sim$4). Due to the flatter GCR spectra in the early Galaxy, the B/Be ratio at [Fe/H]=-2.7 is found to be $\sim$11, to compare with the value of B/Be$\sim$13, obtained with current GCR spectra. This value is quite consistent with the value of B/Be for the star HD140283 (10±4). Notice, however, that the resulting Be vs. Fe relationship is still *not a linear one during the halo phase* (although the obtained slope is smaller than 2).

By normalizing the Be abundance of our model to the solar one at the time of the solar system formation, the interstellar GCR flux of the model (required to produce that quantity of Be) is quantitatively determined. In Fig. 4a,b it is shown as a function of time. This "theoretical" GCR flux today is found to be $\sim$5 protons/cm^2/s, i.e. close to the lower limit of the observed one which is evaluated to 4-20 p/cm^2/s (see, e.g. Reeves in this volume); this is a good "consistency" check of the model. In Fig. 4b it is seen that in the early Galaxy the GCR flux was much higher, up to $\sim$500 p/cm^2/s. Such a high flux may have interesting implications for γ-ray astronomy, as suggested recently by Silk and Schramm (1992). Indeed, the intense pion production through proton-proton collisions (which is found to be proportional to the escape length Λ; see Fig. 5a) results in large fluxes of $\sim$70 MeV photons from π^0 decay (Fig. 5b). If this is a generic feature of the early galaxian phases, a substantial contribution to the cosmic γ-ray background may result (Prantzos and Cassé in preparation).

4 DO COSMIC RAYS ORIGINATE FROM SUPERNOVAE?

In Sec. 2 and 3 it was explicitly assumed that the second component of GCR (CNO nuclei on ISM protons and alphas) evolves as the first one. This is obviously true *if* GCR originate from the surfaces of F-K main sequence stars (see Meyer 1985); their abundances reflect then the one of the ISM at the moment of the formation of the star. In that case component B is completely symmetrical to component A, but *always* smaller (B$\sim$0.2 A), because of the decceleration term ζ <1 in Eq. (1).

Duncan, Lemke and Lambert (1992) suggested that, during the halo phase, component B was more important than component A; this should happen *if the "metallicity"*

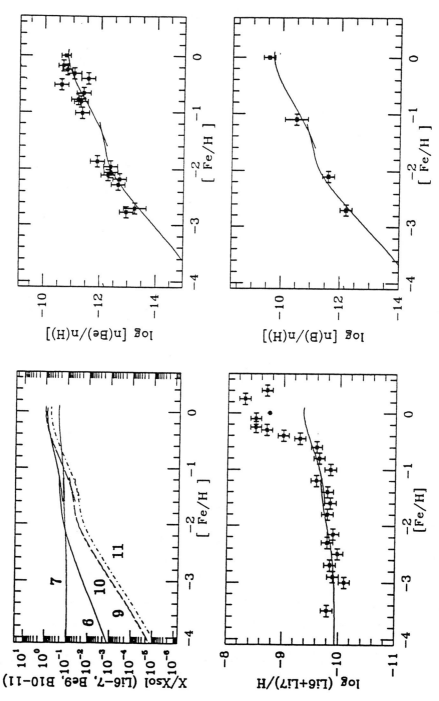

FIGURE 3: Evolution of LiBeB in the framework of a two-zone galactic evolution model with outflow during the halo phase (PCF). (a) Evolution of $X/X_\odot$ for all the isotopes (no stellar source for ^{7}Li is assumed); (b) Evolution of Li/H; (c) Evolution of Be/H; (d) Evolution of B/H.

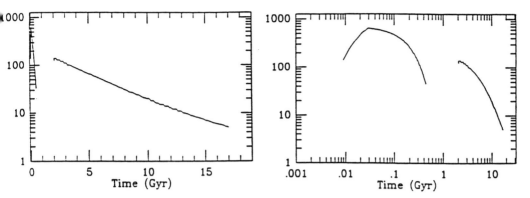

FIGURE 4: Evolution of the GCR flux in our model as a function of time (a) Time is in linear scale, to show the GCR flux variation in the last few Gyr; (b) Time is in logarithmic scale, to show the large GCR fluxes during the halo phase.

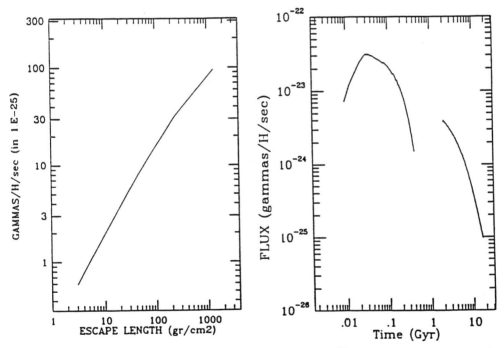

FIGURE 5: (a) Production rate of high energy gamma-rays (in 10^{-25} γ/H atom) by GCR as a function of the escape length Λ; (b) Gamma-ray emissivity in the Galaxy (halo and disk phase), according to the model presented in Sec. 3.

of GCR does not vary with time, and an obvious way to obtain this, is to assume that SN *accelerate their own ejecta*. Then component B in Eq. (1) is $\sim$ *constant*, i.e. slowly decreasing with SFR but not affected by the large increase in ISM metallicity, and dominates BeB production up to $Z \sim 0.2\ Z_\odot$ (Fig. 6a). A *linear* relationship between Be and Fe is naturally obtained then, without any need to invoke large GCR fluxes in the early Galaxy (Fig. 6b).

This interesting suggestion brings, once again, the problem of the *origin of GCR*. The *source composition* of GCR (i.e. the one derived from the observed one after taking into account the propagation effects) is remarquably similar to the solar one, with several important exceptions, however (see, e.g. Ferrando 1992; Prantzos 1992; and Fig. 7):
a) a large depletion in H, He, N (by factors $\sim$10-30);
b) a smaller depletion in C, O, Ne, S, Ar (by factors $\sim$3-5);
c) an anomalous isotopic ratio ^{22}Ne/^{20}Ne$\sim$3 times solar.

Several years ago it has been suggested (Cassé and Goret 1978) that this abundance pattern may (partially) be explained in terms of the first ionisation potential (FIP): species with large FIP (like H, He, N, Ne etc.) are difficult to ionise and, consequently, to pre-accelerate to suprathermal energies, contrary to species with low FIP. The dividing line ($\sim$10 eV) corresponds to temperatures T$\sim$5-10 10^3 K, caracteristic of the photospheres of main sequence stars of F-K spectral type, i.e. sites with $\sim$solar composition (e.g. Meyer 1985). It is now clear that this explanation cannot account for the whole abundance pattern of GCR. Besides the FIP effect (accounting for the moderate depletion of S, Ar, O, Ne), at least two other effects have to be introduced (Silberberg and Tsao 1990):
- a *rigidity* dependent depletion (to account for H, He, N);
- an admixture of $\sim$2% material from WR stars (of the WC type) to account for the C/O ratio [(C/O)$_{GCR}$ >1, contrary to (C/O)$_\odot$ <1] and the ^{22}Ne/^{20}Ne ratio (Cassé and Paul 1982).

Could a SN origin explain that pattern with, at least, the same success? Obviously, the SNI and SNII ejecta are depleted in H, He, N (see Fig. 7) and feature (a) above is naturally explained. However, SNII ejecta are not particularly depleted in Ne, S, or Ar, so one has still to invoke a FIP bias for those species, and this time in a rather "unnatural" environment; notice that WR stars are still needed in order to "bring" C and ^{22}Ne in this model. Another important difficulty for the SN scenario is that no supernova (either SNI or SNII) produces heavy s-elements like Ba; those species should then be depleted in GCR, which is not the case, however (e.g. Ferrando 1992). And it is difficult to assume that it is the r-component of those species that

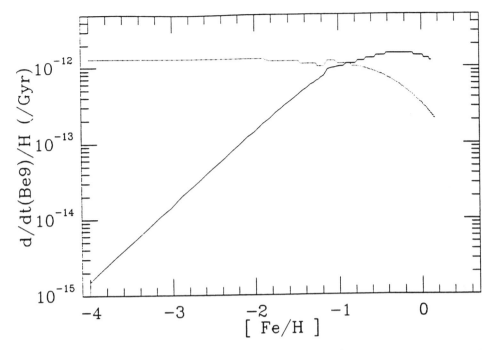

FIGURE 6: (a) Evolution of the production rate of Be/H (in Gyr^{-1}) by components A (solid line) and B (dashed line), respectively (see text) *if GCR originate from supernovae*; (b) Such an assumption leads naturally to a linear relationship between Be and Fe.

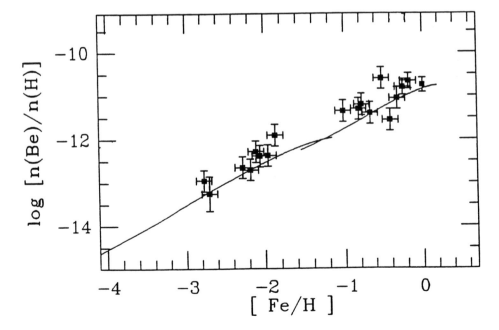

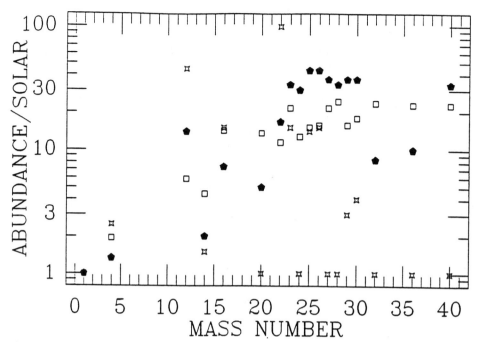

FIGURE 7: Composition of: GCR (from Silberberg and Tsao 1990; full pentagons); SNII ejecta (from Weaver and Woosley 1992; open squares); WR ejecta (from Prantzos et al. 1986; asterisks). $X/X_\odot$ values are given and the three sets of abundances are normalised to H=1.

is observed in GCR, since the pure r-elements should be overabundant then.

Thus, we feel that the attractive (as far as the Be vs. Fe observations are concerned) idea of a SN origin of GCR, is not supported by our current knowledge on GCR source composition. It cannot be excluded, however, that some part of GCR come directly from SN, and that this component played an important role in the LiBeB production in the early Galaxy.

6 CONCLUSION

The recent observations of Be and B in low metallicity halo stars gave a new impetus to the study of the production of LiBeB by GCR. We proposed a new scenario for the propagation of GCR in the early Galaxy suggesting that, during its halo phase, GCR were more efficiently confined than today and had flatter spectra at low energy. Application of this scenario to a consistent model for the chemical evolution of the Galaxy leads to a satisfactory agreement between theory and observations; an associated Li overproduction is avoided because of the peculiar GCR spectra implied by our scenario. Still, the obtained Be vs. Fe relationship is not quite a linear one.

A much better fit to the data is obtained if it is assumed that GCR originate from supernova; in that case, the LiBeB production in the early Galaxy was dominated by the CNO nuclei of GCR impinging on the protons and alphas of the ISM. Our current knowledge on GCR source composition does not support, however, that assumption.

In any case, we think that the observed Be and B abundances in low metallicity stars can still be explained in terms of GCR production, with no need for a less conventional mechanism, like e.g. inhomogeneous big bang nucleosynthesis; the observed B/Be ratio ($\sim$10-12 at all metallicities, i.e. compatible with a GCR origin) is the strongest argument for that hypothesis. However, some of the "classical" ideas on GCR, concerning either their source composition or their confinement in the Galaxy, should be revised in order to account for those recent (and exciting!) observations. Almost twenty years after the works of Reeves, Fowler and Hoyle (1970) and Meneguzzi, Audouze and Reeves (1971), it seems that the final word on the LiBeB production by GCR has not been said yet.

"Bon anniversaire" Hubert!.

REFERENCES
Cassé M., Goret Ph. (1978), *Ap. J.*, **221**, 703
Cassé M., Paul J. (1982), *Ap. J.*, **258**, 860
Duncan D., Lambert D., Lemke D. (1992), *Ap. J.*, in press
Ferrando Ph. (1992), in *Nuclei in the Cosmos*, Ed. F. Kappeler, in press
Gilmore G., Edvardsson B., Nissen P. (1991), *Ap. J.*, **378**, 17
Gilmore G., Gustafsson B., Edvardsson B., Nissen P. (1992), *Nature* , **357**, 379
Hartwick F. (1976), *Ap. J.*, **209**, 418
Meneguzzi M., Audouze J., Reeves H. (1971), *A. A.*, **15**, 337
Meneguzzi M., Reeves H. (1975), *A. A.*, **40**, 110
Meyer J. P. (1985), *Ap. J.*, **57**, 173
Prantzos N., Doom C., Arnould M., de Loore C. (1986), *Ap. J.*, **304**, 695
Prantzos N., Cassé M., Vangioni-Flam E. (1992), *Ap. J.*, in press
Prantzos N. (1992), in *Nuclei in the Cosmos*, Ed. F. Kappeler, in press
Reeves H., Fowler W., Hoyle F. (1970), *Nature*, **226**, 727
Reeves H., Meyer J.P. (1978), *Ap. J.*, **266**, 613
Ryan S., Norris J., Bessell M., Deliyannis C. (1992), *Ap. J.*, **388**, 184
Silberberg R., Tsao C. (1990), *Ap. J. Let.*, **352**, L49
Spite F. and Spite M. (1982), *A. A.*, **115**, 357
Steigman G., Walker T. (1992), *Ap. J. Letters*, **385**, L13
Tinsley B. (1980), *Fund. Cosm. Phys.*, **5**, 287
Vangioni-Flam E., Cassé M., Audouze J., Oberto Y. (1990), *Ap. J.*, **364**, 568

The Saga of the Light Elements

HUBERT REEVES

Dapnia CENS Saclay Gif 91191

Institut d'Astrophysique Paris

1. As a tribute to Ed Salpeter

When I came to Cornell to do graduate studies in 1956, I had the great chance to be accepted as one of Ed Salpeter's graduate students. Those were the days when the stellar energy sources were gradually clarified in terms of nuclear burning of hydrogen, helium, carbon (carbon-carbon reactions were the subject of my thesis) and heavier elements. Following the ingenious intuition of Fred Hoyle, it was already clear that these episodes of thermonuclear reactions in stars could also account for the existence of most of the chemical elements in nature.

There was however a problem with the light species such as D, Li, Be and B. In view of their low Coulomb barrier, these nuclides could not resist the heat of the stellar interiors and hence could not have had a standard stellar origin. In view of this fact, it was not unexpected to find that their abundance are relatively low. But they still had to be accounted for. I remember vividly a moment when Ed discussed the difficulties involved in inventing appropriate mechanisms. There would be a reward , he said . A solution to this problem would undoubtedly tell us something important and unexpected about the build-up of our cosmos. This was the beginning of my interest in this fascinating problem.

More than thirty years later, on the occasion of this Colloquium on " Origin and evolution of the elements", I have had the pleasure to present to Ed Salpeter a number of my students, and also students of my students, who have done their share in the elucidation of this problem. It is a tribute to the quality of his teaching that during this week he will hear about some of their works.

2.Hunting for spallation cross-sections.

It is often said that scientific theories should be "falsifiable". The word originates from the German philosopher Karl Popper. A falsifiable theory is a theory which can make predictions about future observations. If the predictions are not confirmed, the theory is abandoned.

Reality is often quite different. Adopting or rejecting a theory are not the only two possibilities. There is a third one: amending it. A little imagination is required but scientists are in no need of this quality. However there is a price attached to this operation : the theory gradually looses its initial simplicity. The added weight of the proposed amendment makes it less interesting, especially if it involves new parameters whose numerical values must be selected appropriately to match the new data. In this respect I like the "Tooth Fairy Rule" of Mike Turner of Chicago : in a given scenario, one should not appeal more than once to a Fairy Tooth . Each added Tooth Fairy weights heavily against the scenario. Many scientific theories died not when they were falsified but when they became so heavy that everybody lost interest.

Such were my feelings about the FGH theory of the origin of the light elements, fashionable in the early sixties (Fowler et al. 1962). According to these authors, the elements Li, Be, B were generated in the early days of the solar system , by the bombardment of icy meteorites by solar energetic particles .To match all the available data, a scenario had been elaborated in which the number of Tooth Fairies had become far too large. I became interested in finding new ideas about this problem.

One set of data was badly missing. There was almost no measurement of the spallation cross sections for the formation of Li Be B by proton or alpha bombardment of the most abundant cosmic targets, such as carbon , nitrogen and oxygen . In the following years, I contacted many American and Canadian nuclear physics laboratory to advocate this project. I found lots of interest in this program but no one to really get to work on it. There was a problem of purification of the targets to avoid spurious results. Many years of preliminary hard work would be required just to reach the low level of contamination needed. None of these lab could consider spending all that time before coming out with some results.

Such was the situation in 1964, when I was invited to give lectures on Stellar Evolution and Nuclear Astrophysics, in Belgium and in France. I took the occasion to talk about this problem. One evening, after a lecture, at the Institut d'Astrophysique de Paris, a group of physicists from Orsay came to say that they were doing just the sets of measurements I was trying to promote. This group, led by the late René Bernas, had already been involved in the program for many years with the help of a sophisticated mass spectrometer. They would have their first cross-section measurements in a few months!

Needless to say I was overjoyed by this good news. Learning about my interest in this subject, they invited me to work with them on the astrophysical implications of their results. I have remained in France since then...

I am happy to mention the names of these nuclear physicists, to honor their wonderful work, which played a fundamental role in the deciphering of the origin of the light elements : René Bernas, Elie Gradsjtajn , Robert Klapisch, Marcelle Epherre, Françoise Yiou, Grant Raisbeck.

3.   The GCR   origin of LiBeB nuclides in a nutshell.

It turned out that the FGH theory was falsifiable, after all. Inspecting the physics of the proposed scenario, it appeared that the assumed spallation processes required far more power than the gravitational power available during the T Tauri phase of the Sun (Ryter et al 1970).A different astrophysical context was needed to account for the birth of these nuclides.

One evening in 1968, in Copenhagen   I had a long conversation with Bernard Peters , one of the pionneers in the analysis of the chemical composition of galactic cosmic rays (GCR) . Two figures (fig 1 and 2) kept our attention for a long moment : the energy spectrum of the fluxes of the various chemical elements. The most abundant species are, in that order, H, He, C,O, B, N, Li, Be.

One striking difference between these relative GCR   abundances and the stellar abundances is the fact that in the GCR, B, Li and Be are quite comparable to C, O and N , while in stars they are from six to eight orders of magnitude smaller.

Peters had already explained this fact by a very simple idea: the high flux value of LIBeB simply reflects the   effect of the collisions of CNO nuclei with interstellar H and He as they propagate through space, generating, by the spallation process the isotopes of Li Be B. This idea was further corroborated by the recent measurements at Orsay of the corresponding spallation cross sections as a function of proton energy (fig 3 and 4) . The ratios of the cross-sections closely matched the ratio of the flux abundances : B, Li , and Be were in same decreasing order.

We soon realized that the figure was actually trying to tell us something else. We asked: what happens to the LiBeB isotopes which we observe in the GCR? Are we not seeing "live" the very mechanism responsible for all the Li Be B atoms in the galaxy? How much of these atoms have been generated this way since the birth of the galaxy? How does it compare with the stellar and solar system abundances?

A simple calculation   showed that the idea makes sense. The   formation rate of ^9Be , for instance,   relative to H, is   given by the product of the GCR flux of energetic protons (Φ_p ~ 16

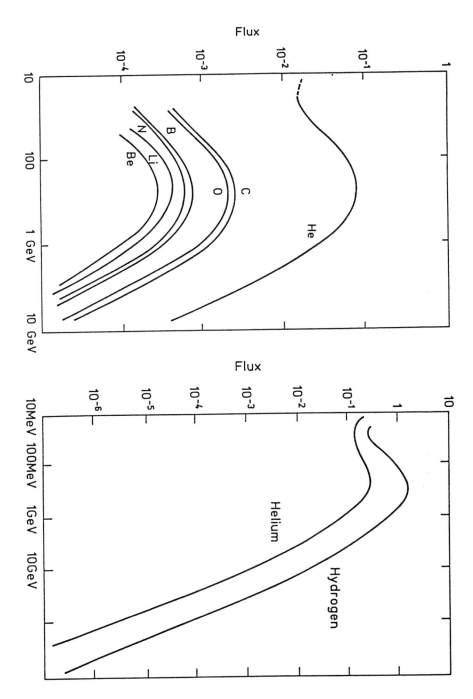

Fig 1 Galactic cosmic ray fluxes of He, O,C, N B, Li, Be as a function of energy per nucleon. Note that B is more abundant than N. Note that B is higher than Li and Li is higher than Be.

Fig 2 Galactic cosmic ray fluxes of H and He as a function of energy per nucleon.Note that most of the particles are found between 100 MeV and 2 GeV.

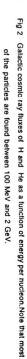

protons cm^{-2}sec^{-1}) times the average cross-section for the formation of Be in the bombardment of the interstellar C, N and O (σ(p+CNO$\Rightarrow$ Be) ~ 5 mb), times the relative abundance of the target nuclides with respect to H (CNO/H~10^{-3}). Assuming a constant rate over the galactic life (T$_g$ ~ 10Gyr), the integrated yield is Be/H $\approx$ 2x10^{-11} , in agreement with the mean stellar value of 1,4x10^{-11}.

The next task was to compare the abundance ratios of the various isotopes in stellar objects with their formation cross-sections in the spallation processes (fig 3 and 4). For this we took advantage of the fact that, comparing the shape of these cross-sections with the proton energy spectrum in the GCR (fig 3,4 and 1; see also Read and Viola 1984), it appears clearly that most of the LiBeB producing collisions are taking take place in the energy range of one hundred MeV to two GeV where the cross-sections are almost constant.

Table : Galactic abundances of Li Be and B in the last few billion years. (see Boesgaard and Steigman 1985; Anders and Grevesse 1989)

Lithium/H = 2x10^{-9} (within a factor of two)
Boron/H = 2x10^{-10} (within a factor of two)
Beryllium/H = 1,4 x 10^{-11}(within a factor of two)
7/6 = 12,5 (within five percent at solar birth; within twenty per cent today)
11/10 = 4,05 (within one percent at solar birth ; unknown today)

The conclusion was that, at least for four of the five stable isotopes: (^{6}Li, ^{9}Be, ^{10}B,^{11}B), the relative cross-sections in this energy range (almost energy independant) are quite comparable with the stellar abundance ratios . This could be considered as another argument in favour of the GCR being a major source of the light nuclides in the galaxy.

A look to the GCR fluxes was already telling us that the GCR origin could not account *for all* the LiBeB observations. While the Be flux is at its expected place at the bottom of the three lists

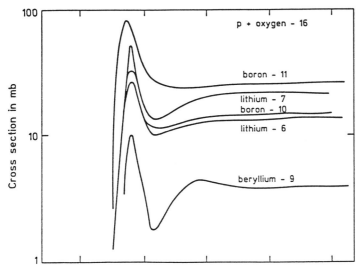

Fig 3 Spallation cross-sections of protons on ^{16}O as a function of energy. Note that from a few
hundred MeV upward, the cross-sections are almost constant. Note the decreasing
sequence of B. Li, and Be.

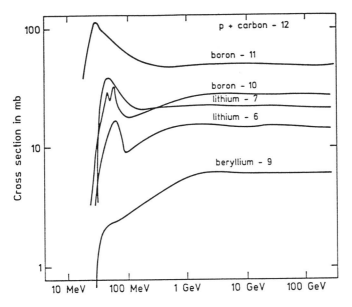

Fig 4 Spallation cross-sections of protons on ^{12}C as a function of energy. Note that from a few
hundred MeV upward, the cross-sections are almost constant. Note the decreasing
sequence of B, Li, and Be.

(GCR, cross-sections and stars), *B is over Li in the GCR and in the cross-sections,* but *under Li in stars* . From the isotopic ratio of lithium it was clear that another source of ^{7}Li was needed to account for the observations. We did not know at this point that we would need two other sources of ^{7}Li...

The same type of computations showed that the amount of D and ^{3}He generated by the spallation of ^{4}He is too small (by factors of 10^{-3} or more) to account for the observed abundances.

In 1969, I presented these conclusions in a seminar at the former IOTA (Institute of Theoretical Astronomy) in Cambridge (UK) . During my seminar , Fred Hoyle kept on talking to Willie Fowler. I could overhear some of his words : "I have been repeating that to you for many years. You should have listened to me" Later on, he told me that he had considered this scenario for a long time. We published a paper together on this subject. (Reeves Fowler and Hoyle 1970)

To pursue this study in a more quantitative way , diffusion models were set up to follow the propagation of GCR in space, including energy deceleration by collisions with electrons and escape of cosmic rays in intergalactic gas. This was done with my thesis students Jean Audouze, Maurice Meneguzzi and later with Jean-Paul Meyer. (Meneguzzi et al 1971; Reeves and Meyer 1978). These models were integrated in models of galactic evolution, taking into account such factors as star formation rates and astration phenomena, to reproduce the expected time variation of these nuclides during the galactic life.(Audouze and Tinsley 1976; Audouze et al 1983; Abia and Canal 1988; Viangoni-Flam and Audouze 1988; D'Antona et al 1991; Browne 1992). Oliver Johns, a nuclear physicist converted to astrophysics, came in Paris as a post-doc and we were able to show that the analysis of abundances of the long-lived and short-lived (fossil) radioactive atoms in the solar system (Wasserburg1985) can be used to constrain the parameters of the galactic models by requiring an almost constant rate of element formation throughout the life of the galaxy. (Reeves and Johns 1976 , Meyer and Schramm 1986 , Reeves 1991, Brown 1992).

The net results of these studies can be formulated in the following way: The GCR origin can account in a satisfactory way for the observed ratios of ^{6}Li, ^{9}Be, and ^{11}B to hydrogen.(Reeves 1974; Austin 1981; Arnould and Forestini 1989) There is a small problem with the boron isotopic ratio (to be discussed later) . As for ^{7}Li, there is clearly a need for (at least) another source.

4. Big bang nucleosynthesis and the abundances of Li Be B.

This part of the story brings me back to my years at Cornell. Once in a conversation , Hans Bethe came to talk about a particular set of nuclear physics data (nothing to do with LiBeB) which was quite cumbersome since it neither proved nor disproved the theory he was working on. In such cases, he added " I store these data in my memory with a little tag. I have a collection of tagged data. Later on , upon reaching a new phase of my research, I pull them out one by one to see how they fit. This is sometimes very useful".

I remembered these words, several years later, when, from his clever aluminum foil left on the moon by the Nasa astronauts and brought back to his laboratory, Johannes Geiss and his team at Berne measured, for the first time the helium isotopic ratio in the solar wind (Geiss this symposium). They obtained the value $^3He/^4He = 4x10^{-4}$. The problem was : how to reconcile this number with the hydrogen isotopic ratio in ocean water (D/H) = $1,5x10^{-4}$, (which, at that time, was considered as the primordial solar system ratio)?

The problem came from the fact that we expect the sun to have burned its initial D nuclides into ^3He by D+p $\Rightarrow ^3He + \gamma$ reactions very early in its history . A (D/H) = $1,5x10^{-4}$,with a standard ratio $H/^4He$ of ten, should have given an $^3He/^4He$ of at least $1,5x 10^{-3}$! (This argument was given weight by the upper limit of D/H=10^{-6} obtained in the solar wind and also by the conviction that the ^3He nuclides present in the solar surface convective zone have not been later burned into ^4He, since ^9Be (which is destroyed at a lower temperature than ^3He) is still present.) . Where was the missing ^3He? I sat on this problem for a long time without a hint of an explanation. I remembered Bethe'tags.

Personal sentimental life sometimes interfere in a positive way on the development of research. One day, I came into the Cornavin station in Geneva in a state of high turmoil because of the sentimental marasm in which I was at that time. I was on my way to Berne to give a seminar at Geiss's Institute. In the hope of finding some mental relief and serenity, I decided to concentrate on this frustrating isotopic problem.

Contingency also often plays a role in research. The night before I had seen on TV the film "La bataille de l'eau lourde"("The battle for heavy water") . It was about a group of Norwegian scientists who, at the beginning of World War II, had succeeded in enriching ordinary water in deuterium by low temperature molecular exchange reactions. The idea is simple: a mixture of H,HD,HDO and H_2O in statistical equilibrium, will see its relative HDO/H_2O content increase when the temperature is decreased.

Could the key of the puzzle lie there? Was it possible that ocean water had already been "naturally" deuterium enriched? Surely low temperature contexts, with mixture of these gases, were easy to imagine, either in the primitive solar nebula or in the interstellar gas where the protosolar nebula came from? I arrived in Berne in a state of high excitation. Johannes was waiting for me at the station. I discovered that he and his collaborators were already trying to solve this problem (Geiss et al 1966)We immediately started to work on this question.

We arrived at an estimation of the protosolar D/H ratio which was about an order of magnitude smaller than the ocean water value. (Geiss and Reeves 1972). This value was later confirmed by the Copernicus satellite and, more recently, by the Hubble Space telescope (Lansky et al 1992). This is discussed in the report of Geiss. His present best estimate of the protosolar D/H ratio is within ten percent of our earlier estimate.

5. Like two pieces of a puzzle.

In 1967, Wagoner, Fowler and Hoyle computed the yields of light elements in the cooling universe (Big Bang Nucleosynthesis : BBN) as a function of the baryon number (ratio of the number of nucleons to the number of photons) or, equivalently , as a function of the nucleonic density. The ^{4}He came out nicely but there was a problem with the deuterium . The ocean water D/H ratio could only be accounted for at a nucleonic density which appeared to be smaller than the nucleonic density of luminous matter in the universe ($\rho_b \approx 10^{-31}$gcm^{-3}). Our lower D/H value pushed the corresponding nucleonic density up at around $\rho_b \approx 3 \times 10^{-31}$ gcm^{-3}, in the range between the lower limit given by the density of luminous matter and the upper limit given by the lack of observable deceleration of galaxies (a few times 10^{-29} gcm^{-3}) . We looked again at Wagoner yields and found with great interest that at $\rho_b = 3 \times 10^{-31}$gcm^{-3}, BBN could account simultaneously for D, ^{3}He, ^{4}He and ^{7}Li. We noticed also that it was hopelessly ineffective in generating ^{6}Li, ^{9}Be ^{10}B, ^{11}B.

At this point it appeared that the two theories of origin matched very well, like two pieces of a puzzle, in accounting for the abundances of all the nuclides with mass less than 12. The Big Bang theory could generate interesting amounts of D, ^{3}He , ^{4} He and ^{7}Li but not ^{6}Li, ^{9}Be, ^{10}B and ^{11}B . The GCR flux could account for the abundances of ^{6}Li, ^{9}Be, ^{10}B and ^{11}B . Actually the case of ^{7}Li turned out to be even more complicated.

I presented these ideas at the Porto Rico APS meeting in 1972 (Reeves 1972) . Ed Salpeter showed great interest in these views. Al Cameron had a brief comment :"I like that" To my knowledge this was the beginning of long-lasting interest of Dave Schramm in this field. Together with Audouze and Fowler we published a paper on the whole subject (Reeves et al 1973),

6.The Spite plateau.

In 1982, François and Monique Spite measured, for the first time , the abundance of Li in very old (PopII) stars, at $^7Li/H \approx 10^{-10}$. The most dramatic aspect of this discovery was the fact that, for stars with very small amount of heavy elements (less than one tenth the solar abundance, also so-called "solar metallicity") , these measured abundances,   are essentially independant of the heavy element content of the star. . The discovery of this lithium plateau , called the "Spite plateau " (fig 5) could be taken as a strong evidence for the existence of a primordial source of ^7Li (Spite and Spite 1982a; 1982b). I consider it also as one of the important confirmations of the Big Bang theory.

There was however the difficult question of the possibility of lithium depletion in the surface of these stars. What fraction of the original lithium was still present after nearly ten billion years? What value should we take for the primordial value? Could it be as large as the recent stars value(PopI) of $^7Li/H \approx 10^{-9}$? (Delyannis et al 1990) Or did we need another source of 7Li to account for the difference between PopI and PopII?

These questions were at the center of a long debate which is still continuing today. My thesis student Sylvie Vauclair (Vauclair 1988a; 1988b) with her thesis student Corinne Charbonnel (Charbonnel et al 1992 and Thesis 1992) have investigated   the depletion of Li and Be, by rotational mixing in the stellar surface layers. The same mechanism has been later used by the Yale group (Pinsonneault et al 1991 to describe the possibility of depletion by a factor of ten or so.

Georges Michaud, with whom I have often collaborated , and his student Paul Charbonneau (Michaud and Charbonneau 1991) have made a global assessment of the situation. They conclude that the depletion must have reached at least (but probably not more) than a factor of two. Their best estimate of the primordial value is $^7Li/H = 3x10^{-10}$. They state that: "more calculations are needed to better establish this value".

Fig 5 Stellar abundances of the elements Li, B, and Be in the old (Pop II stars) as function of their iron content (normalized to solar). Note the Spite plateau in the Li/H curve; the signature of the Big Bang contribution. No such plateau has yet been found for Be and B . They would represent evidence for a first order Q-H phase transition.

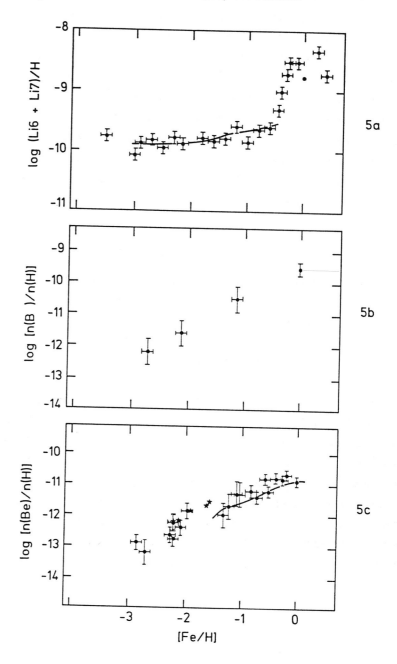

The Wagoner calculations were made in the frame of the so-called standard model of Big Bang Nucleosynthesis (BBN) which makes a certain number of assumptions: homogeneous energy density, zero leptonic number etc. For each of the four cosmological nuclides,   there exist a range of baryonic numbers , or equivalently, nucleonic densities , where this nuclide is generated in satisfactory abundance to meet the observations.   The width of each range is fixed both by the uncertainties in the BBN calculations and in the observations, extrapolated back to the time of BBN. The success of BBN comes from the fact that these ranges have a common part which is thereby selected as the "compatible range" . In the standard BBN this range goes from 2 to $5 \times 10^{-31} \text{gcm}^{-3}$.(Yang et al 1984; Krauss and Romanelli 1990; Smith et al 1992)

With my student Yves David we studied the nucleosynthetic   effect of varying   leptonic numbers of universe (Reeves 1972; David and Reeves 1980; Steigman 1984). No large effect is found until we reach leptonic numbers close to one. With a former student   Gilles Beaudet, we evaluated the uncertainties on the BBN yields from the uncertainties in the nuclear cross-sections (Beaudet and Reeves 1984).

7.The quark-hadron phase transition.

Around 1980, progresses in particle physics have   put into question the standard model used so far to compute BBN yields. Witten (1984)showed that the quark-hadron phase transition occuring around 150 Mev could have generated   density inhomogeneities leading , after weak interaction decoupling at 1MeV, to inhomogeneities in the neutron to proton ratio during BBN at 0,1MeV. This issue is not resolved at the present time.

Models of inhomogeneous BBN were   developped in order to incorporate the possible effects of the Q-H transitions. (Terasawa and Sato 1989; Kurki-Suonio et al 1990; Malaney and Fowler 1989) Some of this work was done in Paris   by Pierre Salat i, Pascale Delbourgo-Salvador, Jean Audouze (Reeves 1988) . Later works were made   with Jacques Richer of Montreal together with Sato and Terasawa from Tokyo (Reeves et al 1990)

Some authors have taken the view that since it is "unlikely " (whatever could be the meaning of this word in this context) that Q-H transition would influence BBN, they just prefer to ignore this possibility . I can not agree with this point of view We are facing here an important issue, related to fundamental physics, which   an acceptable Big Bang theory   must incorporate, even if it complicates life. Said otherwise, the present uncertainties of the parameters of the Q-H transition introduce corresponding uncertainties of the BBN yields, which should be taken into account in making statements about such issues as , for instance, the need (or not) for non-baryonic matter in the universe.

One key-point is the order of the Q-H transition (or transitions since there are both a chiral and a deconfinement transition, taking place at approximately the same energy). If the transition is of second order , no inhomogeneities are created and the standard model is accceptable. If it is of first order, the degree of inhomogeneities depends upon the extent of overcooling and also on other parameters. The effects on BBN yields may or may not be important.

What is the situation in June 1992? The following brief summary comes mainly from a Journée des Particules, held in Orsay in June 1992. Information on the parameters of the phase transition can, in principle, be obtained both from high energy collisions of heavy nuclei in laboratory and from QCD calculations on a lattice. On the laboratory front, it appears likely that the quark-gluon soup has been obtained at CERN but, so far, very little information has been reliably extracted from these experiments. The hope is that the lead-lead collisions, programmed at CERN, in a year or two, will give pertinent experimental results on these processes.

On the calculation front , the interesting new development is the fact that, while previous computations took into account only the presence of gluons , in the last two years quarks have been progressively introduced. From the (apparently non-linear) volume dependance of one of the parameters of the transition (equivalent to a "susceptibilty") Fukugita and Hogan 1991 have concluded that the transition is not of first order. However, according to many specialists, this analysis is far from being conclusive and the order of the transition must still be considered as uncertain. (Bernard, 1991; Toussaint 1992 ; Gottlieb1991).

A second important effect of the inclusion of quarks in the computation is a lowering of the estimated temperature of the transition. While in the pure gauge computations (quark-free), the temperature was found to be around 220-250 Mev, it falls well below 200 MeV when quarks are taken into account . In the case of two light and one heavy quarks, it is close to 150 MeV. The lower the temperature, the higher the density contrast expected in the baryon inhomogeneities potentially created by the transition (if indeed it is of first order) . This is of importance since the difference in the yields between homogeneous and inhomogeneous BBN increases with the contrast parameter .

My favorite approach is to treat the uncertainties on the effect of the Q-H phase transition as uncertainties on the yields. While in the case of the homogeneous Big Bang there is only one free

parameter : the ratio η of the number of nucleons to the number of photons , the Q-H phase

transition introduces three new parameters(see the discussion of Kurki-Suonio 1992) .

The first one is the effective density contrast R between the high density regions (the

clumps) and the low density regions. The word "effective" takes into account the fact that the

contrast created at the beginning of the nucleation process (proportional to exp (-M(proton)/kT$_c$))

is later amplified, by the percolation processes (the bubbles growing to occupy the whole of

space) , to several tens of times its initial value (Alcock et al 1987) . The second one is the fraction

of the mass f$_V$ in the clumps (the clumpiness). The third one is the average distance d between

the clumps. A convenient unit is the present value of d in light-hours (lh) . One lh today

corresponds to 2.5 x 10^5 cm at one billion degrees and approximately one meter at the Q-H

phase transition, when the horizon scale was approximately ten km . At large values of d (d~ 10 ^{4}

lh) the neutron could not diffuse before BBN. Computations made with the assumption of

large values of d give the same results as computations based on a density-inhomogeneous

model. At the lower end of the scale , d <0,1 lh or so, proton diffusion becomes important , the

clumpiness is erased and we are back in the homogeneous model. The most interesting d

range is from 1 to 10 lh.

Calculations , including neutron diffusion during BBN, have been made covering the

whole parameter space corresponding to the uncertainties on the value of these parameters

(Kurki-Suonio et al 1990, Reeves et al 1990) Detailed studies show that the overall situation is not

very different from the homogeneous BBN situation. Again, for each of the four nuclides, a range

of baryonic numbers can be found. And again , a compatible range exists which is quite similar to

the range of the homogeneous BBN.

For a given value of η, one finds a range of possible yields of each nuclide , corresponding to

the variations of the other parameters. When the yields are plotted as a function of η, this range

translates into a width attached to each curve. As a result , the uncertainties in the parameters of

the Q-H transition are reflected in a larger compatibility range for the whole of BBN . This range

appears to practically exclude the possibilty that baryons have the closure density ($\Omega_b < 1$).

What conclusions can we draw from BBN about questions such as the nucleonic density and also the possible existence of dark baryonic matter and non baryonic matter? Several authors have given their views: the differences between them are not large ; they depend mostly on each person's evaluation of the uncertainties attached to the element abundances.

It is highly instructive to give an historical look at the behaviour of scientists in this respect. There is a systematic tendancy to *underestimate* the uncertainties, which appears clearly when the status of a given measurement - the velocity of light for instance- is followed over the years. My personal reaction to this situation is to assign rather wide uncertainties, especially if important conclusions are to be drawn fron the data.

In fact I like to play the game at two simultaneous levels: the reasonable one and the extreme one. For D/H the lower limit (at both levels) is 10^{-5}, corresponding to $\eta_{10}<10$ where η_{10} is the value of the baryonic number in units of 10^{-10}. The upper limits on D/H and also (D+3He)/H are, at the reasonable upper limit: $< 10^{-4}$ corresponding to $\eta_{10} >3$ and, at the extreme upper limit: 2×10^{-4} . corresponding to $\eta_{10} >1,7$. For ^4He, I choose: $0,215<Y<0,245$ as the reasonable limit, corresponding to: $0,8 < \eta_{10} < 5,0$; and $0,21<Y<0,25$ as the extreme limit, corresponding to $0,6 < \eta_{10} < 9,0$. For $^7Li/H : 3\times10^{-10}$ is my reasonable estimate, corresponding to $1,4 < \eta_{10} < 5$; and 10^{-9}, my extreme limit, corresponds to $0,6 < \eta_{10} < 12..$ Given the η range of values compatible with the abundance observations, the corresponding baryonic density is given in units of the closure density by the following expression:

$$\Omega_b(H/100)^2 = 3,7 \times 10^{-3}\eta_{10}$$

where H is the Hubble constant. With the value H =75 we have: $\Omega_b = 6,6 \times 10^{-3}\eta_{10}$. The range of possible H (from 50 to 100) introduces a factor of two on each side for Ω_b

Putting together these requirements of the various nuclides , I find that my reasonable limits require $3<\eta_{10}> 5$ or $0,01< \Omega_b < 0,07$ while ,within the extreme limits ,one could have $0,005 < \Omega_b < 0,12$. In view of these numbers, the following conclusions can be drawn.

1.1) The universe is not closed by baryonic matter $\Omega_b < 1.$ 2) There is probably, but not certainly, a fair amount of dark baryonic matter ($\Omega_{luminous} \approx 0,05$) . 3) At the density level required from dynamic arguments ($\Omega_b \approx 0,1$ to $0,2$), no strong case can be made for the existence of non-baryonic matter.

8. Primordial Be and B ?

One potentially interesting aspect of these Q-H inhomogeneous models is the very large increase in the yields of elements such as   Be , B and also some of the heavy neutron-rich nuclides , which is found for certain values of the parameters of the inhomogeneous BBN.

This possibility has been explored in particular by Kajino and Boyd(1990) and   by Malaney and Fowler(1989). These authors have shown that abundance of Be and B not so much smaller than the PopII observed values   can be obtained for certain choices of the parameters of the Q-H phase transition, thus opening the hope that a plateau of Be and B might be found at very low stellar metallicity , similar to the lithium Spite plateau at one tenth solar metallicity. Such plateaus, if they exist could give interesting information on the properties of the Q-H transition. (Note: later calculations (Terasawa 1992; Thomas 1992) have given appreciably lower values for Be and B)

As discussed before, the recent results of the QCD calculations on a lattice favor a low value ($\approx$ 150 Mev) for the critical temperature , which in turn , implies high   effective density contrast, and increases the difference between the homogeneous and   inhomogeneous primordial nucleosynthesis. In particular it should be noted that all the exotic nucleosynthesis yields (large Be , B and heavier nuclides) are obtained at large   effective R values (R > 100) which the new QCD calculations   seem indirectly to favor.

The most serious   problem posed by this interesting   search is the   large 7-lithium abundances almost always predicted by the "interesting" inhomogeneous BBN (Li/H = 10^{-8} to 10^{-7}) , far above the realistic estimate of 3×10^{-10} by Michaud and Charbonneau (1991). Later on, I will report of a recent measurement of the lithium isotopic ratio in the present interstellar gas which also speaks   against primordial values as large as the Pop I values (10^{-9})

All this, unfortunately, leaves little chance for Be and B to give useful information on the parameters of the Q-H phase transition. But if the probability is low, the expectation value is high.The search for the Be and B plateau should be continued. One should never   trust the theoreticians too much...

9. The lithium isotopic ratio in the galaxy.

The evolutionary abundance curves of the light elements provide important information on the nature of the physical phenomena during the galactic life . I will consider some interesting examples based on recent measurements.

One important piece of evidence for the existence of another source (not BBN or GCR) of ^{7}Li during galactic life has come recently from the observation of the lithium isotopic ratio in the galactic gas , in front of Zeta Ophiucus by Lemoine et al (1992) . Their result : $9,1 < {}^7Li/{}^6Li < 15,5$, with 95% C.L., shows that this ratio did not vary by more than twenty five percent since the birth of the solar system 4,5 Gyr ago , when the ratio was $12,5(\pm0,5)$.

The evidence from meteoritic irradiation shows that the GCR flux has not varied by more than a factor of two during the last few billion years. For the sake of discussion, let us first assume that, since the birth of the sun, the GCR has been the only source of lithium in the galactic gas. Then, from the fact the GCR contribution to ^{7}Li and ^{6}Li are approximately equal, the lithium isotopic ratio should have decreased from its protosolar value of 12,5 to a value appreciably below the lower limit of 9,1 observed by Lemoine et al. Since the mean GCR flux in interstellar space is still uncertain (because of solar modulation effects in the solar cavity where the observations are made), I have recently computed the expected isotopic ratio as a function of the assumed proton flux (Reeves 1992). In the flux range of fluxes estimated with the most recent data from the Voyager satellite at the outer bound of the solar system (22AU) (Ferrando et al 1991 and Ferrando, private communication; Webber et al 1992; see also Cesarsky et al 1977) the isotopic ratio should be smaller than five (fig 6) .

Lemoine et al have also measured the abundance ratios to hydrogen. They find ^{7}Li/H = 3.4×10^{-9} and ^{6}Li/H = $2,7 \times 10^{-10}$ Fig 6 shows also that the ^{7}Li/H expected (always in absence of a stellar source) falls below the observed value in Zeta Oph, while the 6/H is accounted for. However in view of the large uncertainties attached to the Li/H ,this test is far less indicative than the isotopic test.

In order to account for the present isotopic ratio, another source of 7Li is clearly needed . Its yield is evaluated to be $20M_o$ of ^{7}Li per Gyr per $10^{10}M_o$ of galactic gas , averaged over the last 4,5 Gyr .The corresponding increase in ^{7}Li/H ($\Lambda(^7Li/H) \approx 2\times10^{-9}$) is quite model-dependant since

Fig 6 Lithium isotopic abundance expected today as a function of the galactic cosmic ray (demodulated) flux in absence of a stellar source of ^{7}Li. For the best estimate value of the flux the ratio should have decreased from 12,5 (measured value at the solar birth) to less than 5 . The fact that the present ratio is larger than nine (Lemoine et al 1992) is a strong indication of the existence of a stellar source of ^{7}Li.

Also plotted: the ratio of ^{7}Li/H and ^{6}Li/H , normalized to the value of 7/Hss = 1.2x 10^{-9}, ^{6}Li /Hss =10^{-10}. The (7/H)/(7/H)ss shows that without the stellar source this ratio would be too small to match the observed lower limit. The stellar source required to account for the isotopic ratio would lift it up to coincide with the (6/H)/(6/H)ss in better agreement with the data . However the uncertainties are much larger than the uncertainties on the isotopic ratio.

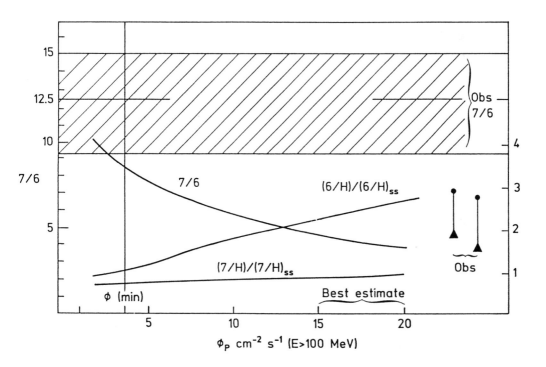

the abundance of H is a function of the adopted galactic evolution parameters (rate of matter rejection by stars, infall of extragalactic matter etc.) . It is of interest to note that this lithium increase is in agreement with the difference between the Pop I and Pop II lithium abundance. This agreement has an incidence on the interpretation of the gradual increase of the stellar lithium abundance as a function of metallicity , at a metallicity above one tenth solar(fig 5a). It shows that the interpretation in terms of a stellar source of lithium is to be preferred to the interpretation as a gradual surface depletion with stellar age (Mathews et al 1990; Brown 1992).

This interpretation has an interesting cosmological corrolary. It leaves little room for the idea that Pop II abundances would be depleted Pop I *and that the primordial values could have been as large as Pop I or more.* The non-detection of lithium in the Magellanic Cloud in front of SN1987 A at a level of $<10^{-10}$ (most of the lithium atoms are stuck in interstellar grains) (Baade and Magain 1988, Sahu, Sahu and Pottash 1989) also points out in the same direction , although the uncertainties are large (Malaney and Alcock 1989).

10. Physical processes in the early galaxy.

In addition to their interest as cosmological observables, the nuclear species lighter than carbon ($Z < 6$) are potentially rich indicators of the physical conditions accompanying the formation of galaxies.

In this respect, the recent observations of B, Be and Li in old PopII stars are particularly worth discussing here(Ryan et al 1990; 1991; 1992) (Rebolo et al 1988; 1990; 1992; Spite et al 1991; Duncan et al 1992; Gilmore et al 1991) .In figure 5 are shown the abundances of Li, Be, and B as a function of metallicity (Fe/H)

While the Li abundance reaches its Big Bang plateau around one tenth of the solar metallicity, the Be and B abundances are still going down. At one part in one thousand of the solar metallicity, they have fallen by a factor of approximately one hundred below their present PopI stellar value. This is qualitatively expected from the fact that they are GCR products. This also can be used to give an upper limit to their hypothetical Big bang contribution as possibly expected in the case of strongly inhomogeneous Big Bang (no plateau similar to the lithium has yet been found).

To obtain quantitative estimates of the GCR expected yields, we cannot simply use the standard GCR models developed for PopI stars. We have to consider several factors by which the ancient situation differed from the present one These factors include the target abundances, the flux intensity of the GCR and also their propagation in the galaxy. The diffusion-like propagation involves, in turn, the matter density, the magnetic configuration governing the mean free-path of the fast particles and also the propagation volume in relation with the changing shape of the galaxy. Michel Cassé (one of my students) Nicolas Prantzos (a student of Michel Cassé) , Elisabeth Flam (a student of Jean Audouze) and Yvette Oberto have been active in investigating these subjetcs (Vangioni-Flam et al 1990 ; Prantzos et al 1992: PCF).

Consider first the evolution of the target abundances. As we move toward the past, the ratio CNO/H decreases progressively while the α/H ratio remains almost constant. In his thesis work, Thierry Montmerle has pointed out the fact that since the CNO group reactions generate the three elements LiBeB while the $\alpha+\alpha$ reactions generate only ^{7}Li and ^{6}Li, one expects the relative ratio of Li/BeB to increase with decreasing metallicity (Montmerle 1977 ; Steigman and Walker 1992) . In this respect we should note the following point: given the shape of the cross sections (fig 3,4 and 7) and the shape of the GCR energy spectrum (fig 2) , the p+ CNO products are mostly created in the hundreds of MeV range, but the $\alpha+\alpha$ products are almost entireley generated in the tens of MeV range. This remark will be of importance later on.

Consider next the evolution of the paleoGCR flux spectrum both at its source and in space . At the source, we expect the injected power , presumably related to acceleration mechanisms in supernova, to have been larger in the past, in relation to the expected higher star formation rate in the early days of the galaxy. This factor is best studied with galactic evolution models aimed at accounting for the variation of abundant elements (Fe,O,C) with galactic age or the number of stars of given metallicity as a function of metallicity (Pagel 1987; 1989;1992; Brown 1992).

The GCR energy spectrum in space is governed by the value of the "escape length" Λ of the galaxy; the amount of interstellar matter met by a cosmic ray before it has a chance to leave the galaxy. The value of the escape length is a function of the density of interstellar matter and also of the magnetic field configurations on which the fast particles diffuse before they reach the border of the galaxy. It depends also upon the geometry of the volume of propagation (disk and halo) of the galaxy at any one time in the past. All these factors are expected to have been different in the early days of the galaxy .

The numerical value of the escape length Λ of the GCR plays an important role on the formation of nuclides through the GCR bombardment of the interstellar gas. Its present value ($\approx$ 5 gcm^{-2}) governs the abundance of secondary nuclides (Li Be B, and also the nuclides with mass number between Fe and Si) in the GCR themselves.

The ratio of tens of MeV to hundred of Mev particles in the interstellar flux of GCR is a key parameter of this discussion. In early times, the low value of the CNO target abundances resulted in an increasingly important relative contribution of the $\alpha+\alpha$ reaction in generating Li (but not in Be and B) . This relative contribution is modulated by the ratio of the fluxes in the tens of MeV (for $\alpha+\alpha$) and hundred of MeV(for p+CNO) : this ratio decreases with increasing Λ as more and more source particles are heavily decelerated before they have a chance to escape the galaxy (see the figure 8 of Prantzos et al 1992).

The escape length (Λ) is the product of the gas density (ρ) times the total trajectory of the particle in the galaxy (d) . This trajectory is dictated by the diffusion of the GCR on magnetic lines of forces in space. It is equal to $d = \lambda N$ where λ the mean free path between each diffusion step and N is the number of steps required to leave the galaxy. Calling h the thickness of the galactic disk, the diffusion theory shows that $N = (h/\lambda)^2$

It is usually believed that the present galactic disk collapsed from an ancient spherical halo . Thus, as we go backward in time, we expect the thickness h of the disk to increase, while the disk surface -density shoud remain approximately the same. The effect on Λ would then be two folds. 1) a decrease in the space density of targets, inversely proportional to h ; 2) an increase in N, proportional to h^2 . The behaviour of λ is much more difficult to assess since we know very little about the time variation of the magnetic field and magnetic configurations in the galaxy. Theoretical estimates (Catherine Cesarsky; private communication) suggest that the mean free path λ should have been smaller (more magnetic irregularities) but a quantitative estimate is not possible;

Putting all this together , we find that Λ varies with h/λ and , according to our previous discussion , this ratio was presumably larger in the past. Increasing Λ corresponds to a relative decrease in the production of Li relative to Be and B. But the main point here is the identification of Λ as the phenomenological parameter most appropriate to integrate the landscape of the early galaxy in discussing the GCR product abundances.

The ratio of B/Be is rather insensitive to the value of the escape length. It never becomes smaller than 10 even for the largest escape length. Its value obtained in PopII stars (10 ± 5) is compatible with the computed production rate of the GCR showing that both species are mostly of GCR origin even in the very early days of the galaxy. (Walker et al 1992 Thomas et al 1992).

The situation is different for the Li/Be ratio. At low interstellar CNO abundances (less than 10% solar) the $\alpha+\alpha$ reaction becomes the main producer of ^6Li and ^7Li (with approximately equal production rate). (Steigman and Walker 1992) The relative contribution are proportional to the metallicity at any given instant of the life of the galaxy.

This effect is best tested at metallicity between 0,01 and 0,1 of solar.The lithium abundance is constant from 0,001 to 0,1. Below 0,01 the observed Li/Be ratio is too high to show the contribution of the GCR Li while above 0,1 the p+ CNO source takes over. According to the model computations of PCF the $\alpha+\alpha$ contribution should lead to a lithium abundance larger than the primordial value for metallicity between 0,03 and 0,1 (in contradiction with the observations). unless one assumes much larger escape length in this period. Values as high as 200gcm^{-2} are required.

Such computations show how the abundances of Be and B can give information on the physics of the early galaxy.

11. An unsolved problem: the boron isotopic ratio.

The observed isotopic ratio of boron in the solar system is 11/10= 4,05($\pm$0,05) (Mason1971). We have no observations of this ratio in other astrophysical contexts.

This value is tantalinzingly close to the value of $^{11}B/^{10}B \approx 2,5$ obtained from the GCR model it is essentially given by the ratio of the O and C cross sections at energies above one hundred MeV). Both the isotopic ratio and the high-energy cross-sections are known accurately enough to convince oneself that the discrepancy is real. We have so far no convincing scenario to reconcile the two values.

The observed value could be related to the existence of fluxes of low energy particles (tens of Mev) located somewhere in the galaxy, perhaps around the GCR acceleration sources (Meneguzzi and Reeves 1975). Because of the high rate at which these low energy particles ionize the interstellar medium, these fluxes are unable to propagate throughout the galactic space. They are expected to remain confined close to their accelerating sources. Furthermore they are excluded from our planetary vicinity by the solar modulation effect. Because of the high peak in the $^{14}N(p,\alpha)$ ^{11}B cross-section around 10 MeV(fig 8) the formation of ^{11}B would be greatly favored in these regions.

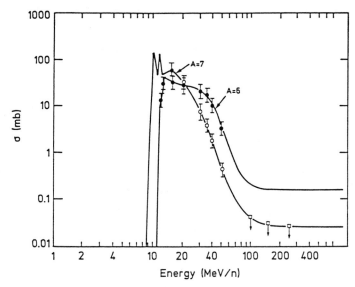

Fig7 The cross-sections for the formation of ^{7}Li and ^{6}Li by α+α reactions as a function of energy
per   nucleon. Note that cross-sections become very small above a few tens of MeV.

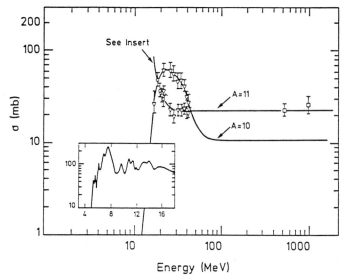

Fig 8 Spallation cross-sections of protons on ^{14}N as a function of energy per nucleon.Note the
high peak of the ^{14}N(p,α)^{11}B around 5MeV , possibly accounting for the boron isotopic
ratio.

One possible argument against this mechanism comes from the recent study of Prantzos et al 1992 which shows that these low energy fluxes could have generated too high Li/Be if they existed in the early days of the galaxy. In view of the low abundance of the CNO species in those times, most of the Li came from the a+a reaction at low energy.

The possibility that this mechanism is the answer to the boron isotopic ratio puzzle can be tested.The same low energy protons would generate gamma rays from the excitation of ^{12}C and ^{16}O. In particular a 4.4 MeV line from ^{12}C and ^{16}O and a 6,1 and 7,1Mev line from ^{16}O would necessarily be associated with these hypothetical fluxes of tens of MeV particles.

Quantitatively, what is the flux of gamma rays expected if we assume that these low energy protons are abundant enough to account for observed boron isotopic ratio? The answer depends upon the degree of clumpiness of the sources of protons and gamma rays, but in all reasonable cases, the expected flux at earth is less than 10^{-6} photons per cm^2 per sec , one or two order of magnitude below the present detection level of the Gamma Ray Observatory. We have to wait for a new generation of gamma ray telescopes before we can test this hypothesis.

I should mention, at this point, the discovery by Paul Pellas in 1968 of meteoritic grains highly irradiated by low energy cosmic rays. On this occasion, we joined forces with him (Pellas et al 1969) to study the caracteristics of this irradiation. This observation gives   credibility to the existence of large fluxes of low energy particles around ordinary young stars,   which are now believed to be the main injectors of GCR particles (Meyer 1985).

It is also of interest to consider other physical mechanisms which are known to produce appreciably higher $^{11}B/^{10}B$ ratios than   nuclear spallation reactions. These processes involve electromagnetic interactions (electrons or photons)   or weak interactions (neutrinos) on ^{12}C.

With Richard Schaeffer, a phycisict   converted to astrophysics, we have studied the electrodisintegration and photodisintegration of ^{12}C (Schaeffer, Reeves and Orland, 1982) by jet-fluxes of electrons in the neighborhood of quasars. Boyd, Ferland and Schramm(1988) have studied the photoerosion of galactic gas by high energy photons. Woosley et al(1990) have considered the neutrinodisitegration of 12C (and other heavy nuclei) by the flux of neutrinos emitted as the core of a massive star collapses to form a neutron star.

The photodisintegration, electrodisintegration or   neutrinodisintegration of ^{12}C could in principle solve the boron isotopic problem if their time-integrated galactic ^{11}B contribution is closely equal (within less than a factor of two) to the GCR contribution. In view of the large uncertainties attached to their galactic yields, these possibilities can neither be confirmed, nor ruled out.   At any rate the comparison between the GCR predicted and observed isotopic values can be used to put *an upper limit* on the   nucleosynthetic importance of electro, photo and neutrino disintegrations processes in the galaxy: they should not have generated more ^{11}B than the GCR themselves!

Since we have no extra-solar system data on the boron isotopic ratio, one should also consider the possibility of chemical or physical fractionation in the   planetary system, with a preferential loss of ^{10}B . Stellar or galactic gas measurements of this ratio are needed to resolve this issue.

A possible solution in terms of a large neutron irradiation (Fowler et al 1962) (the ^{10}B has a large neutron capture cross section) is essentially ruled out by the lack of equivalent isotopic effect on gadolinium, where the isotope ^{154}Gd has a much larger neutron capture cross section.

12. About nature and "naturalness".

Throughout this talk, I have spoken exclusively on things that went right in my research work. This does not give a realistic view of a   scientific career .   It is good for students to know that there seniors have also made mistakes.   I would like to discuss one particular issue , on which my guesses were completely off. It is always a fruitful exercise to ask oneself why ,on a certain issue, your intuition drove you to the wrong conclusion.

When the Spite reported their measurements on the lithium abundance in PopII stars and argued that the low value they had obtained (some ten time less than the Pop I value) was actually the best choice for the   Big Bang yield, I just could not believe that this choice was correct. I had many reasons against this choice, which were all wrong.

First, the surface depletion of lithium abundance   during the Main Sequence stage of stellar evolution was well known . The Sun has lost 99% of its original value after only 4,5 billion years. I remember writing somewhere that, "because of this depletion, there is unfortunately no hope to investigate the Big Bang yield of lithium by looking at very metal poor stars, (just as we do for ^4He)" . In 1982, not only did the Spite find lithium on stars much older than the Sun , but, because of the fact that the abundances measured were quite the same in various stars, they argued that little depletion had taken place and that they had essentially recovered the Big Bang value. This was counterintuitive to everything I believed at that point.

In their first papers, the number of stars was quite small and the statistical argument about the "plateau " was not very strong. I felt that this could be just another trick of statistics and that more data would probably give a different picture. Wrong again! Not only did the Spite come out with more stars with the same lithium abundances but two other groups (Rebolo et al 1988) (Hobbs and Duncan 1987) confirmed the findings with fresh data .

Later on, the explanation for the low depletion of lithium was found . In view of the low metal abundance of these stars, the low opacity of the surface layers reduces considerably the depth of the convective zone with respect to Pop I stars, thereby accounting for the conservation of lithium.

The real issue however, in my mind, was the fact that this interpretation opened the need for a *third* source of ^{7}Li to account for the PopI abundance. In this picture the explanation of the present ^{7}Li abundance requires that *three different mechanisms , completely unrelated, should have contributed amounts of 7Li which differ by less than one order of magnitude!* A very special situation indeed: I know of no other nuclide for which this is the case. My intuition made me suspicious about this situation . I kept thinking that somehow the stellar source would disappear. The recent data of Lemoine et al convinced me that this will not be the case.

The criterion of "naturalness" is most often a fruitful one. But not always. Sometimes, reality is most "unnatural".

References

Abia, C., and Canal, R. 1988 A&A **189,** 55.

Alcock, C.R., Fuller, G.M., and Mathews, G.J. 1987, Ap. J., **320**, 439.

Anders, E and Grevesse, N., 1989 Geochimica and Cosmochimica Acta , **53, 197.**

Applegate, J.H., Hogan, C.,and Sherrer R.J. 1987, Phys Rev., **D35,** 1151.

Applegate, J.H., and Hogan, C. 1985 Phys . Rev. **D30** 3037.

Arnould, M., Forestini, 1989, Proceedings of La Rabida School on Nuclear Astrophysics, Springer Verlag Berlin.

Audouze,J., Boulade, O, Malinie, G., and Poilane, Y.,   1983 A&A **127** 164 1983

Audouze, J and Tinsley, B.M. 1976, ApJ. **192** 487

Austin, S.M., 1981 Prog. in Particle and Nucl. Phys. **7,**1

Baade, D and Magain. P , 1988, A&A **194** 237

Beaudet, G., and Reeves, H.,   1984 A&A **134,** 240

Bernard, C et al , Aug 199 Report on "Hot Summer Daze, Brookhaven National Lab. 1

Boesgaard, A.M. , Steigman, G., 1985 Ann.Rev. Astr. Ap. **23** 319.

Boyd.R.N., Ferland G., and Schramm, D.N. 1988 Fermilab Pub 88 /132A

Boyd, R.N and Kajino, T. 1989 Ap.J. **336** , L55

Brown L.E., 1992 Ap.J. **349,**   251.

Cesarsky, C, J 1992 private communication.

Cesarsky, C.J., Cassé , M., Paul. J 1977, A&A **60** 139

Charbonnel, C.,Vauclair., and Zahn, J.P: 1992 A&A **298**

Charbonnel, C.: 1992 Thesis, Observatoire de Midi-Pyrénées.   Toulouse. France

David, Y., and Reeves,1980 H., Phil. Trans.R. Soc.Lond. A **296**, 415,

D'Antona , F., Matteucci, F. 1991 A&A **248**, 62.

Dearborn, D.S.P., Schramm, D.N., Steigman, G., and Truran, 1989 ApJ **347** 455

Delyannis, C., Demarque, P., and Kawaler, S.D. 1990 ApJS **73** 21

Duncan, D.K. Lambert, D.L. and Lemke, M 1992 The abundance of boron in three halo stars. Preprint Nasa /Esa Hubble Space Telescope.

Ferrando , P. 1992 private communication

Ferrando, P.Lal, N., McDonald, F.B. Webber, W.R. 1991 A&A **247** 163.

Fowler, W.A., Greenstein, J.E., and Hoyle, F. 1962 Geophys.J. Roy. Astron. Soc. **6**, 148

Fukujita, M., Hogan, C., 1991 Nature **354** ,17

Fuller, G.M., Mathews, G.J., and Alcock, C.R. **1988**, Phys. Rev., D37, 1380.

Geiss, J., Eberhardt, P and Signer, P. 1966 Experimental determination of the solar wind
     composition. Proposal to NASA for Apollo experiment S-180.

Geiss, J., and Reeves, H., 1972 A&A **18,** 126,

Geiss, J., 1992 this symposium.

Gilmore, G., Edvardsson,B;, and Nissen, P.E. 1991 ApJ,**378,** 17.

Gottlieb, S., Lattice 90 , Tallahasse, 1991 Nucl. Phys. B. (Proc, Supp.) **20**, 247.

Hobbs, L.M., and Duncan, D.K., 1987 Ap.J. **317,** 796,

Kajino, T.   and Boyd, R.N. 1990 APJ **359,** 267.

Krauss,L.M., and Romanelli, P. 1990 Ap.j. **358**,47.

Kurki-Suonio, H.,1991 Baryon inhomogeneity from the cosmic quark-hadron phase transition in
     "Strange Quark Matter in Physics and Astrophysiscs" Aarhus Denmark

Kurki-Suonio, H.,Matzner, R.A., Olive, K.A. and Schramm, D.N.1990 Ap.J. **353,** p. 406,

Lansky , J.L.,Brown, A., Gayley, K., Diplas, A., Savage, B.D., Ayres, T.R., Landsman, W.,Shore,
     S.N., and Heap, S.H. 1992 D in front of Capella Preprint.

Lemoine, M., Ferlet,R.,Vidal-Madjar,A., Emerich, C.,   Bertin, P., 1992 to appear in AA

Malaney, R.A and Fowler W.A., 1989 ApJ. **345** L5

Malaney, R.A.,and Fowler, W.A., 1988in Origin and Distribution of the Elements, ed. Mathews,
     G., J., World Scientific, Singapore

Malaney, R.A., and Alcock.C.R. 1989 ApJ **351** 31 1990UCRL 101226

Mason, B.1971 , Handbook of Elemental Abundances in Meteorites (New York: Gordon 1
     Breach)

Mathews, G.J. Alcock, C.R. Fuller, G.M. 1990 Ap.J. **349** 449

Meneguzzi, M., Audouze, J., and Reeves, H., 1971 A&A 15, 337,

Meneguzzi, M., and Reeves. H, A&A 1975   **40,** 91

Meyer, J.P. 1985 ApJS **151** 173

Meyer, B., S. and Schramm , D.N.,1986 Ap.J . **406** 150 E1

Michaud, G and Charbonneau,P 1991 Space Science Reviews **57** 1

Montmerle , T. 1977 Ap . J . **217** 872 .

Ormes, J.F., Protheroe , R.J. 1983 Ap.J. **272** 756

Pagel, B.E.J. 1989 in Evolutionnary Phenomena in Galaxies eds J.E. Beckman and B.E.J. Pagel
    Cambridge University Press.

Pagel , B.E.J., 1987 in "A unified view of the macro- and micro-cosmos" First International School
    on Astroparticle Physics, Erice (Sicily, Italy) Ed: A. De Rujula, N.V. Nanopoulos, P.A.
    Shaver , World Scientific Singapore.

Pagel, B.E.J 1992 this symposium.

Pellas, P Poupeau, G., Lorin, J.C. Reeves, H., and Audouze, J 1969 Nature **223** 272

Pinsonneault, M.H., Delyannis, C.P., Demarque, P. : 1992 Ap.J. Supp **78** 179

Prantzos, N., Cassé, M., V angioni-Flam , E., 1992 preprint Institut d'Astrophysique de Paris .
    (PCF)

Read, S.M., and Viola, V.E., Jr 1984 . At.Data . Nucl. Tables, **31,**359

Rebolo , R., Molaro, P., and Beckman, J.E. 1988 A&A **192**, 192.

Rebolo,R., 1990Proceedings of the Cambridge Conference on Elements in the cosmos

Rebolo, R., Garcia-Lopez, R.J., Martin, E.L., Beckman, J.E., McKeith, C.D., Webb, J.K.? and
    Pavlenko, Y.V. 1992 this symposium.

Reeves, H., Fowler, W.A., and Hoyle, F., 1970 Nature **226,** 727

Reeves,   1971, p . 256 , American Physical Society Meeting , Porto Rico Dec
    1971.

Reeves, H ., 1972 Phys Rev D **633** 63.

Reeves, H., Audouze, J., Fowler, W.A. and Schramm, D.N. 1973 ApJ **177** 909

Reeves, H. 1974, Ann. Rev. Astr. Ap, **12,**437.

Reeves, H., and Johns, O.,1976 Ap. J. **206** 958

Reeves,H., and Meyer J.P. 1978, Ap.J. **226,** 613.

Reeves, H., Delbourgo-Salvador, P, Audouze, J., and Salatti, P. 1988, European Journal of
    Physics ,**9,** 179.

Reeves, H., 1990 Phys.Rep , **201,** 335

Reeves,H.,Richer,J., Sato, K., Terasawa, N.: 1990, Ap.J. **355,** 18

Reeves, H 1991 A&A **244**, 294. -

Reeves, H 1992 - to appear in A&A

Ryan, S.G., Norris, J.E., ans Bessel, M.S. 1991 AJ, **102** 303.

Ryan, S.G., Bessell, M.S., Sutherland , R.S., and Norris, J.E. 1990 Astrophys. J. Lett L **348** L57

Ryter, C., Reeves, H., Gradstztajn, E., and Audouze, J., 1970 A&A **8** 387

Sahu, K.C., Sahu, M., and Pottasch,S.R.,1988 A&A 207   L1

Schaeffer, R, Reeves, H and Orland, H 1982 Ap.J.**254** 688 1

Smith, M.S., Kawano, L.H., Malaney, R.A. 1992 Orange Aid Preprint 716. submitted to Ap.J.

Spite, M, and Spite, F. 1982a, Nature, **297,** 483.

Spite , F and Spite M. 1982b A&A . **115,** 337.

Spite, F., Spite, M., Cayrel, R., and Huille, S. 1991 Li on Pop II stars IAU Symposium 149.

Steigman, G , 1986 in Nucleosynthesis and its Implications on Nuclear and Particle Physics 45-55
        J.Audouze and N.Mathieu eds

Steigman, G., and Walker, T.P., 1992 ApJL **385** L13

Terasawa, N., Sato, K., 1989 Prog. Theor. Phys. Lett **385** L13

Terasawa, N., 1992 This symposium.

Thomas, D., Schramm. D.N., Olive, K.A., and Fields, B.D. 1992 UMN-TH 41020/92

Toussaint, D. 1992 Lattice 91 Tsukuba, Nucl. Phys. B. (Proc, Supp.)

Vangioni-Flam   ,E, and Audouze, J., 1988 A&A **193,** 81,

Vangioni-Flam   ,E., Cassé, M., Audouze, J., A., and Oberto, Y., 1990Ap.J. **364**, 568.

Vauclair, S., 1988a, p. 269 , in Dark Matter Ed J. Audouze and Tran Van . Editions Frontières . Gif
        sur Yvette , France

Vauclair, S., 1988b Ap.J., **335** 971

Vidal-Madjar , A., . Adv. Space Res.vol **11** no 11 p 97 -103 1991

Wagoner, R. V., Fowler, W. A., and Hoyle, F. 1967 , Ap.J. **148,** 3,

Walker, T.P., Steigman, G., Schramm, D.N., Olive, K.A., and Fields, B. . Preprint march 1992

Walker,T, Steigman , G, Schramm, D.N., Olive K.A., Kang H;S; 1991 AP J **376** 51

Wasserburg , G.J., 1985 in Protostars and Planets II p. 703 ed D.C.Black and M.S. Matthews
        University of Arizona Press , Tucson, Arizona

Webber, W.R. Ferrando, P., Lukasiak, A., and McDonald, F.B., 1992 ApJ **392 L** ,91

Witten, E., 1984 Phys. Rev. **D30,** 272

Woosley, S.E.,Hartman,D.H.,and Hoffman, R.D. and Haxton,W.C.   1990 ApJ. **356,** 272.

Yang, J., Turner, M.S., Steigman, G., Schramm, D.N. and Olive, K., 1984 Ap. J. **281,** 493,

LITHIUM STORY

Observational status of the lithium abundance

F. and M. SPITE

Observatoire de Paris-Meudon F-92195 Meudon Cedex (France)

1 INTRODUCTION

It is a pleasure to present this review in a symposium in honor of Hubert Reeves, who was always ready to examine kindly the data provided by observers, even if their interpretation turned out to be a headache for all of us.

In the limited space allowed here, we will only try to update the reviews already made in the past (Boesgaard & Steigman 1985, Michaud & Charbonneau 1991, Rebolo 1989, 1991, Boesgaard 1990; a wealth of data and ideas is gathered in D'Antona 1990) but even within this more limited scope, we will have to restrict ourself to a limited number of the numerous papers published recently about the lithium problem, and we beg the pardon of the authors who will not be mentioned here.

Let us briefly recall that in the review of Boesgaard & Steigman (1985), two scenarii are proposed for the evolution of the lithium abundance in the Galaxy. In broad lines, the first scenario proposes a "high" primordial lithium abundance around N(Li)/ N(H) = 10^{-9} or log N(Li) = 3 dex in the scale where log N(H) = 12 dex. This primordial abundance would be found in the presently observed young matter of the Galaxy, explaining the rather uniform abundance in these various objects. Lithium would be progressively destroyed in stars. The second scenario proposes a "low" primordial Li abundance around log N(Li) = 2 dex which would be reflected in the uniform Li abundance found in the old "warm" (Teff > 5500 K) Pop II dwarfs. An efficient production of Li in the Galaxy is then necessary in order to reach the "high" (ten times higher) Li abundance found in the Population I.

It is not clearly established which of these scenarii is near the reality (see however the discussion by Hubert Reeves, this symposium). We will try to concentrate here on the observations which show more or less indirectly (most of the time indirectly !) the signature of the essential processes in the lithium problem.

2 LITHIUM IN POP II STARS

It is now well accepted that the "warm" dwarfs which have both halo kinematics and a metallicity lower than [Fe/H] = −1.4 dex, have a very similar Li abundance. The exact value of this abundance is dependent on the technique of determination of the temperature of the stellar model. Up to now, the temperatures have been determined by using standard calibration of the stellar colors. The new technique of Blackwell et al. (1980) has definite advantages. Its use, through the Magain's formulae (Magain 1987), seems to shift the temperature toward higher temperatures, pushing the Li abundances toward higher values : the increase could be typically 100K for the temperatures and 0.1 dex for the Li abundances. So that the Li abundance of the plateau, rather than the value of Molaro (1991) log N(Li) = + 2.08 dex, should be about 0.1 dex higher, near 2.18 dex (Spite's 1992). The higher value may be more consistent (in the frame of the second scenario) with the combination of both cosmic ray production and primordial nucleosynthesis (Steigman & Walker 1992).

It seems that the interest of the astronomers has shifted towards more subtle questions : is the plateau really flat or with a (small) slope ? Is the scatter around the plateau due only to the determination errors or partly to a real Li abundance scatter? These questions are much more difficult to answer, and are probably pushing the theory of stellar atmospheres to its present limits.

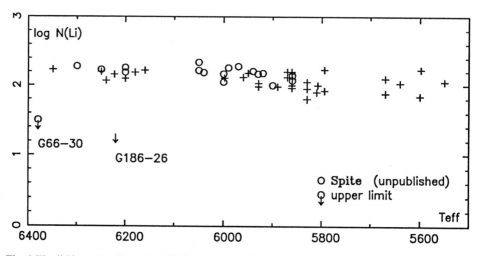

Fig. 1 The lithium abundance log N(Li) versus the effective temperature for the dwarfs more metal-poor than [Fe/H] = -1.4 dex. Two stars have a low lithium abundance by a factor of 6 to 10.

2.1 Strong Differences for Three Exceptional Dwarfs

Three stars have a lithium abundance clearly below the mean. Two such exceptions have been presented : by Hobbs et al. (1991) and Molaro (1991). Very recently, spectra obtained at ESO by Patrick François, show a star with little Li : the star G 66-30 = Wolf 550 : its metallicity (1/40 solar) and its kinematics (high velocity) place it among the Pop II stars. The upper limit of its lithium abundance is about 1.5 dex.

Why such a low Li abundance ? These three stars are a too small proportion of the halo dwarf sample for being the signature of a general destruction process. This could well be an analogy with the conclusions of Lambert et al. (1991) for the Pop I stars, who suggest an occasional and rare depletion process. These three exceptional stars could be spectroscopic binaries but the binarity is not a likely explanation (a large proportion of the stars on the plateau are binaries). Also, this exceptional Li deficiency could be linked to the existence in Pop II of the equivalent of the dip discovered in the Pop I by Boesgaard and Trippico (1986a,b), although not ALL the warmest Pop II stars have this exceptionally large deficiency. Alternately, these stars could be evolved off the main sequence, being now subgiants, with a significant early depletion : it would be very interesting to determine accurately the ionisation equilibrium in the stellar atmospheres of these stars.

2.2 Moderate Differences for most Dwarfs

Some recent observations of "normal" halo dwarfs have been made (Hobbs & Thorburn 1991, Hobbs et al. 1991, Spite et al. 1992a, 1992b) It has been suggested that the scatter of the Li abundance, observed around the mean value of the plateau, is larger than the expected uncertainties of the Li abundance determination (Deliyannis et al. 1992). The quantitative explanation of the scatter is obviously a much more difficult task than the abundance determinations.

Following an interesting idea of Hobbs, the authors looked into the correlation of two direct measurements (the color of the star and the equivalent width of the lithium line), that do not suffer from the uncertainties (and scatter) of the calibration of the colors into temperature and of the translation of equivalent widths into abundances. A scatter is obviously present around the mean relation : color versus equivalent width of the Li line. The scatter is too large to be attributed only to the measurement errors. Another source of scatter has to be present. Deliyannis er al. conclude that the additional cause of the scatter is the real scatter of the Li abundance in the halo dwarfs.

It has to be noted that it is *not* possible to translate the equivalent width W into abundance without the knowledge of the temperature, and we are interested in abundances, not in equivalent widths. The real stars are (unfortunately) much more complex than our simple models, so that there is no perfect relation between color and temperature : this is seen when two independent colors of the same stars are translated into temperatures. Even taking care of the systematic difference, a scatter remains, which seems larger than the scatter computed from the measurement errors.This additional scatter seems large enough for explaining the whole scatter observed in the relation color/equivalent width. It is therefore not yet proven that Li abundance show a real scatter in the warm Pop II dwarfs.

A real scatter of Li abundance could be an argument in favor of a high primordial Li abundance, decreased by the action of a drastic depletion process (the first scenario noted in the introduction). However, it is not the proof of such a process, since other interpretations are possible, for example, in the second scenario, a primordial low Li abundance, will be altered here and there by the ejecta of supernovae, (ejecta which are metal-rich but Li poor), so that the primordial Li content is (slightly) diminished and this decrease may be different from star to star. About supernovae ejecta, if the second scenario is adopted, the remakable stability of the Li abundance, in the halo dwarfs implies that the material ejected from the contaminating supernovae is necessarily very metal-rich and H-poor (so that the little amount of H mixed to the gas does not change the ratio Li/H). This obvious point, necessarily incorporated in models (for example Vangioni-Flam et al. 1990) has been pointed out by Reeves and Pagel, but it is not unuseful to recall it here.

2.3   N-rich Dwarfs

A few dwarfs are enriched in nitrogen. They have a "normal" lithium content in their atmosphere (Spite & Spite 1986). It is therefore very unlikely that this N-enrichment comes from a mixing of CNO processed internal material. Recently it was found that aluminum (Magain 1989) and sodium (Laird & Da Costa 1989) were also enhanced in these stars, which would then be the dwarf counterpart of the anomalous giants in the globular clusters which have strong Al, Na and CN spectral features. Therefore, this kind of anomaly is not linked with mixing, neither on the main sequence (as it had been proposed) nor on the giant branch, but with an abnormal initial chemical composition of the matter which formed the star.

The study of these peculiar stars seems at first glance related only to a minor problem of stellar evolution : it is however directly linked with our understanding of the mixing of the products of the C N O cycle in dwarfs and giants, and this mixing is not yet well predicted by the theory, and remains an important process for our understanding of the depletion of lithium in subgiants and giants. Moreover, the N-rich matter, mixed to the primordial matter, had to be highly concentrated (H-poor) : see preceding section 2.2.

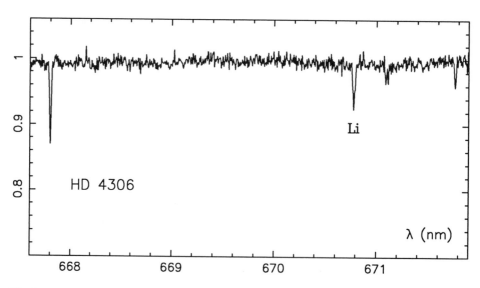

Fig. 2 The spectrum of the very metal-poor giant HD 4306. The lithium line is strong (stronger than the calcium line at the right). The lithium abundance is higher than the abundance expected from the dilution factor.

2.4 Population II Giants with a Li Line

The Pop II giants generally have no detectable Li line, owing to the dilution of Li. A few exceptions are known in globular clusters. A very metal-poor field giant, HD 4306, has a detectable Li line (Spite et al. 1992) : the corresponding lithium abundance is log N(Li) = 1.0 dex. This is significantly larger than the value expected from the dilution (Michaud & Charbonneau 1991) of its previous abundance on the main sequence. A metal-poor giant is also found by Barbuy et al. (1992) with a similar anomaly. These stars could be the Pop II counterpart of the few Pop I giants which are Li-rich (see discussion below, section 3.5).

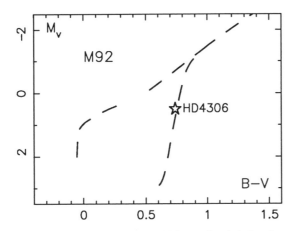

Fig. 3 The giant HD 4306 has about the same metallicity as the globular cluster M92. A cluster star of the same color would have a luminosity as high as Mv =0.5, indicating that the star is highly evolved, and that a large dilution factor is expected.

3 POP I STARS

3.1 Young Stars

Cayrel & Cayrel (1989) analysed a field star which, in spite of its late type and low temperature, has a strong lithium line. The star has to be very young, its kinematics (see G. Cayrel de Strobel, this symposium) shows that indeed it probably escaped from the young Sco-Cen association. In a more general work, Martín et al. (1992) have measured the lithium abundance in the cool companions of massive hot (and therefore young) dwarfs and find in these young stars the classical lithium abundance : the maximum abundance found is log N(Li) = 3.7, that the authors reduce to 3.4 dex, taking into account a NLTE correction. This result is extremely important. I would only propose a word of caution about one of the conclusions of the authors, claiming that this result rules out the inhomogeneous Big Bang theory, which predicts a larger primordial abundance of Li : as said also in the conclusion, the present Li abundance observed in the young objects in the Galaxy is the result of an equilibrium between astration and production : it is not necessarily equal to the primordial abundance (except within the hypothesis of a predominant infall), and this more complex situation is of course taken into account in the models of the Li abundance in the Galaxy (D'Antona & Matteucci 1991, Mathews et al. 1990).

3.2 T Tauri Stars

These are even younger stars, but unfortunately their structures (disk, accretion, activity) are much more complex than the structures of "normal" dwarfs and subgiants. Sometimes, high Li abundances have been claimed for the T Tauri stars (as high as log N(Li) = 4.0 dex or more). We will note here the work of Duncan (1991) who shows by the careful analysis of both Li lines of a T Tau star, that the high Li abundance claimed for this star has to be reduced to a "normal" value. In a very important work, Padgett (1990) measured both the resonance lines of lithium and potassium in pre-Main Sequence stars of the Taurus-Auriga region. The stars with the stronger Li lines, leading to a very high Li abundance, also have stronger K lines showing the signature of atmospheric effects. Moreover, the Li lines are generally lying on the flat part of the curve of growth, and this situation leads to an amplification of the uncertainty in the abundance determination. Obviously the high lithium abundance found sometimes in such stars should not always be taken at face value. Several recent determinations of Li abundances have been made in T Tauri stars (let us quote here : Basri et al. 1991, Magazzù et al. 1992), leading to lithium abundances compatible with the abundance in young stars and young clusters (see also Martín et al., this symposium).

Staude & Neckel (1991) observed systematically an FU Orionis type star, and found a variable Li line, which of course should not be hastily interpreted as a variable Li abundance.

As a conclusion, owing to the recent works in the field, the excess of Li abundance (relative to the normal young stars) sometimes claimed to be found in the T Tauri stars, is no more obvious, and it may be now estimated that the Li abundance found in the T Tau is about the same as the abundance found in the young objects of the Pop I and in the meteorites of the Solar system, possibly on the high side. The final solution will come from the elaboration of good models for the T Tauri stars.

3.3   Active Stars

The large lithium abundance found in active and RS CVn stars (relative to normal stars) has triggered a number of works. Such objects may be young, especially the single dwarfs, but some giants and some binaries are member of an old population. For the typical star FK Com (a high luminosity star), Phillips & Lestrade (1990) propose that the star is a young object retaining the initial high Li abundance of the Pop I. Fekel (1988) had suggested that the active stars, with high rotational velocity, could be young objects, specifically stars evolved from a massive Main Sequence (A type) star, keeping some of the angular momentum of the progenitor. But Fekel & Marschall (1991) observed an active star with the kinematics of an old population, and Fekel & Balachandran (1992) note counter-arguments against this hypothesis of retaining the main sequence angular momentum.

Pallavicini et al. (1992a) observed 65 active or RS CVn stars and found in some of them an abnormally high Li abundance, which has to be attributed to one or several processes, one of them could be the enhancement of Li line by spots. However, Pallavicini et al. (1992b) contest that spots have such a large influence on the strength of the Li line. Is another process acting ? A very important suggestion of Fekel & Balachandran (1992) is a possible Li (Be) production, deep inside the star, combined with a dredge up caused by the transfer of angular momentum between the core and the envelope, transfer which is otherwise a good possible explanation of the rapid rotation of the envelope.

The problem of lithium abundance in the active stars is obviously a complex problem, which will require progresses on both the observational and theoretical sides.

3.4 Field Dwarfs

Numerous measurements were made by Balachandran (1990a,b) in Main Sequence and slightly evolved F-type stars. Her extensive work shows that, when Li depletion is found in evolved stars, it happens that in its previous stay on the Main Sequence, the star was in the temperature range of the Boesgaard's dip. This important point is confirmed by another extensive observation of bright Pop I stars by Lambert et al. (1991). This extremely interesting work brings also a lot of information about the behavior of Li in the Pop I stars an deserves a careful reading.

Stars with a peculiar chemical composition form a wide topic which will not be summarized here : see the recent IAU Symposium 145 (1991, Michaud & Tutukov eds.) and the imminent IAU Colloquium 138 (Trieste, June 1992).

3.5 Field Giants

Extensive observations of giants have been made by Brown et al. 1989 who found that the Li abundance in the giants is generally lower than expected, following the evolution theory, which predicts a dilution by mixing. For investigating this effect, Gilroy (1989) and Gilroy & Brown (1991) observed the $^{12}C/^{13}C$ ratio in cluster giants for an evaluation of the mixing, and found that the mixing was observed earlier than predicted by the theory. But a few giants have a high Li abundance similar to the Li abundance of the young Pop I objects. Is the mixing inhibited in such giants ? Or some Li produced and mixed to the surface ? The case of the giant HD 39853 is puzzling : its kinematics and metal-deficiency (Gratton & D'Antona 1989) points towards a rather great age, and therefore the progenitor should have a moderate mass, so that the star is not expected to produce Li (see however the suggestion of Fekel & Balachandran 1992). Da Silva & De La Reza (1992) analyzed also this star and two other Li-rich giants, using the fainter line at 610.4 nm. One of them (HD 19745) has a large Li abundance : log N(Li) > 4.0 dex, larger than the Li abundance found in the young Pop I objects (see De La Reza, this Symposium).

The high Li abundance found in some giants remains unexplained : some progresses on both observational and theoretical sides are eagerly awaited for.

3.6 The Production of lithium

At least the second scenario, mentioned in the introduction, requires physical processes able to produce lithium. Lithium could be produced in Novae, in neutron stars, in some special processe in supernovae, in the environment of blackholes : the observations of the enigmatic object V 404 Cygni (Wallerstein 1992) could bring essential information on the subject (see the presentation of Rebolo in this symposium). Some exciting progresses have been made about the production of Li in Cosmic Rays ($\alpha + \alpha$ reaction, Steigman & Walker 1992), and last but not least by red giants.

The production of lithium in red giants, which has been suspected, for a long time, as the only viable explanation of the high Li content of the carbon stars, has been caught in the act by Smith & Lambert (1989, 1990) in evolved red giants. This extremely important work brings a new point of view on the whole lithium problem : let us recall that the Li enrichment of the gas by stellar winds requires that these winds have a concentration Li/H higher than the the concentration in the gas.

More information is now available about the amount of Li in the carbon-stars (see the presentation of Abia, this symposium).

4 BINARIES IN POP I AND II

The rotation of stars should play an important rôle in lithium destruction and/or depletion. It could be expected that the analysis of binaries should throw some light on the processes of lithium destruction. Unfortunately, the few well documented observations are rather puzzling.

In a binary of the old disk (Saar et al. 1990) the lithium line is strong in spite of a low temperature estimated to be around $T = 4500K$. The kinematics shows that the star should be old, and the circularization of the orbit points towards the same direction. A depression in the far red spectrum indicates that the star is spotted, and the authors found the signature of a flare during their systematic observing of the star.

The reason for the presence of a strong Li line in such a cool star is unclear : is the line enhanced in the spots? Is the Li abundance really high ? Is the Li depletion inhibited by magnetic fields ? The star seems to be evolved from the main sequence; it may have some similarity with the old disk active binary BD $-0°$ 4234 (Spite et al. 1984).

The binary HD 89499 is different : Ardeberg & Lindgren (1991) find that this cool metal-poor star has a very low metallicity, and a low temperature ($T = 5300K$). The Li abundance is about 1.5 dex (Spite's 1992). The orbit of this binary seems to be circularized; it is not only a binary, it is also an active star (Pasquini et al. 1991), however its Li abundance is about the same as the abundance of the (assumed single) halo dwarfs of the same temperature. The different rotation history of the star did not change the Li abundance.

The influence of rotation, the presence of a strong Li line in some cool old stars (Pop I and II) remains puzzling.

5 ISOTOPIC RATIO IN STARS AND INTERSTELLAR MATERIAL

Since ^{6}Li is more fragile than ^{7}Li, its detection in a halo dwarf would be the proof of the ^{7}Li preservation in this dwarf (Proffitt & Michaud 1989, Deliyannis et al. 1990), and moreover this would be an argument in favor of the "standard model" since ^{6}Li is destroyed in the rotational model. However, only an upper limit has been found, in one of the brightest and the warmest halo dwarfs, by Pilachowski et al. (1989) : more recent analyses provides preliminary results which are rather similar (Smith et al. 1992, Molaro & Spite 1992).

Cayrel (1990) has analysed the isotopic ratio in Hyades stars near the Boesgaard-Tripicco's dip, but this is a very difficult observation, and only an upper limit of the ^{6}Li isotope could be found.

In the interstellar medium, the new results of Lemoine et al. (this symposium) in the direction of the star ρ Oph provide an isotopic ratio for the young interstellar matter : the value of this ratio is in agreement with the meteoritic value. This very important measurement is used by H. Reeves (this symposium) in a discussion providing essential clues about the origin and evolution of Li in our Galaxy.

6 OUTSIDE OUR GALAXY

In our Galaxy, it is well known that some supergiants show a detectable Li line. The Li line has also been detected in some supergiants of the Magellanic Clouds (Spite et al. 1986, Spite 's 1992) but not all of them : perhaps some of them are on their first crossing toward the red, and some are on a "blue loop", but before trying to reach a conclusion, more data are required.

The Supernova 1987 has provided an unrivalled opportunity to observe in detail the interstellar medium in the Large Magellanic Cloud. Even in the combined data of all the observers (Baade et al. 1991), only an upper limit of lithium abundance can be found, and moreover the Li abundance in the interstellar matter depends on a difficult correction : the amount of Li depleted by condensation on the grains : this amount is not well known, and owing to the large final uncertainty, this collective effort unfortunately did not reach a definite conclusion (a high primordial Li abundance is not ruled out).

7 CONCLUSION

Many points have been enumerated, which are not understood or which have several possible interpretations. Future observations will bring more extended material, and also a new kind of information : the future multi-object spectrographs and/or the future large telescopes will enable to describe the lithium behavior in stars of old metal-poor open and globular clusters, in giants of the Magellanic Clouds, will enable the measurements of isotopic ratios etc. providing a useful basis for a better interpretation of lithium, the most elusive (but most fascinating) element.

REFERENCES

Ardeberg A., Lindgren H.1991, A&A 244, 310

Baade D., Cristiani S., Lanz T., Malaney R. A., Sahu K. C., Vladilo G. 1991, A&A 251, 253.

Balachandran S.1990a, ASP Conf. N° 9, Wallerstein G. ed., San Francisco p. 357

Balachandran S. 1990b, ApJ 354, 310

Barbuy B., Gregorio-Hetem J., de Freitas Pacheco, J. A. 1992 (preprint)

Basri G., Martín E. L., Bertout C. 1991, A&A 252, 625

Beers T. C., Preston G. W., Schectman S. A. 1992, AJ 103, 1987

Boesgaard A. M. 1990, in : Sixth Cambridge Workshop, ASP Conf. 9, G. Wallerstein, ed., p. 317

Boesgaard A. M., Steigman G. 1985, Ann. Rev. Astr. Astrophys. 23, 319

Boesgaard A. M., Tripicco 1986a, ApJ 302, L49

Boesgaard A. M., Tripicco 1986b, ApJ 303, 724

Brown J. A., Sneden C., Lambert D. L., Dutchover E., Jr. 1989, ApJS 71, 293

Cayrel R., 1990, Communication at the Montreal CFHT meeting (unpublished).

Cayrel de Strobel G., Cayrel R. 1989, A&A 218, L9

D'Antona F., Matteucci F. 1991, A&A 248, 62

D'Antona F. 1990, editor of "The problem of Lithium", Mem. Soc. Ital. Astr. 62

Da Silva L., De La Reza R. 1992, IAU Sympos. 149, B. Barbuy & A. Renzini, eds., Kluwer, p. 476.

Deliyannis C. P., Demarque P., Kawaler S. D. 1990, ApJ Sup 73, 21

Deliyannis C. Pinsonneault M. H. 1992 preprint

Duncan D. K 1991, ApJ, 373, 250

Fekel 1988 : "A decade of UV astr. with IUE " ESA SP-281, Vol. 1, 331.

Fekel F. C., Quigley R., Gillies K., Africano J. L. 1987, AJ 94, 726

Fekel F. C., Marschall L. A. 1991, AJ 102, 1439

Fekel F. C., Balachandran. S. 1992 (preprint)

Ferlet R., Dennefeld M. 1984, A&A 138, 303

Gilroy K. Krishnaswamy 1989, ApJ 347, 835

Gilroy K. K., Brown J.A. 1991, ApJ 371, 578

Gratton R., D'Antona F., 1989 A&A 215, 66

Hobbs L.M., Thorburn J.A. 1991, ApJ 375, 116

Hobbs L. M., Welty D. E., Thorburn J. A. 1991, ApJ 373, L47

Laird J. B., Da Costa L. N. 1989, BAAS 21,742.

Lambert D. L., Heath J. E., Edvardsson B. 1991, MNRAS 253, 610.

Magain P., 1987 A&A 181, 323

Magain P. 1989, A&A 209, 211.

Magazzù A., Rebolo R., Pavlenko Y. V. 1992, ApJ 392, 159

Martín E. L., Magazzù A., Rebolo R. 1992, A&A 257, 186

Mathews G. J., Alcock C. R., Fuller G. M. 1990, ApJ 349, 449

Michaud G., Charbonneau P. 1991, Space Sci. Rev. 57, 1

Molaro P. 1991, Mem. S. Astr. It. 62, 17

Molaro P., Spite M. 1992, in preparation.

Padgett D. L. 1990 ASP Conf. N°9, Wallerstein G. ed., San Francisco p. 354

Pallavicini R., Randich S., Giampapa M. S. 1992, A&A 253, 185

Pallavicini R., Cutispoto G., Randich S., Gratton R. 1992 (preprint)

Pasquini L., Fleming T., Spite F., Spite M. 1991, A&A 249, L23

Phillips R. B., Lestrade J.-F. 1990, BAAS 22, 832.

Pilachowski C. A., Hobbs L. M., De Young D. S. 1989, ApJ 345, L39

Pilachowski C. A., Sneden C., Hudek D. 1990, AJ 99, 1225

Proffitt C. R., Michaud G. 1989, ApJ 346, 976

Rebolo R. 1989, Astrophys. Space Sci. 157, 47

Rebolo R. 1991, IAU Sympos. 145, G. Michaud & A. Tutukov, eds., Kluwer

Saar S. H., Nordström B., Andersen J.:1990, A&A 235, 291

Smith V. V., Lambert D. L.: 1989 ApJ. 345, L75

Smith V. V., Lambert D. L.: 1990, ApJ 361, L69

Smith V. V., Lambert D. L., Nissen P.: 1992 (in preparation)

Spite M., Spite F. 1992 in preparation.

Spite M., Maillard J. P., Spite F. 1984, A&A 141, 56

Spite M., Cayrel R., Francois P., Richtler T., Spite F. 1986, A&A 168, 197

Spite F., Spite M., Cayrel R., Huille S. 1992a, IAU Sympos. 149, B. Barbuy & A. Renzini, eds., Kluwer, p. 490

Spite F., Spite M., François P. 1992b, ESO Workshop "High resolution with the VLT" (in press)

Staude H. J., Neckel T.1991, A&A 244, L13

Steigman G., Walker T. P. 1992, ApJ 385, L13

Vangioni-Flam E., Cassé M., Oberto Y. 1990, ApJ 349, 510

Wallerstein G. 1992, Nature 356, 569

Lithium in a sample of galactic C-stars

C. ABIA[1]

1: Dpt. Física Teórica y del Cosmos, Universidad de Granada, Avda. Fuentenueva S/N, 18071 Granada (Spain).

INTRODUCTION

Lithium is indeed a striking element. Its cosmic abundance is one of the lowest in the periodic table but, in contrast, it has crucial implications on cosmological theories and on the knowledge of the evolution and structure of the stars. Today it is broadly agreed that the primordial lithium abundance is $\log \epsilon(\text{Li}) \approx 2.1$ (but see Vauclair this volume) [note that $\log \epsilon(\text{Li}) \equiv \log [n(\text{Li})/n(\text{H})]+12$, where n refers to the atoms number density]. Since the actual cosmic Li abundance is found to be $\log \epsilon(\text{Li}) \approx 3.3$, a source of Li is needed in order to explain the increase of its abundance during the life of the galaxy. Spallation reactions in the interstellar medium (Meneguzzi, Audouze and Reeves 1971) was the first nucleosynthetic mechanism proposed for Li production in the galaxy. However, althougth this mechanism accounts for the observed abundance of the other *ligth elements* (Be and B), a mere 20% of the observed Li abundance might be produced in this way. Rouling out the possibility of a high primordial lithium abundance ($\log \epsilon(\text{Li}) \approx 3.0$), we have to appeal to some stellar source for Li production. Novae, supernovae and red giants are the proposed stellar objects (Arnould & Nørgaard 1975; Dearborn et al. 1989; Scalo & Ulrich 1973; Sackman et al. 1974) although up to now, only in red giants we have a clear observational evidence of lithium synthesis. In fact, the existence of some galactic AGB stars (S and C) with strong features of Li I at $\lambda 6707$ Å ($W_\lambda \approx 8$ Å) has been well known for some decades (Torres-Peimbert & Wallerstein 1966). Despite low temperature of S and/or C-stars ($T_{\text{eff}} = 2500 - 3000$ K) which can make the Li line prominent, it seems to be no doubt about the meaning of such strong features (Catchpole & Feast 1971): Li is actually produced in these stars. Recent Li abundance determinations in these stars report abundances in excess to $\log \epsilon(\text{Li}) = 4.0 - 5.0$ (Abia et al. 1991), which are one or two orders of magnitude higher than the abundance observed in the interstellar medium. The consequences on the galactic evolution of Li abundance when this Li-enriched material is put into the interstellar medium by steady mass loss were already pointed by Scalo (1976). However, to estimate correctly the contribution of S and C-stars to the galactic abundance of Li, the determination of Li abundances in a statitiscally significant number of galactic AGB stars is really called for. In this paper we present Li abundance determinations in a homogeneous sample of galactic C-stars.

THE SAMPLE OF STARS

The stars were selected from the Two Micron Sky Survey (those studied by Claussen et al. 1987; hereafter CKJJ). The main characteristics of the sample are the following:
a) They are withim a volumen of radius 1.5 kpc.
b) Their typical luminosity is 10^4 $L_\odot$, spanning a range in bolometric magnitude $-4.8 \leq M_{bol} \leq -5.2$.
c) Masses are in the range 1.2-1.6 $M_\odot$ (see CKJJ for details).
The observations were made at different observatories: La Palma and Calar Alto (Spain) and La Silla (Chile). Spectral resolution was relinquished in order to observe a number of stars as large as possible. Nevertheless, most of the spectra have $\lambda/\Delta\lambda \approx 1.5 \times 10^4$. Signal to noise ratios tipycally excedeed 100. A total number of 215 galactic C-stars were observed.

LI ABUNDANCE DETERMINATION.

The effective temperature was derived from the infrared index $(J-K)_J$ and the callibration by Heng et al. (1985). A gravity of log g = 0.0 was adopted for all the stars. The uncertainty in gravity is not important for the Li abundance derived. The microturbulence was likewise taken as $\xi = 2.2$ km s^{-1} for all the stars. Model atmospheres were interpolated in T_{eff} and CNO abundances from an unpublished grid of models for C-stars with solar metallicity by Eriksson and Gustafsson. Atomic and molecular (CN lines) gf values were taken from the list of Kurucz & Peytremann (1975) and Jørgensen & Larsson (1990), respectively. Synthetic spectra were computed in LTE. The error sources are several. Firstly, the C/O ratio adopted in the analysis: $\Delta C/O = \pm 0.05 \Rightarrow \Delta \log \epsilon(Li) = \pm 0.3$ dex. Secondly, the effective temperature: $\Delta T_{eff} = \pm 200$ K $\Rightarrow \Delta \log \epsilon(Li) = \pm 0.4$ dex. Thirdly, the microturbulence: $\Delta\xi = \pm 0.5$ km s^{-1} $\Rightarrow \Delta \log \epsilon(Li) = \mp 0.2$. Finally, the uncertainty in the continuum location. All these sources of error give a total error in the Li abundance derived typically of $0.4 - 0.5$ dex (see Abia et al. 1991 for details in the analysis).

RESULTS.

Figure 1 shows the main result of this work: the Li abundance distribution for the C-stars sample. The bin size for the abundance groups is 0.5 dex. Four abundance groups can be distinguished from this figure:
1) $\sim 8\%$ of the stars have very low Li abundance: log $\epsilon(Li) < -1.5$ is an upper limit to their Li abundance. It is noticiable that most of these stars present TiO bands, so they are not probably C-stars but S-stars (C/O< 1). This is in contrast with the

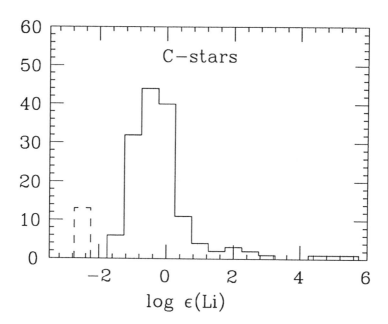

Figure 1. Lithium abundance distribution for the stars in this work. The bin size with dashed line corresponds to stars with $\log \epsilon(\mathrm{Li}) < -1.5$.

finding in the Magallanic Clouds: most of the S-stars observed there have strong Li I features (Smith & Lambert 1989-90).

2) The majority of the stars ($\sim 80\%$) have Li abundances in the range $-1.5 \le \log \epsilon(\mathrm{Li}) \le 0.5$, the mean Li abundance of the sample being $\log \overline{\epsilon(\mathrm{Li})} = -0.3$. This peak value nearly coincides with that found in galactic M and SC-stars (Luck & Lambert 1982; Kipper & Wallerstein 1990).

3) There is a small number of C-stars ($\sim 2\%$) which present very high Li abundance ($\log \epsilon(\mathrm{Li}) > 4.0$). These stars are known as super-rich Li stars. We confirm previous statistic on the matter. Nevertheless, despite this small percentage, they probably represent the main stellar source of Li known in the galaxy. The plume mixing model following a thermal pulse (Scalo & Ulrich 1973) is suggested as the scenario for Li production in these stars. Hot bottom burning would not work given the low temperature expected at the base of the convective envelope in AGB stars with masses lower than 2 $M_\odot$.

4) The most important result of this survey is perhaps the significative number of C-stars ($\sim 10\%$) rich in Li: $1.0 \le \log \epsilon(\mathrm{Li}) \le 3.0$. Some questions inmediatly arise. Are these rich Li C-stars in a previous (or subsequent) stage to the super-rich Li phenomenon? Is there a broad range of Li enrichment in C-stars? Have these stars preserved more efficiently their original Li? and if so, how? Obviously, more observational and theoretical efford is called for in order to answer these questions in

a near future.

This work has been supported in part by the grant PB87-0304 and the PICS *Origin of the ligth elements*. The author was supported by a postdoctoral fellowship of the Spanish Ministry of Education and Science.

REFERENCES

Abia, C., Boffin, H.M.J., Isern, J., Rebolo, R. 1991, A & A 15,337.
Arnould, M., Nørgaard, H. 1975, A & A 42,55.
Catchpole, R.M., Feast, M.W. 1971, MNRAS 154,197.
Claussen, M.J., Kleinmann, S.G., Joyce, R.R., Jura, M. 1987, Ap. J. Suppl. Ser. 65,385.
Dearborn, D.S.P., Schramn, D.N., Steigman, G., Truran, J. 1989, Ap. J. 347,455.
Heng, G., Chen, P., Zhang, Y. 1985, Act. Astrophys. Sin. 44,49.
Jørgensen, U.G., Larsson, M. 1990, A & A 238,424.
Kipper, T., Wallerstein, G. 1990, PASP 102,574.
Kurucz, R.L., Peytremann, E. 1975, SAO Special Report 362.
Luck, R.E., Lambert, D.L. 1982, Ap. J. 256,189.
Meneguzzi, M., Audouze, J., Reeves, H. 1971, A & A 15,337.
Sackman, I.J., Smith, R.L., Despain, K.H. 1974, Ap. J. 187,555.
Scalo, J.M., Ulrich, R.K. 1973, Ap. J. 183,151.
Scalo, J.M. 1976, Ap. J. 206,795.
Smith, V.V., Lambert, D.L. 1989, Ap. J. 345,L75.
Smith, V.V., Lambert, D.L. 1990, Ap. J. 361,L69.
Torres-Peimbert, S., Wallerstein, G. 1966, Ap. J. 146,724.

Lithium Abundances in Strong Li K-Giant Stars

RAMIRO DE LA REZA AND LICIO DA SILVA

Observatorio Nacional - Rio de Janeiro

1 ABSTRACT

Detailed NON LTE and LTE line profile calculations are made of neutral Li lines in strong Li K-giant stars. The stars presenting a stronger secondary Li I line at 6104 Å have a NON LTE Li abundance much larger than the corresponding interstellar abundance. These stars can then be considered as new candidates to sources of Li enrichment in the Galaxy.

2 INTRODUCTION AND METHODS

According to the standard theory of evolution (Iben,1967) Lithium must be strongly depleted in the atmospheres of cool giant stars. Nevertheless, a strong resonance line corresponding to the resonance transition of LiI at 6708 Å has been detected in some giant stars of types S,C and K. It is not clear if the same mechanism of stellar surface enrichment is acting in all these stars. In fact, the low mass K-giants are in a different evolutionary stage than the Li AGB massive S stars recently discovered in the Magellanic clouds (Smith and Lambert,1989). Also, the K-giants do not show the characteristic mixture process of S stars and of the known galactic Super Li Rich C giant stars. Independently of the physical mechanisms producing this enrichment, the presence of an eventual stellar Li overabundance respect to the interstellar medium, will place these K stars as important potencial new sources, by mass loss, of Li in the Galaxy.

To estimate these abundances, we select a group of low rotation strong Li K-giants, in order to perform a detailed NON LTE and LTE analysis of the Li line formation. These analyses are based, not only on the observed resonance LiI line , but also, and on the first time in NON LTE, on the observed secondary line of LiI at 6104 Å.

The observations consist in high resolution optical spectra obtained with the Coudé Auxiliar Spectrometer attached to the 1.4 m telescope of the ESO and low resolution spectra at the UV obtained with IUE satellite. The target objects were two very strong Li K-giant stars HD19745 (K0) and HD39853 (K5) and a strong Li K-giant HD787 (K5). HD19745, the strongest Li star of the group, is a faint object recently discovered by means of an IRAS point sources survey made at the LNA in Brazil (Gregorio-Hetem et al.1992). HD39853 is a bright high velocity and sligthly metal deficient giant star studied by Gratton and D'Antona (1989). HD787 is a star belonging to the group of bright Li rich K-giants discovered by Brown et al.(1989). The relative importance of the Li strength between these objects can be seen comparing the equivalent width of the observed lines. The values in mÅ of the couple (6708,6104) lines are the following: (427,60) for HD787, (479,165) for HD39853 and (487,172) for HD19745. No clear evidence of the presence of ^{6}Li is shown in the spectra.

the NON LTE calculations were made using a 8 level plus continuum neutral Li atom. Explicit calculations (simultaneous resolutions of the statistical and transfer equations) were made for the lines 6708 (2s-2p), 6104 (2p-3d) and 8126 (2p-3s) and for the continua from levels 2p,3d and 3s. Main electron collisional and radiative cross sections were taken from de la Reza and Querci (1978). Due to the importance of photoionization in NON LTE processes, we used our IUE observations to estimate the radiation field shortward than 3500 Å (threshold wavelength from level 2p).
LTE calculations were performed using and independent code. In case of LTE we considered the line structure of the Li lines (Duncan 1991). Bell et al.(1976) atmospheric models were used for the following main atmospheric parameters (T_{eff} in K,log g): HD19745 (4990,2.1); HD39853 (3900,1.16) and HD787 (3980,1.74). A free value constant microturbulence (ξ) was used. A detailed version of this work will be submitted to a current journal of Astrophysics.

RESULTS AND CONCLUSIONS

Our method to obtain the Li abundances consists in fitting the theoretical and absorption profiles taking into account, a posteriori, the supplementary stellar rotation and instrumental effects. The results for the different kinds of line formations considered are the following:

a) In LTE, calculations of the whole blend containing the LiI at 6104 Å together with mainly two FeI and one CaI lines resulted in the following Li abundances deduced only from the 6104 line: log ε = 2.2 for HD787 (log ε_H = 12.00) with ξ =1.7 km/s. ; log ε = 2.92 for HD39853 with ξ = 1 km/s and log ε = 4.08 for HD19745 with ξ = 1.7 km/s. The respective rotational v sin i values were 4.0 ,4.0 and 2.0 km/s. Using these last LTE abundances we were unable to reproduce the observed profiles of the resonant line at 6708 Å. LTE produces cores which are deep and too large for the cooler stars HD39853 and HD787 and on the contrary a not deep enough saturated core for the hotter star HD19745. By means of the FeI lines we found solar metallicities for HD19745 and HD787 and we confirm the mild metal deficiency of HD39853.

b) NON LTE profiles ajust approximately the observed profiles for both lines. This fitting is much better, especially for the hotter star HD19745 , reflecting maybe the existence of better models for values of T_{eff} greater than 4000 K. The main difference between NON LTE and LTE is the general overionization of Li in NON LTE. We must observe however, that the fitting of the resonance line is obtained using a somewhat larger constant microturbulence. Nevertheless the proposed values remain subsonic for the coolest layers of the atmosphere. The 6104 line is much less sensitive to microturbulence. The NON LTE Li abundances are the following: for HD787; log ε = 3.10 ± 0.1 with ξ = 4.5 km/s and v sin i = 4.0 km/s. For HD39853 log ε = 3.9 ± 0.1 with ξ = 3.5 km/s and v sin i = 4.0 km/s and for HD19754 log ε = 4.75 ± 0.1 with ξ = 4.0 km/s and v sin i = 2.0 km/s. Errors estimations cover principally the thermal effects produced by an uncertainty of 200 K in the T_{eff} values. The very strong Li stars HD19745 and HD39853 have Li abundances, as deduced by NON LTE, much larger than the Li abundances of the interstellar medium (log ε = 3.0) and of the solar system (log ε = 3.32). If we consider the NON LTE abundances as the correct Li abundances, these very strong Li K-giant stars can be considered as important candidates of sources, via mass loss, of Li in the Galaxy.

4 REFERENCES

Bell,R.A., Eriksson,K., Gustafsson,B. and Nordlund,A.; 1976 *Astron. Astrophys. Suppl.* 136, 6!

Brown,J.A., Sneden,C., Lambert,d.L., Dutchover,E.Jr.; 1989 *Astrophys. J. Suppl.* 71,293.

Duncan,D.K.,1991 *Astrophys.J.* 373,250.

Iben,I.,1967 *Rev. of Astron. Astrophys.*

Gratton,R.G. and D'Antona,F.; 1989 *Astron. Astrophys.* 215,66.

Gregorio-Hetem,J., Lepine,J.R.D., Quast,G.R., Torres,C.A.O.and de la Reza,R.; 1992, *Astron..* 103,549.

de la Reza, R and Querci,F.; 1978 *Astron. Astrophys.* 67,7.

Smith,V. and Lambert,D.L.; 1989 *Astrophys.J.* 345,L75.

Interstellar Lithium and the $^7Li/^6Li$ Ratio Toward ρ Oph

M. LEMOINE, R. FERLET, A. VIDAL-MADJAR, C. EMERICH, P. BERTIN

Institut d'Astrophysique, CNRS, 98 bis bvd Arago, 75014 Paris, France

1 INTRODUCTION

Up to now, the $(^7Li/^6Li)$ isotopic ratio of lithium has never been measured precisely outside the solar system, as ^{6}Li has never been clearly detected at the surface of stars or in the interstellar medium. However, measurements of the $(^7Li/^6Li)$ ratio in the interstellar medium are crucial in evaluating the relative weights of potential production and destruction mechanisms and thus in reconstructing the lithium history (see Reeves in these proceedings, and Reeves, 1992), thus further in allowing a more precise evaluation of the ^{7}Li primordial abundance.

We report here on new observations toward ρ Oph, which have led to the first actual detection of interstellar ^{6}Li and thus to the first precise measurement of the lithium isotopic ratio outside the solar system (see also Lemoine et al., 1992).

2 OBSERVATIONS

The only resonance line of lithium accessible to ground-based telescopes is the doublet of Li I $(3^2P_{3/2,1/2} - 2^1S_{1/2})$ at 6707.761 Å and 6707.912 Å for ^{7}Li, ^{6}Li having the same structure but shifted by 0.160 Å toward longer wavelengths. These lines are extremely weak and require observations at high spectral resolution, with a very clean instrumental profile, and a very high signal-to-noise ratio (> 1000). We chose to observe ρ Oph at the ESO-La Silla (Chile) 3.6m Telescope, linked via fiber optics to the Coudé Echelle Spectrometer, designed to provide these characteristics with a resolving power $\lambda/\Delta\lambda \simeq 10^5$. ρ Oph combines the following criteria: brightness $(m_v=5.0)$, featureless in the Li region (spectral type B2IV), relatively simple interstellar lines, and the highest possible HI column density ($> 10^{21}$cm^{-2}) together with a high reddening (E(B-V)=0.47).

Raw images were reduced individually using the ESO Image Handling and Processing Software (IHAP). 13 individual spectra of one hour integration time each were then added together, taking into account all the relevant wavelength shifts. Figure 1 shows the final plot of the ρ Oph spectrum in the vicinity of λ6708Å. The signal-to-noise ratio is 2700 per pixel; the spectral resolution is $\simeq$ 3 km/s, implying a limiting detectable equivalent width of 30 μÅ (3σ). The total equivalent width of the Li I doublet is 2.2 mÅ.

Because the ^{6}Li and ^{7}Li lines are partially superimposed, and because the equivalent
width provides only an integrated column density which could be systematically in
error when more than one absorbing region is present on the line of sight, the data
were analyzed with an iterative profile analysis process, which computes a theoretical
Voigt profile for possibly several clouds on the line of sight. Figure 2 shows the final
best fit of the Li I lines, where two interstellar components have been introduced in
^{7}Li, and only one in ^{6}Li, the extra blue component being negligible in ^{6}Li. This fit
of ^{7}Li gives also an excellent fit of the potassium λ7699Å line toward ρ Oph. We
checked on the potassium lines that the ^{6}Li absorption was not due to a high radial
velocity ^{7}Li absorbing cloud.

As a final check, we subtracted from the observed Li I profile the best solution found
for the ^{7}Li absorption alone. This procedure should reveal the actual detection of the
^{6}Li doublet toward ρ Oph. The result, shown in Fig.3 (Top), is extremely convincing :
at the correct position, with the correct separation and the correct oscillator strengths
ratio, appears directly the ^{6}Li doublet well out of the noise. And again, this observed
^{6}Li profile in Fig.3 is well fitted by the ^{6}Li solution found above (Fig.3, Middle); after
subtracting this ^{6}Li absorption from the observed profile in Fig.3, just the noise alone
remains (Fig.3, Bottom) !

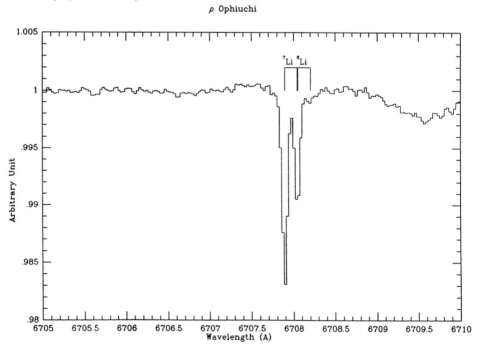

Fig.1: Resulting spectrum of ρ Oph at λ6708; the total equivalent width of the Li I lines is
2.2 mÅ, the signal-to-noise ratio is 2700 per pixel, at $\Delta\lambda/\lambda$=3 km/s, and the limiting detectable
equivalent width is 30 μÅ (3 σ), for a total integration time of 13h toward ρ Oph (m_v=5.0).

ρ Ophiuchi

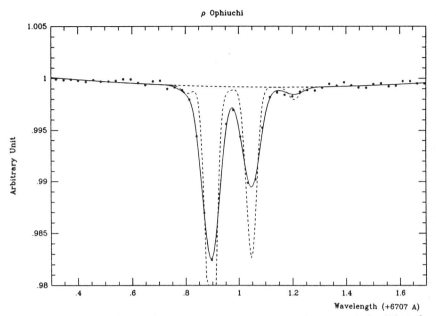

Fig.2: Final best fit for [7]Li and [6]Li. The data points represents the histogram pixels of Fig.1, the dashed lines represent the theoretical Voigt profiles, and the solid lines the Voigt profiles convolved with the instrumental profile. Two interstellar components are needed to fit the [7]Li lines; the second blue component is negligible in [6]Li.

ρ Ophiuchi

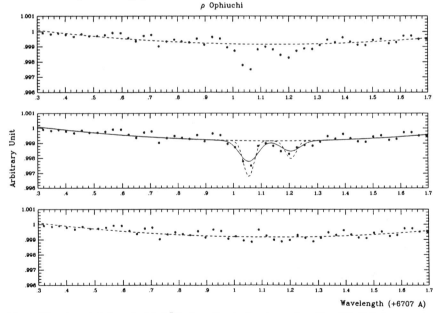

Wavelength (+6707 A)

Fig.3: We subtracted the calculated [7]Li absorption to the observed profile of Fig.2; it thus reveal the [6]Li doublet (Top). This doublet is very well fitted by the calculated [6]Li absorption (Middle): the [6]Li doublet thus appears at the correct position, with the correct oscillator strengths ratio and the correct separation. We checked on the λ7699 K I line that this doublet was not due to a high radial velocity [7]Li interstellar component. When the [6]Li absorption is removed, only the noise remains (Bottom)...

3 DISCUSSION

We derive the error bar on N(^{7}Li) according to the limiting detectable equivalent width, i.e. 20 μÅ with 95% confidence. We derive the error bar on N(^{6}Li) by fitting the ^{6}Li doublet after subtraction of the ^{7}Li doublet for both upper and lower limit of the ^{7}Li error bar, and by taking into account in each case the limiting detectable equivalent width. We thus derive the following lithium isotopic ratio toward ρ Oph: (^{7}Li/^{6}Li) $= 12.5^{+2.8}_{-3.2}$ with 95% confidence.

After correcting for the important ionisation of lithium in the interstellar medium, using N(CaII), N(CaI), and for its depletion onto interstellar grains, using $\delta_{Li} = \delta_K = -0.93$ dex, we derive the resulting effective abundances:

^{7}Li/H $\simeq 3.4 \times 10^{-9}$ ($\geq 2.4 \times 10^{-9}$)

^{6}Li/H $\simeq 2.7 \times 10^{-10}$ ($\geq 1.6 \times 10^{-10}$)

Our value of the isotopic ratio is in strikingly good agreement with the value measured in the meteorites CC I, representative of the solar system formation epoch: ^{7}Li/^{6}Li$=12.5\pm0.2$, and in disagreement with the only previous estimation of Ferlet&Dennefeld (1984): ^{7}Li/^{6}Li$\simeq 38$ (≥ 25) toward ζ Oph, though ^{6}Li was not detected in that case, but only constrained through a profile analysis. Our present measurement is therefore more precise, as we were able to clearly detect the ^{6}Li absorption, and this high value of the interstellar lithium isotopic ratio toward ζ Oph needs to be confirmed. As well, this value of the isotopic ratio toward ρ Oph is not explained by recent models of galactic lithium evolution (Abia&Canal, 1988, D'Antona&Matteucci, 1991) which rather imply an increase of this ratio in the last 4.5 Gyr, perhaps up to a very high value. When related to the "demodulated" energy integrated flux of galactic cosmic rays, our present measurement confirms the existence of a stellar source of ^{7}Li, necessary to maintain the interstellar lithium isotopic ratio to a relatively constant value in the last 4.5 Gyr, and permits to constrain the yield of this stellar source (see Reeves in these proceedings, and Reeves, 1992). Such a stellar source would render unlikely scenarios with a high ^{7}Li primordial abundance, and would instead confirm the choice of the ^{7}Li primordial abundance close to the Pop II value for comparison with the Big-Bang nucleosynthesis calculations.

Therefore, new measurements of the ($^7Li/^6Li$) ratio on different lines of sight are crucial, and now underway, to check for the homogeneity of its value in the interstellar medium. Better constrained, these chemical evolutionary models should shed some light on the ^{7}Li primordial abundance, a fundamental input in the Big-Bang nucleosynthesis models.

REFERENCES

Abia, C., Canal, R.: 1988, A&A **189**, 55
D'Antona, F., Matteucci, F.: 1991, A&A **248**, 62
Ferlet, R., Dennefeld, M.: 1984, A&A **138**, 303
Lemoine, M., Ferlet, R., Vidal-Madjar, A., Emerich, C., Bertin, P.: 1992, submitted to A&A
Reeves, H.: 1992, submitted to A&A

The Initial Lithium Abundance of T Tauri Stars

EDUARDO L. MARTÍN AND RAFAEL REBOLO

Instituto de Astrofísica de Canarias

YAKOV V. PAVLENKO

The Principal Observatory of Ukraine

1 ABSTRACT

Measurements of lithium in different astrophysical sites provide important clues to the origin and galactic evolution of this element. In this paper we present Li abundances for a sample of 24 weak T Tauri stars. The program stars have apparent ages in the H-R diagram less than 3 Myr. The Li abundance calculations use the latest model atmospheres available for cool high-gravity stars and take into account non-LTE effects in the formation of the LiI resonance doublet. The maximum Li abundances found in these WTTS are very close to those found in stars of young open clusters like αPer. We conclude that the initial Li abundance of T Tauri stars is the same as the consensual cosmic value for Population I stars and the interstellar medium.

2 INTRODUCTION

The presence of a strong LiIλ6708 resonance line in the spectra of T Tauri stars has attracted attention since this class of stars began to be studied (cf. Herbig 1962). T Tauri stars (TTS) are the youngest stars were Li can be measured. Their initial lithium content is expected to be the same as that of the molecular cloud where they are formed. In the last few years 3 groups have calculated Li abundances for relatively large samples of TTS taking into account the veiling effects (Strom et al. 1989; Basri, Martín & Bertout 1991 and Magazzù, Rebolo & Pavlenko 1992). Strom et al. claimed that the maximum abundances of TTS could be higher than the maximum Li abundances in young disk stars by at least a factor 2. Basri et al. found maximum Li abundances in TTS of about a factor 10 higher than in the interstellar medium, although they emphasized the large uncertainties of their results. Magazzù et al. obtained Li abundances wich were, on the average, not higher than those of Population I stars, but a few of their TTS had very high abundances.

In this paper we have selected only weak-line T Tauri stars (WTTS), classified as those for wich H$_\alpha$ in emission has equivalent widths less than 10Å. In these stars we do not expect to have optical veiling (cf. Basri et al. 1991) and we can be somewhat more confident on the estimate of atmospheric parameters. We consider only stars that lie as far as possible from the main sequence (MS), sharing the same region in the H-R diagram as the classical T Tauri stars (CTTS). Hence, we expect that the Li abundances of our sample are the same as that of the material that formed the stars.

3 DATABASE AND ANALYSIS

The sample consists of 9 stars observed by us and 15 from the literature. Our observations were made from October 1990 to January 1991 at the telescopes 4.2 m William Herschel and 2.5 m Isaac Newton on the observatory of Roque de los Muchachos. The dispersion of the spectra ranges from 0.22 to 0.37 Å/pixel. We consider here a subset of the sample of pre-main sequence (PMS) stars observed by us with the aim of studying PMS lithium evolution. We chose WTTS with ages less than 3 Myr, according to their positions in the H-R diagram and comparison with conventional isochrones (see Fig. 1). The WTTS selected from the literature have apparent ages similar to our WTTS and LiI equivalent width measurements based on high resolution spectra. The sources of these data are given in Table 1.

The Li abundances for our program stars have been derived from the LiI equivalent widths. The relation of $W_\lambda(LiI)$ with log N(Li) was calculated both in LTE and non-LTE conditions. The method is qualitatively similar as that used by Magazzú *et al.* 1992 and Martín *et al.* 1992. We have improved the calculations by using more recent model atmospheres from Kurucz 1991 (private communication) and a 20-level Li atom model. Details of the estimation of stellar parameters and the abundance calculations will be given in a forthcoming paper. The resulting Li abundances are given in Table 1. The accuracy of the abundance determination is estimated from both the mean uncertainty in T_{eff} ($\pm$100 K) and the error bar in the $W_\lambda(LiI)$.

Figure 1: Our program stars in the H-R diagram. The evolutionary tracks and isochrones (from top to bottom 0.03, 0.3, 1, 1.75, 3, 6, 10 and 20 Myr) are the same as Cohen & Kuhi 1979. The thick solid line is the birthline of low-mass stars from Stahler 1983.

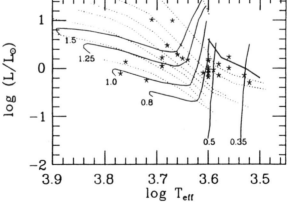

TABLE 1
Lithium abundances for selected Weak-T Tauri stars

Star	Sp.T.	W_λ(LiI) (mÅ)	Source	log N(Li)
035120+3154 SW	G0	200±30	1	3.3:
035120+3154 NE	G5	230±10	1	3.2
Anon 1	M0	480±40	1	3.3
LkCa 3	M1	560	2,3	2.7
LkCa 4	K7	650	2,3	3.2
V410 Tau	K3	420	4	3.2
Hubble 4	K7	610±10	1	3.1
V819 Tau	K7	630±10	1	3.1
LkCa 7	M0	600	2,3,4	2.9:
HD283572	G5	310±30	1	3.1:
IP Tau	M0	500	2,3	2.5
DI Tau	M0	690	3	3.2
V827 Tau	K7-M0	570	3	2.9
GH Tau	M2-3	670±30	1	2.7
V807 Tau	K7	550±50	1	3.0:
V830 Tau	K7	640	4	3.2
LkCa 15	K5	470	2	3.0
IW Tau	K7	440	2	2.5
LkCa 19	K5	440±10	1	2.9
V836 Tau	K7-M0	570	3	2.9
Sz 65	K7-M0	610	5	3.0
Sz 68	K2	420	5	3.3
Sz 82	M0	570	5	2.6
S-R 12	M1	700	6	3.1:

Sources of LiI equivalent widths: 1.- this work, 2.- Hartmann, Soderblom & Stauffer 1987, 3.- Strom et al. 1989, 4.- Basri, Martín & Bertout 1991, 5.- Finkenzeller & Basri 1988, 6.- Magazzù, Rebolo & Pavlenko 1992.

Notes: Abundance of lithium is defined as log N(Li)= 12 + log(N_{Li}/N_H). The average error bar in the lithium abundances is ±0.2 dex, except for the stars marked with ":" where it is about ±0.3 dex.

4 DISCUSSION AND CONCLUSION

The maximum Li abundance in our sample of 24 WTTS is log N(Li)=3.3, and the mean value is log N(Li)=3.0 . This supports the claim of Magazzù et al. 1992 that the initial lithium abundance of T Tauri stars is about log N(Li)=3.2 . We have not found any WTTS with a Li abundance higher than the maximum Li abundances of Population I stars. As a consistency test we used our W_λ(LiI)-log N(Li) computations to derive Li abundances of low-mass stars in the young open cluster αPer (age $\sim$ 50 Myr). Eight stars were selected from Balachandran, Lambert & Stauffer 1988 with $T_{eff} \leq$ 6000 K and log N(Li)$\geq$3.0 . Using their equivalent widths we derived non-LTE Li abundances that are slightly lower than those of Balachandran *et al.* (typically -0.15 dex). The star Ap91 shows the largest discrepancy; Balachandran *et al.* gave log N(Li)=3.6 and we derive log N(Li)=3.0 . The average Li abundance we obtain for these 8 stars is the same as that of our sample of WTTS. This is in contrast with the claim of Strom *et al.* 1989 who suggested a difference between the Li abundances of TTS and that of αPer stars. Our main conclusion is that the average Li abundances in WTTS with H-R diagram ages less than 3 Myr do not exceed the upper values of Li abundances found in MS disk stars.

REFERENCES

Balachandran, S., Lambert, D.L., Stauffer, J.R. 1988, *Ap. J.*, 333, 267

Basri, G., Martín, E.L., Bertout, C. 1991, *Astr. Ap.*, 252, 625

Cohen, M., Kuhi, L.V. 1979, *Ap. J. Suppl.*, 41, 743

Finkenzeller, U., Basri, G. 1987, *Ap. J.*, 318, 823

Hartmann, L.W., Soderblom, D.R., Stauffer, J.R. 1987, *Astron. J.*, 93, 907

Herbig, G.H. 1962, *Adv. Astron. Astroph.*, 1, 47

Magazzù, A., Rebolo, R., Pavlenko, Ya.V. 1992, *Ap. J.*, in press

Martín, E.L., Magazzù, A., Rebolo, R. 1992, *Astr. Ap.*, 257, 186

Stahler, S.W., 1983, *Ap. J.*, 274, 822

Strom, K.M., Wilkin, F.P., Strom, S.E., Seaman, R.L. 1989, *Ap. J.*, 98, 1444

Rotation Induced Mixing and Lithium Depletion in Stellar Surfaces

SYLVIE VAUCLAIR

Observatoire Midi-Pyrénées, 14, avenue Edouard Belin, 31400 Toulouse

ABSTRACT Because of its destruction by nuclear reactions at 2.5 million degrees the lithium abundance that we see at the surface of stars is directly related to the mixing which may occur below the outer convection zone. Several processes can transport the lithium-depleted matter up to the convection zone (Garcia-Lopez and Spruit 1991, Schatzman 1992, Dearborn et al 1992). Here I will focus only on the rotation-induced mixing.

1 ROTATION-INDUCED MIXING

Von Zeipel (1924) demonstrated that a *rotating star cannot be in radiative equilibrium*. The reason is simply that rotating stars are subject to centrifugal potentials which have to be added to the gravitational ones. The level surfaces are no more spherical, but ellipsoidal, with temperature variations. This creates sources and wells of energy (thermal imbalance) leading to meridional circulation (Eddington 1925, Sweet 1950, Mestel 1957, Tassoul and Tassoul 1982).

The lithium advection due to such a meridional circulation and its consequences on the observed lithium abundances have been studied by Charbonneau and Michaud 1988: the lithium-depleted matter goes up towards the convection zone in the upgoing flow, and then backwards in the downgoing flow, so that lithium is eventually completely destroyed in a time-scale which depends on the type of the star.

The situation is however complicated by the occurence of hydrodynamical instabilities in the flow (Zahn 1983, 1987, 1992). The meridional circulation transports angular momentum, leading to shear flow instabilities. The turbulence induced by these instabilities is not isotropic because of the density gradient which stabilizes the vertical shears. Zahn showed that horizontal turbulence developes at large scales, and decays into 3-D motions only at small scales, for which the turnover rate exceeds the angular velocity.

Only vertical motions can mix lithium from the destruction layers up to the convection zone. However the horizontal turbulence has an important effect on the global advection: it reduces the effective particle transport, leading to "truncated advection". The reason for this behavior is that matter which goes up in the upward flow may cross towards the downward flow due to horizontal turbulence before reaching the outer layers.

Chaboyer and Zahn 1992 showed that truncated advection could be treated as a diffusion process. Adding the two effects of vertical turbulence and truncated advection leads to an effective diffusion coefficient which may be approximatly written as :

$$D_{\text{eff}} = \gamma \, u_r \, r$$

where u_r is the radial meridional velocity, r the radius and γ an unknown efficiency factor.

A simple expression for u_r may be obtained in the barotropic case with small differential rotation (Eddington- Sweet velocity):

$$u_r = \frac{L}{M} \frac{\Omega^2 r}{g^2(\nabla_{ad} - \nabla_{rad})} \left[1 - \frac{\Omega^2}{2\pi G \rho}\right]$$

where L, M, g and ρ represent respectively the luminosity, mass, gravity, density at radius r and Ω is the angular velocity.

More sophisticated expressions for the meridional velocities and the diffusion coefficients may be found in the literature (Tassoul and Tassoul 1982, Chaboyer and Zahn 1992, Pinsonneault et al. 1990). They include different physics, like differential rotation inside the stars, but the orders of magnitude and the general behavior of the diffusion coefficients are similar and lead to consistent results on a first order.

The lithium abundance variations are then obtained by solving the diffusion equation in spherical symetry:

$$\rho r^2 \frac{\partial c}{\partial t} = \frac{\partial}{\partial r} \left(\rho r^2 \, D_{\text{eff}} \, \frac{\partial r}{\partial r}\right)$$

2 LITHIUM DEPLETION IN GALACTIC CLUSTERS

The idea that the lithium depletion observed in galactic clusters may be related to rotation-induced mixing is supported by the comparison between the rotation velocities and the lithium abundances in the Hyades (Figure 1).

From our present knowledge on the stellar rotation velocities, F and G type stars are believed to begin their lives on the main sequence with similar equatorial velocities (about 100 km.s^{-1}). Then the G type stars are slowed down due to magnetic braking (Kraft 1965, Skumanich 1972, Schatzman and Baglin 1991, Gray and Nagar 1985).

An example of the results obtained with precise computations of lithium abundance variations due to rotation-induced mixing and nuclear destruction is given in Figure 2 (Charbonnel 1992, Charbonnel et al 1992a and b). The computations have been done with the Geneva stellar evolution code, using Zahn's effective diffusion coefficient. The present average values of the stellar rotation velocities in the Hyades used in the computations are from Gaigé 1992 (see also Charbonnel and Gaigé 1992, this symposium).

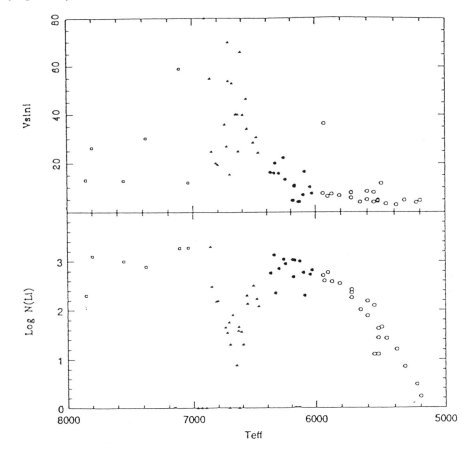

Figure 1. Rotation velocities and lithium abundances in the Hyades versus T_{eff} (after Charbonnel 1992).

The fact that the lithium abundance is normal on the left side of the dip may be due to the separation of the meridional flows in two loops. The observations can be accounted for if the efficiency of the mixing is smaller in the upper loop than in the lower one. This effect can also explain the observations of lithium depletion in subgiants (Charbonnel and Vauclair 1992).

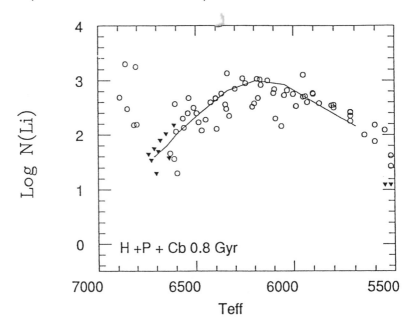

Figure 2. Comparison of the results of lithium depletion due to rotation-induced mixing and the observations of three galactic clusters of the same age: Hyades, Praesepe and Coma Berenices (after Charbonnel et al 1992).

3 LITHIUM ABUNDANCES IN HALO STARS
The observations of lithium in halo stars are compared to those in the Hyades in figure 3 (Charbonnel, Vauclair and Reeves 1992).

We can make the following remarks about this figure:

1) The observations show a very small dispersion in halo stars, as first pointed out by Spite and Spite 1982. The dispersion is small also in the Hyades. Note that the halo stars all have the same age, as it is the case in stellar clusters.

2) The observational points follow average lines which cross at low temperature. This

is due to the depth of the convection zone which is smaller in low metallicity stars than in high metallicity stars due to opacity effects (Figure 4). It is interesting in this respect to plot the lithium abundances as a function of M_{cz}/M^* (figure 3b). Then the lines do not cross anymore.

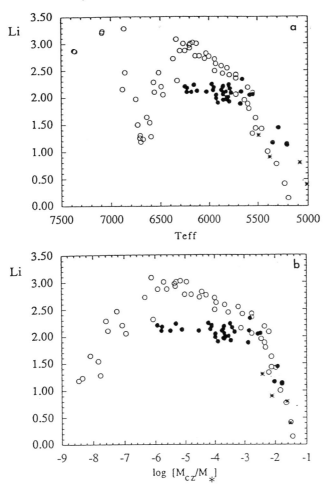

Figure 3. Comparaison of the lithium observations in extreme halo stars (filled circles) and in the Hyades (open circles). The crosses represent upper limits for lithium in halo stars.

Figure 3a: lithium abundances versus T_{eff}

Figure 3b: lithium abundances versus $[M_{cz}/M_*]$, ratio of the convective mass to the total mass of the star. The relation between $[M_{cz}/M_*]$ and T_{eff} in our models is given in figure 4 for three different metallicities.

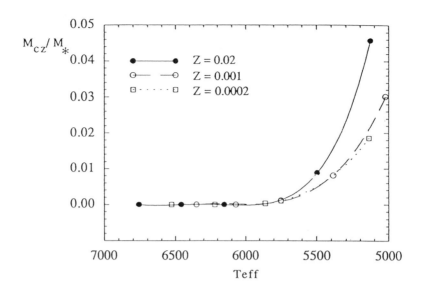

Figure 4. Relation between the ratio of the mass in the outer convection zone to the stellar mass and the effective temperature in our stellar models for three different metallicities.

3) If one takes off from this graph all the stars with effective temperatures below 5500K, all the lithium abundance variations disappear for halo star while most of them remain for young stars. Note that a similar dip as observed in the Hyades could exist for halo stars, at a slightly larger effective temperature, but most of these stars must have evolved away from the main sequence. The very depleted halo star recently observed by Spite and Spite 1992 (this symposium) may be one of them. Figure 5 gives the observations of lithium abundances in all stars (and not only stars with $T_{\text{eff}} > 5500K$ as usually done), as a function of metallicity. Nuclear destruction can be important at all metallicities. The question which arizes concerns the upper envelope of these observations: is the maximum lithium abundance observed in low metallicity stars the primordial one, or has lithium been depleted in all these stars?

The basic argument against the idea that lithium could be depleted in low metallicity stars remains the very small dispersion in the "plateau". Deliyannis et al 1990 computed the lithium depletion in "standard models" and deduced a precise value of the so-called "primordial lithium abundance", which is often used as a constraint for Big-Bang Nucleosynthesis. Unfortunately the standard models for stellar evolution are not realistic, as they include no macroscopic motions outside the convection zones, nor gravitational diffusion. In real nature, gravitational settling must occur unless some macroscopic motions do prevent it. One cannot get rid of both.

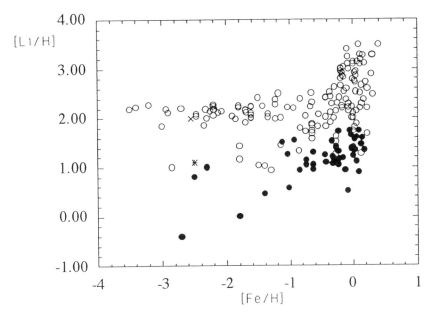

Figure 5. Observations of lithium as a function of metallicity in all the observed stars, including those with effective temperatures smaller than 5500K (see references in Walker et al 1991).

The effect of gravitational settling on lithium in halo stars have been studied by Michaud, Fontaine, Beaudet 1984, Proffitt and Michaud 1991, and also Deliyannis et al 1990. It leads to a non uniform depletion with an average value of about a factor 2. Macroscopic motions can prevent this depletion, but they induce lithium mixing and destruction. Rotation induced mixing could even deplete lithium by a factor 10 and lead to a plateau-shape as observed (Vauclair 1988, Pinsonneault et al 1992). The basic problem for these computations is that we do not know the rotation history of these stars, which has to be postulated.

3 CONCLUSION

Hubert Reeves, in this symposium, stated that "Nature is unnatural", as three different processes are needed to explain the lithium abundance in the Universe. Another unnatural process has to be added to the whole picture: for the lithium abundance in halo stars to represent indeed the primordial one, a fine tuning of turbulence is needed so that gravitational settling may the prevented without lithium being burned. This tuning seems quite difficult to obtain, and lithium is probably depleted by at least a factor 2 in low metallicity stars. More observations of lithium abundances at the two extremes of the plateau, and detailed spectroscopic observations of metals could help giving constraints on the process involved.

REFERENCES

Charbonnel C., 1992, thèse de doctorat
Charbonnel C., Vauclair S. 1992, A&A in press
Charbonnel C. , Vauclair S., Zahn J.-P., 1992a, A&A 255, 191
Charbonnel C., Vauclair S., Maeder A., Meynet G. Schaller G., 1992b, preprint
Charbonnel C., Vauclair S., Reeves H., 1992, preprint
Charbonnel C., Gaigé Y., 1992, this symposium
Chaboyer B., Zahn J.-P., 1992, A&A 253, 173
Charbonneau P., Michaud, G. 1988, ApJ 334, 746
Dearborn D.S.P., Schramm D.N., Hobbs L.M., 1992, preprint
Deliyannis C.P., Demarque P., Kawaler S.D., 1990, ApJS, 73, 21
Eddington A.S., 1925, Observatory 48, 78
Gaigé Y., 1992, preprint
Garcia Lopez R.J., Spruit H.C., 1991, ApJ,377, 268
Gray D.F., Nagar P., 1985, ApJ 298, 756
Kraft R., 1965, ApJ, 142, 681
Mestel L., 1957, ApJ, 126, 550
Michaud G., Fontaine G., Beaudet G. 1984, ApJ, 282, 206
Pinsonneault M.H., Kawaler S.D., Demarque P., 1990 ApJS, 74, 501
Pinsonneault M.H., Deliyannis C.P., Demarque P., 1991, ApJS, 78, 179
Proffitt C.R., Michaud G., 1991, ApJ, 371, 584
Schatzman E., Baglin A., 1991, A&A 249, 125
Schatzman E., 1992 preprint
Skumanich A., 1972, ApJ, 171, 565
Spite F., Spite M., 1982, A&A, 115, 357
Spite F., Spite M., 1992, this symposium
Sweet P.A., 1950, MNRAS 100, 548
Tassoul J.L., Tassoul M., 1982, ApJS 49, 317
Vauclair S., 1988, ApJ, 335, 971
Von Zeipel H., 1924, MNRAS, 84, 665
Walker T.P., Steigman G., Schramm D.N., Olive K.A., Kang H.S., 1991, ApJ, 376, 51
Zahn J.-P., 1983, Astrophys. Processes in Upper Main Sequence Stars (ed. B. Hauck and A. Maeder; Publ. Observatoire de Genève), 253
Zahn J.-P., 1987, The Internal Solar Angular Velocity (ed. B.R. Burney and S. Sofia; Reidel), 201
Zahn J.-P., 1992, A&A (preprint)

The Lithium Abundance and Stellar Evolution

G. MICHAUD and J. RICHER

Département de Physique
Université de Montréal
Montréal H3C 3J7
Canada

1 ASTROPHYSICAL CONTEXT

Since it is a fundamental physical process, atomic diffusion should be included in stellar evolution calculations. A complete calculation would include, atomic diffusion along with physical processes that compete with it. This is generally not done partly because it would complicate calculations but mainly because the competing hydrodynamical processes, such as mass loss, turbulence and meridional circulation are poorly known. This benign neglect seemed vindicated by Eddington's (1926) argumentation that the absence of extreme heavy element abundances ruled out the equilibrium configurations to which diffusion would lead in stellar interiors.

The presence of Fm, Am, Ap and Bp stars however suggests that particle diffusion is important in some stars (Michaud 1970, Watson 1971). In other stars the high precision of observations now justifies including the effects of diffusion even when its effects are small since they may be detected by high accuracy observations. It is then important to first determine the superficial abundances to which stellar evolution leads when diffusion is included. By a comparison to observations, it will then be possible to determine the role it might play in most stars. The progressive introduction of competing hydrodynamical processes should then allow to determine why all stars are not peculiar and which processes are most important.

We first present a brief discussion of the input physics to the diffusion calculations (§2), then the calculations done for the Sun (§3), the AmFm stars and the Li gap (§4), the Halo stars (§5) and show how mass loss could modify the Li abundance (§6). Though probably important, turbulence and meridional circulation will receive only passing remarks. See Michaud and Charbonneau (1991) for a more complete review.

2 DIFFUSION EQUATIONS AND COEFFICIENTS

Various formulations of the diffusion equations and different approximations to the diffusion coefficients have been used over the last few years to calculate the effect of diffusion on stellar evolution. This is reviewed in more detail in Michaud and Proffitt (1992). We here only briefly mention the main points.

The various formulations (see, for instance, Chapman and Cowling 1970 and Burgers 1969) can all be shown to give essentially the same results. The results do depend however on the diffusion coefficients. Approximations have been made by many authors in order to avoid carrying out triple integrals numerically in the evaluation of the coefficients. This leads to approximate analytic solutions which are accurate in the limit of very dilute plasmas. The collision integrals are then proportional to

$$\ln \Lambda_{ij} \simeq \ln \left(\text{constant} \; \frac{2kT\lambda_d}{z_i z_j e^2} \right), \qquad (1)$$

often referred to as the Coulomb logarithm. For very dilute plasmas, it is large and does not depend sensitively on the value of the constant appearing in the logarithmic term. This constant depends on the approximations made in arriving at an analytic expression. In the interior of the sun and halo stars, one is however far from a very dilute plasma. As can be seen from Fig. 1 and 2 of Michaud and Proffitt (1992), $\ln\Lambda_{ij}$ is not much larger than one there. The use of analytic expressions is then unjustified. One should avoid the approximations that they imply and instead use the results of a complete numerical integration of the collision coefficients (Paquette et al. 1986). Michaud and Proffitt (1992) have proposed analytic expressions that they calibrated to the results of Pelletier *et al.* in order to facilitate their use in evolutionary calculations. They are more accurate and as easy to use as the expressions proposed by Bahcall and Loeb (1990).

There remain uncertainties in the calculations of the collision integrals. They come from the physical assumptions involved in the initial definitions of the collision integrals and diffusion equations, not from the mathematical simplifications needed to derive simple analytic expressions. See Michaud (1991) for a discussion of potential uncertainties.

The abundance variations that diffusion leads to depend mainly on the relative value of radiative acceleration and gravity. The evaluations of g_R that currently exist are uncertain but the atomic data banks accumulated by the Opacity project and by Kurucz (1991) should reduce the uncertainty from a factor of 2-3 to a factor of 1.2-1.4. Using the results of the Opacity project Alecian, Michaud and Tully (1992) have obtained accurate g_R for Fe over an important fraction of stellar envelopes. They have shown that Fe can be supported over a large fraction of the envelope and since Fe has been found to be the dominant opacity source over a certain T range in stellar envelopes (Rogers and Iglesias 1992), the local overabundances of Fe created by atomic diffusion could have significant effect on stellar models. One should calculate the atomic diffusion of Fe at the same time as one carries out the stellar evolution calculations. Similar effects may be expected from C, N and O and probably some other Fe peak elements.

3 THE SUN

A number of authors have estimated gravitational settling in the Sun (Aller and Chapman 1960, Noerdlinger 1977, Wambsganss 1988, Cox et al. 1989, Proffitt & Michaud 1991b, Bahcall & Pinsonneault 1992). Disagreements have been analyzed by Proffitt and Michaud (1991b) and, if diffusion is not inhibited below the convection zone, the current surface mass fraction helium abundance of the Sun, Y_s, should be about 0.03 (10%) lower than the initial solar helium abundance. Models with diffusion have slightly higher central densities (0.5%) and temperatures (2%). A slightly larger mixing length is also needed. Predicted neutrino fluxes are slightly increased.

The current solar Li abundance is 100-200 times smaller than the initial value and it appears difficult to account for all of the depletion by pre-main sequence burning. Some mixing appears needed to carry the Li to a temperature where it burns. While the cause and nature of main sequence mixing is controversial, observational data can constrain its magnitude. Figures 1 to 3 of Proffitt and Michaud (1991b) show how two possible distributions of turbulent mixing that approximately reproduce the observed Li depletion affect the surface settling of He. It does not seem possible to reduce the surface settling by more than a factor of two without over-depleting Li, but the shape of the interior He profile and the steepness of the composition gradient can be quite sensitive to the details.

Observations of solar p-modes provide a means for directly measuring the He abundance of the surface convection zone and the sound speed as a function of radius (see the discussion in Dziembowski *et al.* 1992). These measurements are potentially sensitive enough to distinguish between the different profiles shown in Figure 2 of Proffitt and Michaud (1991b) and suggest a surface He abundance consistent with the diffusion predictions noted above (see also Guzik & Cox 1992). Once current uncertainties in other physics (opacities, abundances, and convection theory) are resolved, the oscillations should provide a direct test of turbulent mixing in G dwarfs.

4 Am-Fm STARS AND THE Li GAP

It appears likely that most abundance anomalies in Am-Fm stars (Cayrel Burkhart and Van't Veer 1991) are caused by atomic diffusion in the presence of radiative acceleration (Watson 1971, Michaud *et al.* 1983). Stars are assumed to arrive on the main sequence with normal abundances. A and F stars, have a superficial convection zone but its depth increases by orders of magnitude over the T_{eff} interval from 10000 to 6300 K. The gravitational settling of He occurs below the He II convection zone until it disappears once the He abundance has been sufficiently reduced and the settling then continues below the H convection zone. Overabundances are simultaneously produced for many heavy elements for which $g_R > g$ below the convection zone. For a few heavy elements, underabundances are produced, usually when they are in a rare gas configuration immediately below the convection zone so that $g_R < g$ there. This apparently leads to the signature of the AmFm phenomenon, in the form of underabundances of Ca and Sc. It only occurs over a small T interval at the bottom of the convection zone, that over which Ca and Sc are in the rare gas configuration. This can only happen below the H convection zone, since at temperatures prevalent below the He II convection zone, both Ca and Sc are supported. The observed underabundances of Ca and Sc then imply the underabundance of He in AmFm stars. This is explained by the diffusion model for AmFm stars in so far as over the T_{eff} range where the anomalies are observed, *static* models predict that Ca and Sc are not supported below the H convection zone. The expected anomalies are consistent with observations for most heavy elements (Michaud *et al.* 1976, Michaud *et al.* 1983)

With the observation of AmFm stars in clusters (Burkhart and Coupry 1989), it is possible to go further and test whether evolving stellar models predict the observed abundance anomalies over the T_{eff} range where they are observed in clusters. All stars then have the same age. Given the sensitivity of the radiative acceleration of Sc and Ca to the local T, the observed anomalies in a cluster should test also the effect of evolution on the stellar structure. When does the T at the bottom of the convection zone increase sufficiently for Ca and Sc to be supported and how do the observed anomalies agree as a function of T_{eff}? As one observes cooler stars of the same cluster, what additional abundance anomalies are to be expected? To what extent is the Li gap in clusters linked to the abundance anomalies of the AmFm stars? Or what aspects of the observations of the Li gap stars are explained by the model that explains the AmFm stars?

There have up to now been no systematic calculations of the effect of atomic diffusion on the evolution of A and F stars. Vauclair *et al.* (1974) have determined the effect of atomic diffusion on the disappearance of the convection zone but only in static models. Richer, Michaud and Proffitt (1992) have developed an envelope code that they have coupled to evolutionary calculations to determine precisely the effect of He diffusion on the evolution of A and F stars. Their model is currently being extended to allow inclusion of Fe settling. The Fe abundance modifies the stellar structure through its effect on Rosseland opacities.

Using solar models, it is possible to calibrate the value of α, the ratio of the mixing length to the pressure scale height. This value depends on the opacity tables used and is slightly modified if atomic diffusion is included in the calculations.

The stellar evolution was calculated for models with mass varying from 1.2 to 2.2 $M_\odot$ to cover the T_{eff} range of the Li gap and the Am-Fm stars. The OPAL opacities (Rogers and Iglesias 1992) were used and the value of α was determined using the sun.

In Fig. 1 is shown the time evolution of the T_{eff} in both a model with and without diffusion. The T_{eff} = 7800 K on the zero age main sequence. It goes down to 6300 K just before crossing to the giant branch because of evolutionary effects. The presence of the gravitational settling of He causes the disappearance of the superficial convection zone and decreases the He abundance and so the mean molecular weight over part of the envelope (it is slightly increased close to the centre; see Proffitt and Michaud 1991b). As can be seen by comparing the full and dotted lines on the bottom part of Fig. 1, the two effects combine to reduce the T_{eff} by about 100 K.

On the middle section of the figure, is shown the Li abundance both as a function of time and of position in the star. It is coded in tones of grey as given by the scale. The time evolution of the Li abundance is governed locally by the relative values of the radiative acceleration on Li (g_R(Li)) and of gravity. The evolution of g_R(Li) is shown on part d) of Figure 1. It is larger than gravity over a mass interval covering slightly less than two orders of magnitude where a substantial fraction of Li is in the form of Li III, an hydrogenoid configuration. Deeper in the star, Li has lost all its electrons whereas above it is in the He-like configuration, in which it has no line close to the radiative flux maximum and so absorbs very little radiation. At the bottom of the convection zone, g_R(Li) is increasing outwards due to the increased contribution of neutral Li. Because of the small ionization potential of Li I, this contribution remains small.

As can be seen from parts b) and d) of Fig. 1, the bottom of the H convection zone is progressively deeper during the main sequence evolution. On arrival on the main sequence, Li is not supported below the convection zone since it is mainly in the He like

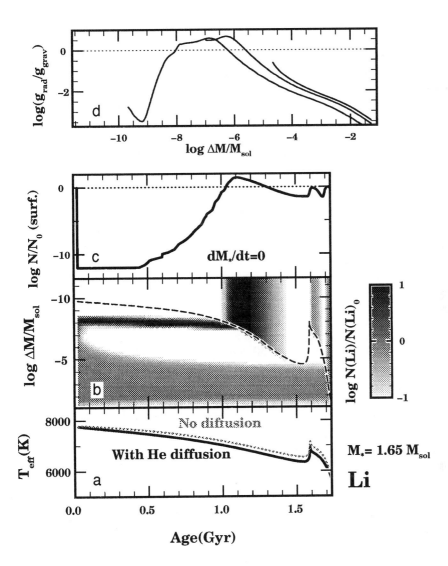

Figure 1: Evolution of the lithium internal distribution in a $1.65\,M_\odot$ Main-Sequence star, without mass loss. (a) Effective temperature of the star when He is settling throughout the envelope (continuous curve). Same star with an homogeneous envelope (dotted). (b) Li concentration (gray shades) and convective zone inner boundary (dashed). (c) Surface Li concentration variations. (d) Variation of the radiative acceleration with depth at age 100 Myr, 1.2 Gyr and 1.5 Gyr (in left to right order) normalized to local gravity. The curves end at the convection zone, on the left.

configuration. As time passes the bottom of the convection zone approaches the Li cloud that has formed during evolution.

Because Li is not supported for $\Delta M/M_\odot < 10^{-6}$ on the zero age main sequence, Li starts settling below that point. Its abundance goes down. Above that point, and up to $\Delta M/M_\odot = 10^{-8}$, its abundance also drops with time because $g_R(Li) > g$ and the radiative acceleration pushes Li into a small cloud centred approximately where $g_R(Li) = g$, slightly above $\Delta M/M_\odot = 10^{-8}$. A thin cloud covering about one order of magnitude in mass forms there. It moves slowly during evolution as it follows the evolution of the position where $g_R(Li) = g$.

During the 1.6 Gyr of main sequence evolution, gravitational settling reduces the Li abundance down to about $\Delta M/M_\odot = 10^{-4.5}$ by a factor of 3 to 30. For $\Delta M/M_\odot < 10^{-1.5}$, the burning of Li by the reaction $Li^7(p,\alpha)He^4$, which is also included in the calculations, reduces the Li abundance by many orders of magnitude. There remains a buffer of about a factor of 100 in $\Delta M/M$ where the Li abundance has not been too affected by either burning or settling.

As evolution proceeds, the bottom of the convection zone, represented by the dashed line on part b of Fig. 1, first approaches and then enters the Li cloud after 10^9 yr of evolution. Dredging up of Li occurs. There is first a 0.25 Gyr interval during which the Li in the cloud is mixed throughout the convection zone which becomes progressively deeper. The superficial Li abundance first exceeds its original value by a factor of 26 but progressively decreases as the convection zone becomes even deeper. An underabundance by a factor of 30 appears as the convection zone mixes with the region which has been depleted by the gravitational settling of Li from the beginning of evolution. Then, a brief stellar contraction phase appears and the T_{eff} increases, the bottom of the convection zone recedes towards the surface and Li is supported again there ($g_R(Li) > g$). The superficial Li abundance increases again. Then the star evolves towards the giant branch and the convection zone rapidly goes through the region from which Li settled (the superficial Li decreases) and then the buffer covering $10^{-4.5} < \Delta M/M_\odot < 10^{-1.5}$ (the superficial Li increases again). Finally the convection zone enters the region where Li was destroyed and the Li abundance decreases again (that part of the evolution was not calculated for that star).

Figure 2 summarizes the results of He and Li diffusion in main sequence A and F stars. The Li gap stars are found in the white patch in the lower right corner of the bottom panel.

A similar time dependence of the abundance is to be expected for other elements in AmFm stars. Calculations are in progress for Be, C, N, O and Fe. For Be, the main difference is that the cloud is deeper since Be has one more electron and so a larger ionization potential before losing its last electron. For heavy elements, the cloud is expected when there is a state of ionization where $g_R < g$. This state has to dominate so that others do not fill the gap. This may not always occur. Only detailed calculations will tell which elements are similarly affected. A comparison to observations of a large number of heavy elements in AmFm stars of clusters should allow a rigorous test of a model for these objects. Because of the potential presence of clouds of different elements, detailed evolutionary calculations are needed. The role of Li, Be and B is

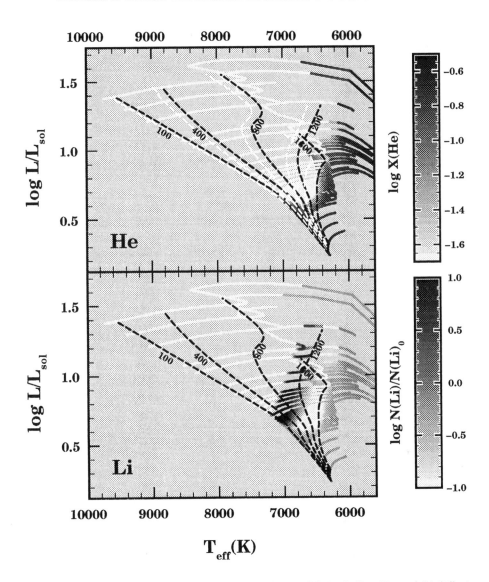

Figure 2: Evolutionary tracks for a family of stellar models including He and Li diffusion, based on OPAL opacities (Rogers & Iglesias 1992). Stellar masses range from 1.2 to 2.2 $M_\odot$, with highest mass resolution in the range where the Li gap forms (around 1.43 $M_\odot$). Along each track, the shade of gray varies according to the current Li surface abundance (lower panel), and Helium surface abundance (upper panel). Isochrones (dashed, black) are labelled with age in Myr. In the top panel, isochrones are also shown for evolution without He diffusion (dashed, white); they illustrate the amount of surface cooling cause by helium settling.

unique in that they test diffusion **and** nuclear burning.

A comparison to observations of Li in clusters (Richer and Michaud, in preparation) shows that the Li gap occurs at the observed T_{eff} but is calculated to be only half as wide as observed.

5 HALO STARS

Pop. II stars are older than most Pop. I stars so that gravitational settling has had more time to operate (Noerdlinger and Arigo 1980, Stringfellow *et al.* 1983). The central settling of He causes core H exhaustion to occur earlier and at a lower luminosity, while surface settling causes an increase in the stellar radius and a decrease in the T_{eff} relative to non-diffusion models. The total shift in the turnoff T_{eff} between diffusion and non-diffusion models is typically about 200 K. Stringfellow *et al.* concluded that this resulted in a 14-25% decrease in the ages of globular clusters. Proffitt and VandenBerg (1991) using Paquette et al.'s diffusion rates concluded that H-He diffusion results in a 10% decrease in the ages of globular clusters as measured by the luminosity of the turn-off (see also Chaboyer *et al.* 1992).

The ^{7}Li is expected to settle at about the same speed as ^{4}He, and can be used to constrain surface settling and mixing. Spite and Spite (1982) discovered that metal poor halo stars with [Fe/H] < -1.3 and T_{eff} > 5500 K all have essentially the same Li abundance, with very little star to star scatter, and no obvious trend with T_{eff} or metallicity (Deliyannis and Pinsonneault 1992 suggest an approximately 10% intrinsic scatter). However, models that include gravitational settling without any extra-mixing predict a noticeable decrease in the surface Li abundance towards the hot end of the plateau, due to gravitational settling of He and Li from the thin convection zones of these stars (see Proffitt and Michaud 1991a). Contrary to main sequence models discussed above, these models were not calibrated using solar properties. One could adjust the mixing length to reduce the settling. However this leads to excessive Li destruction by nuclear burning in cooler halo stars. Models seem to predict an unobserved Pop. II analogue of the Pop. I F-star Li gap (but see G186-26 with Li down by at least 0.9 dex, Hobbs *et al.* 1991; uncertainties, perhaps due to reddening, in the T_{eff} of the stars at the hot end of the halo sequence could erroneously extend it). Deliyannis and Demarque (1991) obtained less settling than Proffitt and Michaud (1991a) but probably because they used smaller diffusion coefficients (see §2 above).

Calculating detailed interaction between gravitational settling and turbulence below the convection zone requires a model of turbulence (see for instance Pinsonneault, *et al.* 1992). It does not however appear possible to have had less than a factor of 1.5 reduction from the original Li abundance in halo stars (see Michaud and Proffitt 1992 for a more detailed discussion).

6 MASS LOSS

On Fig. 3 is shown the evolution of a 2.1 $M_\odot$ star for three mass loss rates. This should be compared to Fig. 1 where the evolution is shown for a 1.65 $M_\odot$ star in the absence of mass loss. A comparison of part b of Fig. 1 with part a of Fig. 3 shows that essentially the same Li distribution prevails in both objects during most of the main

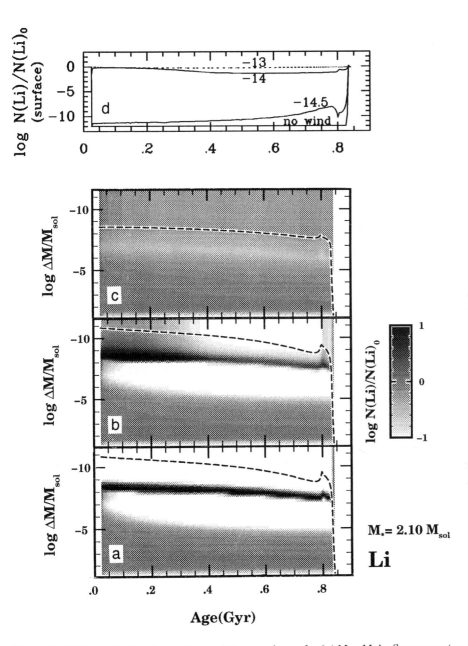

Figure 3: Lithium distribution in the evolving envelope of a 2.1 $M_\odot$ Main-Sequence star, at different constant mass loss rates. In a,b and c, ΔM is the envelope mass integrated from the stellar surface. Only the top $0.05\,M_\odot$ of the envelope is shown. The location of the convection zone inner boundary is indicated (dashed line). Nuclear burning of Li is visible at the bottom of each frame. Mass loss rates are: (a) $10^{-14.5}$, (b) 10^{-14}, (c) 10^{-13} $M_\odot$/yr; (d) change in surface Li concentration in each case (labelled with their corresponding mass loss exponent). Also shown is the evolution without mass loss (no wind).

sequence. The difference occurs only after 60 % of the main sequence life and is caused by the different depths of convection zones. For A and F stars within a cluster this T_{eff} dependence of depth leads to abundance anomalies that depend on T_{eff} not only for Li but also for heavy elements.

Parts a and d of Fig. 3 show that a mass loss rate of $10^{-14.5}$ $M_\odot$ yr^{-1} or smaller does not substantially modify the time dependence of the superficial abundance nor the Li distribution within the star. For a mass loss rate of 10^{-14} $M_\odot$ yr^{-1} the Li abundance as a function of time and position in the star is modified qualitatively above and in the Li cloud. For a mass loss rate of 10^{-13} $M_\odot$ yr^{-1}, the modification also occurs below the Li cloud. This can be understood by a comparison of mass loss and diffusion velocities. This leads to the introduction of a critical mass loss rate at which the diffusion velocity equals the mass loss velocity:

$$\frac{dM}{dt} = \frac{1.6 \ 10^{-15} M \ A \ T_5^{1.5}}{Z^2} \qquad (2)$$

where M is the mass of the star in solar units, T_5 is the temperature in units of 10^5 K, Z is the charge of the ion and A the atomic number of the chemical element. This approximate expression for the critical mass-loss rate was obtained by using an average value for ln Λ. For more accurate work one should include a detailed calculation of the diffusion velocity instead of the average value of ln Λ (see Eq. (1) and §2).

If one applies Eq. (2) to a 2.1 $M_\odot$ star, one can interpret Fig. 3. Immediately above the cloud one finds dM/dt = 6 10^{-15} $M_\odot$ yr^{-1} (Z = 2, T_5 = 1). Immediately above the buffer zone, one finds dM/dt = 3 10^{-14} $M_\odot$ yr^{-1} (Z = 3, T_5 = 5). For mass loss rates smaller than 6 10^{-15} $M_\odot$ yr^{-1}, the matter is not dragged either from the cloud nor from the buffer zone and the solution is not modified (part a of Fig. 3). For mass los rates of 10^{-14} $M_\odot$ yr^{-1}, the matter is dragged from the cloud but not from the buffer (part b). For mass loss rates above 3 10^{-14} $M_\odot$ yr^{-1} (part c), the Li is dragged from both the cloud and the buffer and there remains little effect of diffusion. For detailed solutions, one must solve the continuity equation in detail, since the velocities vary with T_5 and Z, so that it tends to increase as one approaches the bottom of the convection zone where Z = 1 for Li.

Since different chemical elements react differently to mass loss (different Z, A g_R), a detailed comparison of observations should allow a test of the potential effect of mass loss on abundance anomalies of AmFm stars.

REFERENCES

Alecian, G., Michaud, G., and Tully, J. 1992, preprint.
Aller, L. H., and Chapman, S. 1960, ApJ, 132, 461.
Bahcall, J. N., and Loeb, A. 1990, ApJ, 360, 267.
Bahcall J. N., & Pinsonneault, M. H. 1991, ApJ, submitted
Burgers, J. M. 1969, Flow Equations for Composite Gases, (New York: Academic

Press).

Burkhart, C., and Coupry, M. F. 1989, A&A, 220, 197.

Cayrel, R., Burkhart, C., and Van't Veer, C. 1991, in IAU Symposium 145, **Evolution of Stars: the Photospheric Abundance Connection**, Golden Sands, Bulgaria, 25-31 august, ed. G. Michaud and A. Tutukov (Kluwer), pp. 99-110.

Chaboyer, B., Deliyannis, C. P., Demarque, P., Pinsonneault, M. H., & Sarajedini, A. 1992, ApJ, 388, 372

Chapman, S., and Cowling, T. G. 1970, The Mathematical Theory of non-uniform Gases (3d ed.; Cambridge: Cambridge University Press).

Cox, A. N., Guzik, J. A., and Kidman, R. B. 1989, ApJ, 342, 1187.

Deliyannis, C. P., and Demarque, P. 1991, ApJ, 379,

Deliyannis, C. P., and Pinsonneault, M. H. 1992, preprint

Eddington, A. S. 1926, The Internal Constitution of the Stars (New York: Dover [1959] reprint), § 199.

Hobbs, L. M., Welty, D. E., and Thorburn, J. A. 1991, ApJL, 373, L47-L49.

Kurucz, R. L. 1991, in Stellar Atmospheres: Beyond Classical Models, ed. by L. Crivellari, I. Hubeny and D. G. Hummer, NATO ASI Series (Dordrecht, Kluwer), p. 440.

Michaud, G. 1970, ApJ, 160, 641.

Michaud, G. 1991, Annales de Physique Fr., 16, 481.

Michaud, G., and Charbonneau, P. 1991, Space Science Reviews, 57, 1.

Michaud, G., Charland, Y., Vauclair, S., and Vauclair, G. 1976, ApJ, 210, 447.

Michaud, G., and Proffitt, C. R. 1992, in **INSIDE THE STARS**, IAU COLLOQUIUM 137, Vienna, April 1992, ed. A. Baglin and W. W. Weiss, (Conference Series of the PASP), in press.

Michaud, G., Tarasick, D., Charland, Y., and Pelletier, C. 1983, ApJ, 269, 239.

Noerdlinger, P. D. 1977, A&A, 57, 407.

Noerdlinger, P. D. & Arigo, R. J. 1980, ApJL, 237, L15

Paquette, C., Pelletier, C., Fontaine, G., and Michaud, G. 1986, ApJS, 61, 177.

Pinsonneault, M. H., Deliyannis, C. P., and Demarque, P. 1992, ApJS, 78, 179.

Proffitt, C. R., and Michaud, G. 1991a, ApJ, 371, 584.

Proffitt, C. R. and Michaud, G. 1991b, ApJ, 380, 238

Proffitt, C. R., and VandenBerg, D. A. 1991, ApJS, 77, 473.

Richer, J., Michaud, G., and Proffitt, C. 1992, ApJS, in press.

Rogers, F. J., and Iglesias, C. A. 1992, ApJS, 79, 507.

Spite, F., and Spite, M. 1982, A&A, 115, 357.

Stringfellow, G. S., Bodenheimer, P., Noerdlinger, P. D. & Arigo, R. J. 1983, ApJ, 264, 228.

Vauclair, G., Vauclair, S., and Pamjatnikh, A. 1974, A&A, 31, 63.

Wambsganss, J. 1988, A&A, 205, 125.

Watson, W. D. 1971, A&A, 13, 263.

Rotation and Destruction of Li and Be in Main Sequence Galactic Cluster Stars

C.CHARBONNEL[1,2], Y.GAIGE[1]

[1] Observatoire Midi-Pyrénées, France. [2] Geneva Observatory, Switzerland

1 OBSERVATIONAL DATA

Abundances of both ^{7}Li and ^{9}Be in main-sequence stars are complementary tracers of the hydrodynamical processes occuring in stellar interiors. Li is destroyed at $\simeq 2.5$ 10^6K and Be at $\simeq 3.5$ 10^6K. Due to this temperature difference for the destruction of Li versus Be, Be-deficient stars are also Li-deficient, but Li-deficient stars are not necessarily Be-deficient.

The main characteristic of ^{7}Li abundances observed in galactic clusters is the so-called "Boesgaard dip" which was first discovered in the Hyades by Boesgaard and Tripicco 1986 : Main-sequence F stars show a Li depletion for effective temperatures around 6600 K. The dip appears in all the clusters older than 10^8 yr, and the Li depletion increases with time. This feature also exists in field stars, although more spread than in clusters (Balachandran et al. 1988). Observations of subgiants in M67 (Balachandran 1989) show that the Li depletion in the dip persists when the stars leave the main-sequence. In the Hyades, a drop in Be of a factor 2 - 4 has been detected across the temperature region of the Li dip (Boesgaard and Budge 1989). Field main-sequence Be-deficient stars are severely deficient in Li (Boesgaard and Lavery 1986).

As can be seen in figure 1 which presents stellar rotation velocities in the Hyades, the rotation is increasing on the cool side of the dip where the lithium abundances decreases. If the Li depletion in the dip is related to rotation, the dispersion of the rotation velocity should account for the dispersion of the Li observations in the corresponding effective temperature range. We present results of our computations of Li and Be depletion due to the combination of nuclear burning and rotation-induced mixing.

2 ROTATIONAL VELOCITY DISPERSION IN THE HYADES

Assuming a random spatial distribution of stellar rotation axes, Chandrasekhar and Münch 1950 established the integral equations which relate the distributions of the true (v) and apparent (y=vsini) rotational velocities of stars. Simple numerical relations exist between the moments of the frequency functions of both distributions. The mean velocity and the mean square deviation are given respectively by : $\bar{v} = \frac{4}{\pi}\bar{y}$, $\sigma^2 = \overline{(v - \bar{v})^2} = \frac{3}{2}\overline{y^2} - \frac{16}{\pi^2}\bar{y}^2$, and can be deduced directly from observational data.

We used rotational velocity measurements in the Hyades cluster by Kraft 1965 and by Mermilliod 1992 to compute the mean velocity and the mean square deviation of the true distribution as a function of effective temperature. Theoretical $\bar{v}$ and $\bar{v} \pm 1\sigma$ are presented on figure 1 and compared with observational values multiplied by $4/\pi$.

3 ROTATION-INDUCED MIXING

The thermal imbalance in a rotating star generates meridional flows known as "Eddington-Sweet meridional circulation". Zahn 1983 evaluated the turbulence induced by the circulation in terms of a turbulent diffusion coefficient, of the order of $|rU(r)|$, where r is the radius inside the star and U(r) the vertical velocity of the circulation at the radius r. In a uniformly rotating star, the turbulent diffusion coefficient was written (Charbonnel et al. 1992a, hereafter CVZ) : $D_T = \gamma \left| \frac{L}{M^3} \frac{r^6\Omega^2}{G^2(\nabla_{ad}-\nabla_{rad})} \left(1 - \frac{\Omega^2}{2\pi G\rho}\right) \right|$, where Ω is the angular velocity in the star, G the gravitational constant, L, M and ρ the luminosity, the mass and the density at radius r, ∇_{ad} and ∇_{rad} the adiabatic and radiative gradients. This expression was obtained without taking into account the particle transport due to advection nor the mixing due to horizontal turbulence. Chaboyer and Zahn 1992 computed the combined effects of advection and mixing. With the assumption that the horizontal diffusion coefficient varies like $rU(r)$, they obtained an effective diffusion coefficient which can be written : $D_{eff} = \gamma|rU(r)|$ and is similar to D_T, although the physics is somewhat different. The parameter γ which is unknown and depends on the efficiency of horizontal diffusion is calibrated in order to reproduce a Li depletion by a factor 100 during the solar evolution. To compute the solar Li abundance evolution, we used the Skumanich relation (1972) to simulate the evolution of the solar rotation velocity from an assumed value of 100 km.s^{-1} on the zero age main sequence to the present value of 2 km.s^{-1}. The coefficient γ needed to reproduce the Li depletion factor at the surface of the sun in 4.57 Gyr is equal to 1/5. The corresponding Be depletion factor is negligible (^{9}Be$_\odot$/^{9}Be$_0$ = 0.992).

4 LITHIUM AND BERYLLIUM DEPLETION INDUCED BY ROTATION

4.1 Model properties

Using the Geneva stellar evolutionary code in which we have introduced the numerical method described in CVZ to solve the diffusion equation, we have computed the Li and Be abundances evolution due to the combination of nuclear burning and rotation induced mixing on the main sequence of stellar models of 1.0, 1.1, 1.15, 1.2, 1.25, 1.3 and 1.34 $M_\odot$ (Charbonnel et al. 1992b). Stellar models are constructed with the radiative opacities of Rogers and Iglesias 1992 completed at low temperatures (T<6000K) with the opacities of Huebner et al. 1977. The helium abundance Y (0.3) and the mixing length parameter α (1.6) are chosen for a test model of $1M_\odot$ performed with the solar metallicity z=0.0188 (Anders and Grevesse 1989) to reproduce the solar luminosity and radius at the solar age (4.57 Gyr). The metal content for our models is z=0.02. The relative ratios for the elements heavier than hydrogen and helium given by Anders and Grevesse 1989 and used in the opacity tables by Rogers and Iglesias 1992 are adopted.

We assume that all stars reach the main sequence with the same rotation velocity (V_i = 100 km.s^{-1}) and are slowed down according to the Skumanich relation (Skumanich 1972) : $\frac{1}{\Omega_H^2} - \frac{1}{\Omega_i^2} = c.t$ where $\Omega = V/R$. The constant c is adjusted to reproduce the mean velocity $\bar{v}$ and the velocities $\bar{v} \pm \sigma$ corresponding to the effective temperature at the age of the Hyades (figure 1). We neglect the variation of Ω inside the star.

4.2 Lithium and beryllium dispersion in the Hyades.

The influence of the rotational velocity dispersion in the Hyades on the theoretical lithium dispersion at the age of this cluster is shown in figure 2. The initial rotation velocity is 100 km.s^{-1} in each case for all the stars, and the velocities at the age of the Hyades are those given by the curves $\bar{v}$ and $\bar{v} \pm \sigma$; the rotational braking is simulated by the Skumanich relation. Results are compared with the lithium observations in the Hyades, Praesepe and Coma B. Our model leads to improved agreement with the observations. The corresponding dispersion of log(Li/Be) in the Hyades is presented in figure 3, and is in the range present in the observations.

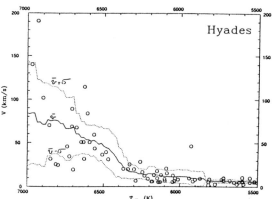

Fig.1 *Theoretical $\bar{v}$ and $\bar{v} \pm 1\sigma$. Circles represent observed vsini multiplied by $\frac{4}{\pi}$; observational data are from Mermilliod (1992) and Kraft (1965).*

Fig.2 *Theoretical lithium abundance dispersion is compared to the lithium observations in the Hyades, Praesepe, and Coma B. The initial rotation velocity is 100 km.s−1 in each case for all the stars; the velocities at 0.8 Gyr are those given respectively by the curves $\bar{v} \pm \sigma$ in Figure 1.*

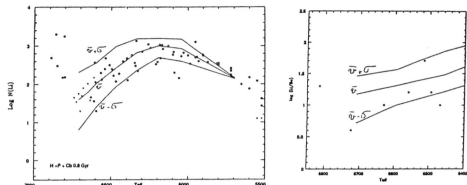

Fig.3 *Theoretical dispersion of log(Li/Be) at the age of the Hyades compared to observations (Boesgaard and Budge 1989).*

REFERENCES

Anders, E., and Grevesse, N., 1989, *Geochim. Cosmochim. Acta*, **53**, 197

Balachandran, S., Lambert, D.L., and Stauffer, J.R., 1988, *Ap.J.*, **333**, 267

Balachandran, S., 1989, in the *6th Cambridge Workshop on Cool Stars, Stellar Systems and the Sun*, Un. of Washington, Seattle

Boesgaard, A.M., and Lavery, R.J., *Ap.J.*, **309**, 762

Boesgaard, A.M., and Tripicco, M.J., 1986, *Ap.J. Letters*, **302**, L49

Boesgaard, A.M., and Budge, K.G., 1989, *Ap.J.*, **338**, 875

Chaboyer, B., and Zahn, J.P., 1992, *Astr.Ap.*, **253**, 173

Chandrasekhar, S., and Münch, G., 1950, *Ap.J.*, **111**, 142

Charbonnel, C., Vauclair, S., and Zahn, J.P., 1992a, *Astr.Ap.*, **255**, 191

Charbonnel, C., Vauclair, S., Maeder, A., Meynet, G., and Schaller, G., 1992b, *Astr.Ap.*, in preparation

Huebner, W.F., Merts, A.L., Magee, N.H.Jr, and Argo, M.F., 1977, *Astrophysical Opacity Library*, UC-34b

Kraft, R.P., 1965, *Ap.J*, **142**, 681

Mermilliod, J.C., 1992, private communication

Rogers, F.J., and Iglesias, C.A., 1992, *Ap.J.*, in press

Skumanich, A., 1972, *Ap.J.*, **171**, 565

Zahn, J.P., 1983, *Astrophysical Processes in Upper Main Sequence Stars*, ed. B.Hauck and A.Maeder, Publ. Geneva Observatory, 253

On the Origin of Lithium in the Black Hole Binary V404 Cyg

RAFAEL REBOLO AND EDUARDO L. MARTÍN

Instituto de Astrofísica de Canarias, E-38200, La Laguna, Spain

1 INTRODUCTION

We have recently discovered the presence of a strong LiIλ6708 resonance line in the spectra of the late type secondary of the X-ray transient binary V404 Cyg (Martín *et al.* 1992). The compact primary has a dynamical mass ≥ 6 M$_\odot$; i.e. higher than the maximum mass for a neutron star and consequently it is currently considered as the firmest black hole candidate in our Galaxy (Casares, Charles & Naylor 1992). The light curve of V404 Cyg in quiescence is dominated by the secondary and its periodic behaviour is suggestive of ellipsoidal variations (Wagner *et al.* 1992). These latter authors propose that the secondary is a solar-mass subgiant which nearly fills its Roche-lobe and is distorted by the gravitational field of the black hole.

In this paper we will discuss some interpretations of the high lithium abundance in the secondary of V404 Cyg in the light of the new work of Wagner *et al.* 1992, which was not known by Martín *et al.* 1992. The lithium abundance derived by Martín *et al.* under the assumption of a main-sequence secondary is not significantly altered, although the uncertainty is higher because the stellar parameters are less constrained. However, the interpretations of Martín *et al.* can be refined, particularly those referring to the evolutionary status of the system.

2 DISCUSSION

2.1. Stellar Evolution

High lithium abundance in a star with spectral type late G or K is the rule rather than the exception when the star is very young. Martín *et al.* 1992 showed that the Li abundance of V04 Cyg is normal if the secondary is younger than the age of the Pleiades (~ 100 Myr). But, if the secondary is a subgiant, as strongly suggested by the photometric variations found by Wagner *et al.* 1992, and the presence of Li is due to youth, the star would have to be an extremely young pre-main sequence (PMS) object. In fact, if we use conventional PMS tracks and isochrones (cf. Cohen & Kuhi 1979) and the stellar parameters adequate for a K0IV star of ~ 7 R$_\odot$, we get an age of less than 1 Myr and a mass of $\sim$ 2 M$\odot$. Martín, Magazzù & Rebolo 1992 have reported on three B-type stars with visual low mass G-type PMS companions. These three low-mass secondaries had high lithium abundance (comparable to that of V404 Cyg). However, if black hole binaries have been forming thoughout the life of the Galaxy, and if they have lifetimes of several hundreds million years (cf. van den Heuvel 1983), it seems highly unlikely that V404 Cyg has been found in its first million year of existence.

It is more probable that the secondary of V404 Cyg is an evolved subgiant. Wagner *et al.* 1992 suggest that it may be a solar-mass star evolving from the main sequence to the red giant branch. At such stage of stellar evolution the dilution caused by the expansion of the outer stellar layers causes the surface abundance of lithium to decrease. If we place the K0IV secondary of V404 Cyg in the post-main sequence evolutionary tracks of Iben 1967, we infer a dilution of a factor $\sim$30, and an age for the star in the range 4 to 11 Gyr, for masses of 1.25 and 1.00 M$\odot$, respectively. Observations of Li in subgiants of the open cluster M67 (age $\sim$ 5 Gyr) by Balachandran 1990 have shown Li depletion greater than predicted by Iben 1967. Such strong depletion can be accounted for by Li destruction during the MS lifetime (Rotationally induced mixing, Gravity waves, Diffusion, Mass loss). In particular, the latter Li depletion mechanism may be very active in the V404 Cyg secondary, as the black hole is accreting matter from it. At this point we remind the reader that the previous considerations do not take into account the possible influence of the black hole on the structure of the solar-type secondary. Martín *et al.* 1992 have discussed that tidal effects do not seem to prevent strong Li depletion in solar-type close binaries. However, the question remains open as we know very little about the influence of the black hole in the structure and evolution of its low-mass companion.

To summarize briefly, interpretations of the high Li abundance in V404 Cyg in the context of standard single-star stellar evolution are not satisfactory because they require the system to be extremely young (1 Myr), which has very low probability. If, as it is more probable, the system is relatively old (a few Gyr) a strong Li depletion is expected (mainly via dilution and mass loss), which is incompatible with the high Li abundance observed, unless there is significant Li production. A possible way of Li generation in V404 Cyg is through high energy spallation reactions during X-ray outbursts. In the following section we will have a closer look to this hypothesis.

2.2. Spallation Reactions

Nucleosynthesis of lithium can take place in spallation reactions when high energy protons and α particles collide with He,C,N,O nuclei. This process takes place in the interstellar medium and is currently the most favoured one for explaining the cosmic abundances of Be, B and the rarer isotope of lithium ^{6}Li, although the dominant isotope ^{7}Li needs other sources of production (cf. Walker, Mathews & Viola 1985). It is very important to find Galactic sites of lithium generation in order to understand the origin and evolution of this element and use it as a constraint to Big Bang nucleosyntesis (cf. Rebolo 1991).

In this context the finding by Martín *et al.* 1992 of high amount of Li in V404 Cyg may be a clue to Li production in the vicinity of the black hole. Ryter *et al.* 1970 have studied the energetics of the spallation reactions leading to Li syntesis and have shown that the creation of Li can be correlated with X-ray emission. The system V404 Cyg has the highest peak X-ray luminosity of all the X-ray binaries so far known in the Galaxy. As a consequence it seems that during the X-ray explosions (which may be recurrent in timescales of ~ 20 yr) there is enough energy available for spallation reactions. The question of how the Li produced in the neighbourhood of the black hole is transferred to the secondary is perhaps more speculative. We can imagine that a fraction of the material accreted onto the black hole is ejected in jets and that some of them may intersect the orbit of the subgiant companion. Some lithium may be generated in the surface of the subgiant star itself as the high energy particles penetrate into its low-density atmosphere. Detailed calculations of these processes are needed but are outside the scope of this paper. We note that Martín *et al.* have proposed a γ-ray test of this hypothesis. If the spallation scenario proves realistic for producing the high Li abundance observed in V404 Cyg it would give support to models of ^{7}Li production in massive black holes that accrete matter from the interstellar medium (Jin 1990).

3 FINAL REMARKS

In this paper we have used the most recent data on the recently recognized black hole binary V404 Cyg in order to interpret the high lithium abundance of the low-mass secondary. From conventional stellar evolution we are led to the highly improbable hypothesis that the companion to the black hole is a pre-main sequence object. If the companion is an evolved subgiant a strong lithium depletion is expected which may be contrarrested by some unknown effect of the black hole on the structure and evolution of the secondary, or by Li generation via spallation reactions. However, we cannot discard that, in a system with such complicated astrophysics as V404 Cyg, there may be some other very different interpretations to those that have been discussed by us. Our views could be criticized as almost prosaic if we did not bear in mind that more spectacular explanations are possible. For example, we may be faced in the future with the unexpected possibility that all the black holes that may be found would show signs of extreme youth, and in such case the high Li abundance of V404 Cyg could be a first hint to unknown physics in the evolution of black holes.

ACKNOWLEDGEMENTS

We thank J. Casares for many helpful discussions. This work is based on observations made with the William Herschel Telescope operated on the island of La Palma by the Royal Greenwich Observatory in the Spanish Observatorio del Roque de los Muchachos of the Instituto de Astrofísica de Canarias.

REFERENCES

Balachandran, S. 1990, in *Cool Stars, Stellar Systems, and the Sun*, ed. G. Wallerstein, A.S.P. 9, 357

Casares, J., Charles, P.A., Naylor, T. 1992, *Nature*, 355, 614

Cohen, M., Kuhi, L.V. 1979, *Ap. J. Suppl.*, 41, 743

Iben, I.Jr. 1967, *Ap. J.*, 147, 624

Jin, L. 1990, *Ap. J.*, 356, 501

Wagner R.M., Kreidl, T.J., Howell, S.B., Starrfield, S.G. 1992, *Nature*, in press

Martín, E.L., Magazzù, A., Rebolo, R. 1992, *Astr. Ap.*, 257, 186

Martín, E.L., Rebolo, R., Casares, J., Charles, P.A. 1992, *Nature*, 358, 129

Rebolo, R. 1991, in *Evolution of Stars: The Photospheric Abundance Connection*, ed. G. Michaud & A. Tutukov, I.A.U. Symp. 145, 85

Ryter, C., Reeves, H., Gradstajn, E., Audouze, J. 1970, *Astr. Ap.*, 8, 389

van den Heuvel, E.P.J. 1983, in *Accretion driven stellar X-ray sources*, eds. W. Lewin & E. van den Heuvel, chapter 8 .

Walker, T.P., Mathews, G.J., Viola, V.E. 1985, *Ap. J.*, 299, 745

NUCLEOSYNTHESIS UP TO THE Fe PEAK

Observational Signatures of Stellar Nucleosynthesis — A Sampler

DAVID L. LAMBERT

European Southern Observatory   and   Department of Astronomy
Karl-Schwarzschild-Str. 2   University of Texas
W–8046 Garching bei München   Austin, Texas 78712
Germany   U.S.A.

1 CIRCA ANNO 0 HR

As early as 1920, Eddington in an address to the British Association declared "I think that the suspicion has been generally entertained that the stars are the crucibles in which the lighter atoms which abound in the nebulae are compounded into more complex elements. In the star matter has its preliminary brewing to prepare the greater variety of elements which are needed for a world of life" (Eddington 1920). Eddington's legendary insight enabled him to support this bold proposal despite the blatant infancy of the crucial disciplines of nuclear physics, stellar structure, and evolution. Insight may have been bolstered by regular contact with a confident Sir Ernest Rutherford for Eddington preceded the above declaration with the suggestion that "what is possible in the Cavendish Laboratory may not be too difficult in the Sun".

The decade 1930–1939 that includes the year 0 HR saw the initial applications of nuclear physics to the questions of stellar energy sources and element synthesis. Rolfs and Rodney (1988) in their superb text "Cauldrons in the Cosmos" note that Atkinson and Houtermans (1929) pioneered studies of energy production in stars and stellar nucleosynthesis of the elements with the first quantitative examination of "how to cook a nucleus in a pot". By the end of the decade thanks to work by von Weizsäcker (1937, 1938), Bethe and Critchfield (1938) and Bethe (1939), "it was clear that two different sets of reactions could convert sufficient hydrogen into helium to provide the energy needed for a star's luminosity, namely, the proton-proton (p-p) chain and CN cycle" (Rolfs and Rodney 1988). Stellar nucleosynthesis had been launched.

In seeking answers to the questions, "How, when, and where were the chemical elements synthesized in stars?", we necessarily provide clues to fundamental questions about the structure and evolution of the stars, the galaxies and the universe. Inevitably, the pursuit of stellar nucleosynthesis leads most often to insights into the structure and evolution of stars. But the appeal of the pursuit is that its impact is wide: stellar nucleosynthesis has been used to test claims for stellar remnants (white dwarfs, neutron stars, and black holes) as major contributors to the dark matter in our Galaxy's halo; accurate specifications of the chemical compositions of stars and gas across the Galaxy together with information on kinematics and ages are the building blocks for a theory of galactic evolution; abundance analyses of the most metal-poor stars in the Galaxy constrain theories of the primordial fireball and the evolution of the early universe as galaxies began to form; observations of H II regions reveal the history of stellar nucleosynthesis in external galaxies and provide data with which to test theories of galactic evolution; analyses of the emission and absorption line spectra of quasars provide estimates of the chemical composition in young systems and so of the objects that inhabited and polluted these systems (normal stars, supermassive objects, ...?). A thorough review of both the observational and theoretical status of the field of stellar nucleosynthesis demands a substantial book. Here, I must be highly selective. Examples of the impact of recent observations on unsolved problems of stellar nucleosynthesis and stellar evolution constitute the body of my essay.

2 OBSERVATIONAL CLUES — AN INTRODUCTION

In this section, I attempt to classify the observations pertinent to stellar nucleosynthesis. An initial broad classification may be made by the mode of analysis: chemical or spectroscopic? By chemical, I mean those observations in which atoms or nuclei are counted. Spectroscopic analysis refers, of course, to the acquisition and interpretation of astronomical spectra.

Under the heading "chemical analyses", I would include the following examples:

● **Carbonaceous chondrites.** These meteorites are presumed to be a record of the composition of the early solar system and even the Sun's parental interstellar cloud. Tables of the elemental and isotopic abundances are assembled at intervals: Suess and Urey's (1956) compilation, which was the first to relate chemical abundances to nuclear structure, was exploited by Burbidge et al. (1957) in their epic theoretical investigation of stellar nucleosynthesis, and the latest table is that by Anders and Grevesse (1989). Volatile elements, among them the astronomically abundant elements H, He, C, N, and O, are not surprisingly underabundant in meteorites and entries for such elements in a table of abundances requires a normalization of a solar, stellar or other astronomical abundance to the meteoritic abundance scale.

● **Interstellar grains in meteorites.** A remarkable recent discovery has been the isolation from meteorites of very small grains that are identified as interstellar grains that survived entry into the solar nebula and incorporation into meteorites

(Wasserburg 1992). Identified grains include microdiamonds, SiC particles, and very small pieces of amorphous graphite. At present, it is studies of the SiC grains that have provided the most information on stellar nucleosynthesis: the majority of the grains came from the circumstellar envelopes of AGB stars and chemical analyses of these grains reveal much about the operation of the s-process in the He-shell of the AGB star. Earlier discoveries of isotopic anomalies in inclusions in meteorites, including excess ^{26}Mg from decay of short-lived ^{26}Al, were also relevant to aspects of stellar nucleosynthesis.

• **Galactic and solar cosmic rays.** Analyses of solar energetic particles and the solar wind reveal a systematic difference between their composition and that of the solar photosphere (as determined spectroscopically). This difference dependent on the first ionization potential (FIP effect) is presumed to arise from fractionation in the low chromosphere. The composition of the observed Galactic cosmic rays differs from their composition at their point of origin: the source abundances are obtained on correction for spallation reactions en route through the interstellar medium. The source abundances resemble closely those of the solar energetic particles leading Meyer (1985) to propose that the galactic cosmic rays come primarily from normal stars and not supernovae — see, however, recent reviews (Cesarsky 1992, Ferrando 1992).

A classification of spectroscopic analyses may be given in terms of the "object" observed. Here I list three subclassifications: stellar systems, stellar surfaces, and stellar ejecta.

• **Stellar systems.** Under stellar systems, I put those observations that are generally (or should be!) considered in theoretical studies of "The Chemical Evolution of the Galaxy, its large subcomponents and other galaxies": subcomponents includes the Galactic disk, globular clusters, and large complexes of gas and stars in the disk. Often, data on kinematics and composition should be integrated. For the disk, spectroscopic analyses of stars and interstellar gas supply information on abundances that is then applied to four traditional tests.

(i)   The age-metallicity relation: the metallicity [X/H] vs stellar age τ_* at a fixed galactocentric radius R_c, where X is, for obvious observational reasons, usually identified with Fe. This test has to be conducted using stars.

(ii) The Galactic abundance gradient: [X/H] vs R_c at a given τ_*. Emission line spectroscopy of H II regions contributes valuable information on abundant light elements (e.g. H, He, C, N, O, Ne, S). Millimetre and radio line spectroscopy of cold interstellar gas provides information on the run of isotopic ratios (e.g. ^{12}C/^{13}C, ^{16}O/^{17}O/^{18}O) with R_c.

(iii)   The variation of the relative abundances with metallicity: most often [X/Fe] vs [Fe/H] is considered.

(iv)   The frequency distribution of the metallicities: $N([Fe/H])\delta([Fe/H])$ for a given R_c and τ_*, a test whose outcome is often referred to as the G-dwarf problem.

The tests sample in different ways the primary influences on the Galaxy's chemical composition such as the star formation rate, the initial mass function, the infall rate, and stellar nucleosynthesis. Since stellar nucleosynthesis is intermingled with several other, generally ill-defined influences, observational studies of Galactic chemical evolution are not a primary provider of direct novel data on stellar nucleosynthesis. Indeed predictions of nucleosynthetic yields are commonly adopted and one or more of the traditional tests then examined in order to constrain the more poorly known influences. Test (iii) has provided the most observational clues about stellar nucleosynthesis.

• **Stellar surfaces.** Stellar spectroscopy is a major source of the abundances considered in unravelling the chemical evolution of stellar systems. In this application, it is assumed that either the compositions of the selected stars are those of their parental clouds or that the elemental abundance under investigation is unaffected by the processes that may have changed the chemical composition of the stellar atmosphere. By identifying "stellar surface" as a source of data on stellar nucleosynthesis, I mean those stellar surfaces whose composition has been changed by processes such as mixing, mass loss by a star, and mass transfer across a binary system. Today, this is an active field of investigation for observers and theoreticians with new major insights into stellar nucleosynthesis appearing steadily. The "stellar surfaces" of relevance appear over almost the entire H-R diagram, from young massive stars (Wolf-Rayet, OBNs) to red giants and supergiants and to lower main sequence stars where loss of surface Li remains an unsolved problem of stellar astration (reverse-nucleosynthesis). Binary stars of interest include Algols and Barium stars.

• **Stellar ejecta.** Products of stellar nucleosynthesis may be observed directly in ejecta from several different types of stars. Obvious examples include supernovae, novae, planetary nebulae and massive young stars such as WR stars with associated nebulae and peculiar objects like η Car.

3 OBSERVATIONAL CLUES — FOUR EXAMPLES

Perusal of this volume and the current journals will show the liveliness of the observational pursuit for clues to stellar nucleosynthesis. Here, I give brief presentations of some recent work and leave to another occasion a comprehensive review. I hope the selected examples illustrate the liveliness and the breadth of the pursuit.

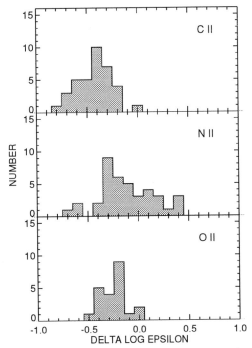

Figure 1: Histograms of the non-LTE abundances relative to solar values.

3.1 Early B-Type Main Sequence Stars

These main sequence stars of masses from 5 to 20 $M_\odot$ might be expected to have atmospheres with compositions identical to those of the nearby H II regions. More massive stars are subject to severe mass loss by winds and modification of the surface abundances: the Wolf-Rayet stars are extreme cases in which the outer layers are stripped to expose first the former H-burning layers (WN stars) and subsequently the He-burning layers (WC stars). OBN stars whose spectra show enhanced N lines have atmospheres enriched in N from the deep H-burning layers but the appearance of the N-rich material may be attributable to mixing induced by rotation rather than to exposure by mass loss. B stars less massive than the early B-type stars infringe upon the strip of the main sequence inhabited by the chemically peculiar stars whose anomalous surface compositions are (presumably) not directly consequences of nucleosynthesis.

Our recent study of a sample of local early-B stars was based on spectra at a resolution $\lambda/\Delta\lambda = 25{,}000$ for blue-green regions containing C II, N II and O II lines — see Gies and Lambert (1992) for the details of the analysis. Abundances of C, N, and O summarized in Figure 1 are non-LTE values derived using tables of predicted equivalent widths of C II, N II, and O II lines (Eber and Butler 1988; Becker and Butler 1988a, 1989, 1990). The contrast between the narrow histogram of the O abundances and the much broader histogram of the N abundances is striking. The width of the former is consistent with the width due to the errors of measurement. Since the three histograms should be broadened to a very similar extent by the errors of measurement, the significantly greater widths of the N and the C histograms are presumably attributable to stellar evolution. The mean O abundance is about 0.3 dex below the solar value and, hence, the local B stars confirm the O deficiency (relative to the Sun) reported previously for local H II regions (Peimbert 1987).

Carbon and nitrogen are also deficient in both B stars and H II regions; the initial C and N abundances of the stars are presumed to be those at the high and low ends of the respective histograms. This presumption is based, of course, on the idea that the N enrichment signals the presence of products of the CN-cycle in which C is processed to N with conservation of the sum of C and N nuclei. If, as appears to be the case, C and N are similarly underabundant in the zero-age B stars, the upper limit for the N abundance is expected to be $\Delta \log \varepsilon(N) \leq 0.45$ from CN-cycling. This limit marks the upper edge of the N histogram. Appearance of the ON-cycle's products at the surface could increase the N abundance to $\Delta \log \varepsilon(N) \simeq 0.9$. The narrow width of the O histogram and the absence of stars with $\Delta \log \varepsilon(N) > 0.45$ shows that the atmospheres are not contaminated with ON-cycled material.

Inspection of the stars in a H-R diagram shows that the N-rich stars are all at least somewhat evolved stars, but not all evolved stars are N-rich. Examination of the N abundances of stars in the mass range $8.5 < M/M_\odot < 13$ as a function of their relative ages (t/t_{ms}) — t_{ms} is the main sequence lifetime — shows that the mean N abundance increases by only 0.3 dex between $t/t_{ms} = 0$ to 1. The increase is less than the dispersion seen at a given age. The noticeable dispersion suggests that the mixing of CN-cycled material into the atmosphere is controlled by factors other than mass, age, and initial composition. Rotation is a candidate: rapidly rotating stars may be mixed more than slowly rotating stars. Mass loss is a second candidate. Mass loss may lead to rotational braking and to a diminution of mixing but severe mass loss may expose layers that previously experienced CN-cycling and, hence, are N-rich. Non-rotating stars may experience little change in the surface N abundance; the minimum N abundance is possibly independent of age. It is also conceivable that the C and N anomalies were inherited from birth and reflect contamination of an interstellar cloud with (for example) ejecta from a WN star.

Examination of a larger sample of stars should provide clues to the agents responsible for the mixing. The present sample specifically excluded the broad-lined stars. One might suppose that mixing effects increase in severity in the rapid rotators. If the observed sample is large, it should be possible to correct statistically for the projection effects and to infer the velocity dependence of the N enrichment. Extension of the survey to M < 8.5 $M_\odot$ is also desirable but, at a spectral type of about B7, one encounters the hottest chemically peculiar stars. A simpler extension of the present results might be more informative: analyse a representative sample of stars with M $\gtrsim$ 13 $M_\odot$. A motivation for analysing massive stars is provided by Sk–69°202 (alias SN 1987A). SN 1987A showed narrow ultraviolet emission lines of C, N, and O lines from the circumstellar shell excited by the soft X-ray burst from the supernova. This shell was presumably ejected when Sk–69°202 was a red supergiant. Analysis of the emission lines gives the shell's composition as N/C $\sim$ 5, N/O $\sim$ 2, and He/H $\sim$ (0.06–0.20). A similar N/O ratio has been reported recently for several Galactic WR ring nebulae: N/O $\sim$ 2 and He/H $\sim$ 0.2 according to Esteban *et al.* (1992) who suggest that these massive stars (M $\sim$ 20–40 $M_\odot$) shed the bulk of the nebular gas as a yellow hypergiant or a red supergiant. These Galactic examples differ in one major way from Sk–69°202: the Galactic stars are Wolf-Rayet stars but Sk–69°202 was a blue supergiant. Continued exploration of OB stars in the Galaxy and the Magellanic Clouds is now needed to define their compositions in order to trace the appearance of products of H-burning in their atmospheres.

The narrow spread of the O abundances in Figure 1 may be ascribed to observational errors and sets a useful constraint on the degree of self-enrichment experienced by the nearby OB associations from which we drew our stars. Reeves (1978) and later Olive and Schramm (1982) suggested that an OB association may be self-enriched as supernovae from earlier generations of massive stars pollute portions of the parental molecular cloud that later undergo star formation. The suggestion was advanced, in part, to explain why the 4.5 Gyr old Sun was more O-rich than local H II regions like the Orion nebula: the Sun formed late in the life of an OB association from polluted/self-enriched gas with an O abundance exceeding the average value for that era. Our results for the O abundance in local B stars (Figure 1) shows that self-enrichment by a factor of 2 is a rare occurrence. In a continuing study of the Orion association, we have adduced evidence for a mild O enrichment in the youngest sub-groups (Ic and Id — see Blaauw's [1964] classifications) relative to the older sub-groups Ia and Ib (Cunha and Lambert 1992). Examination of other OB associations may uncover examples of more severe self-enrichment and so verify Reeves's (1978) pioneering predictions.

3.2 The Asymptotic Giant Branch

Today, no study of stellar nucleosynthesis dare be considered complete without examination of the AGB stars which are likely to be the Galaxy's principal suppliers of ^{7}Li, ^{14}N, ^{19}F, and the s-process ("main component") elements and major suppliers of ^{12}C, ^{13}C and some other elements. Nucleosynthesis occurs primarily during the thermal pulses experienced by the He-shell sandwiched between the electron-degenerate C-O (possibly O-Ne-Mg) core and the extended convective H-rich envelope. After a pulse, products of nucleosynthesis may be dredged up by the convective envelope. Later, as the envelope is shed through the stellar wind and creation of a planetary nebula, the products are dispersed into the interstellar medium. Additional nucleosynthesis occurs through H-burning and associated reactions such as the ^{7}Li-producing ^{7}Be-transport mechanism (Cameron and Fowler 1971) when the base of the convective envelope becomes hot enough, as it is observed and predicted to do in very luminous AGB stars.

The upper limit for the mass of an AGB star is about 8 M$_\odot$; more massive stars do not develop an electron degenerate core. AGB stars of about 3 to 8 M$_\odot$ are termed "intermediate-mass" AGB stars. Observations of the Magellanic Clouds show that "low" mass (M $\leq$ 3M$_\odot$) AGB stars dredge up ^{12}C and other products from the He-shell. But models of these stars do not readily predict this dredge-up; ^{12}C and other observed nuclides are predicted to be synthesized in the He-shell but the transfer of material from the He-shell to the envelope is not predicted. This transfer is, however, predicted for intermediate-mass AGB stars but nature's counterparts of these stars are unexpectedly rare in the Magellanic Clouds and presumably in the Galaxy, too. The question of the dredge-up in AGB stars was aired at IAU Symposium No. 145 by Iben (1991), Lambert (1991) and Sackmann and Boothroyd (1991). The stellar spectroscopist may determine whether an AGB star is a "low" or an "intermediate" mass star. I refer the reader to my 1991 review for a full discussion. First, the neutron source for the s-process is predicted to be ^{22}Ne (α,n) ^{25}Mg in intermediate-mass stars and ^{13}C(α,n) ^{16}O in low-mass stars. After dredge-up, ^{25}Mg and ^{26}Mg are predicted to be overabundant in the atmosphere of an intermediate-mass but not a low-mass AGB star. The searches made to date show no enrichment of these n-rich Mg isotopes, i.e., the observed Galactic stars are low-mass AGB stars. Second, the neutron density in the He-shell controls branches in the s-process path. These branches have long been exploited in classical analyses of the solar system abundances of the heavy nuclides. Two branches are open to scrutiny by stellar spectroscopists: a branch at unstable ^{85}Kr controls the Rb abundance and a branch at unstable ^{95}Zr controls the isotopic abundance ratio of ^{96}Zr to lighter stable Zr isotopes. Our analyses (Lambert and Smith 1992) of Rb and ^{96}Zr show that the s-process operated at the "low" (N(n) $\lesssim$ 10^9 cm^{-3}) densities expected of low-mass AGB stars, not the "high" (N(n) $\geq$ 10^{11} cm^{-3}) densities predicted for intermediate-mass stars.

Surface abundances of the principal elements synthesized by AGB stars may be used to probe physical conditions in and near the He shell. Rb and ^{96}Zr were cited as neutron densitometers. Fluorine is an additional probe of the thermal pulses experienced by the He-shell. Fluorine, an astrophysically elusive element, is represented by the one stable isotope ^{19}F. Various sites for its nucleosynthesis have been proposed: explosive H-burning in novae, He-burning in massive stars or AGB stars, and the Ne-rich layer of a Type II SN. Forestini et al. (1992) and Jorissen et al. (1992) discuss ^{19}F production in AGB stars.

The origin of ^{19}F is uncertain in part because of a paucity of data on its abundance outside the Solar system. Until our survey of the $2\,\mu$m HF lines in red giants (Jorissen *et al.* 1992), the only detections of F were in Betelgeuse where Spinrad *et al.* (1971) detected several HF lines and in high excitation planetary nebulae where a [F IV] emission line was reported by Aller (1978) and Aller and Czyzak (1983). Fluorine is a difficult target for stellar spectroscopy because it is a trace element ($\log \varepsilon(F) =$ 4.5 in the solar system) and the atomic structure is such that only high excitation (therefore, very weak) lines of the neutral atom and ions are present in the optical spectrum. Fortunately, the HF molecule is detectable at $2\,\mu$m in the spectra of red giants and the fundamental V-R band is so placed that several R-branch 1–0 lines are detectable on the high frequency side of the strong CO first-overtone V-R bands.

Our survey investigated the F abundance in normal red giants of spectral types K and M, s-process enriched O-rich giants of spectral types MS and S, C-rich giants of spectral types SC and N, as well as J-type carbon stars and classical Barium giants. Fluorine is clearly enriched in the atmospheres of the AGB stars that have experienced dredge-up from the He-shell. It is also enriched in those stars — the Barium giants and the S stars without Tc — that have accreted mass from a (former) AGB companion star that is now a white dwarf. With respect to ^{16}O, an element whose abundance is not expected to change during AGB evolution, the F enrichment is such that the F/O ratio is increased relative to its solar-like value in normal K and M giants by up to a factor of 10 in S stars and of 30 in carbon stars. At these levels, mass loss from AGB stars is likely to be responsible for Galactic production of F.

Our stellar spectra betray the effects of nuclear reactions in (presumably) an AGB star's He-shell but for a specification of the responsible reactions, one must turn to theory. The proposed scheme for F synthesis uses the neutron source ^{13}C (α,n) ^{16}O that runs the s-process in low mass AGB stars:

$$^{13}C(\alpha,n)^{16}O \text{ followed by } ^{14}N(n,p)^{16}O \text{ and } ^{14}N(\alpha,\gamma)^{18}F(\beta^+\nu)^{18}O(p,\alpha)^{15}N(\alpha,\gamma)^{19}F$$

Forestini *et al.* (1992) examine ^{19}F production in a 3 $M_\odot$ AGB star and, if a dredge-up is imposed, the F enrichment in the atmosphere approaches the modest level observed in S stars. In these calculations, the ^{13}C and ^{14}N are supplied from CNO-cycling in the thin H-burning shell that is active when the He-shell is quiescent. Production of ^{19}F is efficient and can account for the observed correlation for S stars between F enrichment and the increased C/O ratio resulting from the dredge-up. The model predicts that an enrichment of ^{18}O should accompany that of ^{19}F: ^{16}O/^{18}O is reduced by approximately a factor of 5 for a factor of 3 enrichment of F. (The early pulses examined by Forestini *et al.* do not result in significant enrichment of the envelope in s-process products.)

The prediction concerning ^{18}O is open to test using the CO vibration-rotation bands. Note that the isotopic ratios are now also measurable from the pure rotation CO lines for stars with extensive circumstellar envelopes — see, for example Kahane *et al.* (1992) who discuss the $J = 1$–0 and 2–1 emission lines from carbon-rich envelopes. S stars experiencing thermal pulses (i.e., Tc is present in the atmospheres) have an ^{18}O abundance similar to that of K and M giants that have yet to evolve to the AGB — see Smith and Lambert (1990) and Lambert (1992). Certainly, the $^{16}O/^{18}O$ ratio is not reduced by a factor of 5. The issue of the ^{19}F and ^{18}O abundances is presumably a clue to conditions in the He-shell at the time when He is being burnt and during the longer interval when the shell is quiescent and H-burning is occurring just above it. To fit the high F enrichment of extreme S and carbon stars and to avoid simultaneous ^{18}O enrichment, additional production of ^{13}C in the He shell prior to the thermal pulse seems necessary. Jorissen *et al.* (1992), who dub this as "primary ^{13}C production", sketch how ^{18}O enrichment may be suppressed under these conditions. Since ^{19}F is destroyed by α particles in the hot He shells of intermediate mass stars, F enrichment is provided by low not intermediate mass stars. In this sense, ^{19}F is a thermometer for the He-shell.

3.3 Oxygen in the Galactic Halo

That oxygen is overabundant relative to iron in halo dwarfs and giant stars (i.e., [O/Fe] > 0) is not in doubt. What remains uncertain is the magnitude of the over-abundance. The uncertainty was dramatically enlarged by Abia and Rebolo's (1989) analysis of the O I 7774 Å triplet in subdwarfs which showed [O/Fe] in the halo to increase with decreasing metallicity from [O/Fe] $\simeq$ 0.7 at [Fe/H] $\simeq$ -1 to 1.5 at [Fe/H] = -3. In sharp contrast, observations of the [O I] 6300,6363 Å lines in giants give [O/Fe] = 0.4 with no significant dependence on metallicity. Recently, data from ultraviolet OH lines in subdwarfs (Bessell *et al.* 1991, Gilmore *et al.* 1992) has provided [O/Fe] estimates between those from the O I and [O I] lines. Recent reviews summarize the available data on [O/Fe] (Spite 1992, Barbuy 1992). I use this opportunity to sound a warning that lacunae exist in our modelling of line formation in the atmospheres of subdwarfs. If, as is now clear, systematic errors vitiate analyses of the oxygen lines, might they not be present in analyses of other elements: do we truly know the chemical composition of the halo?

Oxygen is an important element in piecing together the chemical evolution of the halo because it is synthesised by the most massive stars. The common interpretation of [O/Fe] > 0 for the halo is that the halo gas was enriched by massive stars but not the SN Ia that provide lots of iron but very little oxygen. Stellar ages derived from evolutionary tracks are sensitive to the adopted O abundance because in metal-poor stars the rate of CNO-cycling is dependent on the O abundance: an increase of [O/Fe] from 0.0 to 0.5 decreases the derived ages of the globular clusters by about 2 Gyr (VandenBerg 1992).

There are strong arguments for supposing that the [O I] lines in dwarfs and giants provide the closest approximation to the [O/Fe]. One expects the [O I] lines to be formed in LTE and to measure in effect the total column density of oxygen atoms through the atmosphere. Ionization and association of O atoms, which are expected to reduce only slightly the number of free O atoms, can be computed with adequate accuracy from LTE. By contrast, the O I 7774 Å lines are sensitive to the temperature profile of the atmosphere and possibly to non-LTE effects. The OH lines are sensitive to the structure of the cool upper photosphere and perhaps also to non-LTE effects.

It is known that the [O I] lines give a systematically lower [O/Fe] value than the O I 7774 Å lines in the same star — see Magain (1988) and Spite and Spite (1991) who analyse subdwarfs with [Fe/H] > −1.7; the 6300 Å line provides a [O/Fe] consistent with that obtained from the same line for halo giants. Since subdwarfs with [Fe/H] < −1.7 are predicted to have a 6300 Å line with an equivalent width of 2 mÅ or less for [O/Fe] = 0.5, direct detection of this line in extreme halo subdwarfs is difficult at best.

Tomkin et $al.$ (1992) have shown from observations of the O I 7774 Å in subdwarfs that the systematic error persists down to at least [Fe/H] = −3 and, more importantly, found that the error is dependent on effective temperature. Tomkin et $al.$ measured the O I 7774 and C I 9100 Å lines which are expected in LTE to give a C/O ratio that is insensitive to the atmospheric structure. The mean ratio from the sample was [O/C] = 0.52 ± 0.12 where the σ is essentially that expected from the measurements of the equivalent widths; this is a non-LTE ratio but a similar value is obtained from the LTE abundances. Tests show that the lines should respond in a similar fashion to changes of atmospheric structure. Clearly, the star-to-star scatter in C/O is small. If [C/Fe] = 0, as may be the case (Wheeler et $al.$ 1989), this analysis of C I and O I lines confirms the result [O/Fe] $\simeq$ 0.5 provided by the 6300 Å in giants.

Close inspection of the [O/Fe] (and [C/Fe]) values reveals that they are correlated with the stellar effective temperatures, as shown in Figure 2. Note that [O/Fe] at high T_{eff}'s tends to the value expected from [O I] analyses of giants. Tomkin et $al.$ surmise that this systematic trend of [O/Fe] with T_{eff} reflects the strengthening of the surface convection zone in the cooler stars; the present models may not predict correctly the temperature profiles of the cooler atmospheres or the atmospheres may be highly inhomogeneous (stellar granulation). The trend may also arise from an inadequate assessment of the non-LTE effects. Kiselman (1991) predicted these effects to be quite large for the O I 7774 Å lines but Tomkin et $al.$'s calculations show them to be 0.1 dex or less. This difference is largely due to Tomkin et $al.$'s inclusion of excitation by collisions between O atoms and neutral H atoms; Kiselman considered only the O-electron collisions. I cannot emphasize too strongly that the adopted cross-sections for the O-H collisions are crude estimates. Reliable calculations are needed of the excitation cross-sections of oxygen and other atoms by the neutral H atoms.

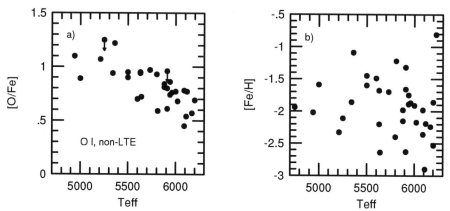

Figure 2: [O/Fe] and [Fe/H] versus effective temperature for the subdwarfs analysed by Tomkin *et al.* (1992).

3.4 Na-rich and O-poor Halo Stars

The [O I] lines in field halo giants show that [O/Fe] = 0.4 ± 0.1 for [Fe/H] < −1. Presently, measurements extend to about [Fe/H] = −3. Inspection of the results for the giants suggests that the star-to-star scatter in [O/Fe] is small and masked by the errors of measurement. By contrast with the field, recent work on globular cluster giants has confirmed earlier suspicions and shown that O-poor giants are common in these clusters (Sneden *et al.* 1991, Kraft *et al.* 1992, Sneden *et al.* 1992). The frequency and severity of O-deficiency varies from cluster to cluster. The O-poor stars are N-rich. Definition of other abundance anomalies correlated with oxygen will surely provide clues to the origins of the O-poor giants. Additional clues will come from the luminosity dependence of the anomalies, e.g., detection of O-poor subgiants and main sequence stars will show that a giant's convective envelope and associated first dredge-up are not solely responsible for the anomalies. It is already known that in clusters where the necessary spectroscopy has been undertaken that the distinction between CN-rich and CN-poor stars is maintained to the stars at the main sequence turn off (e.g., Bell, Hesser and Cannon 1983).

Early work showed that the N-rich stars were Na and Al rich relative to the other giants. Hints of an association between O deficiency and Na-Al enrichment were offered (e.g., Paltoglou and Norris 1989). Now, the association has been defined thanks to Drake *et al.*'s (1992) analysis of high resolution CCD spectra of giants in M4, and the work by Kraft, Sneden and colleagues on M3, M5, M13, and M92. Figure 3 shows the [Na/Fe] vs [O/Fe] correlation for several clusters and some field giants. This gives the impression that the manufacture of Na is correlated with the destruction of O or a related nuclide: [Na/Fe] saturates at about 0.3 as though the source of Na has been exhausted.

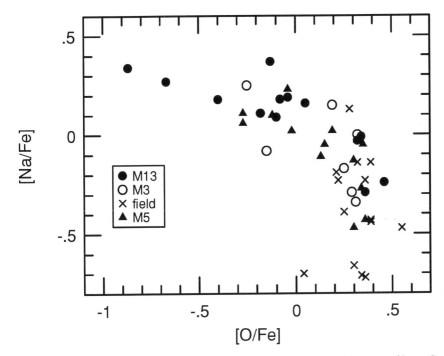

Figure 3: The [Na/Fe] – [O/Fe] correlation for cluster and field giants (from Sneden *et al.* 1992).

It has been widely presumed that a giant can not enrich its atmosphere in Na and Al and, hence, the anomalies must have been present in the star from its birth. This may be so but Denisenkov and Denisenkova (1990) show that synthesis of Na accompanies ON-cycling and, if ON-cycled products are mixed to the surface, Na enrichment must occur. Since observations of C and $^{12}C/^{13}C$ clearly show that nature's giants experience a more severe first dredge-up than current models predict, presence of ON-cycled material cannot be excluded on theoretical grounds. Denisenkov and Denisenkova propose that ^{23}Na, the sole stable isotope of Na, is synthesised by $^{22}Ne(p,\gamma)^{23}Na$ which must accompany the ON-cycle. In a solar mixture, $^{22}Ne/^{23}Na$ = 4. If a similar ratio prevailed in the halo, [Na/Fe] would increase by 0.6 dex as the ON-cycled contamination of an envelope became dominant. It is possibly a coincidence but Figure 3 shows just this magnitude of increase in [Na/Fe]. Of course, the Na-O correlation does not prove that the ON-cycling occurs in the giants themselves. The ON-cycled material may have been present from the birth of the star. A critical test of Denisenkov and Denisenkova's will come when Na abundances are estimated for cluster subgiants and dwarfs. If these stars do not show a spread in the Na abundances, the Na-rich stars may be ascribed to a severe first dredge-up.

The test results are probably already available from the halo field stars. Bessell and Norris (1982) discovered a class of N-rich subdwarfs with N abundances similar to those of the N-rich globular cluster giants. These subdwarfs are Na and Al-rich relative to other subdwarfs of the same metallicity (François 1986, Magain 1989, Zhao 1990, Spiesman 1992). They are also s-process enriched (Magain 1989). The fact that these stars have a normal Li abundance (Spite and Spite 1986) and no obvious deficiency of C implies that they were formed by mixing a small amount of severely ON-cycled (N-rich) material with normal halo gas (Lambert 1987). More detailed analyses of the N-rich subdwarfs are desirable. The rarity of these stars in the field (3 to 5% of subdwarfs) is in striking contrast to the common appearance in globular clusters. This fact suggests that the ON-cycled material was retained by the clusters but rarely by the sites responsible for the field subdwarfs.

Existence of N and Na rich subdwarfs suggests that the abundance anomalies are "primordial", i.e., the stars formed from a contaminated cloud of this anomalous composition. Cottrell and Da Costa (1981) have noted that a N, Na, Al and s-process enrichment of cluster gas could be provided by a preceding generation of AGB stars in the cluster. An alternative explanation may be that the N-rich stars are products of mass transfer between the components of a binary system. This observer will never forget the lesson taught us by McClure (1988) about the Barium stars!

4 BEYOND ANNO 60 HR

The claim that latter decades of the years 0–60 HR (1932–1992) have been the golden years for stellar nucleosynthesis is a readily defensible one. No great insight is needed to see that continued excitement is promised. As an observer, I note that there are large telescopes on the horizon from the Keck in the northern hemisphere to ESO's VLT in the southern hemisphere. These new telescopes will offer fresh opportunities for probing stellar nucleosynthesis. Among these opportunities, I would simply note the extension of high resolution spectroscopy to fainter individual stars in globular clusters and the Magellanic Clouds: spectroscopy of subgiants and main sequence stars in globular clusters will define the relative roles of evolutionary and primordial abundance anomalies (see our discussion of Na-poor stars), and spectroscopy of AGB stars in the Clouds will greatly clarify our understanding of this challenging phase of advanced stellar evolution.

Often — too often — the published chemical compositions of stars are taken by hungry theoreticians and applied with little comprehension of the random and systematic errors of measurement to develop improved theories of stars, stellar clusters, and galaxies. Random errors of measurement are generally well documented in abundance analyses. Although our understanding of the structure of stellar atmospheres and line formation is developing, systematic errors remain a concern. As I tried to show in the discussion of the oxygen abundance in halo stars, observing programs can be designed to undercover systematic errors and to suggest temporary compromise sets of abundance data for use in theoretical investigations. In parallel with observing programs, there must be continued efforts to reduce the systematic errors by improving model atmospheres and the descriptions of line formation; I noted the especial need in the analysis of cool metal-poor stars for calculations of the excitation of atoms and molecules by neutral H atoms.

If the latter decades of the years 0–60HR are a guide, we can expect continued interactions between astrophysicists and those physicists and chemists interested in atomic/molecular spectra and structure and in nuclear physics. These interactions will result in more complete answers to the fundamental questions — How, when and where were the chemical elements synthesised? — that have attracted Hubert Reeves for many years.

My research in stellar nucleosynthesis is supported in part by the US National Science Foundation (grant AST 91-15090), a Texas Advanced Research Project (grant 003658-585) and the Robert A. Welch Foundation of Houston, Texas. I thank Drs. D.R. Gies, C. Sneden, and J. Tomkin for supplying figures.

REFERENCES

Abia C., Rebolo R., 1989, ApJ, 347, 186

Aller L.H., 1978, in Terzian Y., ed., Planetary Nebulae – Obervations and Theory. Reidel, Dordrecht, p. 225

Aller L.H., Czyzak S.J., 1983, ApJS, 52, 211

Anders E., Grevesse N., 1989, Geochim. et Cosmochim. Acta, 53, 197

Atkinson R.d'E., Houtermans F.G., 1929, Z. Phys, 54, 656

Barbuy B., 1992, in Barbuy, B., Renzini, A., eds., The Stellar Populations of Galaxies. Kluwer, Dordrecht, p. 143

Burbidge E.M., Burbidge G.R., Fowler W.A., Hoyle F., 1957, Rev. Mod. Phys., 29, 547

Becker S.R., Butler K., 1988a, A&AS, 76, 331

_____. 1988b, A&AS, 74, 211

_____. 1989, A&A, 209, 244

_____. 1990, A&AS, 84, 95

Bell R.A., Hesser J.E., and Cannon R.D., 1983, ApJ, 269, 580

Bessell M.S., Norris J., 1982, ApJ, 263, L20

Bessell M.S., Sutherland R.S., Ruan K., 1991, ApJ, 383, L71

Bethe H.A., 1939, Phys. Rev., 55, 103 and 434

Bethe H.A., Critchfield C.L., 1938, Phys. Rev., 54, 248 and 862

Blaauw A., 1964, ARA&A, 2, 213

Cunha K., Lambert D.L., 1992, ApJ, in press

Cameron A.G.W., Fowler W.A., 1971, ApJ, 164, 111

Cesarsky C., 1992, this volume

Cottrell P.L., Da Costa G.S., 1981, ApJ, 245, L79

Denisenkov P.A., Denisenkova S.N., 1990, Sov. Astr. Letters, 16, 275

Drake J.J., Smith V.V., Suntzeff N.B., 1992, ApJ, in press

Eber F., Butler K., 1988, A&A, 202, 153

Eddington A.S., 1920, Observatory, 43, 341

Esteban C., Vilchez J.M., Smith L.J., Clegg R.E.S., 1992, A&A, 259, 629

Ferrando P., 1992, in Nuclei in the Cosmos, in press

Forestini M., Goriely S., Jorissen A., Arnould M., 1992, A&A, in press

François P., 1986, A&A, 165, 183

Gies D.R., Lambert D.L., 1992, ApJ, 387, 673

Gilmore G., Gustafsson B., Edvardsson B., Nissen P.E., 1992, Nat, 357, 379

Iben I.Jr., 1991, in Michaud G., Tutukov, A., eds, Evolution of Stars: The Photo-
    spheric Abundance Connection. Kluwer, Dordrecht, p. 257

Jorissen A., Smith V.V., Lambert D.L., 1992, A&A, in press

Kahane C., Cernicharo J., Gómez-González J., Guélin M., 1992, A&A, 256, 235

Kiselman D., 1991, A&A, 245, L29

Kraft R.P., Sneden C., Langer G.E., Prosser C.E., 1992, AJ, in press

Lambert D.L., 1987, in Azzopardi M., Matteucci F., eds., Stellar Evolution and
    Dynamics in the Outer Halo of the Galaxy. ESO, Garching, p. 47

Lambert D.L., 1991, in Michaud G., Tutukov, A., eds, Evolution of Stars: The
    Photospheric Abundance Connection. Kluwer, Dordrecht, p. 299

_____. 1992, in Edmunds M.G., Terlevich R., eds., Elements in the Cosmos.
    CUP, Cambridge, in press

Lambert D.L., Smith V.V., 1992, in preparation

Magain P., 1988, in Cayrel de Strobel G, Spite M., eds., The Impact of Very High
    S/N Spectroscopy on Stellar Physics. Kluwer, Dordrecht, p. 485

_____. 1989, A&A, 209, 211

McClure R.D., 1988, in Johnson H.R., Zuckerman B., eds., Evolution of Peculiar
    Red Giant Stars. CUP, Cambridge, p. 196

Meyer J.-P., 1985, ApJS, 57, 151 and 173

Olive K.A., Schramm D.N., 1982, ApJ, 257, 276

Paltoglou G., Norris J.E., 1989, ApJ, 336, 185

Peimbert M., 1987, in Peimbert M., Jugaku, J., eds., Star Forming Regions. Reidel,
    Dordrecht, p. 111

Reeves H., 1978, in Gehrels T., ed., Protostars and Planets. Univ. Arizona Press, Tucson, p. 399

Rolfs C.E., and Rodney W.S., 1988, Cauldrons in the Cosmos. Chicago Univ. Press, Chicago

Sackmann, I.-J., Boothroyd A.I., 1991, in Michaud G., Tutukov, A., eds, Evolution of Stars: The Photospheric Abundance Connection. Kluwer, Dordrecht, p. 275

Smith V.V., Lambert D.L., 1990, ApJS, 72, 387

Sneden C., Kraft R.P., Prosser C.F., Langer G.E., 1991, AJ, 102, 2001
_____. 1992, AJ, in press

Spiesman W.J., 1992, preprint

Spinrad H., Kaplan L.D., Connes P., Connes J., Kunde V.G., Maillard J.P., 1971, in Lockwood G.W., Dyck, H.M., eds., Proc. Conf. on Late-type Stars. KPNO, Tucson, p. 59

Spite M., 1992, in Barbuy B., Renzini A., eds., The Stellar Populations of Galaxies. Kluwer, Dordrecht, p. 123

Spite F., Spite M., 1986, A&A, 163, 140

Spite M., Spite F., 1991, A&A, 252, 689

Suess H.E., Urey H.C., 1956, Rev. Mod. Phys., 28, 53

Tomkin J., Lemke M., Lambert D.L., Sneden C., 1992, AJ, in press

VandenBerg D.A., 1992, ApJ, 391, 685

Wasserburg G.J., 1992, this volume

Weizsäcker C.F. von, 1937, Phys. Z., 38, 176
_____. 1938, Phys. Z., 39, 633

Wheeler J.C., Sneden C., Truran J.W., 1989, ARA&A, 27, 279

Zhao G., 1990, PhD Thesis, Nanjing University

CNO abundances in the Orion OB association

K. CUNHA, D.L. LAMBERT

University of Texas at Austin

1 INTRODUCTION

Chemical abundances of the stars in OB associations can provide important clues to both stellar evolution and chemical evolution of stellar populations. OB associations have typical lifetimes of $1 - 2 \times 10^7$ years (Reeves 1978) and sometimes consist of distinct stellar generations. Lifetimes of massive O stars are a few million years; such stars end their evolution as Type II Supernovae and can pollute the interstellar environment of the OB associations from which they formed (Olive & Schramm 1982). The Orion association consists of four subgroups , Ia, Ib, Ic and Id of different ages suggesting that star formation in this region happened in progressive steps. The time lag between the formation of the oldest and the youngest subgroups is compatible with the evolution time of massive stars so that one or more stars would have had time to explode (Reeves 1978) and to possibly change the composition of the gas.

Orion seemed thus well suited for trying to detect the possible contamination by Supernovae, in those elements like Oxygen, that are synthesized in the most massive stars/Supernovae. Also, with the study of CNO abundances, one can look for the signatures of CN-cycled material in these main-sequence B stars.

Another interesting aspect of this study is that by analysing the composition of the Orion association B stars, additional constraints can be put on the long standing puzzle about the composition of the Orion Nebula in comparison to the Sun.

In the present work we have derived LTE and NLTE Carbon, Nitrogen and Oxygen abundances for 18 main-sequence B stars belonging to the four subgroups of Orion.

2 OBSERVATIONAL DATA

18 Main sequence B stars belonging to the 4 subgroups of the Orion Association (see table 1) were observed with the 82" and 107" telescopes at the McDonald Observatory at coude, using a CCD detector. The program stars were obtained from Warren & Hesser (1978). Several spectral regions containing lines of OII, NII and CII have been observed with a typical resolution of 0.2A and a typical signal-to-noise of > 200. All reductions were done with the IRAF data package.

3 STELLAR PARAMETERS

Our derivations of T_{eff} and $\log g$ were based on the C_0 color index from the Strom-gren photometric system, which is a measure of the Balmer jump, and also on the comparison of the observed and theoretical pressure broadened line profiles of H_γ Although the Balmer jump is primarily a temperature indicator in early B stars, it is also dependent on gravity; the line profile of H_γ, is primarily a gravity indicator, but is also dependent on temperature so that we employed an iterative scheme to be able to derive self-consistent T_{eff} and $\log g$.

The T_{eff} and $\log g$ calibrations of the color indices and H_γ profile are based on the line-blanketed model atmospheres of Kurucz (1979), which assume local thermodynamic equilibrium.

The derived stellar parameters for the program stars are on Table 1.

MODEL ATMOSPHERES: Line-blanketed model atmospheres were computed for the derived T_{eff} and $\log g$ by means of Kurucz's program ATLAS6 with solar composition.

Table 1

Star	Sub g	T_{eff}	log g
HD 35039	Ia	20550	3.74
HD 35299	Ia	24000	4.25
HD 35912	Ia	19590	4.20
HD 36351	Ia	21950	4.16
HD 36591	Ib	26330	4.21
HD 37744	Ib	24480	4.38
HD 36285	Ic	21930	4.40
HD 36430	Ic	19640	4.36
HD 36512	Ic	31560	4.42
HD 36629	Ic	22300	4.35
HD 36959	Ic	24890	4.41
HD 36960	Ic	28920	4.33
HD 37209	Ic	24050	4.13
HD 37356	Ic	22370	4.13
HD 37481	Ic	23300	4.17
HD 37020	Id	29970	3.92
HD 37023	Id	32600	4.70
HD 37042	Id	31600	4.70

4 LTE ABUNDANCES OF CARBON, NITROGEN AND OXYGEN

Oscillator strengths : The f-values for the OII studied lines were obtained from Becker & Butler (1988a), for the NII lines from Becker & Butler (1989) and for the CII lines from Pradhan (1990).

Microturbulent velocities: The microturbulent velocities for the stars were determined from the OII and NII species. The LTE abundances were calculated for different values of microturbulence. The final microturbulent velocity for each species was the one that represented the best agreement between the abundances of weak and strong lines in the diagram Log ϵ(el) vs. microturbulence. The microturbulence was then determined in this way for our sample of OII and NII lines and we adopted as a final microturbulence value, for the star, the mean microturbulence weighted by the number of lines in each species.

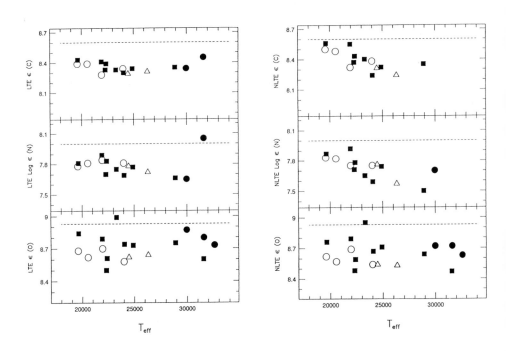

Figure 1. Mean LTE and NLTE Carbon, Nitrogen and Oxygen abundances as a function of effective temperatures. The open circles represent stars belonging to subgroup Ia, the open triangles to Ib, the filled squares to Ic, and the filled circles to Id.

The Carbon, Nitrogen and Oxygen abundances were then calculated for the adopted microturbulent velocities for each star by means of Kurucz's program WIDTH6. The derived C, N and O abundances as a function of effective temperature are shown in Figure 1. Each abundance value for the stars represents the average abundance derived from our sample of lines, but since the abundances from strong lines are the most sensitive to the choice of microturbulence, we have elected to disregard the strongest lines from the mean. In this way we have minimized the effects of uncertainties in the derived microturbulent velocites on the abundance determinations.

5 NLTE ABUNDANCES OF CARBON, NITROGEN AND OXYGEN

Our NLTE abundances are based on the predictions of NLTE equivalent widths for the lines of interest by Becker & Butler (1988b)- OII lines, Becker & Butler (1988c) NII lines and Becker (1992)- CII lines. These published tables contain NLTE equivalent widths as a function of temperature, gravity, microturbulence and abundance, and their calculations are based on line-blanketed LTE models from Gold (1984) and fairly extensive model atoms.

A similar procedure in the determination of the LTE microturbulences was employed for the derivation of the NLTE microturbulences and these were found to be slightly larger than the LTE ones.The OII species were the only ones considered in the microturbulence determination because only weak lines of NII and CII were available and these are independent of the microturbulence.

The derived NLTE C, N and O abundances for the adopted microturbulence for each star as a function of effective temperature are shown in Figure 1. The abundance values for each star represent the average abundance for all the lines available, but as in the LTE case, we have disregarded the strongest lines.

6 DISCUSSION

A few interesting comments can be made on the basis of our results. From inspection of figure 1 it can be seen that the spread in the Oxygen abundances is considerably larger than the spread in the Nitrogen and especially Carbon abundances. Since the errors of measurement and analysis are comparable for Carbon, Nitrogen and Oxygen, this bigger scatter is interpreted as the result of local inhomogeneities in the Oxygen abundance cloud due to the evolution of the massive stars in the association: Oxygen is synthesized in much greater amounts than Carbon and Nitrogen in massive stars (Supernovae of type II). A closer look at the Oxygen abundances shows that the Oxygen abundances are correlated with a star's position in the sky (see Figure 2) and that the most Oxygen rich stars belong to the younger subgroups (Ic and Id) and are therefore closer together in the sky , well separated from the older subgroups (Ia and Ib) which show on average, relatively lower abundances. For a more detailed analysis of the Oxygen results see Cunha & Lambert (1992).

The spread in the Oxygen abundances and the relative constancy of the C and N abundances gives us a hint that we are not seeing major results of CN-cycled matter for the studied main sequence B stars. If this were the case, one would expect to see N enriched stars as a primary indication of CN-cycled matter. The only case of N-rich star we find in our sample is the star HD37042 (Id).

If we simply average the stellar abundances, we obtain mean abundance values that correspond to 8.35, 7.74 and 8.71 for Carbon, Nitrogen and Oxygen respectively. These values are underabundant relative to the Sun, by -0.20, -0.26 and -0.21 dex, and confirm published abundance analysis of the HII regions emission lines which show underabundance relative to the Solar values. If we compare, for example, the CNO abundances we derived for the Orion B stars with the results of Baldwin et al (1991) for the Orion nebula we find fairly good agreement. This important result confirms the study by Gies & Lambert (1992).

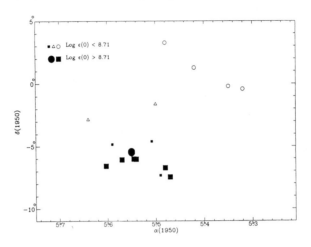

Figure 2. The distribution on the sky of the observed stars from the four subgroups. The symbols used to distinguish the subgroups are the same as in Figures 1 and 2.

REFERENCES
Becker, S. R., and Butler, K. 1988a, A&A, 201, 232
Becker, S. R., and Butler, K. 1988b, A&AS, 74, 211
Becker, S. R., and Butler, K. 1988c, A&AS, 76, 331
Becker, S. R., and Butler, K. 1989, A&A, 209, 244
Becker, S. R., 1992, private communication
Cunha, K., and Lambert, D. L. 1992, ApJ in press
Gies, D. R., and Lambert, D. L. 1992, ApJ, 387, 673
Gold, M. 1984, Diplomarbeit, Ludwig Maximilian Universitat
Kurucz, R. L. 1979, ApJS, 40, 1
Olive, K. A., and Schramm, D. N. 1982, ApJ, 257, 276
Pradhan, A., 1990, private communication
Reeves, H. 1978, in Protostars and Planets, ed. T. Gehrels (Tucson: Univ. Arizona Press), p. 399
Warren, W. H., and Hesser, J. 1978, ApJS, 36, 497

Surface Abundances in Red Giants:
A diagnostic Tool of Stellar Evolution

M.F. EL EID[1] AND N. PRANTZOS[2]

1. Universtitäts–Sternwarte Göttingen, FRG
2. Institut d'Astrophysique de Paris, France

1. INTRODUCTION

The theory of stellar evolution and nucleosynthesis, to which Hubert Reeves has contributed so much, is one of the best established theories in modern astronomy. Despite its success, it still suffers from several theoretical drawbacks, the most important of which is undoubtely the lack of a satisfactory theory of stellar convection; uncertainties in the treatment of mass loss, or in the input atomic and nuclear data, also affect the results of stellar evolution models. Observations of elemental and/or isotopic stellar abundances can help in some cases to constrain stellar models and give some insight into the underlying physical processes. This is the case of e.g. Wolf-Rayet stars, stripped off their hydrogen envelope and revealing directly the content of their hydrogen or helium cores. This is also the case of red giant stars, after their first, second or third "dredge-up" episode, that bring to the surface material processed in the H- or He- burning regions.

In this work, we focus on the chemical abundances of red giants after their first dredge-up episode, which mixes in the stellar atmosphere material affected by H- burning in the core. The resulting surface composition bears the signature of the CNO and, perhaps, Ne-Na and Mg-Al cycles of H- burning. Observations of this composition can then be used as a powerful diagnostic tool of stellar structure (provided the relevant nuclear data are accurate enough, which is not always the case). Several investigations of the CNO isotopic abundances have been performed in the past (e.g. Harris and Lambert 1984a,b; Landré et al. 1990; Dearborn 1992), while the Na abundance has been studied more recently (Denisenkov 1990; Prantzos et al. 1991). Our aim is to revisit these works in the light of up–dated stellar models, incorporating the recent Rogers and Iglesias (1992) opacities, more recent nuclear data, and a different prescription for convection. In Sec. 2 we present briefly the relevant observational material, and in Sec. 3 some results on CNO and Ne-Na abundances for several stars evolved beyond the first dredge-up. In Sec. 4 and 5 we discuss this dredge–up phase and compare our results with observations and with other recent computations.

2. OBSERVATIONS

Table 1 presents a selection of carbon and oxygen isotopic abundances observed in red giants after their first dredge-up episode (from Landré et al. 1990, based on observations by Harris and Lambert 1984a,b). All $^{16}O/^{17}O$ ratios are smaller than the corresponding solar value, indicating an enhancement of ^{17}O; there is, however, a large scatter in those values, contrary to the $^{16}O/^{18}O$ values which are found to lie in a narrow range centered on the solar value of ~442. All $^{12}C/^{13}C$ ratios are smaller than the solar value of ~83, indicating an enhancement of ^{13}C, obviously resulting from the operation of the CN cycle.

TABLE 1

Observed values of carbon and oxygen isotopic ratios in some red giant stars (from Landre et al. 1990, based on work by M. Harris and D. Lambert). The solar values are taken from Anders and Grevesse (1989).

Star	$M(M_\odot)$	$^{16}O/^{17}O$	$^{16}O/^{18}O$	$^{12}C/^{13}C$
α Tau	2–3	530^{-90}_{+180}	430^{-120}_{+180}	8.0 ± 0.9
β And	2–3	150^{-15}_{+25}	370^{-70}_{+140}	10.0 ± 1.3
μGem	2–3	300^{-70}_{+140}	425^{-120}_{+180}	11.5 ± 1.8
α Her	5–7	180 ± 140	490^{-165}_{+210}	15.7 ± 3.7
γ Dra	5	280^{-70}_{+100}	445^{-140}_{+190}	12.0 ± 1.9
α Sco	15	800^{-280}_{+520}	445^{-190}_{+280}	11.0 ± 3.0
α Ori	15	495^{-120}_{+235}	620^{-165}_{+280}	5.5 ± 0.9
solar		2465	442	83

TABLE 2

Sodium overabundance in yellow supergiants according to Boyarchuk et al. (1989). The listed values for [Na/Fe] take into account a correction of 0.05 dex due to sphericity effects (see Prantzos et al. 1991). Note the large uncertainties in the masses of the stars.

Star	$M(M_\odot)$	[Na/Fe]
α Car	8 ± 1.5	0.36
α Umi	6 ± 1.0	0.31
γ Cyg	11 ± 2.0	0.48
ρ Cas	40 ± 15	0.62

Table 2 gives the results of Na/Fe overabundances in four yellow giant/supergiant stars of the F-K type (from Boyarchuk et al. 1989). Sasselov (1986), icluding non-

LTE effects in his analysis, showed that this Na enhancement (by factors 2-5) is a real effect and that it is correlated to the mass of the star. Moreover, it seems that this Na excess is correlated also to the $^{12}C/^{13}C$ ratio, another indication that it is related to H-burning nucleosynthesis.

3. STELLAR MODELS AND RESULTS

The stellar evolution code used to calculate the present stellar models is described in Barraffe and El Eid (1991). The Schwarzschild criterion for convection has been used, and no overshooting is included. The convective energy transport is treated according to the standard mixing length theory with a mixing length parameter of 1.60 pressure scale heights. Mass loss has been included according to the interpolation formula of de Jager et al. (1988). The most recent Rosseland mean opacities from Livermore (Rogers and Iglesias 1992) down to 6000 K are used. Below this temperature we use the opacity tables based on a line library of Kurusz (see El Eid and Höflich 1991). The nuclear reaction network contains all the reactions of the p-p chains, the CNO cycles, the Ne-Na and Mg-Al cycles, and He-burning reactions up to Si. The reaction rates are from Caughlan and Fowler (1988; hereafter CF88), except for the $^{17}O(p,\alpha)$ and $^{17}O(p,\gamma)$ rates which are taken from Landré et al (1990) with correction factors f1=f2=0.2. These recently measured values are much larger than the ones given in CF88 and have a considerable effect on the resulting oxygen isotopic ratios. The $^{12}C(\alpha,\gamma)$ reaction is taken from Caughlan et al. (1985), which is ~3 times higher than that of CF88.

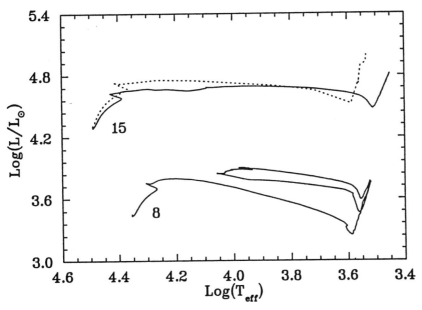

Fig. 1: Evolutionary tracks in the H-R diagram for the 8 $M_\odot$ and 15 $M_\odot$ stars. *solid lines*: this work; *dotted lines*: Schaller et al. (1992), shown for comparison.

The evolution of two stars, of mass 8 $M_\odot$ and 15 $M_\odot$ with initial composition (X,Z)=(0.70,0.02) was followed beyond core helium burning (see Fig. 1). Figs. 2a and 2b show a snap-shot of the internal composition of the stars after core hydrogen burning, i.e. at log $T_{eff} = 4.16$, $\log(L/L_\odot) = 3.78$ for the 8 $M_\odot$ star, and log $T_{eff} = 4.42$, $\log(L/L_\odot) = 3.78$ for the 15 $M_\odot$ star. At this stage, the H–burning shell is located at M~1.0 $M_\odot$ in the 8 $M_\odot$ star, and M~3.0 $M_\odot$ in the 15 $M_\odot$ star. The abundance gradients outside these shells are due to the retreat of the former H–convective core.

The present computations show that during the first dredge–up the convective envelope does not reach the position of the H–shell source. In the 8 $M_\odot$ star, the convective envelope (CE, hereafter) penetrates to ~ 1.6 $M_\odot$, while the H–shell source is located at ~ 1.4 $M_\odot$. In the 15 $M_\odot$ star, these values are ~ 5.0 $M_\odot$ and 4.2 $M_\odot$ respectively. Similar values for the depth of the CE in 15 $M_\odot$ stars were found by Landré et al. (1990) *empirically*, i.e. that depth was found to satisfy the observations of the oxygen isotopic ratios in red giants of that mass, like α Ori and α Sco.

As can be seen in Figs. 2a and 2b, the first affected isotopes starting from the surface inwards are the fragile ^{13}C and ^{15}N, where ^{13}C is enhanced and ^{15}N is depleted. Further inside, ^{12}C is depleted by the CN cycle and is transformed to ^{14}N. The oxygen isotopic abundances are changed in deeper layers, with ^{18}O being depleted, and ^{17}O enhanced. Inside the former H–convective core, the CNO cycle reaches equilibrium, and both ^{16}O and ^{17}O are depleted. The maximum mass of the convective core in the present computations is 2.41 $M_\odot$ for the 8 $M_\odot$ star and 5.74 $M_\odot$ for the 15 $M_\odot$ star. For comparison, the convective core of the 15 $M_\odot$ star had 6.69 $M_\odot$ in the calculations of Schaller et al. (1992) with 0.20 pressure scale heights of core overshooting, and 7.1 $M_\odot$ according to Landré et al. (1990) with 0.30 pressure scale heights for that overshooting. The agreement between the $^{16}O/^{17}O$ ratio obtained in this work and in Schaller et al (1992) (Fig.5) indicates this ratio is not significantly affected by a moderate core overshooting (adopted in the later work). The degree of destruction of ^{17}O depends crucially on the rates of $^{17}O(p, \alpha)$ and $^{17}O(p, \gamma)$. and is much larger in massive stars, due to their higher temperatures.

Finally, the nucleus ^{23}Na is found (Fig. 2a and 2b) to be enhanced in the H–core due to the operation of the Ne–Na cycle. Using the reaction rates of CF88 ^{20}Ne is found to be barely affected, while ^{21}Ne and ^{22}Ne are transfomed into ^{23}Na. In Sec. 5 we shall see that the relatively high observed values of [Na/Fe] are difficult to explain this way.

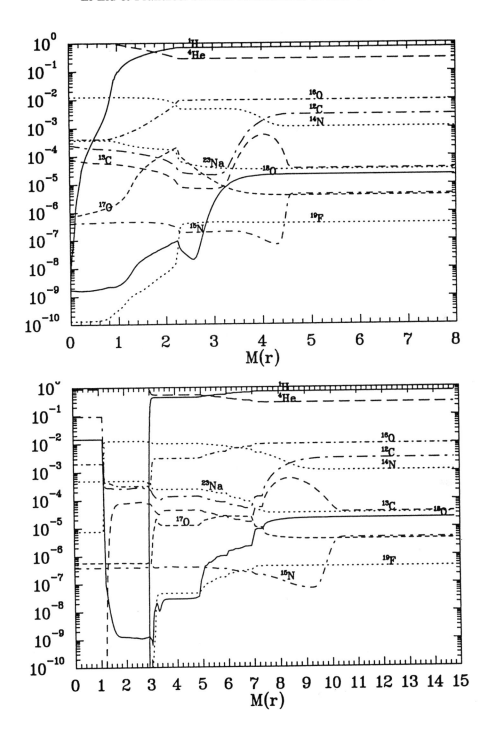

Fig. 2a,b: Abundance profiles as a function of mass coordinate after core H exaustion for 8 $M_\odot$ (2a) and 15 $M_\odot$ stars (2b), respectively.

4. OXYGEN AND CARBON ISOTOPIC ABUNDANCES

In general, the penetration depth of the conective envelope in cool stars is difficult to determine, due to the lack of a satisfactory non–local theory of convection. The confrontation of theoretical predictions with observations, i.e. in particular the observational constraints on the surface abundances of red giant stars, may provide useful information in this respect. It is important to realize that the oxygen isotopes are influenced only inside the H–burning core, while the carbon isotopes are modified outside that core. On the other hand, envelope convection penetrates beyond the ^{13}C enriched region and down to the ^{17}O enriched layers which are depleted in ^{13}C. Therefore, the $^{12}C/^{13}C$ ratio cannot be used to constrain the depth of the convective envelope. The same is valid for $^{16}O/^{18}O$, since ^{18}O is always depleted in the ^{17}O enriched zone (cf. Figs. 2a, 2b).

The behavior of ^{17}O or $^{16}O/^{17}O$ is more interesting. The ^{17}O abundance is modified in the very deep stellar layers and is very sensitive to the reaction rates of $^{17}O(p, \alpha)$ and $^{17}O(p, \gamma)$, as shown by Landré et al. (1990; see also Dearborn 1992). The depletion of ^{17}O is more efficient in massive stars, due to their high temperatures, while in the cooler low mass stars ($M < 3\ M_\odot$) ^{17}O is barely depleted, even with the high $^{17}O(p,\alpha)$ and $^{17}O(p,\gamma)$ values of Landré et al. (1990). Therefore, the observation of $^{16}O/^{17}O$ in low mass red giants may indeed provide useful information about the depth of the convective envelope.

A sketch of the present situation of the carbon and oxygen isotopes in red giants is given in Figs. 3–5. In Fig. 3, the predicted $^{12}C/^{13}C$ ratios show a weak dependence on stellar mass, in line with the observations in Table 1. However, they are systematically too high compared with the observed values. Hence, it seems that standard convective mixing is not enough to reduce these ratios to the observed level. Non–standard mixing already on the main sequence may be invoked, but this process is not yet sufficiently understood (see Dearborn 1992, for a discussion)

In Fig. 4, the predicted $^{16}O/^{18}O$ ratios are in reasonable agreement with the observed ones. They are confined in a narrow range close to the solar value (see Table 1). The difference among the predicted values seems to be related to the adopted $^{18}O(p, \alpha)$ rate In the present computations, this rate is taken from CF 88 with the correction factor f=0.

Fig. 5, shows clearly the extreme sensitivity of the $^{16}O/^{17}O$ ratio to stellar mass, especially for the low mass stars. The difference among the theoretical values reflects probably different treatments of convection in the stellar models, as well as differences in the adopted destruction rates of ^{17}O. This point needs further work. The

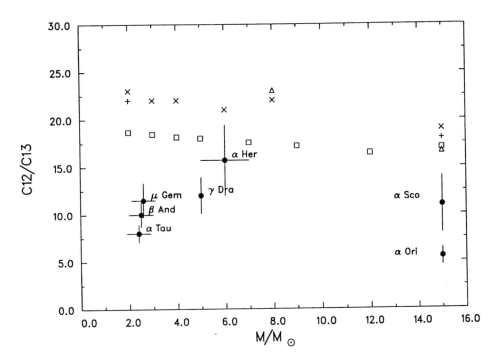

Fig. 3: Surface $^{12}C/^{13}C$ ratios after the first dredge up phase as a function of stellar mass and comparison to observations. *triangles:* present work; X: Dearborn (1992); *squares:* Schaller et al. (1992); *crosses:* Landré et al. (1990).

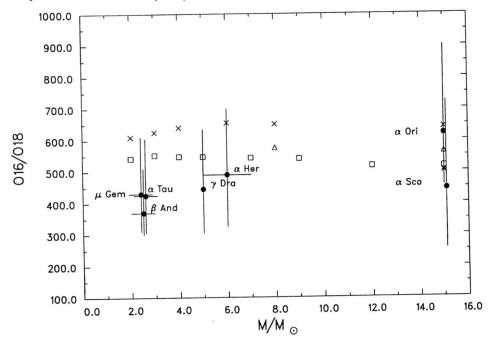

Fig. 4: Same as in Fig.3 for the $^{16}O/^{18}O$ ratio.

observed $^{16}O/^{17}O$ ratio in α Ori and α Sco argue for high destruction rates of ^{17}O, as suggested by Landré et al (1990). Notice that the scatter in the observed values in Fig. 5 probably indicates uncertainties in the estimated masses of the stars. More observations are certainly needed, especially in the mass range 5 to 15 $M_\odot$.

5. THE SURFACE ABUNDANCE OF SODIUM IN GIANT STARS

Our results on the Na/Fe ratio in the surface of our model stars are shown in Fig. 6, together with the observed values of Table 2. If these observations are taken at face value, then three features should be explained: (i) the enhancement of [Na/Fe] by a factor of 2–5, (ii) the increase of the Na excess with decreasing surface gravity, or increasing mass, and (iii) the increase of [Na/Fe] with decreasing $^{12}C/^{13}C$ ratios.

The results in Fig. 6 are obtained by increasing the CF88 rate of $^{20}Ne(p,\gamma)$ by a factor of twenty. The hope is that such an increase will turn a substantial part of the abundant ^{20}Ne into ^{23}Na as suggested in Prantzos et al. (1991), in view of the difficulty to reproduce that excess by some other (plausible) mechanism. Notice that in that paper our current understanding of that reaction is also discussed and it is concluded that such a modification is rather unlikely. Our results show that even with this unlikely modification it is difficult to account for the high values observed. On the other hand, our theoretical values follow the observational trend, i.e. they increase with increasing stellar mass or with decreasing $^{12}C/^{13}C$ values (see Fig. 3); this, in our opinion, clearly indicates that the Ne-Na cycle of H-burning has operated in those stars.

Notice that mass loss in the main sequence can affect significantly the surface abundances of stars more massive than $\sim$15-20 $M_\odot$. In particular, we find that the [Na/Fe] surface ratio is very sensitive to mass loss. The predicted value at M=13.6 $M_\odot$ in Fig. 6 is obtained at the end of core helium burning of the initially 15 $M_\odot$ star which has lost lost 1.4 $M_\odot$. On the other hand, the higher theoretical [Na/Fe] value in Fig. 6 corresponds to a star of 40 $M_\odot$ initially, reduced to M$\sim$25.5 $M_\odot$ when its central helium concentration was $\sim$0.73. At this stage (LogT$_{eff}$=3.48 and LogL/L$_\odot$ = 5.63) the star is on the way to become a Wolf–Rayet star.

In summary, this discussion shows that understanding the Na excess in yellow supergiants is a difficult task. The only encouraging result of the present work is the obtained increase of Na/Fe excess with stellar mass, which supports qualitatively the idea of Ne-Na cycle at work. More observations are clearly needed in order to confirm the values in Table 2, which current models fail to explain quantitatively. If these values are firmly established, and other mechanisms of Na surface enhancement (like e.g. gravitational settling) are excluded, then a very large $^{20}Ne(p,\gamma)$ rate seems the

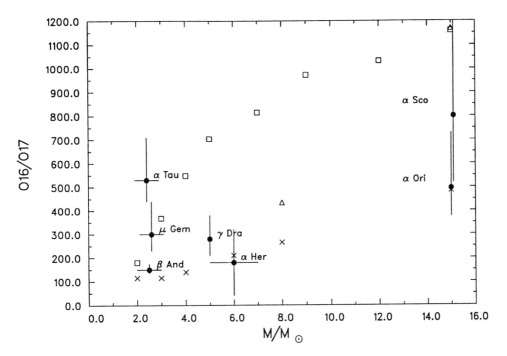

Fig. 5: Same as in Fig.3 for the $^{16}O/^{17}O$ ratio.

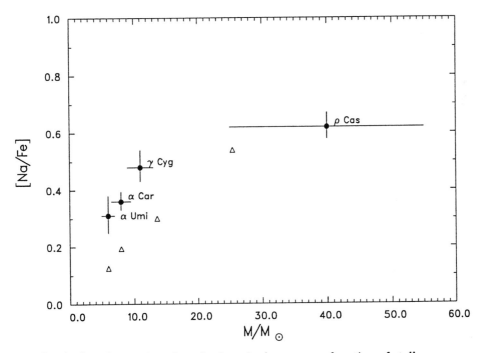

Fig. 6: [Na/Fe] surface ratios after the first dredge-up as a function of stellar mass: theory (*triangles:* this work with the $^{20}Ne(p,\gamma)$ rate 20 times the CF88 value) vs. observations.

only "natural" issue, at least in the framework of standard stellar evolution theory. Indeed, it seems difficult to admit that the isotope ^{22}Ne is selectively enhanced in the original material of those stars, as Denisenkov (1990) suggests.

6. SUMMARY

In this work we report preliminary results of a study of the first dredge-up phase in intermediate and high mass stars. We find that the observed oxygen isotopic abundances in red giants can be reasonably well reproduced with no (or moderate) overshooting and a destruction rate of ^{17}O larger than the CF88 value and closer to the Landré et al. (1990) value. As all previous works in the field, our models fail to reproduce the observed ^{12}C/^{13}C ratio, indicating some extra, non convective, mixing. Finally, we obtain a Na/Fe excess increasing with stellar mass, in qualitative but not quantitative agreement with the recent observations in yellow supergiants. It is clear from the above discussion that a lot of work is still needed before a good understanding of those relatively early phases of stellar evolution is obtained.

AKNOWLEDGEMENTS: M.F. El Eid has enjoyed the hospitality of the Department of Physics and Astronomy of Clemson University, where part of the present work has been done.

REFERENCES

Anders, E., Grevesse, M. (1989), Geochim. Cosmochim. Acta, **53**, 197

Baraffe, I., El Eid, M.F. (1991) A. A. **245**, 548

Boyarchuk A., et al. (1989) Astrophysics **29**, 197

Caughlan, G.R., Fowler, W.A., Harris, M.J., Zimmerman, B.A. (1985) Atomic Data Nucl. Data Tables **32**, 197

Caughlan, G.R., Fowler, W.A. (1988) Atomic Data Nucl. Data Tables 36, 411

Dearborn, S.P.D. (1992) Phys. Report **210**, 367

de Jagger C., Nieuwenhuiizen H., van der Hucht K. (1988), A. A. **173**, 293

Denisenkov P. (1990) Astrophysics **31**, 588

El Eid, M.F., Höflich, P. (1991) ESO/EPIC Workshop on SN1987A and other Supernovae, ed. I.J. Danziger et al. (Garching:FRG), p. 29

Harris M., Lambert D. (1984a) Ap. J. **281**, 739

Harris M., Lambert D. (1984b) Ap. J. **285**, 674

Landré, V., Prantzos, N., Auger, P., Bogaert, G., Levebre, A., Thibaud, J.P.: A. A. **240**, 85

Prantzos N., Coc A. Thibaud J.P. (1991), Ap. J. **379**, 729

Rogers, F.J., Iglesias, C.A. (1992) Ap. J. Suppl. **79**, 507

Sasselov, D.D. (1986) Publ. Astron. Soc. Pac. **98**, 561

Schaller, G., Schaerer, D., Meynet, G., Maeder, A. (1992) A. A. Suppl., in press

Stellar surface abundances and the first dredge-up

M. FORESTINI, Y. BUSEGNIES, N. MOWLAVI

Institut d'Astronomie et d'Astrophysique, Université Libre de Bruxelles, C.P.165;
Av. F.-D. Roosevelt, 50, 1050 Bruxelles, Belgium

1 ABSTRACT

High quality observational data have been recently published concerning the abundances of Li and CNO elements in the surface of red giant stars. While recent theoretical predictions about the surface composition of such objects are lacking, we have undertaken a systematic analysis of that question, on grounds of improved stellar evolution computations. We briefly present here first results for 1, 2.5, 4 and 6 $M_\odot$ Pop I stars.

2 INTRODUCTION

During specific stellar evolution phases, the surface convective envelope enters deep regions where nucleosynthetic processes occured. Such events are called dredge-up phases. The first one is encountered after the central H-burning phase, when stars are climbing the first giant branch in the Hertzsprung-Russell diagram (HRD). In this case, nuclides altered by the central and thick H-burnings are rapidly diluted and mixed into the convective envelope, consequently modifying the surface abundances of the concerned elements.

High quality spectra have been recently obtained for many such objects (*e.g.* Lambert *et al.* , 1980; Tomkin *et al.* , 1976; Lambert and Ries, 1981; Harris *et al.* , 1988), giving us quite accurate constraints on the surface abundances and isotopic ratios of Li and CNO nuclides during the first dredge-up phase. Confrontations between those data and theoretical results provide us interesting informations about our knowledge of the convective envelope depth as well as on the rates of the nuclear reactions involved in the H burning (CNO cycle).

3 INPUT PHYSICS

Our stellar evolution program incorporates 122 nuclear reactions (involving 38 species) for which we use the most recent experimental or theoretical determinations. This allows us to describe in great detail the nucleosynthesis occuring inside the stars. The atmosphere, the treatment of which significantly influences the structure of red giant stars, is calculated from the grid of atmosphere models published by Bell *et al.* (1975). Radiative opacities are taken from Huebner *et al.* (1977) at high temperature and from Alexander (1983) at low temperature. The structure of the convective zones is calculated with the mixing-length theory (with $\alpha = 1.5$). Finally, our equation of state includes partial ionization of H, He, C, N and O, as well as electrostatic corrections calculated following the Debye-Hückel model. We do not yet include microscopic or turbulent diffusion effects in the present calculations.

We always begin our stellar evolution calculations with initial polytropic models of stars arriving on the Hayashi pre-main sequence track of the HRD, with central temperatures around 10^6 K. This allows us to follow the nucleosynthesis as soon as it begins, prior to the main sequence phase. This is quite important for some nuclides, the abundance profile of which already changes at that stage. This will condition their future evolution. On the other hand, light nuclides (such as ^{7}Li) can already be substantially destroyed in low mass stars, before they reach the main sequence.

4 RESULTS

We present here results for four Pop I ($Z = 0.02$) stars of 1, 2.5, 4 and 6 $M_\odot$. Table 1 shows our predictions for the surface abundances of Li and CNO isotopes (relative to their initial value) at different stellar evolution phases up to the top of the first giant branch, where the first dredge-up is complete. Table 2 gives isotopic ratios for the same phases and their corresponding observed ranges for red giant stars (which are experiencing first dredge-up).

Several conclusions can be drawn from these numbers. First, we see that ^{7}Li is already partially destroyed in the solar mass star, during the pre-main sequence phase. On the other hand, the surface Li content of that star becomes negligible

mass	phase	^{7}Li	^{12}C	^{13}C	^{14}N	^{15}N	^{16}O	^{17}O	^{18}O
$1\,M_\odot$	ZAMS	0.31	1.00	1.00	1.00	1.01	1.00	1.00	1.00
	BRG	0.01	1.00	1.03	1.00	1.00	1.00	1.00	1.00
	TRG	0.00	0.87	2.67	1.36	0.75	1.00	1.02	0.96
$2.5\,M_\odot$	ZAMS	1.00	1.00	1.00	1.00	1.00	1.00	1.00	1.00
	BRG	0.05	1.00	1.00	1.00	1.01	1.00	1.00	1.00
	TRG	0.01	0.64	2.69	2.25	0.50	0.99	12.9	0.74
$4\,M_\odot$	ZAMS	1.00	1.00	1.00	1.00	1.00	1.00	1.00	1.00
	BRG	0.06	1.00	1.00	1.00	1.00	1.00	1.00	1.00
	TRG	0.01	0.64	2.80	2.29	0.49	0.97	10.6	0.75
$6\,M_\odot$	ZAMS	1.00	1.00	1.00	1.00	1.00	1.00	1.00	1.00
	BRG	0.18	1.00	1.00	1.00	1.00	1.00	1.00	1.00
	TRG	0.01	0.66	2.95	2.14	0.49	0.99	4.57	0.76

Table 1. Surface abundances, relative to their initial value, at three different stellar evolution phases: ZAMS (Zero Age Main Sequence), BRG (Base of the first Red Giant branch) and TRG (Top of the first Red Giant branch). The initial mass fractions of ^{7}Li, ^{12}C, ^{13}C, ^{14}N, ^{15}N, ^{16}O, ^{17}O and ^{18}O are, respectively, 9.88×10^{-9}, 3.20×10^{-3}, 3.86×10^{-5}, 1.17×10^{-3}, 4.61×10^{-6}, 1.01×10^{-2}, 4.11×10^{-6}, and 2.29×10^{-5}.

after the first dredge-up, which contradicts observations indicating there is again detectable Li in red giant stars. More generally, as seen from Table 2, none of the isotopic ratios observed in stars along the first giant branch could be reproduced by the $1\,M_\odot$ theoretical predictions. In order to confirm or not this fact, we will undertake, in a near future, calculations taking into account diffusion processes in stars of about $1\,M_\odot$.

The three more massive stars show isotopic ratios in good agreement with observations. There is however a noticable exception concerning the oxygen isotopic ratios. It seems that ^{17}O is predicted to be overabundant compared to what observations indicate for the $2.5\,M_\odot$ star. This could be due to the remaining uncertainties on the nuclear reactions involved in the destruction of this isotope.

mass	phase	$^{12}C/^{13}C$	$^{14}N/^{15}N$	$^{16}O/^{17}O$	$^{16}O/^{18}O$
$1\,M_\odot$	ZAMS	83.	252.	2457.	441.
	BRG	81.	254.	2457.	441.
	TRG	29.	461.	2416.	461.
$2.5\,M_\odot$	ZAMS	83.	253.	2457.	441.
	BRG	83.	252.	2457.	441.
	TRG	20.	1148.	188.	590.
$4\,M_\odot$	ZAMS	83.	253.	2457.	441.
	BRG	83.	252.	2457.	441.
	TRG	19.	1186.	225.	574.
$6\,M_\odot$	ZAMS	83.	254.	2457.	441.
	BRG	83.	253.	2457.	441.
	TRG	19.	1116.	530.	569.

Table 2. Surface isotopic ratios (mass fractions) for the same stars and at the same stellar evolution phases as in Table 1. Observations of red giant stars along the first red giant branch indicate that $^{12}C/^{13}C = 10 \longrightarrow 23$, $^{16}O/^{17}O = 240 \longrightarrow 525$, and $^{16}O/^{18}O = 400 \longrightarrow 600$. There is not yet accurate determinations of the nitrogen isotopic ratio.

5 CONCLUDING REMARK

This work consists of the first step in a project we have undertaken, and which aims at calculating the surface abundance changes engendered by the first dredge-up phase occuring along the first giant branch of the HRD. One of our motivations is to give more accurate and consistent predictions than those found in the pioneer work of Renzini and Voli (1981).

6 REFERENCES

Alexander, D.R., Johnson, H.R., Rypma, R.L.: 1983, *Astrophys. J.* **272**, 773
Bell, R.A., Eriksson, K., Gustafsson, B., Nordlund, A.: 1975, *Astron. Astrophys.* **23**, 37
Harris, M.J., Lambert, D.L., Smith, V.V.: 1988, *Astrophys. J.* **325**, 768
Huebner, W.F., Mertz, A.L., Magee, N.H., Argo, M.F.: 1977, *Astrophys. Opacity Library*, UC-346
Lambert, D.L., Dominy, J.F., Sivertsen, S.: 1980, *Astrophys. J.* **235**, 114
Lambert, D.L., Ries, L.M.: 1981, *Astrophys. J.* **248**, 228
Renzini, A., Voli, M.: 1981, *Astron. Astrophys.* **94**, 175
Tomkin, J., Luck, R.E., Lambert, D.L.: 1976, *Astrophys. J.* **210**, 694

Energy Generation in Convective Shells of Low Mass, Low Metallicity AGB Stars

GRANT BAZAN AND JOHN LATTANZIO

McDonald Observatory, University of Texas, Austin, TX 78712, Department of Astronomy, University of Illinois, Urbana, IL 61801, Department of Mathematics, Monash University, Australia, and Institute for Geophysics and Planetary Physics, L-413, Lawrence Livermore National Laboratory, CA 94550.

1 INTRODUCTION

In low mass stars of low metallicity, a sizable amount of ^{13}C in a small region ($\Delta m \leq 5 \times 10^{-3}$ $M_\odot$) can be made by semi-convection mixing proton-rich and carbon-rich material during interpulse periods (Iben and Renzini 1982). This material is ingested by an expanding convective shell and burned at high temperatures (Gallino et al. 1988). In all previous stellar evolution calculations of thermally pulsing AGB evolution, the energy output of the neutron emission and capture reactions and beta decays has been ignored. We have calculated this energy and included it in AGB evolutionary models.

2. CALCULATIONS

Because of the impracticality of implementing a full neutron capture network into a stellar evolution code, we have iterated between calculations of the energy associated with neutron generation and captures and calculations of the stellar evolution resulting from the consideration of the energy. We derive time dependent energy generation rates through the solution of a charged particle/neutron capture nucleosynthesis network of 605 nuclei. The ^{13}C is initially assumed to lie outside the convective shell in a mass distribution similar to that of Hollowell (1988). The ingestion rate, which was shown by Hollowell and Iben (1990) to determine the magnitude of the neutron density, is (initially) taken from the fifteenth pulse of Hollowell (1988). The neutron density is assumed to take its equilibrium value. Initially, the temperature and density behaviors are also taken from Hollowell (1988). Temperature and density values in later iterations are taken from the new AGB models with mass dependencies according to the isentropic conditions in the convective shell. Nucleosynthesis via neutron irradiation is determined according to the usual equations (see Bazan and Lattanzio 1992 for details). The resulting energy generation is then employed in an

AGB model of C-O core mass 0.59 $M_\odot$ and metallicity $Z = 0.001$ calculated from a ZAMS mass of 1.5 $M_\odot$ using the same code outlined in Lattanzio (1986).

3. RESULTS AND DISCUSSION

3.1 Altered AGB Evolution

The integrated luminosity through the convective shell from neutron captures, beta decays, and the $^{13}C(\alpha, n)^{16}O$ reactions varies between 10^3 and $10^4 L_\odot$, according to the model parameters for the 15th thermal pulse from Hollowell (1988). For comparison, the total energy in the convective shell (radiative + convective) peaks at $\sim 1500\ L_\odot$ at this time. This extra energy causes the upper convective boundary to be pushed out in mass (the lower convective boundary is also advanced in mass). The more rapid growth of the upper convective boundary means that the actual ^{13}C ingestion rate should be higher than described by evolution calculations not considering the extra energy, which from Hollowell and Iben (1990) means that the neutron generation and capture proceeds at a faster rate. The resultant energy generation should then evolve faster with a higher peak value.

From the nucleosynthesis code we determine the energy generation as a function of time for our assumed ingestion rate of the ^{13}C pocket. The evolution code then determines the new time dependence of the boundary of the convective shell and a new ^{13}C ingestion rate. We use this new ingestion rate in the nucleosynthesis code to determine the time dependence of the energy generation, which is then used to determine the new convective shell boundary behavior. The procedure 'converges' (as determined by the ingestion rate) after about 3 iterations. Figure 1 shows the detailed evolution of the shell due to the changes in the rate of ^{13}C ingestion, which begins at $t \sim 11540$ years in the figure. We show the changes in the convective boundaries for 3 iterations, with the solid line being the standard case. The first iteration is shown by the dashed line and gives the new evolution assuming added energy according to the ingestion rate of the Hollowell (1988) model. The dot-dashed line represents our 'converged' model, after we have performed two iterations of energy calculations and stellar modelling. As the rates of the last two iterations were very nearly equal, we stopped calculations here.

There are still differences between the runs, however, as shown clearly in figure. The shell is almost quenched due to the energy from neutron captures and assorted other reactions. In run '3', the maximum extent of the convection is still enough to engulf all the ^{13}C (which extends over $\Delta m = 5 \times 10^{-4}\ M_\odot$ only). But we note that the details of the processes leading to the ^{13}C pocket are not well understood, and if the pocket extended much further, we could have the situation where the temporary quenching of the convection and its subsequent growth and renewed ^{13}C ingestion could lead to mini-pulses of s-processing *within* a single thermal pulse. Details, of course, will depend on the dynamics of the convection and nuclear burning.

3.2 Nucleosynthesis Considerations

A larger ingestion rate leads to higher peak and average neutron densities experienced by material in the shell (Hollowell and Iben 1990). In the usual case we find a peak neutron density of 5×10^9 cm^{-3}, while the final iteration has a peak of 2×10^{10} cm^{-3}. (The latter value would be higher still if not for the retreat of the convective shell and the corresponding decrease in temperature at the base of the shell.) Following the arguments of Hollowell and Iben (1990), the peak neutron density in run '3' should be roughly three times that in run '2', since the mass expansion rate in run '3' is roughly three times that in run '2'. The fact that the neutron densities differ by only about 0.05 dex is the result of the decrease in the base temperature. The magnitude of the neutron density in the converged model seems to conflict with previous work on parameters governing the s-process. Parameter studies of the s-process (Kappeler et al. 1990, Bauer et al. 1991) have hinted that the s-process average neutron density should be $\sim 10^8$ cm^{-3} in order to reproduce the relative abundances of the s-only isotopes. Another problem is the lack of any measurable ^{96}Zr in MS and S stars (Smith and Lambert 1985, 1986, 1990). This mandates that the neutron density be no more than a few times $\sim 10^8$ cm^{-3}.

Another descriptive parameter of the s-process, the integrated neutron flux or exposure, changes due to the new energy source. A simple model of s-processing in AGB stars relates exposures received by material in individual thermal pulses to the mean exposure of the 'classical' model exposure distribution (Ulrich 1973). Initially, our models give a per pulse exposure of 0.30 mb^{-1}, while in our final run it reaches only 0.19 mb^{-1}. Taking into account the convective shell overlap fractionand the dilution fraction of ^{13}C s-processed material gives an actual mean exposure of 0.13 mb^{-1}.

4. CONCLUSION

We have shown that the inclusion of energy from neutron production, capture, and nuclear decay reactions significantly alters the development of the flash-driven convective shell in low-mass, low-metallicity AGB stars. Our exploratory analysis indicates the possibility of mini-pulses of s-processing within a single thermal pulse. A significant change is the more rapid growth of the outer boundary of the convective shell, leading to a much more rapid ingestion of the ^{13}C pocket. This leads to a neutron density higher than cases where the extra energy is ignored. As a result of the altered evolution, mean exposure values fall, which leads to a corresponding change in the expected s-process abundances.

5. ACKNOWLEDGEMENTS
The authors wish to thank Jim Truran, Grant Mathews, Michael Howard, and Stan Woosley for their comments. This work was performed under the auspices of the U.S. Department of Energy by the Lawrence Livermore National Laboratory under contract No. W-7405-ENG-48, for the University of Illinois under NSF contract AST-8611500, and for the McDonald Observatory, University of Texas at Austin.

6. REFERENCES
Bauer, R. W., Bazan, G., Becker, J. A., Howe, R. E., and Mathews, G. J. 1991, *Phys. Rev. C*, **43**, 2004.

Bazan, G., and Lattanzio, J. C., 1992, *Astrophys. J.*, submitted.

Gallino, R., Busso, M., Picchio, G., Raiteri, C. M., and Renzini, A. 1988, *Astrophys. J. Lett.*, **334**, L45.

Hollowell, D. E. 1988, *Ph. D. Thesis*, University of Illinois.

Hollowell, D. E. and Iben, I. Jr. 1990, *Astrophys. J.*, **349**, 208.

Iben, I. Jr. I. Jr. and Renzini, A. 1982, *Astrophys. J. Lett.*, **259**, L79.

Kappeler, F., Gallino, R., Busso, M., Picchio, G., and Raiteri, C. M. 1990, *Astrophys. J.*, **354**, 630.

Lattanzio, J. C., 1986, *Astrophys J.*, **311**, 708.

Smith, V. V. and Lambert, D. L. 1985, *Astrophys. J.*, **294**, 326.

Smith, V. V. and Lambert, D. L. 1986, *Astrophys. J.*, **311**, 843.

Smith, V. V. and Lambert, D. L. 1990, *Astrophys. J. Suppl.*, **72**, 387.

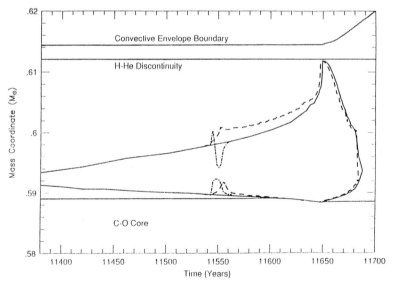

Figure 1:   The convective shell behavior under the influence of added energy. The models shown include an unaltered model (solid line), run '1' (dashed line), and run '3' (dot-dashed line).

Explosive Nucleosynthesis in Supernovae

F.-K. THIELEMANN[1], K. NOMOTO[2], M. HASHIMOTO[3]

[1]Harvard-Smithsonian Center for Astrophysics, Cambridge, MA 02138, USA
[2]Department of Astronomy, University of Tokyo, Bunkyo-ku, Tokyo 113, Japan
[3]Department of Physics, Kyushu University, Rapponmatsu, Fukuoka 810, Japan

1 INTRODUCTION

In this article we discuss type I and type II supernovae (SNe I and SNe II) with a special emphasis on nucleosynthesis processes. SNe Ia are explained by exploding white dwarfs in binary systems. This class seemed homogeneous until recently, when variations in the early spectra before maximum light became apparent. Such variations might be tied to varitions in the explosion mechanism and progenitor history. All other supernova types (SNe Ib, Ic, II-L and II-P) seem to be linked to the gravitational collapse of massive stars (main sequence mass $M>8M_\odot$) at the end of their hydrostatic evolution. Results for nuclei with $A<70$ produced in 13, 15, 20 and $25M_\odot$ stars are presented, as well as the average nucleosynthesis of SNe II, integrated over the progenitor mass range. The amount of Fe-group nuclei, especially ^{56}Ni (decaying to ^{56}Co and ^{56}Fe), is directly linked to the explosion mechanism and the mass cut between ejecta and the central compact object (neutron star or black hole). With self-consistent calculations being not available yet, we make an attempt to derive constraints from observational information.

2 TYPE IA SUPERNOVAE (SNE IA)

2.1 Explosive Burning Conditions

SNe Ia can be explained by the central carbon ignition and total disruption of a cold (but not solid) degenerate white dwarf, when exceeding the Chandrasekhar mass via mass transfer in a binary system. This agrees with the total energetics, light curves and spectra of SNe Ia. There exists also the possibility that ignition occurs before M_{Ch} is reached via accretion (Shigeyama et al. 1992), but this outcome is still uncertain. In the following we want to discuss in detail the burning conditions as they occur in SNe Ia. We take the model W7 by Nomoto, Thielemann, and Yokoi (1984) and Thielemann, Nomoto, and Yokoi (1986) as a typical example. There is still considerable uncertainty in the physics of propagating flame fronts and open questions remain (see e.g. Woosley and Weaver 1986b, 1992, Müller and Arnett 1986, Zeldovich et al. 1985, Nomoto and Shigeyama 1990, Khokhlov 1991ab, Wheeler and

Harkness 1990, Yamaoka et al. 1992). The propagation of the burning front after central C-ignition has only been treated in a parametrized way. These parameters can be tuned to either obtain deflagrations (subsonic) or detonations (supersonic burning front propagation) or also delayed detonations or late detonations (see e.g. models W7 and C8 in Nomoto, Thielemann, Yokoi 1984). Deflagrations, which turn into detonations in the outer mass zones of the white dwarf, could explain the recent variety of spectra before maximum light (SN 1990N, 1991T, Leibundgut et al. 1991, Filippenko et al. 1992, Phillips et al. 1992, Yamaoka et al. 1992), while spectra after maximum light seem to follow closely the model W7.

Temperatures in the burning front are increased by about a factor of 10, in comparison to the intitial values. The outer zones experience explosive C and Ne-burning, where first the carbon fusion produces ^{20}Ne which photodisintegrates back to ^{16}O. Further towards the center, zones undergo explosive O-burning where also ^{16}O is burned by fusion reactions to ^{28}Si and ^{32}S. Even higher temperatures lead to the burning of Si. In incomplete Si-burning doubly-magic ^{56}Ni is produced together with intermediate nuclei like ^{40}Ca. Inside 0.8M$_\odot$ only Fe-group nuclei exist with the dominant abundance of ^{56}Ni, which has the highest binding per nucleon for N=Z nuclei. In case of a late detonation, where the transition from deflagration to detonation occurs at $\approx$1.2M$_\odot$, the outer zones experience a somewhat different behavior. Towards the very center the densities become high enough to cause Fermi energies of the electron gas in excess of several MeV, which enables appreciable amounts of electron captures on free protons (and to a minor extent on heavy nuclei $\approx$40%). This changes the total proton to neutron ratio and the most abundant nuclei become first ^{54}Fe and ^{58}Ni and finally ^{56}Fe. The region of complete Si-burning is devided into an inner zone of 0.35M$_\odot$ with a normal freeze-out and an outer region with an alpha-rich freeze-out where ^{54}Fe is transformed by alpha captures into ^{58}Ni.

2.2 Abundances in the Ejecta

One of the major aspects of nucleosynthesis calculations is to understand the chemical evolution of galaxies and especially the present abundances in our Galaxy, with the solar system abundances being a snapshot in time at a specific location. If every SN Ia event starts from almost the same configuration (a white dwarf with $M=M_{Ch}$), very similar nucleosynthesis products are expected from each event and the comparison with solar abundances is actually meaningful without averaging over a complete sample. The production of Fe-group nuclei in comparison to their solar values is a factor of 2 larger than the production of intermediate nuclei from Si to Ca. This shows that SNe Ia are the dominant production sites of Fe-group nuclei, while SNe II have to fill in the intermediate nuclei.

An undesirable aspect is a large scale deviation from solar abundances within the Fe-group, when SNe Ia are the main contributors of these nuclei to the interstellar medium. In this respect ^{54}Fe, ^{58}Ni, and ^{62}Ni (decay product of ^{62}Zn) are overproduced in deflagration models with solar metallicities. All these nuclei come from a chain in the nuclear chart which is displaced by two units to the neutron-rich side from the N=Z chain and measures therefore the neutron excess of the material. Outside of $0.3M_\odot$, where electron capture is not effective, the neutron excess is only determined by the ^{22}Ne admixture to ^{12}C and ^{16}O in the original white dwarf, coming from ^{14}N in He-burning, which in turn originated from all CNO-nuclei in H-burning, i.e., the metallicity. Using time-averaged metallicities would reduce the overproduction of these nuclei (see colums W7 and W07 in Table 1 for deflagration models with solar (W7) and zero (W07) metallicities – recalculated with present day reaction rates [Thielemann, Arnould, Truran 1987], see also Bravo et al. 1992). As most SNe Ia occur in spiral galaxies on scale heights larger than 300pc (Della Valle and Panagia 1992), where [Fe/H] values are 0.0 to -0.5 (Fig.11.5 in Gilmore 1989), one has average metallicities of 0.5-0.6 times solar. Then the ^{54}Fe/^{56}Fe, ^{57}Fe/^{56}Fe and ^{58}Ni/^{60}Ni ratios are within a factor of 2 to 3 of solar, respectively, which lies within the present uncertainty range of thermonuclear reaction rates. Important is also that the propagation of the burning front is not fully understood yet. A burning front which starts with a small velocity and then accelerates (Woosley and Weaver 1986b, 1992, Khokhlov 1991ab), could reduce the amount of material in the mass zones between 0.05 to $0.3M_\odot$, where ^{54}Fe, ^{58}Ni, and ^{62}Ni are produced predominantly.

TABLE 1

MASSES IN SNe Ia MODELS

	W7	W07	W7DN	W7DT
C	0.048	0.051	5.8(-5)	4.8(-5)
O	0.143	0.133	0.047	0.004
Ne	0.005	0.002	2.9(-5)	2.0(-5)
Mg	0.009	0.016	0.002	6.3(-5)
Si	0.153	0.138	0.202	0.143
S	0.086	0.092	0.122	0.086
Ar	0.016	0.020	0.024	0.017
Ca	0.012	0.019	0.021	0.017
Ti	2.5(-4)	3.3(-4)	4.7(-4)	5.1(-4)
^{54}Fe	0.744	0.775	0.788	0.934
^{56}Fe	0.104	0.082	0.107	0.114
^{57}Fe	0.613	0.672	0.627	0.792
^{58}Ni	0.025	0.020	0.029	0.029
^{60}Ni	0.128	0.097	0.128	0.132
	0.011	0.014	0.011	0.012

In order to judge the uncertainty of the total amount of ^{56}Ni and Fe-group elements in deflagrations, we present in Table 1 also the results of late detonations (Yamaoka et al. 1992) which might be possible explanations of the variations among SNe Ia.

3 TYPE II SUPERNOVAE (SNE II)

3.1 Shock Wave Generation

A self-consistent treatment of the problem would include hydrodynamic calculations, following the Fe-core collapse, the bounce at nuclear densities, the delayed formation of the shock wave, and explosive nucleosynthesis in the envelope, which will be ejected. There exists recent progress in the understanding of the supernova mechanism of massive stars (see e.g. Bruenn 1989ab; Cooperstein and Baron 1989; Baron and Cooperstein 1990; Myra and Bludman 1989; Wilson and Mayle 1988; Mayle and Wilson 1988, 1990; Bethe 1990; Bruenn and Haxton 1991; Wilson 1992; Janka 1992; Herant, Benz, Colgate 1992), but complete self-consistent hydro and nucleosynthesis calculations have not been performed yet. Here we still make use of the fact that typical energies of 10^{51} erg are observed and light curve as well as explosive nucleosynthesis calculations can be performed with an artificially induced shock wave of appropriate energy, applied to the pre-collapse stellar model (see Woosley and Weaver 1986a; Shigeyama, Nomoto, Hashimoto 1988; Woosley 1988; Woosley, Pinto, Weaver 1988; Hashimoto, Nomoto, Shigeyama 1989; Arnett et al. 1989; Thielemann, Hashimoto, Nomoto 1990; Thielemann, Nomoto, Hashimoto 1992). There will, of course, be a variation of this value with progenitor mass, but only by a small factor. The temperatures obtained during the propagation of the shock wave are proportional to $E^{1/3}$ (see section 3.2). Thus, we expect that this uncertainty introduces negligible errors. Larger uncertainties are expected from the missing knowledge of the exact core structure at the time when the shock wave starts propagating outward. The lack of knowledge of the shock structure introduces an additional uncertainty, because different ways of initiating the explosion (piston, thermal bomb, kinetic bomb) can lead to different results. Aufderheide, Baron, and Thielemann (1991) have studied these questions in detail and came to the conclusion that errors in the bulk composition of up to 30% are introduced. While this is not negligible, it is small enough for a first order approach to nucleosynthesis studies in SNe II.

One remaining problem, when performing calculations with initiated shock waves, cannot be avoided. The location of the mass cut between neutron star and ejected envelope is unclear and can only be deduced from observational constraints. In case of SN1987A the knowledge of its distance and the light curve, powered by decaying ^{56}Ni and ^{56}Co, could be utilized to determine the mass of ^{56}Ni produced. ^{56}Ni is located in the innermost region of the ejecta and therefore provides information about the mass cut. For progenitors of different mass, where such information is not available, we have to search for other sources of information. In addition we cannot make judgements about the ejection of matter which experienced stronger neutronization via electron captures during core collapse, because this effect was not included in the pre-explosion models.

Haxton (1988), Woosley and Haxton (1988), Woosley et al. (1990) and Kolbe et al. (1992) examined the possible effect of inelastic neutrino scattering on explosive nucleosynthesis, an idea which was already introduced earlier by Domogatsky and Nadyozhin (1977). Inelastic neutrino scattering can populate excited states which are unstable against particle emission and produce neighboring nuclei. Outside the neutrino-sphere the scattering events will be rare and therefore this process will be mostly of importance for nuclei with very small abundances, which are not produced otherwise. This effect is not included in the present calculations. Kolbe et al. (1992) showed that with more realistic neutrino energy distributions (Janka and Hillebrandt 1989ab, Myra and Burrows 1990) the influence will be reduced.

3.2 Burning Products

As discussed above, the most significant parameter in explosive nucleosynthesis is the temperature, and a good prediction for the composition can already be made by only knowing T_{max}, without having to perform complex nucleosynthesis calculations. Weaver and Woosley (1980) already recognized, that matter behind the shock front is strongly radiation dominated. Assuming an almost homogeneous density and temperature distribution behind the shock (which is approximately correct, see Fig.3 in Shigeyama, Nomoto, and Hashimoto 1988), one can equate the supernove energy with the radiation energy inside the radius R of the shock front

$$E_{SN} = \frac{4\pi}{3} R^3 a T^4. \tag{1}$$

This equation can be solved for R, and with $T = 5 \times 10^9 \mathrm{K}$, the lower bound for explosive Si-burning with complete Si-exhaustion, and $E_{SN} = 10^{51} \mathrm{erg}$, the result is $R \approx 3700 \mathrm{km}$ (see Woosley 1988). For the evolutionary model by Nomoto and Hashimoto (1988) of a $20 \mathrm{M}_\odot$ star this radius corresponds to $1.7 \mathrm{M}_\odot$, in excellent agreement with the exact hydrodynamic calculation (see Thielemann, Hashimoto, Nomoto 1990). Temperatures which characterize the edge of the other explosive burning zones correspond to the following radii: incomplete Si-burning ($T_9=4$, $R=4980\mathrm{km}$), explosive O-burning (3.3, 6430), and explosive Ne/C-burning (2.1, 11750). This relates to masses of 1.75, 1.81, and $2.05\mathrm{M}_\odot$ in case of the $20\mathrm{M}_\odot$ star. The radii mentioned are model independent and vary only with the supernova energy. The results of explosive burning for the $20\mathrm{M}_\odot$ star (Thielemann, Hashimoto, Nomoto 1990) compared favorably with observed abundances for SN1987A (Danziger et al. 1990). This gave a reliable calibration for extending the same procedure to other SN II progenitor models by Nomoto and Hashimoto (1988), which all made use of the Caughlan et al. (1985) $^{12}C(\alpha,\gamma)$-rate and the Schwarzschild criterion of convection without over-

shooting, ranging over the SN II progenitor mass range and assuming an average
supernova energy of 10^{51}erg in the ejecta.

TABLE 2
MAJOR NUCLEOSYNTHESIS YIELDS

Element	$13M_\odot$	$15M_\odot$	$20M_\odot$	$25M_\odot$
C	0.060	0.083	0.115	0.148
O	0.218	0.433	1.480	3.000
Ne	0.028	0.039	0.257	0.631
Mg	0.012	0.046	0.182	0.219
Si	0.047	0.071	0.095	0.116
S	0.026	0.023	0.025	0.040
Ar	0.0055	0.0040	0.0045	0.0072
Ca	0.0053	0.0033	0.0037	0.0062
Fe	?0.150?	?0.150?	0.075	?0.050?

The main results, focusing just on a few important elements, are presented in Table 2 (a more detailed account, discussing also all minor elements and isotopes, is
given in Thielemann, Nomoto, Hashimoto 1992a). The content of Table 2 indicates
an interesting behavior. While the heavier intermediate mass nuclei originate only
from explosive O and Si-burning, which contribute similar amounts for all progenitor
masses, the lighter elements, C through Si, have dominant or essential contributions
from hydrostatic burning (C/Ne-core) or explosive Ne-burning. For both latter cases
we see a tremendous mass dependence in Table 2. The mass cuts between ejecta and
neutron stars were taken in such a way that the $0.07M_\odot$ of ejected ^{56}Ni for SN1987A
(a $20M_\odot$ star) could be reproduced. For 15 and $13M_\odot$ stars we followed the arguments by Shigeyama et al. (1990), Nomoto et al. (1990), and Hachisu et al. (1990)
that one can reproduce the light curves of SNe Ib and SNe Ic, respectively, when these
progenitors lost (almost all) their H-envelope by mass transfer in a binary system.
Thus, SN Ib/c events are interpreted as explosions of massive stars in binary systems
which lost the complete H-envelope to the binary companion (Ib) or retained a minor
amount of the H-envelope (Ic), which is sufficient to suppress the spectral He-lines
and leads to only minor Hα-features. The need for the relatively low mass progenitors
is due to the fact that they also possess small He-cores (3.3 and $4M_\odot$, respectively)
which result in the steep observed slope of their light curves (Ic similar to Ia). The
early escape of x-rays and gamma-rays from ^{56}Co-decay for small He-cores, steepens
the light curves in comparison to the pure ^{56}Co exponential decay slope. There is no
direct observational reason for the low Fe-yields in the case of the $25M_\odot$ star, except
for a recent observation (Schmidt 1992) which indicates for the first time that such
small amounts of Ni/Fe ejecta are possible in massive SNe II.

The main feature, which we learned from this exercise, is that the elements for which
SNe II are responsible in the chemical evolution of galaxies, fall into two categories:

(1) O, Ne and Mg are originating predominantly from hydrostatic burning shells and their ejected amount rises sharply with progenitor mass, while (2) S, Ar and Ca are mostly due to explosive burning and their ejected mass is varying very little. Si plays an intermediate role. This conclusion is based on models applying the Schwarzschild criterion for convection, which gained recent support from modelling P-Cygni (El Eid and Hartmann 1992, see also Weaver and Woosley 1992) and has to be tested further with respect to abundance observations in supernova remnants (Hughes, private communication).

3.3 Neutron Star Masses

When making use of the observed ^{56}Ni in SN1987A and arguments about the light curves of Type Ib/c supernovae, we obtained the results in Table 2. ^{56}Ni is produced inside the explosive Si-burning zone, where temperatures exceed 5×10^9 K. Accordingly the mass cut has to be at $M_{Si-ex} - M_{Ni}$. This is given in Table 3. ΔM_E is the uncertainty in the mass cut when 50% uncertainty in the explosion energy is assumed. All this is correct for prompt explosions. For delayed explosions accretion will occur for about 1s after the bounce. This time interval contains the sound travel time to an outer zone before infall can be triggered and an (almost) free fall time afterwards (for details see Thielemann, Nomoto, Hashimoto 1992b). Zones which can reach the proto-neutron star during $\approx$1s after the bounce contribute to ΔM_{acc}.

TABLE 3
MASS CUT IN SN II EVENTS

$M/M_\odot$	M_{core}	M_{Si-ex}	M_{cut}	ΔM_E	ΔM_{acc}
13	1.18	1.43	1.28	0.03	0.14
15	1.28	1.47	1.32	0.03	0.15
20	1.40	1.70	1.60	0.03	0.17
25	1.61	1.82	1.72	0.03	0.21

The baryonic masses of Table 3 will be reduced by the gravitational binding energy, which is released in neutrinos. Lattimer and Yahil (1989) give a relation with about 10% uncertainty. This relation is used to obtain the numbers in Table 4. Whether such neutron star masses are stable or result in a collapse to a black hole at the upper end of the mass range, will have a decisive impact on the average ejecta of SNe II and can only result from self-consistent calculations.

TABLE 4
NEUTRON STAR MASSES

$M/M_\odot$	M_b	M_g
13	1.25-1.45	1.13-1.32
15	1.29-1.50	1.16-1.36
20	1.57-1.80	1.39-1.60
25	1.69-1.96	1.49-1.73

3.4 Early Galactic Chemical Evolution

Galactic chemical evolution calculations take into account the continuous enrichment of the interstellar medium by SNe I and SNe II, stellar winds (planetary nebulae), etc.. In the very early evolution of the Galaxy only the most massive stars can contribute because of their short lifetime. Following Matteucci (1987), Matteucci and Francois (1989) and Mathews, Bazan and Cowan (1992), the elements listed in Table 2 can only derive from supernovae (with the exception of C) and can be contributed solely to SNe II for $[Fe/H]<-1$, $[Fe/H]=\log_{10}[(Fe/H)/(Fe/H)_\odot]$. Therefore, it is not surprising that observations show a constant $[x/Fe]$, x being O, Mg, Si, S, Ca, Ti, Cr, or Ni below -1. The integrated yields of SNe II should result in an abundance pattern such as found in low metallicity stars.

TABLE 5

INTEGRATED ABUNDANCE YIELDS

Element	[x/Fe]	$< M_x >$	[x/Fe] for [Fe/H]$<$-1
C	-0.13	0.067	0.0
O	0.68	1.369	0.5
Ne	0.76	0.295	-
Mg	0.63	0.082	0.4
Si	0.40	0.052	0.4
S	0.21	0.020	0.5
Ar	0.13	0.004	-
Ca	0.25	0.003	0.3
Fe	0.00	0.038	0.0

In a first attempt we have tried a crude method to test whether these considerations are consistent with the results from our explosion calculations. We utilized the coarse grid given by a set consisting of 13, 15, 20, and 25$M_\odot$ stars – neglecting the contribution from stars less massive than 13$M_\odot$ – and performed an integration over the abundance yields of the individual elements $M_{ej,x}(M)$ from Table 2, weighted with a Salpeter initial mass function (IMF)

$$M_x \propto \int_8^{100} M_{ej,x}(M)M^{-2.35}dM. \tag{2}$$

We extrapolated the yields smoothly up to $M = 40$ $M_\odot$, from where they were kept constant. This relates to the fact that more massive stars will lose large amounts of envelope mass, in form of stellar winds, during their early evolution, so that the later evolution resembles that of less massive stars. The procedure differed only for Fe, which was kept constant for $M>25M_\odot$. The integration over an IMF gives the same abundance ratios as full scale galactic chemical evolution calculations, provided that the IMF and the star formation rate (SFR) are constant in time. The resulting [x/Fe] ratios and the average mass ejected for the elements are given in Table 5 and compared to the [x/Fe] values observed in low metallicity stars. The latter are taken from Gehren (1988); Wheeler, Sneden, and Truran (1989); Lambert (1989); Gratton and Sneeden (1991); Sneden, Gratton, Crocker (1991).

The results are encouraging, considering the crude method and the fact that the observations probably have errors of 0.1-0.2 dex. They agree within the observational errors for all elements, with the exception of Mg and S, but they are by no means perfect. In order to obtain the results for the Fe-yields it was also necessary to assume mass cuts for the 13, 15, and $25M_\odot$ stars. If one wants to obtain a [Si/Fe] value of 0.4 and believes in the given Si yields, where no mass cut uncertainty enters (except for a possible lower value for the upper integrand limit due to black hole formation), then the average Fe-mass produced in SN II+Ib events has to be $0.038M_\odot$ (if our predictions for Si are correct and results do not change drastically, when using an initial mass function different from a Salpeter IMF). We were only able to achieve this goal when assuming for the $25M_\odot$ star and more massive stars the ejection of only $0.02M_\odot$, thus changing the respective values given in Table 2. This value is small by any standards and less than previously expected. The statistically still insignificant sample of bolometric type II plateau supernova light curves (Young and Branch 1989, Phillips et al. 1990) indicates that the average type II-P supernova (being less massive than $20M_\odot$) produces more Fe than SN1987A (which has the dimmest tail by about a factor of 2) and therefore predicts decreasing Fe-yields with increasing progenitor mass. Recently Schmidt (1992) even found a massive SN II remnant with $M(Ni)<0.05M_\odot$.

We neglected contributions by SNe II in the mass range 8–10 $M_\odot$, which will undergo collapse to neutron star densities as well as C- through Si-burning in one continuous burning stage initiated by e-capture in a strongly degenerate core. 10–13 $M_\odot$ stars are also strongly affected by core degeneracy and have a very steep density gradient at the edge of the Fe-core. In both cases minor amounts of explosive nucleosynthesis ejecta are expected (see e.g. the Crab nebula), although it is not completely clear whether these are negligible.

3.5 Constraints for Self-Consistent Models

Taken the above results, there are several points which are of concern for obtaining the observed abundance ratios in low metallicity stars:

(1) The Ni/Fe ejecta have to decrease drastically for stars in the range 20-25$M_\odot$.
(2) The [O/Fe] and [Mg/Fe]-ratios are too high. It would be helpful if very massive SNe II, which contribute large amounts of O and Mg (from hydrostatic burning, see

Table 2), would not be taken into account. (3) The ratios [S/Fe] and [Ca/Fe] are too small. These are the elements which are almost entirely produced in explosive burning (which gives roughly constant yields over the progenitor mass range), while elements like O and Mg originate almost fully and Si partially from hydrostatic burning, which increases with progenitor mass. (4) r-process elements seem only to be ejected from low mass SNe II (Mathews and Cowan 1990; Cowan, Thielemann, Truran 1991; Mathews, Bazan, Cowan 1992). (5) The observational $\Delta Y/\Delta Z$-values (gradient of the helium mass fraction with respect to metallicity) are much larger than can be reproduced with present stellar yields. In case the burning products of a $25M_\odot$ star are not ejected anymore (but rather end in a black hole?), observations and theory can be made consistent (Maeder 1992). (6) With the inclusion of hyperons the equation of state for neutron stars becomes softer and gives maximum neutron stars masses of $1.5\text{-}1.7M_\odot$ (e.g. Glendenning et al. 1992). We reach such values between 20 and $25M_\odot$ when consulting Table 4.

Do we wittness here the following transition in the supernova mechanism as a function of progenitor mass: (a) formation of a neutron star and a delayed explosion powered by neutrino heating of the high entropy bubble plus explosively produced Ni/Fe and r-process elements, (b) with increasing neutron star mass an accretion fall-back of matter containing r-process material, (c) formation of a hot neutron star which survives accretion long enough (seconds) to cause neutrino heating and explosive ejection of the envelope, but accretes enough matter that it finally collapses to a black hole with substantial fall-back of explosively produced Ni/Fe, and (d) straight collapse into a black hole and no ejection of hydrostatic nor explosive yields?? This might still sound hypothetical and could be wrong, but it is a strong reason to obtain an answer from self-consistent supernova calculations with varying equation of state (EOS), in order to see the possible effect of encountering the maximum neutron star mass during a supernova collapse.

4 CONCLUSIONS

Explosive nucleosynthesis in supernovae has matured from a pure nucleosynthesis approach with peak temperatures and densities and an expansion time scale to a topic clearly tied in with observational and theoretical astrophysics. There are still many open questions in supernova physics, ranging from the progenitor system and burning front propagation in SNe Ia to the details of multidimensional effects and the equation of state in SNe II. But progress is inevitable and new observational constraints give the guidance. SN1990N and SN1991T seem to indicate that delayed (or late) detonations occur in SNe Ia and allow for the variations of spectra at early times. New self-consistent calculations will make it possible to find a unique relation between light curves, spectra and explosion models (Höflich 1992, Eastman and Pinto

1992).

A similar improvement for SNe II models is expected. After (i) Wilson (1992) gets clear explosions based on the delayed mechanism with neutrino heating which, dependent on his convection parameter, reach energies as high as 1.5×10^{51}erg, (ii) Janka (1992) obtains clear explosions without convection with energies of the order 7×10^{50}erg, and (iii) Herant, Benz, Colgate (1992) obtained a strong indication for convective turnover in 2D and 3D calculations, the supernova mechanism as such seems to be solved. It is now time to explore the fate of massive stars over the whole SN II mass range and analyze the connections to the chemical evolution of galaxies and the central remants (neutron stars or black holes).

Acknowledgements

We want to thank our collaborators T. Shigeyama and H. Yamaoka, who contributed to the material presented here. This research was supported in part by NSF grant AST 89-13799. The computations were performed at the National Center for Super-computer Applications at the University of Illinois (AST 890009N).

,

REFERENCES
Arnett, W.D., Bahcall, J.N., Kirshner, R.P., Woosley, S.E. 1989, *Ann. Rev. Astron. Astrophys.* **27**, 629

Aufderheide, M., Baron, E., Thielemann, F.-K. 1991, *Ap. J.* **370**, 630

Baron, E., Cooperstein, J. 1990, in *Supernovae*, ed. S.E. Woosley (Springer-Verlag, New York), p. 342

Bethe, H.A. 1990, *Rev. Mod. Phys.* **62**, 801

Bravo, E., Isern, J., Canal., R., Labay, J. 1992, *Astron. Astrophys.*, **257**, 534

Bruenn, S.W. 1989, *Ap. J.* **340**, 955

Bruenn, S.W. 1989, *Ap. J.* **341**, 385

Bruenn, S.W., Haxton, W.C. 1991, *Ap. J.* **376**, 678

Caughlan, G.R., Fowler, W.A., Harris, M.J, Zimmerman, G.E. 1985, *At. Nucl. Data Tables* **32**, 197

Cooperstein, J., Baron, E. 1989, in *Supernovae*, ed. A. Petschek (Springer-Verlag, New York), p. 213

Cowan, J.J., Thielemann, F.-K., Truran, J. W. 1991, *Phys. Rep.* **208**, 267

Danziger, I.J., Bouchet, P., Gouiffes, C., Lucy, L. 1990, in *Supernovae*, ed S.E. Woosley, (Springer-Verlag, New York), p.69

Della Valle, M., Panagia, N. 1992, *Astron. J.* **104**, 696

Domogatsky, G.V., Nadyozhin, D.K. 1977, *M.N.R.A.S.* **178**, 33

Eastman, R.G., Pinto, P. 1992, *Ap. J.*, submitted

El Eid, M., Hartmann, D. 1992, *Ap. J.*, submitted

Filippenko, A.V. et al. 1992, *Ap. J. Lett.* **384**,L15

Gehren. T. 1988, in *Rev. in Mod. Astronomy* **1**, p.52

Gilmore, G. 1989, in The Milky Way as a Galaxy, eds. G. Gilmore, I. King, P. van
  der Kruit (Geneva, Geneva Observatory), 227

Glendenning, N.K., Weber, F., Moszkowski, S.A. 1992, *Phys. Rev.* **C45**, 844

Gratton, R.G., Sneden, C. 1991, *Astron. Astrophys.* **241**, 501

Hachisu, I., Matsuda, T., Shigeyama, T., Nomoto, K., 1990, *Ap. J. Lett.***358**, L57

Hashimoto, M., Nomoto, K., Shigeyama, T. 1989, *Astron. Astrophys.* **210**, L5

Haxton, W.C. 1988, *Phys. Rev. Lett.* **60**, 1999

Herant, M., Benz, W., Colgate, S. 1992, *Ap. J.*, in press

Höflich, P. 1992, Habilitation Thesis, MPI Munich

Janka, H.-T. 1992, private communication

Janka, H.-T., Hillebrandt, W. 1989 *Astron. Astrophys.* **224**, 49

Janka, H.-T., Hillebrandt, W. 1989 *Astron. Astrophys. Suppl.* **78**, 375

Khokhlov, A.M. 1991, *Astron. Astrophys.* **245**, 114

Khokhlov, A.M. 1991, *Astron. Astrophys.* **245**, L25

Kolbe, E., Krewald, S., Langanke, K., Thielemann, F.-K. 1992, *Nucl. Phys.* **A540**,
  599

Lambert, D.L. 1989, in *Cosmic Abundances of Matter*, ed. C.J. Waddington, AIP
  Conf. Proc. 183, p.168

Lattimer, J.M., Yahil, A. 1989, *Ap. J.* **340**, 426

Leibundgut, B. et al. 1991, *Ap. J. Lett.* **371**, L23

Maeder, A. 1992, *Astron. Astrophys.*, in press

Mathews, G.J., Bazan, G., Cowan, J.J. 1992, *Ap. J.* **391**, 719

Mathews, G.J., Cowan, J.J. 1990, *Nature* **345**, 491

Matteucci, F. 1987, in *Stellar Evolution and Dynamics of the Outer Halo of the
  Galaxy*, eds. M. Azzopardi, F. Matteucci, *ESO Conf. Proc.* **27**, p.609

Matteucci, F., Francois, P. 1989, *M.N.R.A.S.*, **239**, 885

Mayle, R.W., Wilson, J.R. 1988, *Ap. J.* **334**, 909

Mayle, R.W., Wilson, J.R. 1990, in *Supernovae*, ed. S.E. Woosley (Springer-Verlag,
  New York), p. 333

Müller, E., Arnett, W.D. 1986, *Ap. J.* **307**, 619

Myra, E.S., Bludman, S. 1989, *Ap. J.* **340**, 384

Myra, E.S., Burrows, A. 1990, *Ap. J.* **364**, 222

Nomoto, K., Hashimoto, M. 1988, *Phys. Rep.* **163**, 13

Nomoto, K., Shigeyama, T. 1990, in *Supernovae*, ed S.E. Woosley, (Springer-Verlag,
  New York), p.572

Nomoto, K., Thielemann, F.-K., Yokoi, K. 1984, *Ap. J.* **286**, 644

Phillips, M.M. et al. 1992, *Astron. J.* **103**, 1632

Schmidt, B. 1992, private communication

Shigeyama, T., Nomoto, K., Hashimoto, M. 1988, *Astron. Astrophys.* **196**, 141

Shigeyama, T., Nomoto, K., Tsujimoto. T.,Hashimoto, M. 1990, *Ap. J. Lett.* **361**, L23

Shigeyama, T., Nomoto, K., Yamaoka. H., Thielemann, F.-K. 1992, *Ap. J. Lett.* **386**, L13

Sneden, C., Gratton, R.G., Crocker, C. 1991, *Astron. Astrophys.*, **246**, 354

Thielemann, F.-K., Arnould, M., Truran, J.W. 1987, in *Advances in Nuclear Astrophysics*, eds. E. Vangioni-Flam et al., (Editions frontières: Gif sur Yvette), p. 525

Thielemann, F.-K., Hashimoto, M., Nomoto, K. 1990, *Ap. J.* **349**, 222

Thielemann, F.-K., Nomoto, K., Hashimoto, M. 1992, in *Supernovae, Les Houches, Session LIV*, eds. J. Audouze, S. et al. (Elsevier, Amsterdam), in press

Thielemann, F.-K. Nomoto, K., Hashimoto, M., 1992 in *The Structure and Evolution of Neutron Stars*, ed. R. Tamagaki, S. Tsuruta (Addison-Wesley, Redwood City), in press

Thielemann, F.-K., Nomoto, K., Yokoi, K. 1986, *Astron. Astrophys.*, **158**, 17

Weaver, T.A., Woosley, S.E. 1980, *Ann. N.Y. Acad. Sci.* **366**, 335

Weaver, T.A., Woosley, S.E. 1992, *Phys. Rep.*, in press

Wheeler, J.C., Harkness, R.P. 1990, *Rep. Prog. Phys.* **53**, 1467

Wheeler, J.C., Sneden, C., Truran, J.W. 1989, *Ann. Rev. Astron. Astrophys.*, **27**, 279

Wilson, J.R. 1992, private communication

Wilson, J.R., Mayle, R.W. 1988, *Phys. Rep.* **163**, 63

Woosley, S.E. 1988, *Ap. J.* **330**, 218

Woosley, S.E., Hartmann, D., Hoffman, R.B., Haxton, W.C. 1990, *Ap. J.* **356**, 272

Woosley, S.E., Haxton, W.C. 1988, *Nature* **334**, 45

Woosley, S.E., Pinto, P.A., Weaver, T.A. 1988, *Proc. Astron. Soc. Australia* **7**, 355

Woosley, S.E., Weaver, T.A. 1986, *Ann. Rev. Astron. Astrophys.* **24**, 205

Woosley, S.E., Weaver, T.A. 1986, in *Radiation Hydrodynamics*, IAU Colloq. No 89, eds. D. Mihalas, K.H., Winkler (Reidel, Dordrecht), p. 91

Woosley, S.E., Weaver, T.A. 1992, in *Les Houches, Session LIV, Supernovae*, eds. J.Audouze, et al. (Elsevier Science Publ.), in press

Yamaoka, H., Nomoto, K., Shigeyama, T., Thielemann, F.-K. 1992, *Ap. J. Lett.* **393**, L55

Young, T.R., Branch, D. 1989, *Ap. J. Lett.* **342**, L79

Zeldovich, Ya.B., Baerenblatt, G.I., Librovich, V.B., Makhviladze, G.M. 1985, *The Mathematical Theory of Combustion and Explosions*, (Plenum, New York)

Nucleosynthesis and Light Curves of Exploding Very Massive Stars

K. HERZIG[1], M. F. EL EID[1], P. HÖFLICH[2]

[1] Universitätssternwarte Göttingen, Germany
[2] Max-Planck-Institut für Astrophysik, Garching, Germany

ABSTRACT

We have investigated the evolution of stars of initial masses of about 100 to 140 $M_\odot$ from the main sequence up to supernova stage. These stars become dynamically unstable due to electron-positron pair creation prior to central oxygen ignition, which then proceeds explosively. The nucleosynthesis has been followed with an extended nuclear reaction network to especially determine the quantity of synthesized radioactive nuclei which power the light curve. Preliminary results of the nucleosynthesis and the light curve of such explosions are presented in this short contribution. We find that oxygen cores with masses lower than approximately 60 $M_\odot$ do not suffer total disruption, and will likely lead to the formation of black holes of stellar masses.

1 INTRODUCTION

In various recent studies (cf. Herzig et al., 1990; Herzig 1992), the evolution of VMS has been followed from the main sequence through the explosive oxygen burning phases, which led to the explosion of such stars. Here, such computations have been revised by improving the calculations of nucleosynthesis and light curve in many respects (Herzig, 1992; Herzig et al., 1993, for details).

2 EVOLUTION TOWARD DYNAMICAL INSTABILITY

The initial masses examined are 100 $M_\odot$ and 120 $M_\odot$ having initial solar composition. In the case of the 100 $M_\odot$ star, the Schwarzschild criterion for convection has been used, and convective overshooting from the core has been included by assuming an overshoot distance of 1.9 pressure scale heights. The star in this case evolved to a WO type star retaining 61 $M_\odot$. In the case of the 120 $M_\odot$ star, no overshooting has been assumed. During the LBV phase, the mass loss rates are artificially increased to prevent the star from evolving to the red supergiant stage, where no such luminous star are observed. For the WR phase (surface hydrogen mass fraction smaller 0.35), an average mass loss rate of $3 \cdot 10^{-5}$ $M_\odot$ a^{-1} has been assumed. As a result of this simulation, a 56 $M_\odot$ WR star of type WNL has been obtained. At this stage we artificially increased the mass to 65 $M_\odot$, corresponding to initially 140 $M_\odot$.

The two models briefly described above differ mainly in the structure and composition of the layers outside the carbon/oxygen cores. The radius of the WNL model is about ten times larger than that of the WO star, and its atmospheric composition is dominated by nitrogen, helium (85%) and hydrogen (13%), which is completely absent in the WO model.

3 ONSET OF PAIR-INSTABILITY AND EXPLOSIVE OXYGEN BURNING

At the end of central neon burning, temperatures and densities are achieved which favour the creation of electron-positron pairs by the radiation field. This causes the adiabatic index localy and globally to fall below the critical value of 4/3. A collapse phase follows during which oxygen burning proceeds explosively. The nucleosynthesis and energy production during this phase have been followed by using an extended network of nuclear reactions containing 72 nuclei linked by 483 reactions. The reaction rates up to Mg are taken from Caughlan & Fowler (1988), for the higher elements we used a compilation by M. Rayet (1990, priv. com.).

In particular we found, that all reactions, especially (α,p) and (p,γ) reactions, have to be taken into account to determine the abundances of the heavier elements properly. The total amount of nickel, for instance, may differ by a factor of 30 if only (α,γ) reactions are considered.

The final chemical composition for the two WR-models of type WNL and WO respectively is shown in Fig. 1 and 2. Since the temperature in the WO model rose higher (up to $3.73 \cdot 10^9$ in the central parts) than in the WNL model, less nickel (0.009 $M_\odot$ compared to 0.015 $M_\odot$) was synthesized in the latter and the mass of the burned oxygen was smaller (4.1 $M_\odot$ instead of 7.8 $M_\odot$).

4 SUPERNOVA STAGE AND LIGHT CURVE

Light curves for the supernova models have been calculated using the radiation-hydrodynamical code developed at Göttingen Observatory (Herzig, 1988; Herzig et al., 1990). Besides the luminosity we also calculated colour magnitudes in the Johnson UBV bands to compare our results with observations. PCSN light curves in the visual band are shown in Fig. 3 together with typical light curves of standard SN I and SN II events.

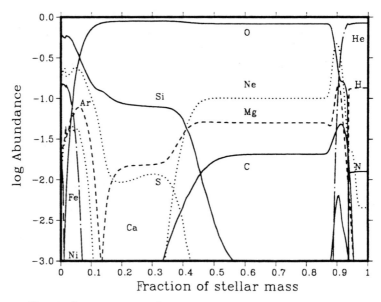

Fig. 1 *Chemical composition after explosive oxygen burning for model WNL65.*

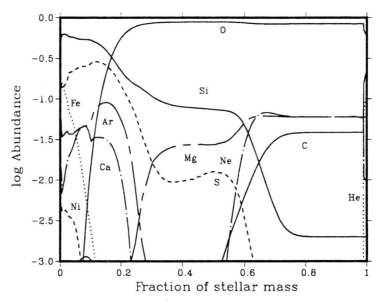

Fig. 2 *Chemical composition after explosive oxygen burning for model WO61.*

Typically they are much broader and less luminous, the former resulting from the large mass of the exploding core, the latter due to the small quantity of radioactive material and the absence of a strong shock wave. An important energy source during the first 50 days is the recombination energy of oxygen. Details of the calculations will be published elsewhere (Herzig et al., 1993).

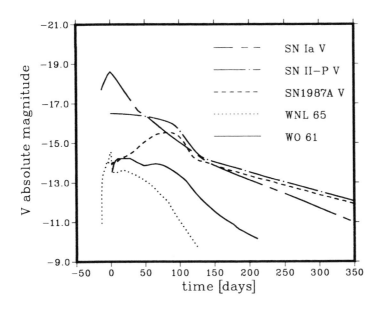

Fig. 3 *Light curves for the PCSN models WNL65 and WO61 in comparison with standard light curves of SN I and SN II (taken from Doggett & Branch, 1985) and of SN 1987A (Catchpole et al, 1989).*

In the case of the WNL model kinetic energy ($4.4 \cdot 10^{51}$ erg compared to $7.3 \cdot 10^{51}$ erg in the WO model) is not sufficient to disrupt the star. 71 % of the stellar mass reaches escape velocities, therefore about 19 $M_\odot$ remains as stellar remnant. This core is too small to become pair unstable again and will eventually become a black hole.

REFERENCES
Catchpole, R.M. et al: 1989, MNRAS 237, 55p
Caughlan, G.R.; Fowler, W.A.:1988, Atom Data and Nucl. Data Tables 40, 283
Doggett, J.B.; Branch, D.: 1985, Astron. J. 90, 2303
Herzig, K.: 1992, Ph.D. thesis, Universität Göttingen
Herzig, K.; El Eid, M.F.; Fricke, K.J.; Langer, N.: 1990, Astron. Astrophys. 233, 462
Herzig, K.; El Eid, M.F.; Höflich, P.: 1993, in preparation

Fe-group yields of SNIa and metallicity

E. BRAVO[1,4], J. ISERN[2,4], R. CANAL[3,4], J. LABAY[3,4]

1: Dpt. Física i Enginyería Nuclear, U.P.C., Av. Diagonal 647, 08028 Barcelona (Spain). 2: Centre d'Estudis Avançats, C.S.I.C., Blanes (Spain). 3: Dpt. Astronomía i Astrofísica, U.B., Barcelona (Spain). 4: Laboratori d'Astrofísica, I.E.C., Barcelona.

Fe-group nuclei are thought to be synthesized mainly in supernova (SN) explosions, even though some of them may be built also in intermediate-mass stars (^{58}Fe, ^{59}Co). The contribution of the different types of SN is of the same order in view of the present SN rates (Van den Bergh & Tamman 1991) and total mass of iron ejected. In order to follow the evolution of the ratios between the different nuclei of the Fe-group during the galactic life, and compare them with the observational data for stars (Bravo et al. 1992c), it is interesting to know the yields of these nuclei in type Ia SN (SNIa). Much work has been done up to now in the standard model (Thielemann et al. 1986, Woosley 1991, Khokhlov 1991), and the nucleosynthesis has been calculated for a range of initial conditions. Alternative models (Livne & Glasner 1990) must yet be developed. The purpose of this paper is to study the dependence of the yields on some unknowns of the standard model. These are due to:

I) possible occurrence of SNIa with lightly different initial conditions (central density at ignition, metallicity).

II) lack of knowledge of some physical processes involved that affect the thermal structure at ignition, and the combustion propagation.

The influence of the central density of the white dwarf at ignition has been studied in a previous paper (Bravo et al. 1992b), and we adopt in all the present calculations the value of $\rho = 4\,10^9$ g/cm^3 (Iben 1982). The combustion propagation is simulated in two groups of models as a deflagration (models R) and as a delayed detonation (models D) respectively. As to the metallicity, the effect on the nucleosynthesis can be factorized in the ^{22}Ne abundance prior to the explosion. We calculated the yields for a SNIa coming from a Population I star, $X_{Ne} = 0.025$ (models 2), as well as a SNIa coming from a Population II star, $X_{Ne} = 0$ (models 0), and an intermediate metallicity case, $X_{Ne} = 0.01$ (models 1). The results are presented in Table 1 and Figs. 1, 2. In deflagration models (R) the only elements affected by the initial presence of ^{22}Ne are ^{54}Fe, ^{58}Ni, and ^{62}Ni, because these elements are synthesized in a zone of the star that has undergone little electron captures (Bravo et al. 1992a). Indeed, those elements are only slightly affected. In delayed detonation models (D) the influence of the initial ^{22}Ne is much more evident, leading to an overproduction of

^{53}Cr, ^{54}Fe, ^{55}Mn, and ^{58}Ni in the models with a higher metallicity. Other nuclei (^{50}Ti, ^{54}Cr, ^{58}Fe, and in models D also ^{64}Ni) are extremely overproduced in our calculations as a result of the electron captures that lead to a high neutron excess in the center of the star. This is in part due to the high central density, and has been also found by other autors (c.f. Khokhlov 1991), specially for the overproduction of ^{54}Cr. The final yield of those nuclei depends only on the behaviour of the burning front near to the center, and so depends on the central thermal gradient at ignition, and on the conductive velocity of the burning front.

Table 1. $\text{Log}(M(^A Z))$ for the Fe-group elements (in $M_\odot$)

$^A Z$	R0	R1	R2	$^A Z$	R0	R1	R2	$^A Z$	R0	R1	R2
^{46}Ti	-7.05	-6.36	-5.84	^{54}Cr	-2.37	-2.37	-2.38	^{62}Ni	-2.39	-2.36	-2.34
^{47}Ti	-7.26	-7.11	-7.11	^{55}Mn	-2.02	-1.98	-1.95	^{64}Ni	-3.10	-3.10	-3.10
^{48}Ti	-3.61	-3.77	-3.86	^{54}Fe	-1.04	-0.99	-0.95	^{63}Cu	-4.62	-4.62	-4.62
^{49}Ti	-4.36	-4.34	-4.33	^{56}Fe	-0.23	-0.25	-0.27	^{65}Cu	-5.32	-5.33	-5.36
^{50}Ti	-3.09	-3.09	-3.09	^{57}Fe	-1.68	-1.66	-1.62	^{64}Zn	-3.84	-3.96	-4.19
^{50}V	-7.87	-7.88	-7.90	^{58}Fe	-2.24	-2.25	-2.25	^{66}Zn	-3.22	-3.22	-3.22
^{51}V	-4.69	-4.10	-4.08	^{59}Co	-2.72	-2.72	-2.71	^{67}Zn	-5.29	-5.29	-5.29
^{50}Cr	-4.01	-3.86	-3.74	^{58}Ni	-0.94	-0.92	-0.87	^{68}Zn	-4.84	-4.85	-4.84
^{52}Cr	-2.08	-2.14	-2.19	^{60}Ni	-1.55	-1.57	-1.58	^{70}Zn	-6.17	-6.17	-6.17
^{53}Cr	-3.04	-2.99	-2.97	^{61}Ni	-3.52	-3.53	-3.57				

$^A Z$	D0	D1	D2	$^A Z$	D0	D1	D2	$^A Z$	D0	D1	D2
^{46}Ti	-6.49	-5.58	-5.24	^{54}Cr	-1.87	-1.88	-1.87	^{62}Ni	-2.31	-2.31	-2.24
^{47}Ti	-6.81	-6.42	-6.33	^{55}Mn	-1.96	-1.85	-1.71	^{64}Ni	-2.71	-2.71	-2.70
^{48}Ti	-2.88	-2.94	-3.11	^{54}Fe	-1.17	-0.98	-0.76	^{63}Cu	-4.80	-4.83	-4.77
^{49}Ti	-4.14	-4.00	-3.99	^{56}Fe	-0.15	-0.21	-0.14	^{65}Cu	-5.41	-5.42	-5.33
^{50}Ti	-2.57	-2.57	-2.56	^{57}Fe	-1.95	-1.90	-1.69	^{64}Zn	-5.13	-5.07	-4.16
^{50}V	-7.60	-7.64	-7.54	^{58}Fe	-1.67	-1.68	-1.67	^{66}Zn	-4.06	-4.09	-3.97
^{51}V	-3.56	-3.46	-3.38	^{59}Co	-3.14	-3.15	-3.07	^{67}Zn	-5.29	-5.43	-5.27
^{50}Cr	-3.70	-3.30	-3.10	^{58}Ni	-1.43	-1.41	-1.22	^{68}Zn	-6.24	-6.37	-6.21
^{52}Cr	-1.52	-1.58	-1.65	^{60}Ni	-2.08	-2.09	-2.00	^{70}Zn	-6.50	-6.68	-6.47
^{53}Cr	-2.63	-2.53	-2.46	^{61}Ni	-3.86	-3.90	-3.64				

The flame front advances through the star with a velocity which is the highest of three quantities: the spontaneous flame velocity, the conductive velocity, and the turbulent front velocity. Turbulence is triggered in the front as a result of the Rayleigh-Taylor instability, but it isn't efficient near to the center. The spontaneous flame velocity depends on the thermal gradient. The above mentioned models were made with the

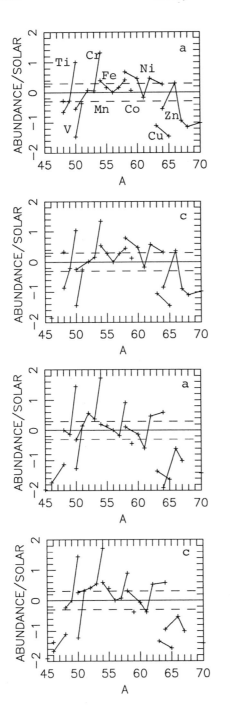

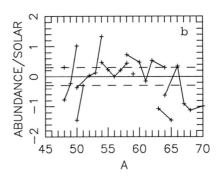

Figure 1. Log of the abundance of the Fe-group elements normalized to ^{56}Fe, in deflagration models: a) R0, b) R1, c) R2

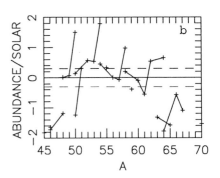

Figure 2. Log of the abundance of the Fe-group elements normalized to ^{56}Fe, in delayed detonation models: a) D0, b) D1, c) D2

assumption of a nearly isothermal white dwarf, except in the central region which was instantaneously incinerated. This made the spontaneous flame velocity unimportant out of the initially burnt zone, that had 0.014 $M_\odot$. If one assumes instead an adiabatic thermal gradient, the spontaneous flame velocity is greater than the speed of sound in the central 0.03 $M_\odot$, and is greater than the conductive velocity up to $\simeq 0.15\ M_\odot$. We have repeated the calculation increasing the size of the initially burnt zone. The result was a reduction of the overproduction of ^{50}Ti, ^{54}Cr, and ^{58}Fe but in a small amount, while the production of the other nucli remained practically unchanged. The sensitivity of nucleosynthesis to the conductive velocity has benn tested, repeating the calculation with a faster conductive velocity. Again the changes were small, but in this case ^{50}Ti, ^{54}Cr, and ^{58}Fe production was increased.

CONCLUSIONS

The nucleosynthesis of the standard model of SNIa has been tested against changes in the initial metallicity of the white dwarf, the thermal gradient near to the center of the star, and the conductive velocity. The main results are:

I) Moderately neutronized nuclei production, like ^{54}Fe and ^{58}Ni, varies only slightly for a deflagration model.

II) In delayed detonation models an increase in the initial metallicity leads to a large overproduction of ^{54}Fe, and increases considerably the amount of ^{58}Ni, and ^{50}Cr.

III) The large overproduction of the nuclei ^{50}Ti, ^{54}Cr, and ^{58}Fe is due to the high central density ($\rho = 4\ 10^9$ K), and can not be removed changing the velocity of the burning front or the initial amount of mass incinerated.

This work has been supported in part by the CICYT grants PB87-0304 and PB90-0912, and by a EASI/CESCA project.

REFERENCES

Bravo E., Isern J., Canal R., Labay J., 1992a, A&A 257, 534

Bravo E., Dominguez I., Isern J., Canal R., Höflich P., Labay J., 1992b, in prep.

Bravo E., Isern J., Canal R., 1992c, in preparation

Iben I., 1982, ApJ 259, 244

Khokhlov A.M., 1991, A&A 245, L25, and 245, 114

Livne E., Glasner A.S., 1990, ApJ 361, 244

Thielemann F.K., Nomoto K., Yokoi K., 1986, A&A 158, 17

Van den Bergh S., Tamman S., 1991, ARA&A 29, 363

Woosley S.E., 1991, in: Proceedings of the Les Houches School on Supernovae, eds. J. Audouze, S. Bludman, R. Mochkovitch and J. Zinn-Justin, Elsevier, Amsterdam, in press

Fe-peak elements in SNIa : Constraints from Spectral Modeling

P. RUIZ-LAPUENTE [1,2] AND A.V. FILIPPENKO [3]

[1] Dpto. de Astronomía y Meteorología. Facultad de Física. U.de Barcelona, [2] Institut d'Astrophysique de Paris, [3] Department of Astronomy. U.of California, Berkeley.

ABSTRACT

Theoretical modeling of the atmospheres of Type Ia SNs at nebular phases gives a clear diagnostic on the amount of ^{56}Ni synthesized in these events and puts constraints on the abundance of other Fe-peak nuclei. Such an analysis can help to discriminate among different models proposed to explain this type of explosions and should be taken into account when studying the contribution of Type Ia SNs to the chemical evolution of galaxies. Results obtained by the application of this analysis to different Type Ia events are presented and discussed.

1. INTRODUCTION

Spectral modeling of SNIa at late phases gives a clear idea of the yields of different nucleosynthetic products ejected in these explosions. In that phase (which begins about 6 months after explosion) the supernova shows a nebular-like spectrum where the emission comes out in forbidden lines of the main ions present at this stage: Fe and Co ions (Axelrod 1980, Meyerott 1980). In this late phase continuum emission is negligible. The treatment of radiative transfer can therefore be simplified.

In a previous work (Ruiz-Lapuente and Lucy 1992, hereafter RL) the use of a least-squares approach has been explored as a useful method for a mathematically-consistent determination of the nucleosynthetic yields of supernovae explosions, since it provides statistically-founded best estimates of the amount of the different products synthesised in the explosion. This method, as applied to SNIa, is in addition a way to achieve a simultaneous determination of nucleosynthesis, reddening and distance towards SNIa (RL). The method consists in an iterative determination of the best estimates of those quantities, obtained by minimizing the residuas of the theoretical spectrum with respect to the observed one. To that purpose, one calculates the spectral formation in the nebular conditions of type Ia at late phases (taking into account nonthermal processes in the ionization and solving the rate equations

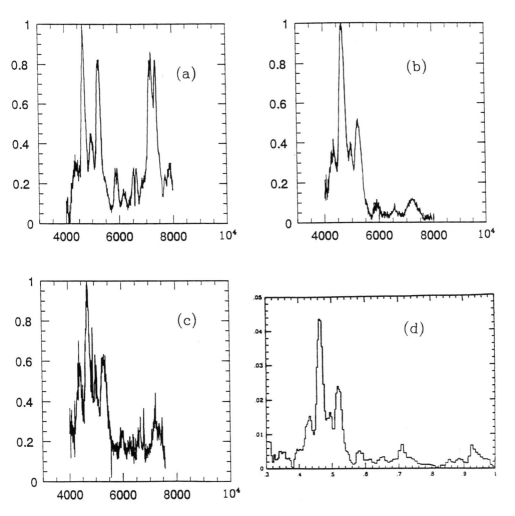

Figure 1. Spectra of SN 1986G in Cen A (a), SN 1991T in NGC 4527 (b) SN 1991aa in anonymous galaxy (c) and SN 1972E in NGC 5253 (d). See references in the text. The first three spectra have been normalized to the flux of the [FeIII] forbidden line at λ 4600.

for the populations of the energy levels of the different ions) and then applies a correcting reddening law to compare with the observations. In the following pages we
come back to the spectral information recently collected on SNIa and on the infer-
ences concerning nucleosynthesis obtained from the application of spectral modeling
techniques.

2. NEBULAR SPECTRA OF SNIA

Very few Type Ia supernovae have been observed in the nebular phase. Among those
with published spectra, two SNIa are in the Centaurus Group: SN 1986G in Cen A
(Cristiani et al. 1992) and SN 1972E in NGC 5253 (Kirshner and Oke 1974, Meyerott
1980). Recently, two other SNIa can be added to this list: SN 1991T in NGC 4527,
which showed Fe-dominated spectra at premaximum (Filippenko et al. 1992; Ruiz-
Lapuente et al. 1992) and has been subject of ample discussion as to the mechanism
of explosion; and also SN 1991aa, which is a SNIa discovered at premaximum (Mc
Naught, IAUC 5263) which showed a premaximum spectrum similar to the one of
SN 1990N at a similar phase (Phillips, IAUC 5267). SN 1991aa is located in an
anonymous galaxy beyond the Virgo Cluster, with a redshift of 0.01, thus being up to
now the farthest SNIa observed in the nebular phase (an account of the observations
of these SNIa will be given elsewhere).

It is our hope that a significant increase in the number of SNIa observed at late phases
will be achieved in the next future. In the next pages the cases of SN 1991aa and SN
1991T, SN 1986G and SN 1972E will be used for comparison between different SNIa.

3. RADIOACTIVE NI

Our analysis of SN 1986G (RL) pointed to a small quantity of ^{56}Ni synthesized in this
event, that being around 0.4 $M_\odot$. Recently Filippenko et al. (1992) have reexamined
this event and compared it with SN 1991bg another faint SNIa, and thus suggest that
those might correspond to "lazy deflagrations", which have not achieved complete
burning to the extention found in normal events.

The amount of ^{56}Ni derived from the least-squares analyses is larger for SN 1991T. It
is of the order of 0.7-0.8 $M_\odot$. In normal Type Ia such as SN 1972E this mass is about
0.5-0.6 $M_\odot$ as inferred from our analysis (RL). The case of SN 1991aa will correspond
to a "normal type Ia" following our results. The electron temperature is around 6800
K around 270 days after explosion in the case of clasical SNIa events and as high as
8000 K in the case of SN 1991T.

In Fig. 2 we present synthetic spectra for SN 1991aa and SN 1991T.

Thus from the set of supernovae analysed here our results indicate different amounts
of ^{56}Ni synthesized in the explosion. Whereas ^{56}Ni in normal Type Ia SNs is spread
out in the inner mass fraction covering up to 7000 km s^{-1} (see Table I), in SN 1991T
the ^{56}Ni mass is found up to velocities of 9700 km s^{-1}.

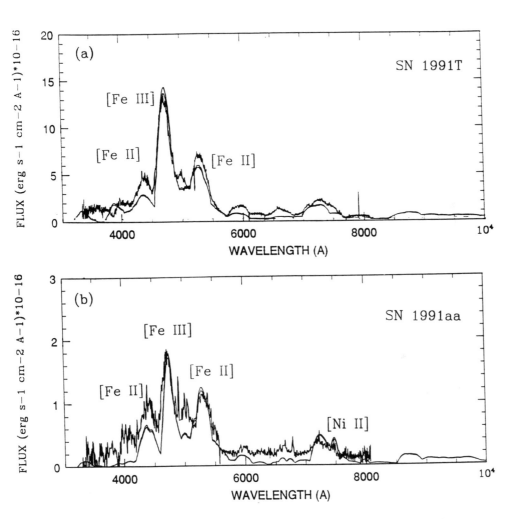

Figure 2. Synthetic spectra for SN 1991T 275 days after explosion (a) and for SN 1991aa, 250 days after explosion (b). The ionization stage and the electron temperature are higher for SN 1991T.

Table I

SNIa

	SN 1972E	SN 1991AA	SN 1991 T	SN 1986 G
v_{56}^*	$\simeq 7000.$	$\simeq 7100$	$\simeq 9800$	$\simeq 6800$
$^{56}Ni\ (M_\odot)$	0.5-0.6	$\simeq 0.6$	$\simeq 0.7$-0.8	0.4

[*] v_{56} stands for the maximum velocity (km s^{-1}) of the layers containing ^{56}Ni

4. STABLE NI

As discussed in RL, the spectra of SNIa can also give a clear indication of the amount of stable Ni synthesized in the explosion. The forbidden line at λ 7379.5 of NiII should be present if the amount of ^{58}Ni is of the order of that predicted for Type Ia events. An amount of about 0.04 $M_\odot$ is fitted with the use of a least-squares procedure to derive the amount of the different products ejected. In the case of SN 1991aa consistency with a NiII line has been obtained and the quantity derived would again correspond to a ratio of Ni/Fe $\simeq$ 0.1. A similar proportion is consistent with SN 1991T, although in this SNIa the large broadening of the lines prevents to discriminate the [NiII] emission from the rest of forbidden emission. The abovementioned ratio constitutes, to our best knowledge, the first determination of the Ni/Fe ratio in a sample of Type Ia SNs. This ratio is very important for the discussion of nucleosynthetic models of SNIa (see Table II for predictions of different models) and of chemical enrichment of the Universe, since the abovementioned Ni / ^{56}Fe ratio is approximately 2 times the solar one. This means that the excess ratio produced by SNIa should be compensated by a lower ratio in other SN types. The inspection of the Ni/Fe ratio should also be done in remnants of SNIa, SNIb, and SNII, with the use of a similar nebular analysis technique.

Table II

Stable Ni in SNIa

*	C6 [a]	W7N [b]	Kh2 [c]	Kh5 [c]
^{58}Ni	4.0e-02	1.3e-01	5.5e-02	6.5e-02
^{60}Ni	3.9e-03	1.4e-02	1.0e-02	1.2e-02
stable Ni	4.4e-02	1.44e-01	6.5e-02	7.7e-02
Ni/^{56}Fe	8.9e-02	2.2e-01	7.8e-02	8.4e-2
[Ni/^{56}Fe]	1.56	3.86	1.37	1.47

[a] From Nomoto, Thielemann and Yokoi (1984) [b] From the revised W7 nucleosynthesis (Thielemann, Nomoto and Yokoi 1986) [c] From Khokhlov (1990), models 2 and 5.

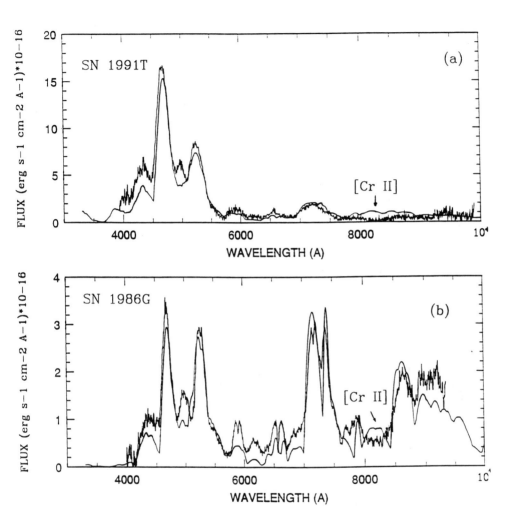

Figure 3. Synthetic spectra for SN 1991T (a) and SN 1986G (b) using the $Cr/^{56}Fe$ ratio from delayed detonation models (Khokhlov 1991). The lack of [CrII] lines at λ, λ 8000-8200 put constraints on this abundance.

5. WHAT ABOUT CR?

Nebular spectra of SNIa do also provide constraints as to the abundances of other elements present in the γ-ray deposition region. Among the elements of the Fe-peak group, Cr might be detected in the nebular phase, since the presence of small amounts of CrII might be revealed by forbidden emission of this ion, which would appear in the λ, λ 8000-8200 range. Table III shows the prediction of Cr abundances for different SNIa models and Fig. 3 shows the spectra obtained in the case of a $Cr/^{56}Fe$ ratio corresponding to delayed detonation models. The least-squares fits of these spectra point to a ratio $Cr/^{56}Fe \leq 0.01$, thus suggesting that this ratio is not higher than the solar one in Type Ia SNs.

Table III

Cr in SNIa ($M_\odot$)

*	C6 [a]	W7N [b]	Kh2 [c]	Kh5 [c]
^{50}Cr	2.4e-04	2.3e-04	3.3e-04	3.1e-04
^{52}Cr	5.4e-03	5.2e-03	2.6e-02	1.9e-02
^{53}Cr	2.1e-03	6.6e-04	1.8e-03	1.4e-03
^{54}Cr	2.4e-07	3.8e-05	7.7e-03	5.5e-03
Cr	7.7e-03	6.1e-03	3.5e-02	2.7e-02
$Cr/^{56}Fe$	0.015	9.7e-03	4.4e-02	2.9e-02
$[Cr/^{56}Fe]$	1.05	0.68	3.06	2.02

[a] From Nomoto, Thilemann and Yokoi (1984) [b] From the revised W7 nucleosynthesis (Thielemann, Nomoto and Yokoi 1986) [c] From Khokhlov (1990), models 2 and 5.

6. CONCLUSIONS AND FUTURE PROSPECTS

It thus seems very interesting to try to enlarge the sample of SNIa supernovae observed at late phases. The application of our method reveals that different amounts of ^{56}Ni can be synthethised in the explosion. The higher electron temperature found in SN 1991T as compared with that in SN 1986G and in normal Type Ia cleary points to a higher amount of ^{56}Ni synthesized in the explosion. The difference in the distribution of ^{56}Ni along the envelope reveals also that the ignition density can differ from one event to another, that giving rise to different burning conditions and final products. The nucleosynthesis predicted by deflagration models of SNIa, in what concerns the amount of ^{56}Ni synthesized, seems to be in agreement with the amounts derived by modeling the nebular phases of SNIa to reproduce the observations. Concerning the case of SN 1991T , the burning propagation could have been faster, giving rise to larger amounts of ^{56}Ni. Such behavior is predicted by delayed detonation models

(Khokhlov 1991, Woosley α Weaver 1991), late detonation models (Yamaoka et al. 1992) and fast deflagration models (see next contribution on this topic). A point to be taken into account are the velocities observed at premaximum on SN 1991T, which are lower than those predicted by some of these models and can, on the contrary, be in agreement with fast deflagration models (see discussion in this volume).

We are grateful to the Berkeley SN Team for observations of SN 1991T, and to R.Canal, F.Sánchez and A.Mampaso for their useful collaboration in the programme of SNIa observations at nebular phase in the William Herschel Telescope. P.R.L would like to thank R. Pelló and J.F.Le Borgne for implementing data analysis algorithms for reduction of spectroscopic data at the Department of Astronomy and Meteorology of Barcelona, and for helpful teaching of those procedures.

REFERENCES

Filippenko, A.V. et al., 1992. Ap.J. 384, L15.

Filippenko, A.V. et al., 1992 (submitted to AJ).

Khokhlov, A.M., 1991. Astron. Ap. 245, L25.

Kirshner, R.P. and Oke, J.B., 1975. Ap.J. 200, 574.

Meyerott, R.E., 1980. Ap.J. 259, 257.

Nomoto, K., Thielemann, F.-K., and Yokoi, K., 1984. Ap.J. 286, 644.

Ruiz-Lapuente, P. and Lucy, L.B., 1992. Ap.J. in press.

Ruiz-Lapuente, P. et al., 1992. Ap.J. 387, L33.

Thielemann, F.-K., Nomoto, K., and Yokoi, K., 1986. Astron. Ap., 158, 17.

Gamma-Ray Spectra from Fast Deflagration Models of SNIa

P.RUIZ-LAPUENTE [1], R.LEHOUCQ [2], R. CANAL [3] AND M. CASSÉ [2]

[1] Institut d'Astrophysique de Paris, [2] Service d'Astrophysique. CEA. Saclay, [3] Dpto. de Astronomía y Meteorología. Facultad de Física. U. de Barcelona.

ABSTRACT

We present the results of Monte Carlo calculations of γ-ray spectra for a grid of fast deflagration models of SNIa. The models here explored produce large amounts of ^{56}Ni as compared with standard deflagration models and close to the surface they give rise to incineration of the material up to NSE. The case of SN 1991T could be well explained by these models. Contrary to previous calculations of delayed detonation models (Khokhlov 1991) and late detonation models (Yamaoka et al. 1992), the velocities obtained in the material are consistent with that observed at premaximum, thus making such models better candidates to explain this SNIa. A discussion is made in view of the constraints obtained from GRO observations of SN 1991T.

1. INTRODUCTION

The chemical composition of the outer layers of SN 1991T, as revealed by premaximum spectra (Filippenko et al. 1992; Ruiz-Lapuente et al. 1992), poses a new challenge to model builders: it appears to be dominated by ^{56}Ni and its decay products and a quantitative analysis by means of a NLTE model atmosphere (Ruiz-Lapuente et al. 1992) indicates relative abundances consistent with the radioactive Ni isotope being synthesized at the time of the explosion. Furthermore, the layers immediately below, seen around maximum, show the intermediate-mass elements composition (Si, S, Ca) characteristic of partial burning and they overlay a core made again of material that has been in NSE (Filippenko et al. 1992; Phillips et al. 1992). Such a stratification of the chemical composition is in contrast with that shown by SN 1990N, another recent SNIa well observed before maximum (Leibundgut et al. 1991), where a mixture of Si, Ca, Fe, and Co was detected in the outermost layers.

Incineration to NSE up to the surface should happen when a detonation (supersonic burning) initiated at the centre of a C+O white dwarf (the standard SNIa progenitor) makes its way up to the outermost layers (Woosley & Weaver 1986). No intermediate-mass elements would, however, be synthesized anywhere. Delayed-detonation models (Khokhlov 1991; Woosley & Weaver 1991; Bravo 1991; Canal et al. 1991) can explain the presence of intermediate-mass elements close to the surface, with a NSE core below (and they can thus, in principle, fit a stratification such as that observed in SN 1990N), but they do not predict any composition inversion (Fe on top of Si) such as that observed in SN 1991T. A somewhat different approach, labelled "late detonation", has recently been proposed for that (Yamaoka et al. 1992): a carbon deflagration (subsonic burning) produces a central NSE core and an intermediate Si/S/Ca layer and later it changes into a detonation which incinerates the outermost layers. Depending on the density at which the transition from deflagration to detonation takes place, the surface layers can be either Fe-dominated (the SN 1991T case) or just Si/Fe-rich (the SN 1990 case). The Fe-dominated composition, however, corresponds to layers moving at velocities ranging from $\sim 15,000$ km s^{-1} up to $\sim 40,000$ km s^{-1} whilst the bulk of the matter in the outer Ni/Co/Fe layers of SN 1991T was moving at velocities $v_{exp} \sim 13,000$ km s^{-1} only.

2. FAST DEFLAGRATIONS

Here we explore another type of models, where detonation of the outer layers is induced by the shock waves propagating ahead of a "fast" (that is: only slightly subsonic) deflagration initiated at the centre of the electron-degenerate C+O core. Such a behaviour had been already obtained in model C8 of Nomoto et al. (1984) and in model 3 of Woosley & Weaver (1984), as well as in models of Canal (1986, unpublished).

We start from C+O cores ($X_C = X_O = 0.50$), in hydrostatic equilibrium and with masses $\simeq 1.4$ M$_\odot$. Models with central densities 1×10^9 (FD1), 2×10^9 (FD2) , and 3×10^9 g cm^{-3} (FD3) have been considered. This density range still only partially covers the full range that can result from the evolutionary history of mass-accreting C+O white dwarfs and it is just within the uncertainties associated with the effects of conductively driven Urca neutrino cooling in the standard scenario (Woosley & Weaver 1986). Explosive ignition is started by artificially rising the temperature of the central layer (before that it is $\simeq 2.5\times10^8$K). The subsequent evolution is followed by means of a one-dimensional, implicit hydrocode (Canal et al. 1990). The velocity of the burning front is here taken to be a constant fraction of the local sound speed: $v_{burn} = 0.5$ c$_s$. Details on the hydrodynamics and nucleosynthesis of these explosions will be given elsewhere (Canal et al., in preparation).

All three models detonate their outermost layers. It is interesting to note that the velocities of the incinerated material are in the range $v_{exp} = 13,000 - 28,550$ km s^{-1}, which is closer to the velocities actually observed in SN 1991T than those obtained by Yamaoka et al. (1992) with their "late detonation" scheme.

3. CALCULATIONS OF GAMMA-RAY SPECTRA

We have computed different spectra with a Monte Carlo approach based on the one developed by Ambwani and Sutherland (1988) and has been already applied by Lehoucq, Cassé and Cesarsky (1989) to calculate gamma-ray spectra of SN 1987A. In the present calculations we assume formation of positronium by the positrons originated in the ^{56}Co decay and we take into account the two and three photons decay (Bussard, Ramaty and Drachman 1979). The spectra of fast deflagration models of SNIa are calculated for day 73 and day 183 after explosion, which correspond to the middle point of the period in which GRO made some observations of SN 1991T (Lichti et al. 1992, hereafter L92). In Fig. 1 we show the spectra which correspond to model FD2, which is up to now our best proposed model to explain SN 1991T. This model synthesizes about 0.75 $M_\odot$ of ^{56}Ni in the inner core and about 0.01 $M_\odot$ of ^{56}Ni in the external layers. The velocities reached by the external layers are from 13000 km s^{-1} to 28550 km s^{-1}.

The observations done with COMPTEL (L92) failed to detect this SNIa, and this lack of detection put constraints to possible models proposed to explain SN 1991T for given distance estimates. Our calculations predict the photon flux for the observing period. In Figure 2 we present the integrated flux predicted for the 847 keV line and the 1238 keV line for fast deflagration models FD1, FD2, and FD3 for 73 days after explosion. For comparison, the same quantities are presented for other SNIa models, among them the W7 model of Nomoto, Thielemann, & Yokoi (1984), the delayed detonation model FDEFA1 from Woosley & Weaver (1991), and the late detonation model W7DT of Yamaoka et al. (1991). We normalize to a distance of 13.5 Mpc (bottom), which is the distance quoted by Tully (1988) to the host galaxy NGC 4527, and to 10 Mpc (top), which is a distance scale for which the current flux limits would give a clear discrimination among models (L92).

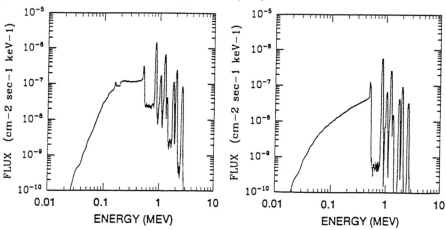

Figure 1: Theoretical spectra predicted by model FD2 for 73 days after explosion (left) and 183 days after explosion (right).

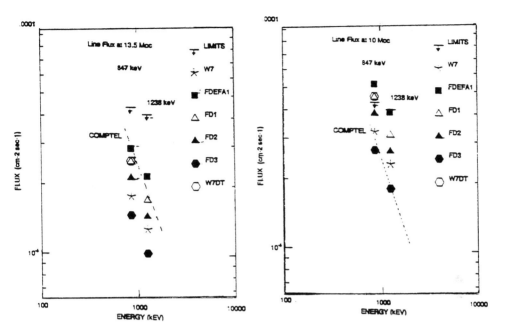

Figure 2.   Integrated flux predicted for the 847 keV line and the 1238 keV line adopting a distance to NGC 4527 of 13.5 Mpc.   The preliminary GRO flux limits (L92) are shown by an arrow. All predictions are consistent with these limits. The same quantities are given for a distance to NGC 4527 of 10 Mpc.   As can be seen some models such as the delayed detonation model FDEFA1 of Woosley & Weaver (1991) (see Burrows, Shankar, & Van Riper 1991, for predictions of the γ-ray line flux), the late detonation model W7DT of Yamaoka et al. (1992) (see Shigeyama et al. 1992 for predictions of the γ-ray line flux), and the fast deflagration model FD1 presented in this work are incompatible with a distance to NGC 4527 of 10 Mpc. On the contrary the fast deflagration models FD2 and FD3 presented in this work are compatible with the preliminar GRO upper limits for such a short distance.

4. DISCUSSION AND CONCLUSIONS

The critical question of the distance to NGC 4527 is restricting the possibilities of interpreting the lack of detection of gamma photons in the 847 keV line and in the 1238 keV line (L92). It has been shown that for a distance of 13.5 Mpc the current limits by GRO can not serve to discriminate among different fast deflagration models. For a shorter distance (around 10 Mpc), however, they would provide interesting constraints.

One of the main differences concerning the predicted gamma-ray spectra of the fast deflagration models here presented and the alternative delayed-detonation (DD) and late detonation (LD) models is the lower gamma-ray flux obtained for a similar ^{56}Ni mass. This fact is due to the lower velocities predicted in the external layers in fast deflagration models, which reduce the escape probability of gamma-ray photons in comparison with that of DD and LD models. A pitfall of DD and LD models as potential models for SN 1991T are the high velocities predicted for the intermediate-mass elements in the external layers. which are in the range of 40000 km s^{-1} and 50000 km s^{-1} for Ca and Si in LD models (Yamaoka et al. 1992), and of the order of 40000 km s^{-1} in DD models (Khokhlov 1991). These large velocities favour the escape of γ-ray photons. giving rise to a larger gamma-ray flux.

In counterpart fast deflagration models produce velocities in the range of 13000-28550 km s^{-1}. These models can be in better accordance with the range of velocities deduced from the blueshift of the P-Cygni absorption features observed at premaximum. and thus can be better potential models for SN 1991T.

ACKNOWLEDGEMENTS

We would like to thank G.G.Lichti for very valuable information and discussions concerning GRO results. P.R.L. wants to thank the hospitality and funding from the Institut d'Astrophysique de Paris. where this work was developed.

REFERENCES

Ambwani, K., & Sutherland, P., 1988, ApJ., 325. 820
Bravo, E. 1991, PhD Thesis (Barcelona: Univ. Barcelona)
Burrows, A., Shankar, A., & Van Riper, K. 1992. ApJ., 379, L7
Canal, R., Garcia, D., Labay, J., & Isern, J. 1990. ApJ, 356, L51
Canal, R., Isern, J., Bravo, E., & Labay, J. 1991, in SN 1987A and other Supernovae.
ed. L.J. Danziger & K. Kjär (Garching: ESO), 153
Filippenko, A. et al. 1992. ApJ, 384, L15
Khokhlov, A.M. 1991, A&A, 245, 114
Lehoucq, R., Cassé, & Cesarsky, C.J. 1989. A&A. 224, 117
Leibundgut, B. et al. 1991, ApJ, 371, L23
Lichti, G.G. et al. 1992. L92. To appear in A&A Supp.Ser.
Phillips, M.M. et al. 1992, AJ (in press).
Ruiz-Lapuente, P. et al. 1992, ApJ, 387, L33.
Shigeyama, T., Kumagai, S., Yamaoka, H., Nomoto, K., & Thielemann. F.-K. 1992.
To appear in A&A Supp.Ser.
Tully, R.B. 1988. Nearby Galaxies Catalog (Cambridge: Cambridge Univ.Press)
Woosley, S.E., & Weaver, T.A. 1984, in Stellar Nucleosynthesis, ed. C.Chiosi & A.
Renzini (Dordrecht: Reidel), 263
Woosley, S.E., & Weaver, T.A. 1986, ARA&A, 24. 205
Woosley, S.E., & Weaver, T.A. 1991, in Supernovae. ed. J. Audouze, S.Bludman. R.
Mochkovitch. & J. Zinn-Justin (Amsterdam: Elsevier), in press
Yamaoka, H., Nomoto, K., Shigeyama, T., & Thielemann. F.-K. 1992. ApJ, in press

Dense Matter and the Supernova Mechanism

JAMES M. LATTIMER AND F. D. SWESTY

Dept. of Earth & Space Sciences, SUNY–Stony Brook, Stony Brook, NY
11794-2100, USA

1 INTRODUCTION

Despite nearly three decades of attempts to model the mechanism for producing the
brilliant optical displays of type II supernovae, and in spite of considerable advances
in our understanding of gravitational collapse that have occurred in the last fifteen
years, the underlying mechanism remains elusive.

These supernovae are thought to originate from the collapse of the core of a massive
star at the end of its normal life. The iron core of the star becomes unstable and
collapses due to electron captures on the heavy nuclei. Capture reactions continue
during the collapse until the increasing density prevents the neutrinos from escaping
the core on collapse timescales. Even with this neutrino "trapping", which impedes
the consumption of electrons, the pressure of the matter remains insufficient to halt
the collapse until supernuclear densities are reached and nucleon degeneracy and
repulsive forces become dominant. The inner 2/3 of the core collapses faster than
the outer core (Yahil 1983); when the collapse is halted, the core rebounds, forming
a shock at the interface between the inner and outer cores. The rebounding core
is stabilized and becomes the bulk of a proto-neutron star. The shock propagates
into the infalling outer core, and dissociates the heavy nuclei in it. However, the
9 MeV per baryon energy of dissociation, together with neutrino losses from the
high-temperature shocked material, serve to weaken this shock as it propagates.

In a purely hydrodynamic theory known as the bounce-shock hypothesis, this shock
advances through the infalling matter beyond, reverses the infall, and ejects the
matter. However, work in the last decade showed that this process can only succeed
if the initial iron core is extremely small, if electron captures during the collapse are
relatively small, and if the equation of state (EOS) above nuclear density is quite
soft (see, *e.g.*, Cooperstein and Baron 1990). However, when certain observational
constraints are placed on the nuclear parameters which go into the EOS, it now

seems that the bounce-shock mechanism fails even under the most optimistic conditions (Swesty, Lattimer and Myra 1992). Current thinking is that late-time heating processes, due to neutrino heating and/or convective overturn, drive the supernova mechansim. In all the proposed processes, the EOS plays an important role. In this contribution, we discuss the constraints observations impose upon the underlying nuclear parameters of the EOS, and the results of applying these constraints to collapse calculations.

2 THE EQUATION OF STATE

2.1 Nuclear Parameters

For clarity, we summarize the most important nuclear parameters in the EOS. The energy can be conveniently be expressed in terms of a liquid drop expansion: $E_{Nucleus} = E_{Volume} + E_{Surface} + E_{Coulomb} + \ldots$, where

$$E_{Volume} = -B + (K_s/18)\left[1 - n/n_s\right]^2 + S_V\left(1 - 2Z/A\right)^2 + a_V T^2 + \ldots,$$
$$E_{Surface} = 4\pi r_N^2 \sigma/A = A^{-1/3}\left[4\pi r_o^2 \sigma_s - S_s\left(1 - 2Z/A\right)^2 + a_S T^2 + \ldots,\right] \quad (1)$$

and $E_{Coulomb} = 3Z^2 e^2/5r_o A^{4/3}$. Here, the nuclear radius is $r_N = (3A/4\pi n_s)^{1/3} = r_o A^{1/3}$. Typical values of the parameters are shown in Table 1 (n is the baryon density and T is the temperature).

Table 1: Nuclear Parameters ($n = n_s, Z/A = 0.5, T = 0$)

Symbol	Parameter	Approximate Value
n_s	saturation density	0.16 fm^{-3}
B	bulk binding energy	16 MeV
K_s	bulk incompressibility	160–400 MeV
S_V	bulk symmetry energy	27–35 MeV
a_V	bulk level density	0.06 MeV^{-1}
σ_S	surface tension	1.1 MeV fm^{-2}
S_S	surface symmetry energy	30–160 MeV
a_S	surface level density	0.1–0.25 MeV^{-1}
m^*/m	nucleon effective mass	0.7–1.1
T_c	critical temperature	12–20 MeV

2.2 Parameter Correlations

Several of these parameters are highly correlated and cannot be chosen independently, a fact often overlooked in some equations of state used for supernovae. Here, we discuss three: correlations between T_c and K_s, a_S and T_c, and S_V and S_S.

The critical temperature of bulk nuclear matter is determined by $\partial P/\partial n|_{Z/A=0.5} = \partial^2 P/\partial n^2|_{Z/A=0.5} = 0$, which generally implies

$$T_c \approx 20\sqrt{K_s/360\ \text{MeV}}\,(0.16\ \text{fm}^{-3}/n_s)^{1/3}\sqrt{m/m^*}\ \text{MeV}.$$

Therefore, a small value of K_s is only consistent with a correspondingly small value of T_c. This correlation affects the nuclear specific heat a through the surface level density parameter, defined by

$$a_S = -\left(4\pi r_o^2/2T\right)\partial\sigma/\partial T|_{x=0.5,T=0,n_s} \approx \left(54/T_c^2\right)\ \text{MeV}^{-1}.$$

The total nuclear specific heat is the sum of volume and surface contributions:

$$a = a_V + a_S A^{-1/3} \approx m^*/15m + .2A^{-1/3}\ \text{MeV}^{-1}.$$

Some of the discrepancy between the experimental values of a and the liquid drop values historically used can be resolved if T_c (and thus K_s) are relatively small.

In the *Liquid Drop Model*, the total nuclear symmetry energy is $E_{sym} = (1 - 2Z/A)^2(S_V - S_S A^{-1/3})$. It is straightforward to show that variations of S_V and S_S, in order to preserve the observed values of E_{sym} for ordinary nuclei, must be highly correlated. That is, if (S_{Vo}, S_{So}) is a typical set of symmetry energy coefficients, a new set (S_V, S_S) is related by

$$S_S \approx S_{So}(S_V/S_{Vo})^\alpha; \qquad \alpha = S_{Vo}A^{1/3}/S_{So} \approx 3 - 4.$$

If the nuclear symmetry energy is, instead, parametrized by the *Liquid Droplet Model*, in which $E_{sym} = (1 - 2Z/A)^2 S_V/[1 + S_S A^{-1/3}/S_V]$, the value of α is larger than that for the Liquid Drop model by exactly 2. This severely constrains the form of the symmetry energy. Some effort (Meyers *et al.*1977, Lipparini and Stringari 1982) has been made to extract the symmetry energy from the giant dipole resonance (GDR), but these efforts are hampered by complexities in the macroscopic models of the GDR.

2.3 Determination of Nuclear Incompressibility

The two major, as yet, uncorrelated parameters may thus be taken as K_s and S_V. The value of the incompressibility parameter is important both for its role above nuclear density and for its affect on the nuclear specific heat below nuclear density. The best hope for its evaluation seems to be from giant monopole resonance (GMR) data. The scaling model, which has been argued to give a reasonable representation of the GMR, predicts that the total nuclear incompressibility may be expanded in a liquid-drop fashion. Furthermore, it predicts that the Coulomb and surface contributions to the incompressibility, K_S and K_C, satisfy

$$
\begin{aligned}
K_C &= \left(3e^2/5r_o\right)\left(1 - 3K'_s/K_s\right), \\
K_S &= 4\pi r_o^2 \sigma_s \left(9\sigma'_s + 4 + 6K'_s/K_s\right),
\end{aligned} \tag{2}
$$

where $K'_s = n_s \partial K_s / \partial n |_{n_s, Z/A=0.5}$ and $\sigma'_s = (n^2/\sigma)(\partial^2 \sigma/\partial n^2)|_{n_s, Z/A=0.5}$. Pearson (1991) has shown that the linear relations

$$
K_C \approx 15.4 \pm 2 \text{ MeV} - 0.065K_s, \qquad K_S \approx 230 \pm 50 \text{ MeV} - 3.2K_s, \tag{3}
$$

may be established from the GMR data. The Coulomb relations in Eqs. (2) and (3) thus give a linear relation between K'_s and K_s, but not a value for K_s itself, since the data is so meager. However, by combining the surface relations in Eqs. (2) and (3), and using the values $r_o = 1.143$ fm and $\sigma_s = 1.15$ MeV fm^{-2}, we find

$$
K_s = 137.4 \pm 23.2 \text{ MeV} - 26.36\sigma'_s.
$$

The value of σ'_s can be calculated from any particular nuclear interaction, and will be a function of K_s and K'_s. Thus, in the scaling model, surface energy calculations, combined with experimental data, will yield an estimate of the incompressibility. For all Skyrme-type interactions this procedure yields $K_s \approx 312 \pm 40$ MeV. This is somewhat larger than the value established by Blaizot (1980), 210 ± 30 MeV, from RPA calculations and may show that the scaling model is an incomplete description.

3 THE PROMPT PHASE OF COLLAPSE

With an EOS (Lattimer and Swesty 1991) in which the nuclear parameters are consistently varied as discussed above, we have performed a series of collapse calculations using a 1-d relativistic radiation hydrodynamic code (based upon the work of Myra *et al.*1987 and Myra and Bludman 1989). One major result is that when the appropriate nuclear parameter correlations are included, and when the EOS is constrained by the PSR1913+16 mass limit (1.44 M$_\odot$), the shock generated by core bounce consistently stalls near 100 km, independently of the assumed nuclear incompressibility in the range 180–375 MeV and the volume symmetry energy in the range 27–35 MeV. This result holds even for equations of state which have maximum neutron star masses as small as 1.4 M$_\odot$. This result is in contrast to earlier simulations which did not include the correlations or constraints. Only when substantially softer equations of state, such as that of Cooperstein and Baron (1990), hereafter CB, with $K_s = 180$ MeV and $\gamma = 2.5$, are used does the shock initially propagate to larger radii. In addition, the final trapped lepton fraction is independent of variations in both K_s and S_V. The neutrino luminosities vary somewhat with K_s, however. Since K_s and a are inversely correlated, the temperature during collapse increases with K_s which leads to a larger electron capture rate and ν_e luminosity.

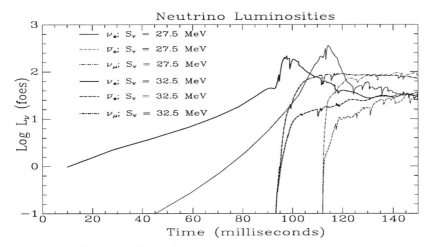

Figure 1. Neutrino lightcurves in the prompt phase.

In fact, the only significant differences observed are S_V variations in the entropy profiles and neutrino luminosities as a function of time (shown in Figs. 1 and 2). For comparison, we also plotted results for the CB EOS. With the CB EOS, the central temperature reaches 25 MeV, but only 15 MeV in the other cases, resulting from specific heat and EOS softness differences.

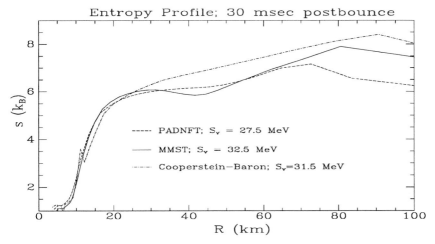

Figure 2. Entropy per baryon versus radius 30 ms after bounce.

Increasing S_V decreases $\hat{\mu} = \mu_n - \mu_p$ and increases the electron capture rate and ν_e luminosity; however, the collapse rate is also increased, neutrino trapping is acheived sooner, and the final lepton fraction doesn't change. It is possible that large neutrino experiments such as Super-Kamiokande or Sudbury may be able to observe such differences as those indicated in Fig. 1 from galactic supernovae.

The entropy profiles are shown in Fig. 2. Neutrino heating and convection are sensitive to the post-shock density, temperature and entropy profiles (Herant, Benz and Colgate 1992). Thus, for example, note the convectively unstable regions in which $ds/dr < 0$. These regions are noticeably larger in the large S_V cases.

We gratefully acknowledge support for this research from USDOE (grant No. DE-FG02-87ER40317) and the Pittsburgh and NERSC Supercomputing centers.

REFERENCES

Blaizot, J. P. 1980, *Phys. Rep.* **64**, 171.

Cooperstein, J. and Baron, E. 1990, in *Supernovae*, ed. A. Petschek (New York: Springer-Verlag), p. 323.

Herant, M., Benz, W. and Colgate, S.A. 1992, *Ap. J.*, in press.

Lattimer, J. M. and Swesty, F. D. 1991, *Nucl. Phys.* **A535**, 331.

Lipparini, E. and Stringari, S. 1982, *Phys. Lett.* **B112**, 421.

Möller, Meyers, W., Swiatecki, W.J. and Treiner, J. 1988, *At. Data Nucl. Data Tables* **39**, 225.

Myers, W.D., Swiatecki, W. J., Kodama, T., El-Jaick, E. L., and Hilf, E. R. 1977, *Phys. Rev.* **C15**, 2032.

Myra, E.S. and Bludman, S. 1989, *Ap. J.* **340**, 384.

Myra, E.S., Bludman, S.A., Hoffman, Y., Lichtenstadt, I., Sack, N. and Van Riper, K.A. 1987, *Ap. J.* **318**, 744.

Pearson, J. M. 1991, *Phys. Lett.* **B271**, 12.

Pearson, J.M., Aboussir, Y., Dutta, A.K., Nayak, R.C., Farine, M. and Tondeur, F. 1991, *Nucl. Phys.* **A528**, 1.

Swesty, F. D., Lattimer, J. M., and Myra, E. S. 1992, in preparation.

Yahil, A. 1983, *Ap. J.* **265**, 1047.

Nucleosynthesis in Nova Explosions

S. STARRFIELD[1], J.W. TRURAN[2], W.M. SPARKS[3],

M. POLITANO[1], I. NOFAR[4], G. SHAVIV[4]

[1] Department of Physics and Astronomy, Arizona State University, Tempe, AZ 85287-1504
[2] Department of Astronomy and Enrico Fermi Institute, University of Chicago, Chicago, IL 60637
[3] Applied Theoretical Physics Division, Los Alamos National Laboratory, Los Alamos, NM 87545
[4] Department of Physics and Asher Space Institute, The Technion, Haifa, 32000, Israel

ABSTRACT

This paper examines the consequences of accretion of hydrogen rich material onto ONeMg white dwarfs with masses of $1.0 M_\odot$, $1.25 M_\odot$, and $1.35 M_\odot$. Our results demonstrate that novae produce both ^{22}Na and ^{26}Al in astrophysically interesting amounts. Hot hydrogen burning on ONeMg white dwarfs can produce as much as 2% of the ejected material as ^{26}Al and 3% as ^{22}Na. The largest amount of ^{22}Na is produced by the highest mass ONeMg white dwarfs nova systems which are predicted to be the fastest and most luminous of all types of novae. The largest amount of ^{26}Al is produced in the lowest mass white dwarfs, which according to our evolutionary calculations, should eject the largest amount of material. In support of this prediction, the observations of QU Vul, a slow ONeMg nova, indicate that it has ejected about $10^{-3} M_\odot$. However, such a large amount of ejected mass is inconsistent with a typical ONeMg white dwarf mass $>$ $1.25 M_\odot$.

1. INTRODUCTION

Urey (1955) was the first to suggest that radioactive decay of ^{26}Al was important for the heating of the small bodies in the solar system. However, it was not until Lee, Papanastassiou and Wasserburg (1977) found that large excesses of ^{26}Mg were correlated with the ratio of ^{27}Al/^{24}Mg in the Allende meteorite, that it was demonstrated that ^{26}Al, the radioactive parent of ^{26}Mg, was present in the early solar system. Because its half life, 7.2×10^5yr, is short compared to a Hubble time, this result implied that the ^{26}Al must have been produced by some astrophysical process shortly before the formation of the solar system and then mixed with the pre-solar nebula just prior to its collapse. Its presence in the interstellar medium (ISM) was confirmed by the discovery, by HEAO-3, of the 1.809 Mev γ-ray line, which results from the decay of ^{26}Al to the first excited state of ^{26}Mg (Mahoney et al. 1982). Mahoney et al. also reported that the γ-ray photons appeared to originate from the general direction of the galactic plane. They later (Mahoney et al. 1984) used their detection to estimate that there was about $3 M_\odot$ of ^{26}Al in the ISM and examined the possibility that novae could produce this isotope. The Mahoney et al. results were confirmed by measurements with the SMM γ-ray spectrometer (Share et al. 1985). The discovery of the existence of ^{26}Al, both in meteorites and the ISM, has stimulated a variety of studies to determine both the astrophysical site and the mechanisms which could produce this isotope. An excellent review of the ^{26}Al problem can be found in Clayton and Leising (1987).

Outbursts of classical novae are caused by thermonuclear runaways (hereafter, TNR) proceeding in the accreted hydrogen-rich envelopes of the white dwarf components of nova binary systems (Truran 1982; Starrfield 1989; 1990). For the physical conditions of temperature and density that are expected to obtain in this environment, thermonuclear burning will proceed by means of hydrogen burning from either the proton-proton chain (early in the accretion phase) or the carbon, nitrogen, and oxygen (CNO) bi-cycle (late in the accretion phase and through the peak of the outburst). If there are heavier nuclei present in the nuclear burning shell, they will contribute to the nucleosynthesis and only a small amount to energy production. For solar composition material, energy production and nucleosynthesis from the CNO hydrogen burning reaction sequences imposes interesting constraints on the energetics of the runaway: in particular, the rate of nuclear energy generation at high temperatures (T > 10^8K) is limited by the timescales of the slower *and temperature insensitive positron decays*, particularly of ^{13}N (τ = 600s), ^{14}O (τ = 102s) and ^{15}O (τ = 176s). The behavior of the β^+-unstable nuclei holds important implications for the nature and consequences of classical nova outbursts. For example, significant enhancements of envelope CNO concentrations are required to insure higher levels of energy release on a hydrodynamic timescale (seconds) and thus produce a more violent outburst (Starrfield, Truran, and Sparks 1978; Truran 1982; Starrfield 1989).

The large abundances of the positron unstable nuclei have important and exciting consequences for the evolution. (1) Since the energy production in the CNO cycle comes from proton captures followed by a β^+-decay, the rate at which energy is produced, at temperatures exceeding 10^8K, depends only on the half-lives of the β^+-unstable nuclei and the numbers of CNO nuclei initially present in the envelope. (2) Since the convective turn-over time scale is usually about 10^2 sec near the peak of the TNR, a significant fraction of the β^+-unstable nuclei can reach the surface without decaying and the rate of energy generation at the surface will exceed 10^{12} to 10^{13} erg gm^{-1} s^{-1} (Starrfield 1989). This will produce a burst of γ-rays prior to optical maximum light (Starrfield et al 1992). (3) Since convection operates over the entire accreted envelope, it brings unburned CNO nuclei into the shell source, when the temperature is rising very rapidly, and keeps the CNO nuclear reactions operating far from equilibrium. (4) These nuclei decay when the temperatures in the envelope have declined to values that are too low for any further proton captures to occur, yielding isotopic ratios in the ejected material that are distinctly different from those ratios predicted from studies of equilibrium CNO burning.

2. THEORETICAL STUDIES OF O-NE-MG NOVAE

The existence of ONeMg white dwarfs (Starrfield et al. 1992) and the success of the one-zone nucleosynthesis calculations (Weiss and Truran 1990; Nofar, Shaviv, and Starrfield 1991), allow us to examine the consequences of accretion of hydrogen rich material onto such a white dwarf. We have done this, using our one-dimensional hydrodynamic computer code which incorporates a large nuclear reaction network to follow the changes in abundance of 78 nuclei (Kutter and Sparks 1972; Weiss and Truran 1990; Nofar, Shaviv, and Starrfield 1991; Politano et al. 1992). A detailed description of the current version of this code will appear elsewhere (Politano et al. 1992). Here, we present the results of one set of calculations.

In work reported in Politano et al. (1992), we evolved TNR's in accreted hydrogen rich layers of white dwarfs with masses of $1.0 M_\odot$, $1.25 M_\odot$, and $1.35 M_\odot$. The abundance results are given in Table 1. For all three white dwarf masses, we assumed that the rate of accretion onto the white dwarf was 10^{17}gm s^{-1} (1.6×10^{-9} $M_\odot$yr^{-1}). In all three sequences, we assumed an initial abundance of ONeMg nuclei equal to 50% of the envelope material (by mass). The remaining 50% consisted of a solar mixture of the elements. It is assumed that this composition results from the mixing of the accreted layers with core material.

The abundances of some of the nuclei, produced in our hydrodynamic simulations, are given in Table 1. Note, first, that the abundance of ^{26}Al declines while the abundance of ^{22}Na increases as the mass of the white dwarf increases. This suggests that the novae which exhibit the largest production of ^{26}Al may not be the same novae that produce enhanced ^{22}Na. In addition, as we proceed to higher white dwarf masses, the abundances of ^{31}P, ^{32}S, and ^{36}Ar increase to very large values. All of these nuclei must be produced as the result of "slow" (relative to the rates which dictate energy generation) proton captures on ^{24}Mg over the few minute lifetime of the explosion. We also call attention to the behavior of the light nuclei in our results. ^{12}C increases in abundance with white dwarf mass, but ^{16}O decreases in abundance. In contrast, the total neon abundance remains virtually constant. This could be the explanation of the puzzling feature that all of the observed ONeMg novae always show strong neon lines even when the Mg or Al lines are weak. When the actual abundances become available for some of the recent ONeMg novae (Her 1991, LMC 1990 #1, Sgr 1991, Nova Pup 1991, and Nova Cyg 1992), we will have much more information with which to test these predictions. However, the analyses of Nova Her 1991 imply that oxygen is depleted in this ONeMg nova while sulfur is strongly enhanced. We interpret these data as implying that the explosion occurred on a very massive white dwarf.

3. ABUNDANCES OF ^{22}NA AND ^{26}AL IN THE ISM

In this section we summarize the determinations of the amount of ^{22}Na and ^{26}Al in the ISM (see Weiss and Truran [1990] and Nofar, Shaviv, and Starrfield [1991] and references therein).

3.1 ^{22}Na

^{22}Na has a half-life of $\tau_{1/2} = 2.6$yr and produces a γ-ray of energy $E_\gamma = 1.275$ MeV in its decay to ^{22}Ne. The possibility exists that the γ-rays emitted by the ^{22}Na decay may be detectible by **GRO** from nearby ONeMg nova explosions. Based upon the results for ^{22}Na production that we obtained from our simulations, we find that the luminosity of a nova at this energy is:

$$L_\gamma = 9.4 \times 10^{33} \times (\frac{M_{env}}{10^{-5}M_\odot}) \times (\frac{X_{22_{Na}}}{10^{-3}}) \times e^{-\frac{t}{3.75}yr} \ erg \ s^{-1}$$

where $X(^{22}$Na$)$ is the mass fraction of ^{22}Na in the ejecta and M_{env} is the ejected envelope mass ($M_\odot$). An examination of the ^{22}Na abundance given in Table 1 indicates that this luminosity can reach $\sim 100L_\odot$ (Politano et al. 1992). Our results also predict that a $1.35M_\odot$ ONeMg nova can emit as much as 10^{34} erg s^{-1} in γ-rays if it ejects a mass of $10^{-5}M_\odot$. Here, we note that the presence of large amounts of ^{22}Na nuclei, in a barely optically thick (to γ-rays) expanding envelope, would Compton scatter a small fraction of them down in energy to hard X-ray wavelengths. The X-rays in both Her 1991 and N Cyg 1992 were observed at a time when the expanding envelope was optically thick and we suggest that some of the X-ray emission in this nova was caused by Comptonization of ^{22}Na γ-rays (Truran, Starrfield, and Sparks 1978; Starrfield et al. 1992).

The determinations of the amount of material ejected in the outburst of QU Vul 1984 range up to $10^{-3}M_\odot$ (Saizar et al. 1992). If the higher value is verified by other methods, then this nova, near maximum light, could have been as bright in γ-rays as it was in X-rays (Ögelman, Krautter, and Beuermann 1987). In any case, the abundance of this nucleus is high enough for us to include its decay as an energy source in the expanding nebula.

3.2 ^{26}Al

^{26}Al has a half-life of $\tau_{1/2} = 7.3 \times 10^5$ yr and we can reasonably expect this nuclei to be mixed through the ISM. Calculations of the abundance of this nucleus in the ISM required to reproduce the observed flux can be found in Mahoney et al. (1984), Weiss and Truran (1990), and Nofar, Shaviv, and Starrfield (1991). There are significant differences, however, in the predictions of the latter two papers: Weiss and Truran do not find more than 2×10^{-3} of ^{26}Al while Nofar, Shaviv, and Starrfield reported that the abundance of ^{26}Al reached 5%. Note, that the higher value is supported by our calculations which are given in Table 1. In fact, under some circumstances, virtually every ^{24}Mg nucleus present at the beginning of the evolution can be converted to ^{26}Al. Following Weiss and Truran (1990), the equation for estimating the amount of ^{26}Al in the ISM is:

$$M_{26}(Novae) = 0.4M_{\odot} \times (\frac{R_{nova}}{40yr^{-1}}) \times (\frac{F_{ONeMg}}{0.25}) \times (\frac{M_{ej}}{2\times10^{-5}M_{\odot}}) \times (\frac{X_{26}}{2\times10^{-3}})$$

where the various terms are the nova rate in the galaxy, the fraction of those novae that are ONeMg novae, the mean mass ejected per event, and the fraction of ejected mass in the form of ^{26}Al. The fraction of novae that are ONeMg novae has been discussed in a number of papers (Truran and Livio 1986; Ritter et al. 1991) with a result that ranges from 0.25 to 0.6. We predict that a lower mass white dwarf can accrete and, ultimately, eject more mass than a high mass dwarf. The range is probably from $\sim 10^{-4}M_{\odot}$ (1.0$M_{\odot}$ white dwarf) down to $10^{-8}M_{\odot}$ (1.38$M_{\odot}$ white dwarf). Observations suggest that the fastest ONeMg novae eject $\sim 10^{-6}M_{\odot}$ while QU Vul may have ejected about $10^{-3}M_{\odot}$. Finally, the amount of ^{26}Al produced in the outburst also depends upon white dwarf mass (see Table 1) with the lower mass white dwarfs producing the most ^{26}Al. This is a potentially important result, since it means that low mass novae, which eject the most mass, are the same novae which produce the largest amount of ^{26}Al. A nova such as QU Vul, with Al determined to be about 35 to 50 times solar (Saizar et al. 1992), and an ejected mass of $10^{-3}M_{\odot}$ could produce almost $10^{-6}M_{\odot}$ of ^{26}Al by itself. In contrast, the very fastest and most luminous ONeMg novae such as V693 CrA, LMC 1990 #1, and Sgr 1991, which we assume occur on very massive white dwarfs, may only eject $10^{-6}M\odot$ and the ^{26}Al mass fraction in the ejecta would be far smaller than in QU Vul. *Therefore, the entire amount of ^{26}Al in the galaxy could come from novae of the QU Vul type.* However, we do not wish to imply that we understand the entire picture since, for example, Nova Her 1991, which was very fast, probably ejected more material than the other fast ONeMg novae.

4. SUMMARY AND DISCUSSION

In this paper we have examined the consequences of accretion of hydrogen rich material onto ONeMg white dwarfs with masses of 1.0$M_{\odot}$, 1.25$M_{\odot}$, and 1.35$M_{\odot}$. These results, in combination with the one zone nucleosynthesis studies of Nofar, Shaviv, and Starrfield (1991), and Weiss and Truran (1990), have demonstrated that novae produce both ^{22}Na and ^{26}Al in astrophysically interesting amounts. Specifically: 1) Hot hydrogen burning on ONeMg white dwarfs can produce as much as 2% of the ejected material as ^{26}Al and 3% as ^{22}Na. 2) The largest amount of ^{26}Al is produced in the lowest mass white dwarfs, which according to our evolutionary calculations, should eject the largest amount of material. The observations of QU Vul, a slow ONeMg nova, indicate that it has ejected about $10^{-3}M_{\odot}$. 3) The largest amount of ^{22}Na is produced by the highest mass ONeMg white dwarfs nova systems. These novae are predicted to be the fastest and most luminous of all types of novae and the abundance of ^{22}Na is sufficiently high for its presence to be included in the energy budget of the ejected material. 4) There cannot be large numbers of QU Vul type novae in the galaxy since, if there were, the observed γ-ray emission from ^{26}Al would be far higher.

TABLE 1: EJECTED ABUNDANCES (MASS FRACTION)

SEQUENCE	1	2	3
MASS($M_\odot$)	1.00	1.25	1.35
X	.33	.30	.27
Y	.17	.19	.20
$^{12}C+^{13}C$ (10^{-2})	.94	4.3	3.4
$^{14}N+^{15}N$ (10^{-2})	2.0	2.5	8.8
$^{16}O+^{17}O$ (10^{-2})	11.8	7.0	1.1
$^{18}F+^{19}F$ (10^{-4})	1.7	4.0	25.9
$^{20}Ne+^{21}Ne+^{22}Ne$	.25	.23	.17
$^{22}Na+^{23}Na$ (10^{-3})	.53	6.4	35.2
$^{24}Mg+^{25}Mg+^{26}Mg$ (10^{-2})	5.9	3.8	4.4
^{26}Al (10^{-3})	19.6	9.4	7.4
^{27}Al (10^{-2})	1.4	1.6	1.9
$^{28}Si+^{29}Si+^{30}Si$ (10^{-2})	1.6	5.8	5.3
^{31}P (10^{-4})	.02	40.2	202.
^{32}S (10^{-4})	1.1	29.4	289.
^{36}Ar (10^{-5})	1.9	2.1	40.9

We emphasize that it is the integrated contribution over white dwarf mass that will determine the ^{26}Al returned to the ISM by nova outbursts. While higher mass ONeMg novae appear to produce less ^{26}Al and eject less material, they recur more frequently and the frequency weighted distribution rises very steeply toward higher white dwarf masses. Therefore, there must be a white dwarf mass in which the maximum amount of ^{26}Al is returned to the ISM and that mass has yet to be determined.

This work was supported in part by NSF and NASA grants to the University of Illinois and Arizona State University and by the DOE.

REFERENCES

Clayton D.D., and Leising M.D. 1987, *Physics Reports*, **144**, 1.
Kutter, G. S., and Sparks, W. M. 1972, *Ap. J.*, **175**, 407.
Lee T, Papanastassiou D.A., and Wasserburg G.J. 1977, *Ap. J. Lett.*, **211**, L107.
Mahoney W.A., Ling J.C., Jacobson A.S., and Lingenfelter R.E. 1982, *Ap. J.*, **262**, 742.
Mahoney W.A., Ling J.C., Wheaton, W. M, and Jacobson A.S. 1984, *Ap. J.*, **286**, 578.
Nofar, I., Shaviv, G., and Starrfield, S. 1991, *Ap.J.*, **369**, 440.
Ögelman, H., Krautter, J., and Beuermann, K. 1987, *A. & A.*, **177**, 110.
Politano, M., Starrfield, S., Truran, J. W., Sparks, W. M., and Weiss, A. 1992, in preparation.
Ritter, H., Politano, M., Livio, M., and Webbink, R. 1991, *Ap. J.*, **376**, 177.
Saizar, P. Starrfield, S., Ferland, G. J., Wagner, R. M., Truran, J. W., Kenyon, S. J., Sparks, W. M., Williams, R. E., and Stryker, L. L. 1992, *Ap. J.*, in press.
Share G.H., Kinzer R.L., Kurfess J.D., Forrest J.D., Chupp F.L., Rieger E. 1985, *Ap. J. Lett.*, **292**, L61.
Starrfield, S. 1989, in *Classical Novae*, ed. N. Evans and M. Bode (New York: Wiley), p. 123.
Starrfield, S. 1990, in *Evolution in Astrophysics: IUE Astronomy in the Era of New Space Missions*, ed. E. Rolfe (ESA SP-310; Noordwijk), p. 101.
Starrfield, S., Shore, S. N., Sparks, Sonneborn, G., W. M., Politano. M., and Truran, J. W. 1992, *Ap. J. Letters*, **391**, L71.
Starrfield, S., Truran, J. W., and Sparks, W. M., 1978, *Ap. J.*, **226**, 186.
Truran, J. W. 1982, in *Essays in Nuclear Physics*, eds. C. A. Barnes, D. D. Clayton and D. N. Schramm (Cambridge: Cambridge U. Press), p. 467.
Truran, J. W. and Livio, M. 1986, *Ap. J.*, **308**, 721.
Truran, J. W., Starrfield, S., and Sparks W. M. 1978, in *Gamma Ray Lines from Novae*, ed. T.L. Cline, and R. Ramaty, (NASA Tech. Memorandum #79619), p. 315.
Weiss, A. and Truran J. W. 1990, *A&A*, **238**, 178.

The Li production in novae revisited

H.M.J. BOFFIN, G. PAULUS, M. ARNOULD

Institut d'Astronomie et d'Astrophysique, C.P.165, Université Libre de Bruxelles
50 av. F.D. Roosevelt, B-1050 Brussels, Belgium

1 INTRODUCTION

This paper reexamines the possibility of ^{7}Li production in thermonuclear explosions, with special emphasis on novae. It updates the parametric study of Arnould and Norgaard (1975) and the detailed model calculations by Starrfield et al. (1978a), from which ^{7}Li was predicted to be produced in novae. We address this problem again because the reaction ^{8}B$(p, \gamma)^9$C was not taken into account in those previous calculations, and because new reaction rates have since become available.

2 THE MODEL

We use the same formalism as Arnould and Norgaard (1975). An exploding one zone model is adopted, the density and temperature of which increase suddenly to peak values ρ_o and T_o, and then decrease in time following

$$\rho = \rho_o \exp(-t/\tau_{ex})$$
$$T = T_o \exp(-t/3\tau_{ex}), \tag{1}$$

where τ_{ex} is the expansion time, which is varying between about 300 to more than 1000 seconds in the nova models of Starrfield et al. (1978b). According to those models, ρ_o varies in the range 10^3 to about $5 \ 10^4$ gcm^{-3}, while T_o lies between 10^8 and $5 \ 10^8$ K. We have checked numerically that the thermodynamic evolution to peak conditions has no influence on the ^{7}Li produced.

3 LI PRODUCTION IN EXPLOSIVE HYDROGEN BURNING

In realistic explosive hydrogen burning conditions, ^{7}Li is made as ^{7}Be, which decays into ^{7}Li by electron capture with a half-life of about 53 days after freeze-out of the nuclear reactions. The main reactions involved in the production and destruction of ^{7}Be are

$$\begin{array}{c} {}^3\mathrm{He}(\alpha,\gamma) \\ \end{array} \begin{array}{c} {}^7\mathrm{Be} \xrightarrow[\gamma \quad \mathrm{p}]{\mathrm{p} \quad \gamma} {}^8\mathrm{B} \begin{cases} (\beta^+) \ 2 \ {}^4\mathrm{He} \\ (\mathrm{p},\gamma) \ {}^9\mathrm{C} \\ (\alpha,\mathrm{p}) \ {}^{11}\mathrm{C} \end{cases} \\ {}^7\mathrm{Be} \ (\alpha,\gamma) \ {}^{11}\mathrm{C} \ . \end{array} \qquad (2)$$

The final yield of ^{7}Be thus results from a balance between its production by alpha capture on ^{3}He, and its destruction by either proton or alpha captures. As discussed in detail by Arnould and Norgaard (1975) , the proton captures on ^{7}Be are, in most realistic cases, in equilibrium with the ^{8}B photodisintegration, leading to a so-called hot p-p nuclear pattern. In such conditions, any ^{8}B destruction necessarily leads to an equivalent ^{7}Be depletion. From that equilibrium, and taking into account the ^{8}B destruction channels indicated in (2), one can define an effective ^{8}B destruction timescale by

$$\tau_{\mathrm{eff}}(^8\mathrm{B}) = \left(1 + \frac{\tau_{\mathrm{p}}(^7\mathrm{Be})}{\tau_{\gamma}(^8\mathrm{B})}\right)\left(\frac{1}{\tau_{\mathrm{p}}(^8\mathrm{B})} + \frac{1}{\tau_{\alpha}(^8\mathrm{B})} + \frac{1}{\tau_{\beta}(^8\mathrm{B})}\right)^{-1}, \qquad (3)$$

where $\tau_{\gamma}(^8\mathrm{B})$ and $\tau_{\beta}(^8\mathrm{B})$ are the ^{8}B lifetimes against photodisintegration and β-decay, the other τ's indicating lifetimes against proton or α-particle captures.

On such grounds, Fig. 1 delineates the area in the $T_o - \rho_o$ plane where the ^{7}Be production should be most efficient. The way that area is constructed is discussed in great detail by Arnould and Norgaard (1975). The main differences with the results reported here concern changes in the rates of the reactions of sequence (2), due consideration of the $^8\mathrm{B}(\mathrm{p},\gamma)$ and $^8\mathrm{B}(\alpha,\mathrm{p})$ channels, as well as different choices of τ_{ex}, that was adopted proportional to the hydrodynamic free expansion timescale $\tau_{\mathrm{HD}} = 446\rho_o^{-1/2}$ s by Arnould and Norgaard (1975).

4 NUMERICAL RESULTS

4.1 General exploration of the T_o, ρ_o plane
Numerical calculations of the ^{7}Be yields have been performed for the thermodynamic conditions described in Sect. 2. Use has been made of a reaction network including nuclei from H to Ar linked by 313 nuclear reactions or β decays. The latest available reaction rates have been adopted. For $^8\mathrm{B}(\mathrm{p},\gamma)^9\mathrm{C}$ and $^8\mathrm{B}(\alpha,\mathrm{p})^{11}\mathrm{C}$, the rates of Wiescher et al. (1989) have been selected.
The numerical predictions are in very good agreement with the qualitative analysis presented in Sect. 3. In particular, no significant ^{7}Be production is obtained for densities above roughly 1000 gcm^{-3} .

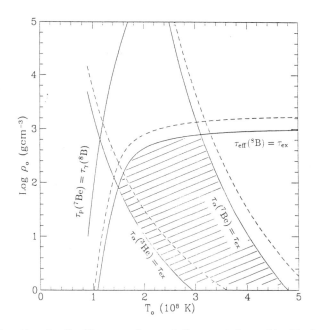

Fig. 1.   *Delineation in the $T_o - \rho_o$ plane of the most favorable (dashed) area for ^7Be production. Solid and dashed curves correspond to $\tau_{ex}=300$ s and $\tau_{ex}=100$ s, respectively.*

4.2 Importance of the ^{3}He abundance

In their evaluation of the possible role of novae in the galactic ^{7}Li enrichment, Starrfield et al. (1978a) assumed that the nova ^{7}Be yields are just proportional to the (quite uncertain) initial ^{3}He concentration. From calculations performed for ^{3}He abundances in the range $2\ 10^{-5} \leq X(^3He) \leq 2\ 10^{-3}$, we obtain that, independently of τ_{ex}, the ^{7}Li overabundance x is given by

$$\frac{x}{x_o} = 1.56 \log \frac{^3He}{^3He_\odot} + 0.88, \tag{4}$$

where x_o is the lithium overabundance derived with initial solar ^{3}He. The non-linear dependence of x on ^{3}He is due to $^3He(^3He, 2p)\ ^4He$, which acts as a more and more important leak as the initial ^{3}He abundance increases.

4.3 Importance of the ^{8}B (p, γ) ^{9}C reaction

The main reason for the destruction of ^{7}Be at high density is the prominence of $^8B(p, \gamma)\ ^9C$. Figure 2 shows the impact of that reaction on the final ^{7}Be yields for a given set of thermodynamic conditions. It is seen that ^{7}Be is destroyed very soon (after ~ 1 s) when $^8B(p, \gamma)\ ^9C$ is considered while its survives for ~ 100 s (which is of the order of τ_{ex}) when that reaction is neglected. In this case, the late ^{7}Be destruction is due to ^{7}Be$(p, \gamma)^8$B$(\beta^+)2\,^4$He. That reaction sequence develops because ^{8}B$(\gamma, p)^7$Be freezes out earlier than the ^{7}Be proton capture (such an effect cannot be predicted by the qualitative analysis of Sect. 3).

It has to be emphasized that the ^{8}B$(p, \gamma)^9$C rate is still uncertain. A microscopic cluster model (Descouvemont 1992) predicts a rate that is ~2.5 times lower than the evaluation of Wiescher et al. (1989). The results of Figs. 1 and 2 derived with the latter rate are not drastically affected by such a change. It would be important to accertain the ^{8}B$(p, \gamma)^9$C rate through further theoretical or experimental work.

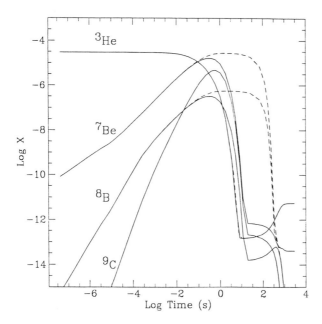

Fig. 2.   *Time evolution of the abundances of several nuclei for a given set of T_o (3 10^8 K) and ρ_o (10^4 gcm^{-3}). Solid and dashed lines are obtained when ^{8}B$(p, \gamma)^9$C is considered (with the rate of Wiescher et al. 1989) and neglected, respectively.*

4.4 Real nova models

We have also examined the ^{7}Be synthesis that could develop in detailed nova models. This has been done with the help of the one zone model approximation discussed in Sect. 1, the corresponding thermodynamic conditions mimicking those at the base of the thermonuclear runaway layers predicted by the full models 1,2, 4 and 8 of Starrfield et al. (1978b).

The qualitative considerations of Sect. 3 argue against a significant ^{7}Be production in those cases, the predicted peak densities being too high. This is fully confirmed by our numerical calculations, the derived final mass fractions lying in the $-13.5 \leq \log X(^7Be) \leq -11$ range. Taken at face value, such a result forces to conclude that those model novae cannot contribute in a significant way to the galactic ^{7}Li enrichment.

5 CONCLUSION

Our numerical reexamination of the possibility of ^{7}Be synthesis in explosive hydrogen burning, complemented with a general qualitative analysis of the situation, leads us to conclude that low enough peak densities and/or short enough timescales to avoid as much ^{8}B destruction as possible are required in order to produce a ^{7}Be yield that is a significant fraction of the initially available ^{3}He amount.

Those constraints imply in particular that one zone models approximating realistic nova models yield negligible ^{7}Be abundances. In order to see if the conclusions of Starrfield et al. (1978a) concerning the possible important contribution of novae to the galactic ^{7}Li have to be drastically revised, predictions from "complete" nova models are urgently called for.

Acknowledgements. This work has been supported in part by the SCIENCE Program SCI-0065, and by a Programme International de Collaboration Scientifique (PICS). M.A. is Chercheur Qualifié F.N.R.S. (Belgium).

References:

Arnould M., Norgaard H., 1975, A&A 42, 55

Descouvemont P., 1992, preprint

Starrfield S., Sparks W., Truran J.W., Arnould M., 1978a, ApJ 222, 600

Starrfield S., Truran J.W., Sparks W., 1978b, ApJ 226, 186

Wiescher M., Görres J., Graff S., Buchmann L., Thielemann F.-K., 1989, ApJ 343, 352

Nuclear Gamma-Ray Line Astronomy

M. CASSÉ[1] AND N. PRANTZOS[1,2]

1. Centre d' Etudes de Saclay, CEA, France
2. Institut d'Astrophysique de Paris, CNRS

1. INTRODUCTION

Gamma-ray lines constitute the most genuine diagnostic tool of nuclear astrophysics, since they allow for an unambiguous identification of isotopic species. Theoretical development in the field started more than 25 years ago, and H. Reeves left his mark here also by estimating the (still unobserved!) γ-ray line emissivity of the interstellar medium nuclei, excited by low-energy cosmic rays (Meneguzzi and Reeves 1975). Progress in the seventies has been quite slow, mainly because of experimental difficulties due (among other things) to the small number of γ-photons reaching the Earth. The situation improved dramatically in the eighties, due to several remarkable discoveries (see next Section) and perspectives for the nineties are even brighter, after the launch of the Compton (GRO) Observatory in April 1991, which has already provided a wealth of new data.

In this short review we focus on the nuclear gamma-ray lines that have been observed up to now: the 511 keV line of positron annihilation and the 1809 keV line of ^{26}Al decay, that have been detected in the plane of our Galaxy; and the ^{56,57}Co decay lines at 847 keV, 1238 keV and 122 keV that have been observed in SN1987A. After a short historical introduction (Sec. 2), the production sites of positrons and ^{26}Al are presented in Sec. 3 and 4, respectively, while the corresponding flux profiles as a function of galactic longitude and latitude are discussed in Sec. 5. In Sec. 6 we comment on the recent observational results on SN1987A and other candidate sources.

2. HISTORICAL BACKGROUND

The detectability of gamma ray lines from the decay of radioactive nuclei produced in supernova explosions was investigated more than 25 years ago (Clayton and Craddock 1965): it was based on the hypothesis that the decay of radioactive ^{254}Cf (an r-process isotope with a lifetime of $\sim$2 months) powers the light curves of supernovae. Subsequent nucleosynthesis studies (Bodansky et al. 1968) showed that the abundant nucleus ^{56}Fe is likely to be synthesized in supernova interiors as radioactive ^{56}Ni, decaying into ^{56}Co in ~7 days; this gave support to the idea of Pankey (1962) that ^{56}Co (lifetime ~77 days)rather than ^{254}Cf powers the late light curves of supernovae. These developments led to the first calculation (Clayton et al. 1969) of the escaping gamma ray spectra resulting from the decay of ^{56}Ni and ^{56}Co from Type I supernovae (SNI). Several years later, similar calculations performed for Type II supernovae (SNII) showed that the expected signal was likely to be much fainter than for SNI, due to the larger mass and the slower expansion of the ejecta of SNII (Woosley et al. 1980).

The appearence of SN1987A in the Large Magellanic Cloud (LMC) confirmed in a spectacular way those early ideas, bringing at the same time new challenges to modelers of stellar explosions (e.g. Arnett et al. 1989 for a review). The shape of the light curve after day 120, combined with the well known distance to LMC (50±5 kpc), convincingly showed that ~ 0.075 M$_\odot$ of ^{56}Co were the main source of light of the supernova, at least until day 900 (see Sec. 6). The definitive confirmation came from the detection of the 0.847 and 1.238 MeV lines of ^{56}Co decay (Matz et al. 1988), slightly preceded by the hard X-rays of the corresponding Comptonization continuum (Dotani et al 1987; Sunyaev et al. 1988); their appearance several months earlier than predicted suggested that mixing and/or fragmentation had taken place in the mantle of SN1987A during the explosion or shortly after. To complete the panorama, the detection of the 122 keV line of ^{57}Co in SN1987A has been reported recently by the GRO team (Kurfess et al. 1992).

SN1987A is not the first observed manifestation of cosmic radioactivity: the story started ten years earlier, with the identification of the 511 keV line in the direction of the galactic center (Leventhal et al. 1977). Observations of that emission with different instruments in the eighties have shown an increase of the recorded flux with the detector's aperture (Gehrels 1991). This result is now commonly interpreted in terms of a two-component galactic emission: a compact and variable component, located in the galactic center region, and a diffuse one, distributed essentially in the galactic plane (e.g. Ramaty and Lingenfelter 1991). The compact source is probably one of the sources recently resolved by the SIGMA experiment, like 1E1740.7-2942 (Paul et al. 1991), presumably a black hole producing positrons by high energy processes in its vicinity. As for the diffuse component, which has been recently "mapped" by GRO (Purcell et al. 1991; see Sec. 5) radioactivity of unstable nuclei produced in supernova explosions seems to be the most important source of the ~2 10^{43} e$^+$/s suggested by the observations, although the parent nuclei of the positrons have not yet been identified.

The first unambiguous identification of a cosmic radioactivity came in the early eighties, when the 1809 keV line of ^{26}Al was detected by the HEAO-3 satellite (Mahoney et al. 1984), at a level of $\sim$4 10^{-4} ph/cm^2/s/rad from the galactic center direction. The observations of the last few years have clearly demonstrated that the 1809 keV emission is diffuse in the galactic plane (Schoenfelder and Varendorff 1991), like the second component of the 511 keV emission. This is due to the timescales of the corresponding processes ($\sim$$10^5$ years for positron annihilation and $\sim$$10^6$ years for ^{26}Al decay), which are large compared to the timescale of occurence of supernovae (<100 years) or other potential sources of e^+ or ^{26}Al in the Galaxy. On the other hand, those timescales are short compared to the timescale of galactic evolution (a few billion years), a clear indication that nucleosynthesis is currently active in the Galaxy. Despite the many theoretical efforts, however, it is still difficult to justify the observationally derived $\sim$2-3 M$_\odot$ of ^{26}Al/10^6 years in the Galaxy, or to identify the parent nuclei of the galactic positrons.

3. GALACTIC SOURCES OF POSITRONS

Positron production can take place in various astrophysical sites (see Ramaty and Lingelfelter 1991 for a review). Subsequent annihilation with electrons proceeds either directly (giving two 511 keV photons) or through positronium formation (in which case, either two 511 keV photons or three continuum γ-ray photons are obtained). The complex physics of positronium annihilation in various astrophysical media has been recently reviewed by Guessoum et al. (1991). The positronium fraction (derived from the 511 keV/continuum ratio) of the observed diffuse 511 keV flux indicates an annihilation rate of $\sim$2 10^{43} e^+/sec. The most important source of this activity seems to be positron emission from unstable nuclei produced in supernova explosions (TABLE 1).

The decay chain ^{56}Ni $\longrightarrow$^{56}Co $\longrightarrow$^{56}Fe in SNIa (exploding white dwarfs) is favoured by the large mass of ^{56}Ni produced in such explosions (typically $\sim$0.6 M$_\odot$ per SNIa, according to theoretical models). Its actual role is, however, obscured by the uncertainties affecting the escape of positrons from the supernova during the relatively short lifetime of the parent nuclei ($\tau_{1/2}$ $\sim$6.1 days for ^{56}Ni and $\sim$77.3 days for ^{56}Co). The escaping fraction is very difficult to estimate, since it depends on various unknown factors, like the strength and topology of the ambient magnetic field etc. In the less favourable case, only 0.5% of the released positrons are found to escape from an SNIa (Chan and Lingelfelter 1990), but this fraction could be as high as 10-15 %. Positrons from the same decay chain in SNII (exploding massive stars) are less interesting, because of the larger envelope mass of SNII and the smaller Ni quantity produced in these explosions (typically $\sim$0.1 M$_\odot$ per SNII).

The radioactive chain ^{44}Ti $\longrightarrow$^{44}Sc $\longrightarrow$^{44}Ca is another interesting candidate, since the relatively long lifetime of ^{44}Ti ($\tau_{1/2}$ ~47 years) allows the positrons to be emitted in a rarefied environment, facilitating their escape before annihilation. There is, however, a considerable uncertainty on the amount of ^{44}Ti produced by supernovae: the estimated yields span the range 10^{-5} - $5\ 10^{-4}$ M$_\odot$ for SNIa, depending on the nature of the explosion (He-detonation models give larger amounts than C-deflagrations), and 0.5-2 10^{-4} M$_\odot$ for SNII (see Mahoney et al. 1992 for references).

Assuming that the current frequency of supernova in our Galaxy is ~3. per century for SNII and 0.6 per century for SNIa (Tutukov et al. 1992), one may evaluate the corresponding rate of injection of ^{56}Ni to ~0.6 M$_\odot$ per century; more than half of that amount comes from SNIa and should give positrons detectable through their 511 keV line. Assuming, furthermore, that supernovae (SNIa+SNII) produce a solar ^{44}Ca/^{56}Fe ratio (~ $1.2\ 10^{-3}$ by mass), leads to ~$0.7\ 10^{-3}$ M$_\odot$ of ^{44}Ti per century. With an escape factor of ~1 for positrons the decay of ^{44}Ti can account for almost half of the observed annihilation rate (~$2\ 10^{43}$ e$^+$/s), while the decay of ^{56}Co can explain all of it *if* the escape fraction is larger than ~0.04.

The only radioactive nucleus that is definitely known to be currently synthesized in the Galaxy, at a rate of ~2-3 M$_\odot$ per 10^6 years is the long lived ^{26}Al (see next Sec.) but the positrons from its decay can account for less than ~30% of the observed diffuse 511 keV emission. Finally, all the other candidate sources of positrons (other radioactive nuclei like ^{22}Na, pair production in pulsars or γ-ray bursts etc.) suffer from large theoretical uncertainties, and their estimated positron production in the Galaxy is smaller than the one of ^{26}Al. Considering that the observed diffuse 511 keV flux results only from the three decay chains presented here (i.e. ^{26}Al, ^{44}Ti, and ^{56}Ni) and taking into account the relative supernova frequency in the Galaxy, Lingelfelter et al. (1992) derived recently an interesting upper limit to the current rate of galactic nucleosynthesis of Fe: ~1 M$_\odot$/100 years.

4. PRODUCTION OF ^{26}AL IN THE GALAXY
There is no lack of candidate astrophysical sources for the production of ^{26}Al in the Galaxy. The main production mechanism is the reaction ^{25}Mg(p,γ)^{26}Al, which can take place in various astrophysical environements (see Clayton and Leising 1987, Prantzos 1987, 1991 for reviews).

Hydrostatic core H-burning may produce important amounts of ^{26}Al if the central temperature of the star is T$_c$ > 35-40 10^6 K, i.e. for stars with mass M>30 M$_\odot$. ^{26}Al is then ejected by the strong stellar winds during the so called *Wolf-Rayet* (WR) phase. Several independent calculations have shown that a WR star can produce a few 10^{-5}-10^{-4} M$_\odot$ of ^{26}Al, depending of its mass; these numbers increase somewhat if the effects of the initial stellar metallicity (z) are taken into account (the yield of a WR star scaling roughly as $\propto z^{1.7}$). On the basis of recent theoretical models it turns out that the galactic population of WR stars can produce 0.2-0.3 $M_\odot$ of ^{26}Al in the last 10^6 years (e.g. Prantzos 1991), and even up to 0.5 $M_\odot$ if the preliminary results of Meynet and Arnould (this volume) are confirmed.

Hydrostatic shell H-burning, at temperatures T$\sim$70-90 10^6 K, can produce substantial quantities of ^{26}Al in intermediate mass stars, during their *asymptotic giant branch* (AGB) phase (a poorly known phase of stellar evolution). According to the only full-scale calculation available up to now (Paulus and Forestini 1991), a 3 $M_\odot$ AGB star may eject, on average, $\sim$5 10^{-9} $M_\odot$ of ^{26}Al/10^6 years. The contribution of AGB stars to the galactic ^{26}Al is then evaluated to $\sim$0.03 $M_\odot$ per 10^6 years.

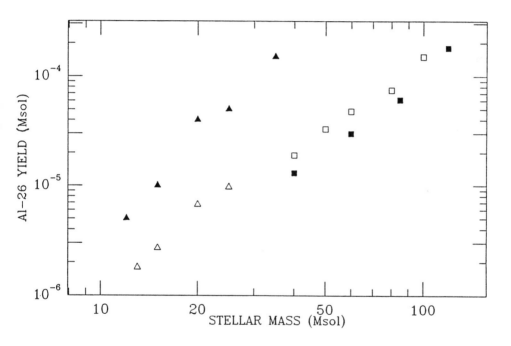

Figure 1: Mass of ^{26}Al ejected by massive stars as a function of stellar mass, according to various calculations. *Triangles*: ejection at the end of the stellar life, with the SNII explosion; *Squares*: ejection by the strong stellar wind during the Wolf-Rayet phase of the more massive stars (from Prantzos 1991).

Explosive H-burning at peak temperatures T~200-400 10^9 K during *nova* explosions can also produce large amounts of ^{26}Al (e.g. Leising 1991 for a review). The main uncertainty here is the treatment of convection, since ^{26}Al can be convectively transported in cooler layers and ejected in the ISM before destruction. The most recent and self-consistent calculations (Starrfield et al. this volume), including a nuclear network coupled to hydrodynamics, indicate a large production of ^{26}Al by low-mass O-Ne-Mg novae, considerably larger than previous estimates. In that case the galactic nova population could produce all the observationally derived ^{26}Al in the Galaxy (see, however, Sec. 5).

Finally, ^{26}Al can be produced during *hydrostatic*, or *explosive C-* or *Ne- burning* in massive stars, ending their lives as *SNII* . Notice that the treatment of convection is again crucial in evaluating the amount of ^{26}Al produced before the explosion, in the convective C and Ne shells; but, equally important are other parameters, like e.g. the ^{12}C(α,γ) reaction rate, determining the position and extent of those shells inside the star. Adopting the most recent and complete set of calculations (Woosley 1991, 1992 for stars with mass between 12 and 35 $M_\odot$) one finds that the galactic SNII can produce up to 0.5 -1.0 $M_\odot$ in the last 10^6 years, the larger number being obtained when the yields are extrapolated up to 100 $M_\odot$ stars.

It seems then that none of the candidate sources can provide the 2-3 $M_\odot$ of ^{26}Al required by the observations (baring the recent results of Starrfield; see next Sec.). Notice, however, that the uncertainties affecting the yields of novae and AGB stars are much larger than the ones of SNII (which are uncertain by a factor of at least ~3-4), while those for WR stars are rather small (~50%). In our opinion, it is premature to conclude that there is a real problem of "missing sources" of ^{26}Al; perhaps the problem lies in the, still poor, modelisation of the relevant sites.

5. DIFFUSE GALACTIC EMISSIONS AT 511 KEV AND 1809 KEV

The discussion of Sec. 3 and 4 shows that relatively old objects (i.e. of age >10^9 years) like SNIa, are the most probable sources of positrons in the Galaxy, while both young objects (massive stars of age ~10^7 years) and old ones (novae) are the most serious candidates for the production of ^{26}Al. In view of the existing uncertainties in the relevant yields it has been suggested that observations with good angular resolution could help, mapping the galactic distribution of the 511 keV and 1809 keV emissions, and revealing the distribution and nature of the corresponding sources.

Unfortunately, the galactic distributions of all those candidate sources are not well known. Direct estimates, based on extrapolations of local distributions (or on observations of novae or WR stars in nearby galaxies) are quite uncertain. Indirect attempts have thus been made, and several tracers of old or young populations have been proposed, based on observations in different wavelengths (e.g. Leising and Clayton 1985, Skibo et al. 1992). Such observations, combined to some straightforward theoretical considerations, lead to the following general expectations:

a) The emissivity profile as a function of galactic longitude (l) is expected to be considerably "peaked" in the GC direction in the case of old objects, because of the predominant contribution of the galactic bulge. On the other hand, the longitude profile of a young population is not expected to be peaked in the GC direction: indeed, current star formation seems to be rather weak in the inner galactic regions (despite the large quantities of molecular hydrogen there; Gusten 1989) and there is no bulge contribution. Moreover, the spiral pattern of a young stellar population should give rise to asymmetries (or spikes in the direction of the spiral arms) in the corresponding longitude profile (see Prantzos 1991, 1993)

b) the latitude profile of a young stellar population should be very peaked around $l=0°$, since the corresponding scale height is quite small (h_z <80 pc); old objects have a larger scale height (>200 pc) and the corresponding latitude profile should be considerably broader.

Notice also a potentially important difference between the distribution of the positron sources and the corresponding flux distribution: the intensity of the annihilation line depends on the properties of the annihilating medium (e.g. Guessoum et al. 1991), so that the resulting flux distribution may not directly reveal that of the underlying sources; this is clearly not the case for the photons of ^{26}Al decay.

During the last year, the OSSE detector aboard the Compton Observatory has given, for the first time, results on the longitude and latitude profiles of the galactic 511 keV emission (Purcell et al. 1991). The contribution of the central compact source was rather low ($\sim$2 10^{-4} ph/cm^2/s), giving a nice opportunity to study the diffuse component. The OSSE data show a longitude profile considerably peaked in the GC direction, while the latitude profile at $l=0°$ corresponds to a scale height $h_z > 150$ pc. Taken together the two profiles strongly suggest a contribution from the galactic bulge and favour an old population at the origin of the diffuse 511 keV emission, in line with theoretical ideas on the sources of galactic positrons (see also Prantzos 1993, Purcell et al. 1993, Skibo et al. 1992).

The situation is less satisfactory for the diffuse 1809 keV line. A preliminary map, due to the COMPTEL detector aboard the Compton Observatory, has been presented recently (Toulouse Symposium on High Energy Astrophysics, March 1992 and

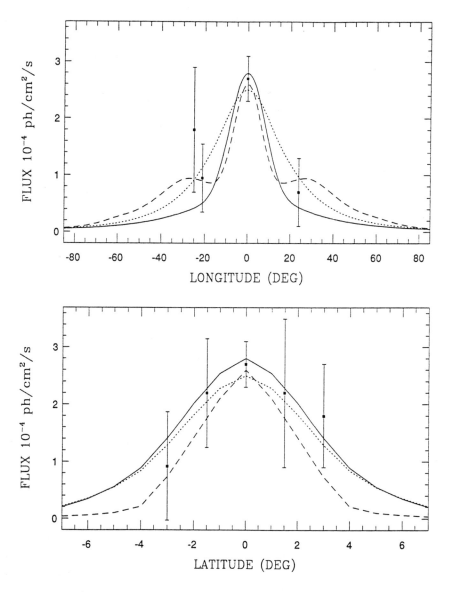

Figure 2: Observations of the 511 keV flux from the galactic plane, as a function of galactic longitude l (2a) and latitude at $l=0°$ (2b), respectively, with the OSSE instrument aboard the Compton Observatory. Source distributions with an enhanced population in the central regions of the Galaxy (or, equivalently, with a contribution from the galactic bulge), can easily fit the longitude data. The latitude data are better fitted with source distributions with rather large scale-height ($h_z > 150$ pc). Taken together those two profiles suggest a relatively old population (SNIa) at the origin of galactic positrons (from Prantzos 1993).

Compton Observatory Workshop, October 1992); the data seem to indicate a rather flat profile and some asymmetry between the first and fourth galactic quadrants. If confirmed, those features clearly indicate a young stellar population with a defficiency in the central Galaxy, as already suggested (Prantzos 1991; Ramaty and Prantzos 1991). Notice that in that case it should be rather difficult to accept novae as the main sources of ^{26}Al in the Galaxy, despite the large yields recently obtained in the calculations of Starrfield et al. (Sec. 4); otherwise one should explain why the nova distribution in our Galaxy is quite unlike the very peaked one in M31.

6. SN1987A AND OTHER SUPERNOVAE

The emergence of high energy emission from supernovae has been studied many years before the appearance of SN1987A in LMC. We briefly remind here its main features (see e.g. Burrows 1991; Cassé and Lehoucq 1993 for a thorough presentation).

Positrons and γ-ray photons are expected to be produced by radioactive decays of species with mass number 56, 57, and 44. Positrons transfer their kinetic energy to the medium by electromagnetic interactions and most of them annihilate. Gamma photons are comptonized by continuous transfer of energy to the ambient electrons, and they finally escape from the supernova or are photoelectrically absorbed. The enegy injection to the ejecta can be calculated by a Monte-Carlo simulation. The procedure is usually applied with the simplifying assumptions of a spherical, free-coasting and homologuous expansion (which seems to be legitimated by hydrodynamic calculations). The density and velocity profiles of the ejected material, which serve as initial conditions are taken from detailed numerical models.

The Monte-Carlo simulation allows to calculate three main observables: i) the bolometric light curve, through the energy deposition factor, ii) the X-ray flux in different energy bands, as a function of time, and iii) the gamma-ray emission (both continuous and discrete) as a function of time, as well as γ-ray line profiles.

As mentionned in Sec. 2, the light curve of SN1987A was powered by 0.075 $M_\odot$ of ^{56}Co, at least until day 900. The late light curve shows a slower decrease, although there is some discrepancy between the ESO and CTIO data between days 991 and 1030 (Bouchet et al. 1991; Suntzeff et al. 1991). Radioactivity of nuclei with lifetimes >1 year (like ^{57}Co and ^{44}Ti), or a pulsar inside the supernova, may be responsible for that behaviour. Adopting the supernova structure of Nomoto et al. (1988; Model 11E1), Cassé, Lehoucq and Cesarsky (1991) obtain a reasonable fit to the data of Suntzeff et al. (1992) with the following set of parameters: ^{57}Co/^{56}Co$\sim$2 times solar, M(^{44}Ti)=9 10^{-4} and $v_{expansion} \sim 11\,000$ km/s (Fig. 3). This amount of ^{57}Co is consistent with the infrared and X-ray upper limits and the recent GRO observation

TABLE 1

Characteristics of the main radioactivities in SN1987A: mean lifetime, energies of the main γ-ray photons, branching ratio, total γ-ray energy, mean energy of the positron (electron).emitted, branching ratio of β-decay, energy deposited by the positron (electron) and ejected mass of the nucleus of interest.

Decay	τ	E_r (keV)	%	ΣE_r	$\bar{E}_e$ (keV)	%	E_e (keV)	$M/M_\odot$
^{56}Co	113.7 d	847	100	3670	β^+ 660	19	125	0.070
$\downarrow$								
^{56}Fe		1238	68					
^{57}Co	391 d	122	89	136				$4.3\ 10^{-3}$
$\downarrow$								
^{57}Fe		136	11					
^{44}Ti	78 yr	62	100	2263	β^+ 597	94	561	$1.2\ 10^{-4}$
$\downarrow$								
^{44}Sc		78	100					
$\downarrow$								
^{44}Ca		1156	100					
^{60}Co	7.6 yr	1173	100	2505	β^- 96	100	96	10^{-5}
$\downarrow$								
^{60}Ni		1132	100					

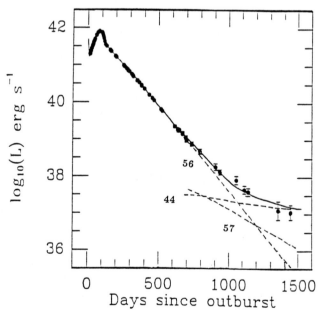

Figure 3: Bolometric Light curve of SN1987A (from Suntzeff et al. 1992), and fit (from Lehoucq 1993), based on: $M_{56}=0.075\ M_\odot$, $M_{57}=3.6\ 10^{-3}\ M_\odot$, $M_{44}=9.\ 10^{-4}\ M_\odot$. The dotted curves labelled 57 and 44 indicated the contribution of ^{57}Co and ^{44}Ti, respectively.

of the 122 keV line of ^{57}Co decay (Kurfess et al. 1992).

The amount of ^{44}Ti required in this model is, however, much larger (a factor of 3) than what is obtained in detailed nucleosynthesis calculations (Nomoto et al. 1991; Woosley 1991), which give ^{44}Ti/^{56}Ni $\leq$ 3 (^{44}Ca/^{56}Fe)$_{\odot}$. The crucial test as to the exact amount of 44Ti in SN1987A will come from the detection of (or some firm upper limits on) its caracteristic decay lines, at 68, 78 and 1156 keV; indeed, due to the long mean life of ^{44}Ti ($\sim$78 years) for many decades in the future these lines will have the same intensity, predicted at the $\sim$a few 10^{-5} ph/cm^2/s level and detectable with an instrument like INTEGRAL.

The emergence of hard X-ray and γ-ray lines $\sim$6 months earlier than expected (Sunyaev et al. 1988; Dotani et al 1987) is a clear indication that some ^{56}Co has been transported and mixed in the external layers, instead of remaining confined in the innermost layers of SN1987A. The mixing mechanism, presumably hydrodynamical instabilities of e.g. the Rayleigh-Taylor type, is usually introduced artificialy in the models, i.e. the radial distribution of the radioactive species is adjusted in order to explain the main features of the evolution of the light curve, the X-ray spectra and the ^{56}Co decay line intensities (Fig. 4 and 5). On the other hand, hydrodynamical 2- and 3-D calculations, performed by various groups (e.g. Herant and Benz 1991), do indeed find some mixing, and even some fragmentation in the ejecta.

All these calculations fail, up to now, to reproduce one puzzling feature of the high energy emission of SN1987A: the profile of the 847 keV and 1238 keV lines, resolved by the GRIS experiment (Tueller et al. 1990; Fig. 6). These lines, observed 433 and 613 days after explosion, are redshifted by 500-800 km/s, contrary to what is expected from an optically thick source. Moreover, the line width is larger than predicted, indicating that some fraction of ^{56}Co has penetrated deeply in the H-rich envelope. This result is corroborated by the infrared observations of an asymmetric and redshifted FeII line (λ = 1.257 μm) after day 700 (Spyromilio et al. 1990). Such a broad profile could, perhaps, be interpreted in terms of fragmentation in the ejecta, resulting in few large optically thick fragments, allowing γ-rays from the opposite side to filter through them. Unfortunately, hydrodynamic calculations fail up to now to produce the small ^{56}Ni quantities at velocities as high as $\sim$3000 km/s, as indicated by the GRIS data. No doubt, however, that observations of that kind are an inavaluable tool for supernova diagnostics.

The detection of gamma-ray lines from SN1987A opened a new window on supernova research. Indeed, each type of supernova is expected to produce different amounts of radionuclides and consequently a high energy emission (both in the lines and the

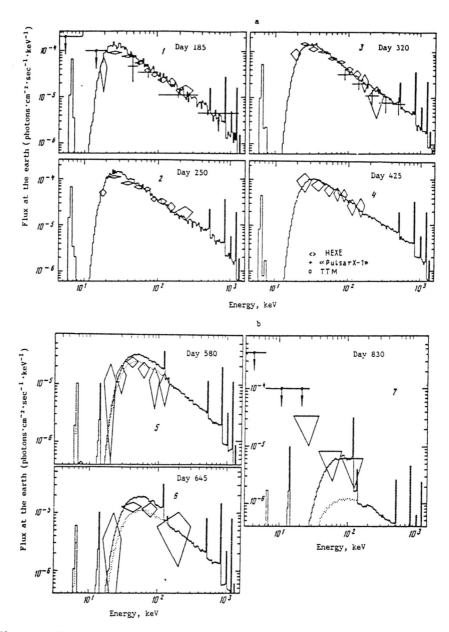

Figure 4: X-ray spectra of SN1987A at different times and Monte Carlo fits (Sunyaev et al. 1990); the ^{56}Co contribution is shown by the dotted line and the 57/56 ratio is twice solar.

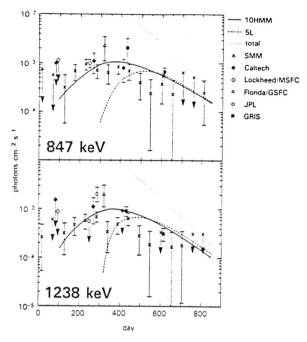

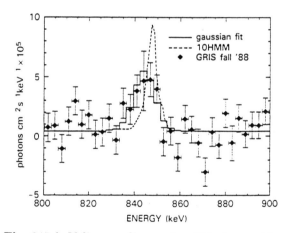

Figure 5: Gamma-ray line light curves of SN1987A at 847 keV and 1238 keV (Tueller et al. 1990); theoretical models are from Pinto and Woosley (1988).

Figure 6: The 847 keV line profile on day 613 observed by GRIS (Tueller et al. 1990). The solid line is a gaussian fit to the data and the dashed line the predicted profile from the 10HMM model of Woosley and collaborators.

continuum) that can be used to probe the supernova structure, i.e. reveal the mass of the debris, the amount of heavy elements produced, the explosion energy, velocity profiles, etc. (see Burrows 1991 for a review). The most promising candidates turn out to be Type Ia supernovae; due to their small envelope, large expansion velocities and large amounts of ^{56}Ni produced, they should be much brighter (in the dominant ^{56}Co lines) than other supernova types. But observational results are very poor up to now, and the upper limits published on a few cases are not very constraining. Uncertainties on distance determinations are one of the main difficulties. An interesting upper limit has been given recently by OSSE for the Co line of SN1991T in the periphery of the Virgo cluster. This supernova has the peculiarity of showing (in the optical band) Ni and Fe in its outer layers, before maximum light; this seems to indicate that nuclear statistical equilibrium has affected those layers. Because of uncertainties in the distance of the host galaxy, however, no model can be firmly dicarded (see Ruiz-Lapuente et al. these proceedings).

7. SUMMARY

In this short review we presented the main features of the astrophysical γ-ray lines related to nucleosynthesis, that have been detected up to now outside the solar system. The main points of the discussion can be summarized as follows:

– The diffuse 511 keV emission in the galactic plane should be attributed to positrons emitted from radioactive nuclei produced in SNIa explosions. This conclusion is supported both by theoretical arguments and the recent OSSE observations of the galactic distribution of that emission. The radioactive progenitor nuclei of the positrons, however, are not yet identified.

– On theoretical grounds, massive stars during their hydrostatic or explosive evolutionary stages, seem to be the most probable sources of the galactic ^{26}Al at the origin of the diffuse 1809 keV emission. The recent COMPTEL mapping of that emission (albeit a preliminary one), strongly corroborates that view and does not favour a major contribution from relatively old objects, like e.g. novae.

– The radioactive decay of ^{56}Co powered the lightcurve of SN1987A until day 900 after the explosion, while long lived radioactivities, like ^{57}Co or ^{44}Ti, are presumably at work now. The role of ^{44}Ti should be clarified by future observations of its caracteristic decay lines.

In any case, γ-ray line astronomy has been firmly established in the past few years as a new and powerful diagnostic tool of nuclear astrophysics.

AKNOWLEDGEMENTS: We gratefully aknowledge support for this work from the PICS N° 114 (*Origine des Eléments Légers*)

REFERENCES

Arnett D., Bahcall J., Kirshner R., Woosley S., *A.R.A.A.*, **27**, 629

Bodansky D., Clayton D., Fowler W. (1968), *Ap. J.*, **16**, 299

Bouchet P., Danziger J., Lucy L. (1991), *A. J.*, **102**, 1135

Burrows A. (1991), in *Gamma-Ray Line Astrophysics*, Eds. Ph. Durouchoux and N. Prantzos, (AIP: New York), p. 297

Cassé M., Lehoucq R., Cesarsky C. (1991), in *Gamma-Ray Line Astrophysics*, Eds. Ph. Durouchoux and N. Prantzos, (AIP: New York), p. 218

Cassé M., Lehoucq R. (1993), in *Supernovae*, Les Houches Lectures, ed. R. Mochkovitch in press

Chan K., Lingenfelter R. (1990), *21st Int. Cosmic Ray Conf. Papers*, **3**, 253

Clayton D., Craddock W. (1965), *Ap. J.*, **142**, 189

Clayton D., Colgate S., Fishman G. (1969), *Ap. J.*, **155**, 755

Clayton D., Leising M. (1987), *Phys. Rep.*, **144**, 1

Dotani I. et al. (1987), *Nature*, **330**, 230

Gehrels N. (1991), in *Gamma-Ray Line Astrophysics*, Eds. Ph. Durouchoux and N. Prantzos, (AIP: New York), p. 3

Gehrels N., Barthelemy S., Teegarden B., Tueller J., Leventhal M., Mac Callum C. (1991), *Ap. J. Letters*, **375**, L13

Guessoum N., Ramaty R., Lingenfelter R. (1991), *Ap. J.*, **378**, 170

Gusten R. (1989), in *The Center of the Galaxy*, Ed. M. Morris (IAU Symp.), p. 89

Herant M., Benz W. (1991), *Ap. J. Let.*, **370**, L81

Kurfess J., et al (1992), *Ap. J. Let.*, in print

Lehoucq R. (1993), *Ph. D. Thesis*, Université de Paris (unpublished)

Leising M. D., Clayton D. D. (1985), *Ap. J.*, **294**, 591

Leventhal M., MacCallum C., Stang P. (1978), *Ap. J.*, **225**, L11

Lingelfelter R., Chan K., Ramaty R. (1992), *Physics Reports*, in press

Mahoney W. A., Ling J. C., Wheaton W. A., Jacobson A. S. (1984), *Ap. J.*, **286**, 578

Mahoney W., Ling J., Wheaton W., Higdon J. (1992), *Ap. J.*, **387**, 314

Matz S., et al (1988), *Nature*, **331**, 416

Nomoto K., et al. (1988), *Proc. Astr. Soc. Austr.*, **7**, 490

Nomoto K., Kumagai S., Shigeyama T. (1991), in *Gamma-Ray Line Astrophysics*, Eds. Ph. Durouchoux and N. Prantzos, (AIP: New York), p. 236

Pankey T., (1962), *Ph.D. Thesis* (unpublished), Howard University

Paul J. *et al.*, (1991), in *Gamma-Ray Line Astrophysics*, Eds. Ph. Durouchoux and N. Prantzos, (AIP: New York), p. 17

Paulus G., Forestini M. (1991), in *Gamma-Ray Line Astrophysics*, Eds. Ph. Durouchoux and N. Prantzos, (AIP: New York), p. 157

Prantzos N. (1987), in *Nuclear Astrophysics*, Eds. W. Hillebrandt, R. Kuhfuss, E.
Mueller, J. Truran (Springer-Verlag), p. 250

Prantzos N. (1991), in *Gamma-Ray Line Astrophysics*, Eds. Ph. Durouchoux and N.
Prantzos, (AIP: New York), p. 129

Prantzos N. (1993), in *Advances in High Energy Astrophysics*, Ed. P. Mandrou, *A.
A. Suppl.* in press

Purcell W. et al. (1991), in *The Compton Observatory Science Workshop*, NASA
Conf. Publ. 3137, Eds. C. Shrader, N. Gehrels, B. Dennis, p. 431

Purcell W. (1993), in *Advances in High Energy Astrophysics*, Ed. P. Mandrou, *A. A.
Suppl.* in press

Ramaty R., Lingenfelter R. (1977), *Ap. J.*, **213**, L5

Ramaty R., Lingenfelter R. E. (1991), in *Gamma-Ray Line Astrophysics*, Eds. Ph.
Durouchoux and N. Prantzos, (AIP: New York), p. 67

Ramaty R. and Prantzos N. (1991), *Comments on Astrophysics*, **XV**, p. 301

Schoenfelder V., Varendorff M. (1991), in *Gamma-Ray Line Astrophysics*, Eds. Ph.
Durouchoux and N. Prantzos, (AIP: New York), p. 129

Skibo J., Ramaty R., Leventhal M. (1992), *Ap. J.*, **397**, 135

Spyromilio J., Meikle W., Allen D. (1990), *M.N.R.A.S.*, **242**, 669

Suntzeff B., et al. (1992), *Ap. J. Let.*, **384**, L33

Sunyaev R., et al (1988), *Sov. Astr. Let.*, **14**, 247

Sunyaev R., et al (1990), *Sov. Astr. Let.*, **16**, 403

Tueller J. et al. (1990), *Ap. J. Let.*, **351**, L41

Woosley S. E. (1991), in *Gamma-Ray Line Astrophysics*, Eds. Ph. Durouchoux and
N. Prantzos, (AIP: New York), p. 270

Woosley S. (1992), in *Advances in High Energy Astrophysics*, Ed. P. Mandrou, to
appear in *A. A. Suppl.*

Woosley S., Axelrod T., Weaver T. (1980), *Comments Nucl. Part. Phys.*, **9**, 185

511 keV Galactic distribution : a signature for nucleosynthetic yields.

Ph.DUROUCHOUX[1], P.WALLYN[1], Z. HE[2], S. YANAGITA[3] .

1: C.E Saclay, Service d'Astrophysique, 91191 Gif sur Yvette Cedex, France
2: on leave at Southampton University, Southampton, U.K.
3: Ibaraki University, Mito, Japan.

1- ABSTRACT

Since the 1970's, more than 25 observations of the annihilation radiation from the galactic center disk have been performed. Previous analysis have shown indication of two components to this emission:

i) a variable compact source(s) of 511 keV annihilation radiation within the galactic center region (Ramaty &-Lingenfelter 1989, Ramaty & Lingenfelter 1990, Paul & al 1991, Durouchoux & Wallyn 1993).

ii) a steady diffuse interstellar 511 keV source (Guessoum et al 1991, Ramaty et al 1991, Wallyn & Durouchoux 1993)

Considering the diffuse emission, many scientists have tried to compare the 511 keV distribution with the distributions of different types of stellar objects along the galactic plane, but none of this distributions fit well the data.

We consider here a multiprogenitor unidimensional model, were we take into account different progenitor distributions, leaving the percentage of each progenitor as a free parameter within an interval given by theoretical calculations. These intervals are estimated, taking into account the ranges of the yields of different radioactive elements giving β^+ decays. The frequencies of the occurence of this events are also taken into account.

We find that the best fit between measurements and multiprogenitor distributions is obtain with a SN Ia / Nova contribution of 68.1±5% (assuming .05% for the e^+ SN Ia escaping factor). The contribution of SN II is 21.3±5% whereas the possible other contributors (WR, pulsars, AGB) are within the statistical errors and do not allow to consider any substantial contribution of this objects.

We apply this results to the recent COMPTON/OSSE GC measurements and find a quite well fit with our composite distribution.

2- β^+ DECAYS

Among the different elements synthesized during explosive nucleosynthesis, some of them present β^+ decays in their radioactive chain, and therefore inject positrons in the interstellar medium.

After energy losses, thermalization and charge exchange, radiative combination with electrons or direct annihilation with electrons or hydrogen, 511 keV radiation is formed, and correlations might exist between the measurements of the 511 keV distribution and the β^+ progenitors.

We present in table 1, the four elements emitting β^+ in their decay chain, as well as γ ray lines (^{56}Co, ^{22}Na, ^{44}Ti, ^{26}Al)

Decay (Parent - Daughter)	Gamma-ray energy (keV) and Branching (%)	Half-life
$^{56}Ni \rightarrow ^{56}Co$	163(85),276(34), 427(34), 748(51), 812(85)	6.1d
$^{56}Co \rightarrow ^{56}Fe$	847(100), 1030(16), 1240(67), 1760(14), 2600(17), e^+(20)	77 d
$^{22}Na \rightarrow ^{22}Ne$	1275(100), e^+(90)	2.6 y
$^{44}Ti \rightarrow ^{44}Sc$	68(100), 78(100)	46 y
$^{44}Sc \rightarrow ^{44}Ca$	1156(100), e^+(94)	3.9 y
$^{26}Al \rightarrow ^{26}Mg$	1809(100), 1130(4), e^+(85)	7.4(5) y

Table 1 : Nucleosynthetic elements producing e^+ in their decay chain.

3- SOURCES OF POSITRONS

If we associate the radionuclides which produce e^+ in their decay chains, we see -table 2- that the principal progenitors are Type I (I_a and I_b) supernovae, Type II Supernovae, Novae, Wolf-Rayets, and possibly Red Giants and AGB (asymptotic giant branch). Pulsars, which can also be considered, in some circunstances, as e^+ emitters, are also taken into account.

In table 3, we indicate, for each event (SN, Nova) the frequency and the mass element yields (from Nomoto, private communication).

Radionuclide Decay Chain	Mean Life (years)	Principal Sources
$^{56}Ni \rightarrow ^{56}Co \rightarrow ^{56}Fe$	0.31 (^{56}Co)	Type I Supernovae (0.5 M$\odot$/SN)
	0.024 (^{56}Ni)	
$^{22}Na \rightarrow ^{22}Ne$	3.8	Novae (6×10^{-8}M$\odot$/Nova)
$^{44}Ti \rightarrow ^{44}Sc \rightarrow ^{44}Ca$	68	Type I Supernovae (7×10^{-5}M$\odot$/SN)
$^{26}Al \rightarrow ^{26}Mg$	1.1×10^6	Novae, Wolf-Rayet, Red Giant & Type II Supernovae

Table 2 : Principal sources of the four radionuclides producing e^+ in their decay chain

	frequency	$^{56}Ni^{**}$	$^{44}Ti^{**}$	$^{26}Al^{**}$	$^{22}Na^{**}$
SNII	$1.2 \times 10^{-2}y^{-1}$	0.076-0.1958M$\odot$	$(0.7492-1.22) \times 10^{-4}M\odot$	$(2.197-2.312) \times 10^{-6}M\odot$	$(0.9678-2.844) \times 10^{-8}M\odot$
	e^+ yield sec^{-1}	0	$(7.37-12.0) \times 10^{-4}$	$(3.16-3.33) \times 10^{40}$	$(1.916-5.34) \times 10^{38}$
SNIb	$.34 \times 10^{-2}y^{-1}$	0.07-0.25M$\odot$	$(0.6-3.6) \times 10^{-41}M\odot$	$(1.8-9.7) \times 10^{-6}$M$\odot$	$(0.9-2.6) \times 10^{-7}$M$\odot$
	e^+	$3.068-10.96) \times 10^{43}$	$(1.67-10.04) \times 10^{41}$	$(0.734-3.961) \times 10^{40}$	$(0.479-1.38) \times 10^{39}$
SNIa	$.36 \times 10^{-2}y^{-1}$	0.48-0.58M$\odot$	$(6.3-8.2) \times 10^{-5}$M$\odot$	$(0.38-3.8) \times 10^{-6}M\odot$	$(0.22-1.6) \times 10^{-7}M\odot$
		$(2.23-2.69) \times 10^{44}$	$(1.86-2.42) \times 10^{41}$	$(0.164-164) \times 10^{40}$	$(1.24-9.01) \times 10^{38}$
Nova*		-	-	$6.429 \times 10^{40}-6429 \times 10^{42}$	$3.177 \times 10^{39}-2.28 \times 10^{43}$

Table 3 : Estimations of e^+ production rate based on 1991 Nomoto model
(Private communication)

* using canonical numbers :
   40 novae/year/galaxy
   10^{-4} M$_\Theta$ ejecta/novae
   Yields of ^{22}Na and ^{26}Al are taken from Paulus and Forestini 1991

** β^+ branching ratio:

	^{56}Ni	^{44}Ti	^{26}Al	^{22}Na
ratios	0.19	0.95	0.821	0.9057

Using the yields and lifetime of ^{56}Ni, ^{44}Ti, ^{26}Al and ^{22}Na and taking into account of the frequency of this events, we calculate what might be the positron production each progenitor, in a range of yields (e.g for SNII, we used calculation provided by K. Nomoto performed for 13 M$_\Theta$ and 15 M$_\Theta$ presupernova).
We then calculate the positron productions (Table 3) and deduce the total number of e$^+$ per type of object (Table 4).

	^{56}Ni	^{44}Ti	^{26}Al	^{22}Na	Total
SNII 13MΘ	0	12.0E41	3.16E40	18.16E38	1.233E42
15MΘ	0	7.3741	3.33E40	5.34E38	7.708E41
SNIb min	3.068E43 *	1.67E41	0.734E40	0.479E39	1.70E42 **
max	10.96E43	10.04E41	3.96E40	1.38E39	6.525E42**
SNIa C6	2.23E44	1.86E41	0.164E40	1.24E38	1.13E43 **
W7	2.69E44	2.42E41	1.64E40	9.01E38	1.371E43 **
Nova min	-	-	6.429E40	3.177E39	6.747E40
max			6.429E42	2.28E43	2.923E43

Table 4 : Total e$^+$ production per type of event.

  *   should be multiplied by escape factor e
  ** calculed with e = 0.05

In the Table 5, we transformed this e$^+$ production in percentage of the total 511 keV flux.

	^{56}Ni	^{44}Ti	^{26}Al	^{22}Na	Total
SN II (15MΘ)	0	7.37E41	3.33E41	5.34E38	7.708E41 (1.53%)
SN Ib (max)	5.48E42	10.04E41	3.96E40	1.38E39	6.525E42 (13%)
SN Ia (W7)	1.345E43	2.42E41	1.64E40	9.01E38	1.371E43 (27.3%)
Nova (max)	-	-	6.429E42	2.28E43	2.923E43 (58.2%)

Table 5 : Fraction of e$^+$ production versus the different types of events.

Finally, the percentages are applied to the respective progenitor distributions : SN Ia and SN Ib are associated with "old progenitor", as well as nova and we will use one of the different nova distributions (Leising & Clayton 1985, Mahoney & al 1985)
For SN II, considered as "new population" distributions such as CO and 6.7 keV iron line, will be used; Finally we included possible contributions of Wolf-Rayets, AGB and pulsars.

4- PROGENITOR DISTRIBUTIONS AND FREQUENCIES OF OCCURENCE

As input distributions, we used the following distributions of progenitors:
- CO : Dame et al (1987),
- Molecular clouds : Leising & Clayton (1985)
This two distributions are characteristic of young population, and are considered as SNI distribution, in our model, without distinction between *deflagration model* (where the propagation of the burning front after the central carbon ignition is subsonic and which describes satisfactorily the optical observations of SNIa) and the *detonnation model* (which is less common and where the burning front is supersonic and characterized by a detonnation)
- Iron line -6.7 keV- iron produced in SN explosions, and then heated via shock processes to the temperature of several keV and forms an optically thin hot plasma emitting 6.7 keV X-rays which might represent more or less the SNR distribution , Koyama & al (1989)
- Nova distribution (based on visual luminosity of the older disk) Mahoney & al (1985), which are tought to be the progenitors of SNIa
- Nova distribution (based on radial distribution of novae in M31) Leising & Clayton (1985)
- Wolf Rayet distribution , Van der Hucht & al (1988)
About frequencies of occurence, one choose $0.36 \ 10^{-2}$ year^{-1} for the SNIa, compared to $0.34 \ 10^{-2}$ year^{-1} for the SNIb and $1.2 \ 10^{-2}$ year^{-1} for the SNII.

5- DATA ANALYSIS, FIT PROCEDURES AND RESULTS

Among the 25 observations of the galactic center region we separated the data into two sets of data :
i) the high flux data -where we assume a central source(s) "ON", and therefore we included a Dirac distribution with a free parameter (flux intensity)
ii) a low flux data - corresponding to an "OFF" state for the central object
and we calculate, for each set, a *composite distribution* by minimizing the χ^2 and deduce the masses of elements produced by the different progenitors.
We listed in Table 6 the percentage of each possible progenitor, for the "OFF" data (corresponding to a 511 keV flux coming mainly from the diffuse component)

PROGENITORS	"OFF" %
CO	$21.3 \pm 5\%$
NOVA	$68.1 \pm 5\%$
PULSAR	$4.4 \pm 5\%$
Wolf-Rayets	$1.7 \pm 5\%$
AGB	$4.4 \pm 5\%$

Table 6 : Contribution of different progenitors for the "OFF" observations of the 511 keV annihilation line.

One can see that the SN II, which are the major component of the ^{44}Ti contribute for about 20% of the total amount of the yield of positrons, whereas the SN Ia / Nova (which cannot be separated by this technics since their signatures are similar) are the major contributor of e$^+$; If we assume a mean yield of 0.5 M$_\Theta$ of ^{56}Ni per SNIa, one can put constraints on the ^{26}Al produced by Novae: of the order of 5.10^{-7} M$_\Theta$ year^{-1}.

The calculated distribution presented in Fig 1, based on "OFF" observations of the galactic plane, fits quite well with the measurements obtained out of the galactic center direction (GRIS & COMPTON/OSSE)

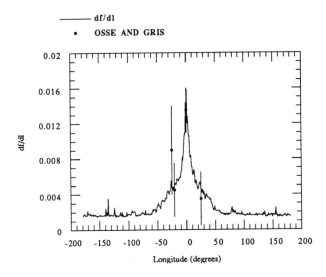

Fig 1: Diffuse component distribution based on "Off point source" observation of the Galactic Center region (with different fields of view)

More data are strongly needed in order to use, in the same way, a two dimension composite distribution and therefore put more stringent constraints on nucleosynthetic yields.

REFERENCES

1- R. Ramaty & R. Lingenfelter "Origin of the annihilation radiation from the galactic center region. Nuclear Physics 13, Proc. Suppl., (1989).

2- R. Ramaty and R.E. Lingenfelter."The positron annihilation radiation from the Galactic Center Region" 1991, in proceedings on Gamma-Ray Line Astrophysics, eds. Ph. Durouchoux and N. Prantzos (New-York : AIP 232, p. 67).

3- J. Paul, B. Cordier, A. Goldwurm, F. Lebrun, P. Mandrou, J.P. Roques, G. Vedrenne, L. Bouchet, E. Churazov, M. Gilfanov, A. Kuznetzov, R. Sunyaev, I. Chultov, A. Dyachtov, N. Khavenson and B. Novikov." The soft Gamma-ray Sources Identified by Sigma in the galactic center region" 1991, in proceedings on Gamma-Ray Line Astrophysics, eds. Ph. Durouchoux and N. Prantzos (New-York : AIP 232, p. 17).

4- Guessoum N., Ramaty R., Lingenfelter R.E.. Ap.J., 378, 170 (1991).

5- R. Ramaty, M. Leventhal, K.W. Chan and R.E. Lingenfelter.1992, Ap. J.(Letters), 392, L63.

6- P.Wallyn & Ph.Durouchoux accepted for publication in Ap.J. (1993).

7- Ph. Durouchoux & P. Wallyn to appear in Ap.J .(1993).

8- T.M. Dame, H. Ungerechts, R.S. Cohen, E.J. de Gens, I.A. Grenter, J. May, D.C. Murphy, L.A. Nyman, P. Thaddeus. Ap. J. 322, 706 (1987).

9- G. Paulus & M. Forestini "Radionuclides of interest for gamma-ray line Astrophysics" 1991, in proceedings on Gamma-Ray Line Astrophysics, eds. Ph. Durouchoux and N. Prantzos (New-York : AIP 232, p. 183).

10- M.D. Leising and D.D. Clayton. Ap. J. 294, 591 (1985).

11- W.A. Mahoney, J.C. Hidgon, W.A. Wheaton, A.S. Jacobson. Proc. of 19th International Cosmic Ray Conference (La Jolla) 1, 357 (1985).

12- K.A. van der Hucht, B. Hidayat, A.G. Admirauto, K.R. Suppelle and C. Doom. Astron. Astrosphys. 199, 217 (1988).

13- K. Koyama, H. Awaki, H. Kunedia, S. Takano, Y. Tawara, S. Yamanchi, I. Hatsukade, F. Nagasc. Nature 339, 603 (1989).

Solar Neutrinos

E.SCHATZMAN

Observatoire de Meudon

Abstract. Here is presented a discussion of the possible interpretation of the recent data on the solar neutrino flux (GALLEX collaboration). The possible modulation by solar magnetic fields of the energy flux of internal waves, down to the center of the Sun, is very appealing for connecting neutrino flux variation and its correlation with variations of the solar radius. However, the production of electronic ripples present a number of difficulties.which have not been overcome here.

1.INTRODUCTION

The search for solar neutrinos begins with the discovery by Fowler (1958) and Cameron (1958) of the PPII and PPIII reaction cycles. The first attempt by Davis to measure the

solar neutrino flux (1955) was based on a very optimistic estimate and just gave

$$\sum_i (\Phi_i \sigma_i) - \leq 4.10^4 \text{ SNU's.}$$ It was nevertheless important, as it gave for the first time an

experimental estimate of an upper limit for the central temperature of the Sun.(see the historical bibliography in Bahcall and Sears, 1972).

Since this early time the information has become more important and more reliable. Two other kinds of detectors have been at work, Kamioka (Hirata *et al*, 1988, 1990, 1991) which detects directly the neutrinos passing through a pool of about 2 000 tons of water, and two detectors with Gallium, SAGE in Baksan (Russia) and GALLEX in Gran Sasso (Italy). These diferent experiments are sensitive to different ranges of enery of the neutrinos. To the problems raised by the values of the neutrinos flux in these different experiments, we should add the problem of the variability found by Davis and the possible correlation with the solar cycle of activity, the most remarkable being the correlation between the solar radius and the neutrino flux (Ph. Delache *et al* , 1992).

Explaining the well known flux deficiency found by Davis in his experiment has been an important question for many years. There were two possible explanations, one being an astrophyscial one: the so called "standard model" is wrong and some important physical process has been forgotten, the other one being a particle physics explanation: the neutrino has a mass and through its interaction with the solar plasma a fraction of the e-neutrinos is transformed into μ-neutrinos (MSW effect : Wolfenstein, 1978, Mikheyev and Smirnov, 1986, Smirnov, 1990). The difficulty is that, with the helioseismological data, we have to fullfill other constraints on the solar model, and the range of possibilities which is offered is rather small.

It is imposible in this brief review to get deeply into the details of the problem. What is the most important is to reach consistency between the different data and the different theoretical processes which can be used in order to reach an acceptable explanantion.

2.CONSTRAINTS

2.1. Astrophysical constraints.

These have been presented several times (Christensen-Dalsgaard,1990). From the point of view of the neutrino flux, the most important one comes from the frequency difference between modes (n,l) and $(n+1,l+2)$,

$$v_{n,l} - v_{n-1,l+2} = \delta v_{n,l} = \frac{(4l+6)\,A}{n+\frac{1}{2}l+\varepsilon} \tag{1}$$

with

$$A = \frac{-1}{2\pi^2} \int_0^R \frac{1}{r}\,\frac{dc}{dr}\,dr \tag{2}$$

which depends strongly not only of the sound velocity but also of its derivative. ε is a small quantity which is not important here. This dependance is at the origin of the failure of the explanantion of the neutrino flux deficiency proposed by Schatzman and Maeder (1981). Mixing can carry ^3He near the center. Then, the energy production taking place with more fuel, the luminosity of the Sun can be obtained with a lower temperature; the importance of PPIII, which is highly temperature dependant, becomes smaller and the number of high energy neutrinos is smaller. However, the effect of mixing is to change appreciably $\nabla\mu$.wheras with a strong μ gradient, $(dc\text{-}/dr\text{-})$ changes sign close to the solar center, the effect of mixing is to cancel this change of sign. Consequently, the quantity $\delta\nu_{n,l}$ jumps from 9.5 μHz to 16.5 μHz, wheras the observationnal uncertainty is of the order of 0.5 μHz. A decrease of the central temperature would certainly make $\delta\nu_{n,l}$ smaller. To give an order of magnitude, a 5% decrease of the central temperature would bring the frequency difference to 9μHz, and this could be compensated by a mixing due to a diffusion coefficient which should not exceed something like 100 (pseudo Reynolds number less than 10).

The variability, if it is real, could not be due to a change of the density distribution inside the Sun, as shown by Schatzman and Ribes-Nesmes (1987). In the model which was used, even the large perturbation of the density, incompatible with any consistent solar model, did not give a variability of the order of the observed one. Perhaps, other possibilities are open, if we take into account the suggestion of Haxton (1991). We shall come back on this point later (sections 4 and 5).

2.2 Neutrinos.

As mentionned above, with the recent results of GALLEX (P. Anselmann*et al*, GALLEX Collaboration, 1992) we have now a measurement of the neutrino flux by three different

Predicted neutrino flux

		Rate ^{37}Cl			Rate ^{71}Ga		
		CESAM[1]	T-Ch[2]	Bahcall[3]	CESAM	T-Ch	Bahcall
ppI	pp	0	0	0	70	70.6	70.8
	pep	0.22	0.2	0.2	3	2.8	3.1
pp II		1.06	1.0	1.2	32.5	30.6	35.8
pp III		5.73	4.1	6.2	13.1	9.3	13.8
CN0		0.42	0.5	0.4	9.1	10.4	7.9
Total		7.43	5.8±1	8.0±1	127	124±5	131.5±$^{7}_{6}$
measured		2.10±0.30			GALLEX : 83±19±8		

(1) Morel *et al* (1992) ; (2) Turck-Chièze *et al* (1988); (3) Bahcall *et al* (1988)

methods. The interest of the Gallium experiment is that it is directly related to the solar luminosity, the pp-neutrinos representing about 55% (70.5 SNU) of the prediction of the standard solar models (SSM) for gallium detectors. We shall follow closely here the discussion presented by the GALLEX collaboration (1992). The general comparison with the observationnal data can be found in table 1, which summarizes the table given by P.Anselmann *et al.*. (1992).

Let us first consider the effect of a reduction of the central temperature. The temperature dependance $(T_c)^{n_i}$ can be obtained on the basis of the well known self-similar argument (see, for exemple Cox and Giuli, 1968), or by using the empirical approach of Bahcall and Ulrich (1988), which provides for the different neutrinos sources, $n_{pp}= -1.2$, $n_{Be} = 8$ and $N_B = 18$. With a temperature reduction of 5 and 10% %, one obtains the following results (Table 2).

Table 2

Reduction of the neutrino flux by a temperature decrease

	Davis	Davis	GALLEX	Kamioka..[g]	Kamioka.[f]
Observed	2.10±0.2	2.10±0.2	83±19	0.46±0.0	0.70±0.0
5% decrease	3.62[a]	3.35[b]	110[c,d]	120[a,b]	
10% decrease	2.2		102	310[a,b]	

The reduction factor is applied to the models of Bahcall and Ulrich (a,c,e,g) and to the CESAM model (b,d,)(Morel *et al* , 1992). For the Kamiokande II experiment, is given the ratio of the observed flux to the flux predicted by the TurckChièze model (f)(1988).

It is clear that, as a temperature reduction by a factor 0.9, which brings the neutrino fluxes respectively to 2.2 and 102 for the Chlorine and the Gallium experiments, give an unacceptable reduction by a factor 0.15 for the Kamioka II experiment, instead of a ratio (Observation/SSM) of 0.46 (Bahcall-Ulrich) or 0.7 (Turck-Chièze). This difficulty can be expressed in terms of confidence level (less than 5%). Anselmann *et al* (1992) then conclude that it is necessary to take into account the MSW effect.

2.3. Variability. If the correlation between the neutrino flux observed by Davis and the solar diameter measured by Laclare is real (Delache *et al*, 1992), this implies two conditions :(1) from the physical point of view, to find a physical process which could produce a change of the flux of ν_e arriving on Earth, and (2) from the astrophysical point of view, a change of the physical structure of the Sun, having the solar scale, which could allow changes of the proper magnitude of the neutrino flux. The whole problem will be discussed later (sections 4 and 5)

3.NEUTRINOS OSCILLATIONS AND MSW EFFECT.

It is useful for further discussion to have under hand the classical description of the MSW effect. (Bouchez *et al* , 1986). The local length of oscillation in the plasma is

$$l_m = l_0 \left(1 - 2\frac{l_v}{l_0}\cos2\theta + \left(\frac{l_v}{l_0}\right)^2 \right)^{-1/2}$$ (3)

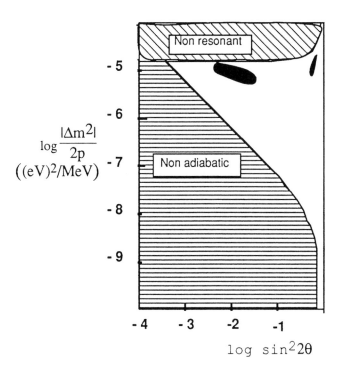

Fig.1. In white, the area where the adiabatic and the resonant conditions are fullfilled. It is possible to plot on the same diagram the areas where the MSW effect is valid by chosing the value $2p = 1$ MeV. They are represented by the black areas (confidence level 90%)

is the oscillation length in vacuum, and

$$l_0 = \frac{2\pi}{\sqrt{2}GN} = \frac{1.624 \ 10^9 \ cm}{(N/6.10^{23})} \tag{5}$$

The characteristic oscillation length l_0, near the center of the Sun, is of the order of 100 km. The vacuum oscillation length is of the order of 200 km for an energy of the neutrinos of 1 MeV and a mass difference Δm^2 of 10^{-5} eV^2. The most interesting cas is when $\Delta m^2 < 0$. Then there is a value of the density which yields maximal mixing ($\theta_m = \pi/4$) and an oscillation length (resonance length) given by

$$l_R = \frac{l_v}{sin2\theta} \tag{6}$$

This means that in a medium of constant electron density $N_R = \sqrt{2}/ G(|\Delta m^2|/4p).cos2\theta$ a neutrino produced as a v_e would propagate with maximal oscillation between the two states. If the density gradient is small enough, (adiabatic condition) the neutrinos keep on the instantaneous eigenstate, and the probability of finding a v_e outside the Sun will then be $sin^2\theta$ as found by Mikheyev and Smirnov (1985, 1986) and Smirnov (1990).

The resonant condition and the adiabatic condition define in the plane ($|\Delta m^2|/p$), $sin^22\theta$ a triangular area where the resonant and the adiabatic conditions are fullfilled (fig. 1). Anselmann *et al* (1992), using a χ^2 method, in order to rely on the statistical significance, have fitted the three experimental results to the reduction flux prediceted by the MSW mechanisme in the ($\Delta m^2, sin^22\theta$) plane. For $2p = 1$ MeV, it is possible to plot the two areas selected at 90% confidence level on fig 1. They fall near the boundary of the adiabatic region but are in the resonance region. This suggests that the use of the resonance approximation is valid in order to get an estimate of the reduction of the neutrino flux.

4. VARIABILITY

4.1.Effect of density perturbations
Several authors have explored the effect of solar density fluctuations: the adiabatic propagation of a neutrino mass eigenstate can be destroyed by a small perturbation if the

frequency of that perturbation is close to the local oscillation frequency. Essentially, non-adiabatic behaviour can arise at that point where the perturbation wave length matches the local oscillation wave length. Haxton and Zhang (1991) have suggested to introduce the effect of solar currents. But, instead of considering the motion of the solar matter, we shall look at the possible effect of electric currents in the solar plasma (Schatzman and Tsintsadze, 1992).

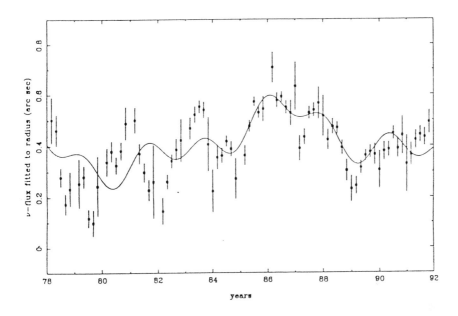

Figure 3 : $R_\odot$ and neutrino variations 4-mode model

One of the most important problem is due to the fact that, wheras the ^{37}Cl experiment seems to show a variation of the neutrino flux with time (fig.3) the KamiokandeII experiment does not show any apprecialbe variation. We should take into account the fact that the ranges of energies are different, Kamiokande II being sensitive only to neutrinos of energy higher than 7.5 MeV. With sin2θ of the order of 0.1, and an energy of 7.5 Mev, the oscillation length at resonance will be of the order of 18000km, wheras it would be of the order of 2000 km for the neutrinos at chlorine threshhold (0.814 MeV), and only of the order of 500 km for the PPI neutrinos.

Haxton and Zhang (1991) notice that, even if $|(d/dx)j_x|$ is too small to break the non-adiabatic condition, non-adiabatic behaviour can arise at that point where the perturbation frequency matches the local oscillation frequency. To give an exemple, let us write the value of the resonant density N_R and the local resonant wave length,

$$\frac{N_R}{6.10^{23}} = 3.27 \ 10^6 \ \cos 2\theta_v \ \frac{(eV)^2}{MeV} \tag{7}$$

$$L_R = 248 \ \frac{MeV}{(eV)^2} \tag{8}$$

from which we can derive the place where the local resonant wave length matches the perturbation wave length. Assuming, as we shall see later (section 5) that the ripples are induced by internal waves, a characteristic lenght is the vertical scale height, we then define the scale of the ripples by $(2\pi/\omega) = (\alpha H_{P\ 0})$, where the scale height is taken at the boundary of the convective zone,

$$\frac{N_R}{6.10^{23}} = 8.11 \ 10^9 \ ctg \ 2\theta_v \frac{2\pi}{\alpha H_{P\ 0}} \frac{R}{r} \tag{9}$$

With $\alpha = 2$, $\sin \theta_v = 0.1$, this defines two points where this matching is satisfied, for $r/R = 0.047$ and 0.22. This corresponds, for $|\Delta m^2| = 10^{-5} \ eV^2$ to an energy of the neutrinos respectively of 1.9 MeV and 8.9 MeV. It is clear that, if these electronic currents exist, the interaction between the plasma and the neutrinos can take place at at different distances from the solar center.

Another way of looking at the problem is the following. If the ripples of the electron current have a scale of the order of a few thousand kilometers, the average effect on neutrinos with a large wave length will be negligible. The difficulty is that the largest fraction of the Boron neutrinos detected by the chlorine experiment are high energy neutrinos, and that neutrinos detected by the Kamiokande II are also high energy neutrinos. It is difficult to imagin that a perturbation which changes the neutrino flux detected by the Chlorine experiment will not also change the flux detected by Kamiokande II. On the contrary, a change in the properties of the electron ripples will hardly alter the PPI neutrinos,, their wave lenght being much smaller than the perturbation wave length. Before drawing any conclusion, a complete numerical investigation should be carried. Altogether, we should remember that there is no reason for the wave length spectrum of the collective motions to be monochromatic : it would be necesary in any case to take into consideration the wave length spectrum of the plasma turbulence.

4.2.Plasma waves.

We shall consider now the question of the possible existence of electron currents which satisfy to the conditions we have just mentionned. (Schatzman and Tsintsaze, 1992). We have to consider here two different properties of the plasma waves (1) their amplitude, and (2), their wave length. The basic condition for the presence of plasma waves is that the plasma frequency should be larger than the collision frequency. With a collision frequency written as usual:

$$V_{coll} = \frac{n_e \, e^{\,4}}{m_e^{1/2} \, (kT\,)^{3/2} \, \ln\Lambda} \tag{10}$$

we find a condition

$$\frac{\rho}{T^{\,3}} \leq 10^{-12} \tag{11}$$

This condition is fullfilled everywhere in the Sun. We shall conclude here that the presence of plasma waves inside the Sun is a property which certainly has to be taken into account (fig. 4).The frequency of a plasma wave is given by

$$\omega^2 = \omega_{pe}^2 + \frac{3kT}{m_e}k^2 = 2.145 \, E35 \left(\frac{\rho}{150}\,\frac{1+X}{1.5}\right) + 4.544 \, E18 \left(\frac{T}{10^7}\right) k^2 \tag{12}$$

where k is the wave number and k the Boltzmann constant. The damping constant is

$$\gamma = -\omega_0 \sqrt{\frac{\pi}{8} \frac{1}{(kd)^2}} \exp\left(-\frac{1}{2(kd)^2}\right) \qquad (13)$$

and d, the Debye length is

$$d = 9.750 \text{ E-9} \left(\frac{T}{10^7} \frac{150}{(1+X)\rho}\right) \qquad (14)$$

The number of electrons in the Debye sphere is

$$n = 7.87 \left(\frac{T}{1.5.10^7}\right)^{3/2} \left(\frac{150}{\rho(1+X)}\right)^{1/2} \qquad (15)$$

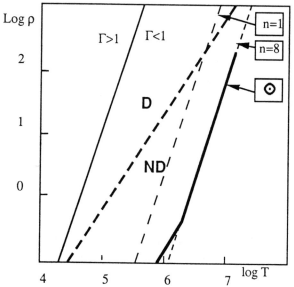

Fig. 4. Properties of the solar plasma. It degenerate, The solar plasma is weakly correlated, non-degenerate, and the energy density of plasma waves is of the order of 10%.

and a rule of the thumb is that the amount of energy of the Langmuir waves is of the order of (1/n). This shows (fig. 4) that plasma waves carry an appreciable amount of energy and have to be taken into account.

The group velocity is

$$\frac{d\omega}{dk} = \frac{3kT}{m_e} \frac{k}{\omega} = 9.81 \left(\frac{T}{10^7}\right) \left(\frac{150}{\rho}\right) \left(\frac{1.5}{1+X}\right) \tag{16}$$

The group velocity is very slow and the plasma wave is almost a standing wave. Plasma waves are present if their damping constant is small. This is the case for any wave length larger than a few Debye lengths.

We have now to consider the questions of the velocity and the number of electrons involved in these collective motions and of their wave number. Haxton and Zhang (1991), assuming that all the solar plasma is moving, suggest velocities of the order of 0.005 c, where c is the velocity of light. This is difficult to accept, as the sound velocity near the center of the Sun is of the order of 500 km.s^{-1}. However, we should remember that the velocity of the electrons is already 43 time larger, of the order of 13 000 km s^{-1}, or of the order of 0.04 c. In other words, in order to fullfill the condition of Haxton and Zhang, it is just necessary that 12 % of the electrons take part to these collective motions. The energy content of these collective motion will be of the order of 10 % of the total kinetic energy. This comparable to the situation in Tokomaks. But this is not a rarefied plasma. Is this beyond the possibilities? It should be noticed that with a frequency of the order of 10^{18} a neutrino will move over 3 microns over one period of oscillation. This raises the question of the nature of the interaction of the neutrinos with the Langmuir waves.Furthermore, there remain the question of the size of these collective motions and their generating mechanism. We shall consider this in the next section.

5. INTERNAL WAVES

5.1 Generation of internal waves.

Let us remind briefly here the properties of internal waves. Internal waves, or *gravity waves*, are well known by the geophysicists. They can be present in radiatively stable regions. They are generated at their boundary by the motions in the convective zone. Thei importance has been underlined by Zahn (1990) and by Knobloch (1990). Press (1981) has presented an approach of the description of the propagation of gravity waves and Schatzman (1991a,b,c) has applied these results to the problem of mixing in radiative zone. They carry efficiently angular momentum (Schatzman b) through wave action (Dewar, 1970, Grimshaw, 1984, Goldreich and Nicholson 1990 a,b)) and explain well the negative value of $\nabla\Omega$ (Brown *et al* 1989, Thomson, 1990), contrary to transport by turbulent viscosity (Endal and Sofia, 1978). The amplitude of the radial component of the velocity, obtained at the WKB approximation, varies like :

$$u_v \propto k_H^{3/2}\, \rho^{-1/2} \left(\frac{N}{\omega}\right)^{-1/2} \exp\!\left(i \int k_v\, dr - i\omega t \right) \exp(-A/2) \tag{17}$$

with

$$-(A/2) = \left(-\frac{1}{2} \int \left(\frac{D_{th}\, k_H^2}{N}\right) \left(\frac{N}{\omega}\right)^4 \left(\frac{N^2}{N^2-\omega^2}\right)^{1/2} k_H\, dr \right) \tag{18}$$

$$N^2 = -g\left(\frac{d\ln\rho}{dr} - \frac{1}{\Gamma_1}\frac{d\ln P}{dr}\right) = \frac{g}{H_P}\left(\nabla_{ad} - \nabla_{rad}\right) = \frac{g}{H_P}\Delta\nabla \tag{19}$$

and the thermometric diffusivity D_{th} can be written $D_{th} = \dfrac{L}{4\pi GM\rho}\dfrac{\nabla_{ad}}{\nabla_{rad}}$. For the damping factor we have made the same choice as Press (1981) for the wave number k_H and write :

$$k_H = \frac{2\pi}{\alpha H_{PI}}\frac{r_I}{r}$$

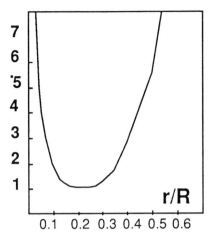

Fig.5. The square of the horizontal velocity as a function of depth. The unit is $u^2 = 10^3$ cm^2 s^{-2}

where Hp_I is the scale height at the boundary of the convective zone.

It easy to show that the damping factor remains finite when we approach the center of the Sun. Then as the Brunt-Väissälä frequency vanishes like r , it is clear that the radial component of the velocity diverges like r^{-4} .The curve given fig 5 shows the variation of the mean value of the square of the vertical component of the velocity as a function of the radius, assuming a Kolmogoroff spectrum of the turbulence at the boundary of the convective zone. In fact, the linear approximation, which assumes $u_v < (\omega/ k_v)$ is not valid inside $(r / R_\odot) < 0.05$, and some kind of turbulence is produced in the solar core, its importance being limited by the effects of $\nabla\mu$. The characteristic scale at this depth is of the order of $(r / r_I) Hp \approx 2500$ km. However, the velocities of the motion of the internal waves, even near the solar center are strikingly small. It is clear that internal waves cannot generate directly the collective motion of electrons with the amount of energy mentionned above.

5.2. Thermal diffusivity.
Press (1981) and Press and Ribicky (1981) have shown that, due to entropy transport, the thermal diffusivity is changed in the presence of internal waves. The heat flux is given by

$$F = -\left(K + \frac{\nabla^* - \nabla_{ad}}{\nabla^*} K \frac{\varepsilon^2}{2}\right)\frac{\partial T}{\partial x} \qquad (20)$$

where K is the radiative thermometric diffusivity and $\varepsilon = \left(\frac{u_H\, k_H}{\omega}\right)$ is the measure of the linear approximation describing the motion ($\varepsilon \ll 1$). In the presence of a μ-gradient, isotherma internal waves can propagate. There is nevertheless advection of entropy. We can think that when we approach the region where the linear approximation is not valid anymore, a new instability is generated. We are used to the fact that in that case a convective zone apeears, but we are here in a situation such that with a proper mixing internal waves will not propagate anymore and the entropy advection by internal waves will stop; then the modification of the radiative thermal diffusivity ceases also. A μ-gradient will be established such that the Brunt-Väissälä frequency remains close to zero, generating a new kind of semi-convective zone. The motions which will be induced by this instability borrow their energy to the radiative flux and this is of a completely different order of magnitude. This will induce a new structure in the central regions and we can also speculate at this point that enough energy can be found then for the generation of electrons collective motions.

5.3. Effect of magnetic field on internal waves.

The analysis of heliosismological data by Dziembowsky *et al* (1989, 1990) clearly shows the presence of a strong magnetic field at the bottom of the convective zone (abour 10^6 Gauss). The variations of the strenght of the magnetic field with the solar cycle suggest a modulation of the flux of internal waves with the solar cycle. The change of flux of internal waves near the solar center could then explain the modulation of the neutrino flux. In that sense, the time of propagation of internal waves from the bottom of the convective zone to the center of the Sun could introduce a phase lag between the neutrino flux and the measure of solar activity, e.g. the solar diameter. Fig 6 give an idea of the time of propagation from the surface to the center (Schatzman, 1992).

6. CONCLUSIONS.

The recent results of the GALLEX experiment put such constraints on the solar models that they are difficult to accept. This suggests strongly that the neutrino flux deficiency is due to the MSW effect.It comes out from the discussion carried by the GALLEX

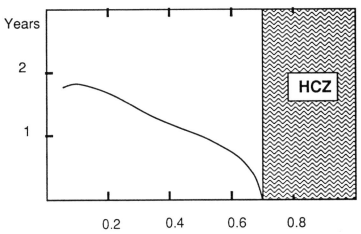

Fig. 8. Average lag time between production of internal waves and arrival at the group velocity at level r in the Sun.

collaboration that the variations of the neutrino flux by interaction between neutrinos and the solar magnetic field is very unlikely. The fact that internal waves produced at the boundary of the convective zone penetrate deeply in the Sun and can be modulated by variations, related to the solar cycle, of the magnetic field, is very appealing for an explanation of the variations of the neutrino flux. Collective motions of the electrons of the solar plasma could well replace the velocity field considered by Haxton and Zhang. However, a number of difficulties appear : it is difficult to find the energy necesary to feed the collective motions of the electrons, with a total amount of energy which is at least a few per cents of the internal energy of the plasma; the effect of the ripples can hardly affect differently the neutrino flux detected by Davis ans by the Kamiokande II collaboration : this is contradictory with the fact that that no variations have been observed by the Kamiokande collaboration. The present achievement of the model is such that it is not yet possible to conclude. Internal waves remain however an important field for further studies.

REFERENCES

Anselmann P. *et al* (1992) GALLEX collaboration *Phys. Lett.* **B 285** 390

Bahcall J. N. and Sears (1972), *Ann. Rev. Astron. Astrophys.* **10**, 25

Bahcall J. N. and Ulrich R. K.(1988) , *Rev. Mod. Phys.* **60** 297

Bouchez J. , Cribier M. ,Hampel W. , Rich J. , Spiro M. , Vignaud D. (1986), *Z. Phys. C, Particles, Physics and Fields,* **32**, 499

Brown T. M. *et al* (1989),*Astrophys. J.* **343** 526

Cameron A. G. W. (1958), *Ann. Rev. Nucl. Sci.,* **8**, 299

Christensen-Dalsgaard J., (1990) in *Inside the Sun,* G.Berthomieu and M. Cribier Eds., Kluwer Academic Press, p. 305

Cox J. P. and Giuli T. (1968) *Principles of Stellar Strucutre,* Gordon and Breach Publishers.

Davis R. Jr. (1955) *Phys. Rev.* **97**, 766

Delache Ph., Laclare F., Gavryusev V. Gavryuseva E.(1992) , in *Proceedings of neutrino conference* (to be published).

Dewar R. L. (1970)*Phys. Fluids.* **13**, 2710

Durney B. R. and Latour J.(1978), *Geophys. Astrophys. Fluid Dynamics,* **9**, 241

Dziembowski W. A. and Goode P. R. (1989), *Astrophys. J.,* **347**, , 540

Dziembowski W. A. and Goode P. R. (1990), in *Inside the Sun* , G. Berthomieu and M. Cribier Eds., Kluwer Academic Press, 341

Endal A. S. and Sofia S.(1978), *Astrophys. J.* , **220**, 279

Fowler W., (1958), *Astrophys. J.* , **127**, 551

Goldreich G. and Nicholson, P. (1989 a) *Astrophys. J.* **342** (1989) 1075

Goldreich G. and Nicholson P. (1989 b), *Astrophys. J.* **342** (1989) 1079

Grimshaw R. (1984), *Ann. Rev. Fluid Dynamics* , **16**, 11

Haxton W. C. and Zhang W. M.(1991), *Phys. Rev.* **D 43** 2484

Hirata K.S. *et al.,* (1989), Phys. Rev. Lett. **63**, 16

Hirata K.S. *et al,* (1990), Phys. Rev. Lett. **65**, 1297

Hirata K.S. *et al* (1991), Phys. Rev. **D44**, 2241

Knobloch E.(1990), in *Rotation and Mixing in Stellar Interiors* M.J. Goupil and J.P. Zahn Eds, Lecture Notes in Physics N° 366, Springer, 1990,p.109

Mikheyev S. P. and Smirnov A. Yu.(1986), *Nuovo Cimento* **9C** 17

Morel P., Berthomieu G., Provost J. and Lebreton Y (1992), *Astron. Astrophys.* (1992)

Press W. H. (1981), *Astrophys. J.*, **245**, 286

Press W. H. and G.B. Ribicki G. B. (1981), *Astrophys. J.* **248**, 751

Schatzman E. (1991) in *The Solar Interior and Atmosphere* , A.N. Cox Eds., Tucson ,
    University Press, Arizona, p.192

Schatzman E.(1991) in *Angular Momentum Evoliution of Young Stars* , S. Catalano and
    R.J. Stauffer Eds, Kluwer Academi Press, p.223

Schatzman E., (1991) *Mem. Soc Ast. It.* **62**, 111

Schatzman E. (1992) (*to be published*)

Schatzman E., Maeder A., (1981)*Astron. Astrophys.* **96** 1

Schatzman E. and Ribes-Nesmes E., (1987) in *New and Exotic Phenomena*, O. Fackler
    and J. Tran Than Van Eds., Editions Frontières, p. 365

Schatzman E. and Tsintsadze N. (1992)(*in progress*)

Smirnov A. Yu., (1990) in *Inside the Sun,* G. Berthomieu and M. Cribier Eds. Kluwer
    Academic Publishers, Dordrecht, 231

Thomson M. F. (1990) *Solar Physics* **125** 1

Turck-Chièze S. *et al* , (1988)*Astrophys. J.* **335** 415

Wolfenstein L. (1978), *Phys. Rev.* **D17**, 2369

Zahn J. P. (1990), in *Inside the Sun* G. Berthomieu and M. Cribier Eds, Kluyver, 425

Screening effect in solar conditions

H. DZITKO[1], P. DELBOURGO-SALVADOR[2], CH. LAGRANGE[2], S. TURCK-CHIEZE[1]

[1] Service d'Astrophysique, DAPNIA, CE Saclay, 91191 Gif sur Yvette, FRANCE

[2] Service PTN, CE Bruyères le Châtel, BP 12, 91680 Bruyères le Châtel, FRANCE

ABSTRACT

As four neutrino experiments are running, the solar modelling must give more precise neutrino predictions and therefore Solar microscopic physics is now more precisely studied. This paper deals with the different prescriptions used to calculate the screening factors, the enhancement of nuclear reaction rates due to the plasma electrons. As known since several years, intermediate screening is justified for reactions involving nuclei with a number of protons greater than 3. We discuss here the different prescriptions available for such a treatment and suggest that the Mitler's one is more correct than the generally used Graboske et al's intermediate screening.

1 THE PRESENT SOLAR NEUTRINO PROBLEM

The difference between theoretical predictions and experiments seems smaller in the case of the Kamiokande and gallium experiments than in the case of the chlorine one (see Table 1) and justifies to reconsider each ingredient of the calculation.

experimental results	Predictions	ratio $\frac{exp}{th}$ (R_H)
Chlorine experiment 2.33 ± 0.25 SNU	6.4 ± 1.4 SNU	$(36.4 \pm 10 \pm 8)\%$
Kamiokande experiment 0.28 ± 0.03 ev/d ($E \geq 7.5$ MeV)	0.48 ± 0.12 ev/d	KII= $(60 \pm 6 \pm 13)\%$
Gallium experiments (1991) SAGE : $85^{+22}_{-32} \pm 20$ SNU GALLEX : $83 \pm 19 \pm 8$ SNU	123 ± 7 SNU	$(68.5^{+30}_{-38} \pm 4)\%$ $(68 \pm 17 \pm 4)\%$

Table 1 : Comparison between experiments (Davis 1989, Hirata 1991, Anselmann 1992, Gavrin 1992) and theory for the solar model of Turck-Chièze and Lopes (1992).

Neutrino oscillations as well as physical or astrophysical solutions may explain these results. For the second point, improvements have been noticed in the determination of the equation of state, the photospheric abundances, and the opacity calculations (see for a review, Turck-Chièze et al 1992). We consider here the screening effect on the nuclear reaction rates.

2 SCREENING EFFECT IN NUCLEAR REACTIONS

In the central region of the Sun, the atoms are, except for iron, completely stripped of their atomic electrons and the nuclei are immersed in a sea of free electrons which generally cluster in their vicinity. This shielding effect reduces the Coulomb potential and increases the thermonuclear reaction rates within the star by a factor

$$f = exp\Lambda = exp(\frac{Z_1 Z_2 e^2}{R_D kT}), \quad R_D = \sqrt{\frac{kT}{4\pi e^2 \rho N_a \xi}}, \quad \xi = \sqrt{\sum_i \frac{(Z_i(Z_i + \theta_e))X_i}{A_i}}$$

where, R_D is the Debye-Huckel radius, ξ the rms charge of the plasma and θ_e the electronic degeneracy factor. This expression is actually only valid if $\Lambda \ll 1$, which is the case called by Salpeter (1954) weak screening (WS). This means that the electrostatic interaction between the two reacting nuclei is small compared to the mean thermal energy per particle. The other limiting case $\Lambda \gg 1$ is the strong screening which applies to high density cases (that is $\rho \geq 10^4 - 10^8$ g/cm^3). In the central region of the Sun (T$\approx$ 15 10^6 K and $\rho \approx$ 150 g/cm^3) the Λ ratio may reach 0.2, 0.4 for interactions between proton and ^7Be or ^{14}N. Therefore the **"intermediate screening"** hypothesis is better adapted. It has been developed by DeWitt et al in 1973 and a tentative of generalization to a two components plasma has been made by Graboske et al (1973), latter called GWGC. They have considered a statistical-mechanical theory for the screening function f, suitable for most astrophysical cases without assumption on the charge mixture and the charge of the reacting particles. They propose an expression in terms of chemical potentials which is made to retrieve weak, intermediate, and strong screening results :

$$\log f = k_b \eta_b \Lambda_0^b [(Z_1 + Z_2)^{1+b} - Z_1^{1+b} - Z_2^{1+b}], \quad \Lambda_0 = 1.88 \ 10^8 (\rho/\mu_I T^3)^{1/2}$$

where Λ_0 is a charge independent variable, which characterizes the plasma. The exponent b lies between 1 and 2/3 corresponding to weak or strong screening and for intermediate screening, b=0.860. η_b depends on the appropriate charge plasma average ξ. The difficulty for intermediate screening is to find the intermediate average charge which is, in GWGC theory, equals to $< z^{3b-1} > /(\xi^{3b-2}z^{2-2b})$. For weak screening $\eta_b = \xi$, $k_b = 1/2$ giving so the classical WS formula.

The GWGC prescription is generally used in the astrophysical community but the limitation of such approach comes from the non relativistic treatment and the simplified one or two specific components plasma. This prescription is not accurate enough for the Sun and we have tested Mitler's formalism (1977) which was carried out by doing the two fluid approximation to the plasma. The advantage of this calculation is to consider the distortion of the electron clouds of the nuclei when they approach each other and to perform more accurate integrals.

The previous authors have usually written $Log f = H(0) = \Delta U/kT$ where ΔU was the electrostatic energy difference between the two reacting nuclei when they are widely separated and when they are fused. Mitler has shown that ΔF—the difference in Helmholtz free energy between the final and initial configuration of the two particles—is better adapted than ΔU. This prescription recovers the previous Salpeter and GWGC results for $H(0)$ in the weak screening limit and is valid for all regimes. We call it latter the standard assumption. A more precise approach introduces a radial dependence of ΔF. Therefore, in a few cases, we have calculated the penetration factor for a screening potential with $\Delta F(r)$, using numerical methods developed by one of us (Lagrange 1991). The reaction rates were then calculated and the enhancement factor f deduced. The differences between values obtained with the standard assumption and the numerical methods were smaller than 2.5 %, so we recommend the simpler standard assumption for solar conditions. Table 2 presents the different screening factors for the various prescriptions. In the specific case of the Sun, the Mitler's prescription gives results between the classical weak screening and the GWGC intermediate screening (see figure below).

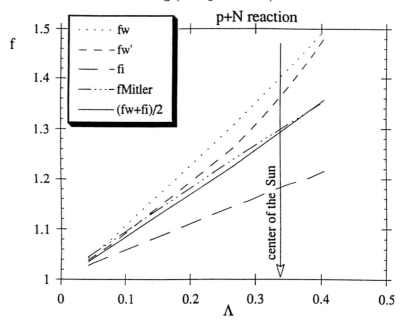

Figure 1 : Different screening prescriptions for the p+N reaction and for various Λ. In the center of the Sun $\Lambda \approx 0.34$ for this reaction.

It seems that the weak screening prescription (fw) overestimates the f value out of its validity domain ($\Lambda < 0.1$). However, we can remark that the extented weak screening (quoted fw'), which is a second order approximation in Λ for the pair distribution function involved in the calculation of f, gives results closer to Mitler's f (fMitler). On the contrary, GWGC intermediate screening (fi), seems to underestimate this factor. In the weak screening regime, all the prescriptions merge to the same values as expected.

density, temp.	147.8 g/cm^3	15.5 10^{6}K	22.0 g/cm^3	8.04 10^{6}K
Reaction	p+N	p+Be	p+N	p+Be
Λ	0.34	0.19	0.37	0.21
fw	1.41	1.22	1.45	1.24
fw'	1.38	1.18	1.42	1.20
fi	1.19	1.11	1.20	1.12
(fw+fi)/2	1.30	1.17	1.33	1.18
fMitler	1.29	1.17	1.32	1.19

Table 2 : Estimation of the different screening factor prescriptions for two parts of the present Sun, the central region and the external part of the nuclear area.

3 CONCLUSION

So far, the screening formalisms which have been generally used in stellar codes seem to be not accurate enough for the solar modelling. The Mitler's standard assumption appears more carefully determined and relatively easy to introduce in the solar case to compute the f factor within 3 % accuracy. Therefore, we suggest to use it in evolution codes since it should contribute to have a more accurate determination of the solar neutrino fluxes.

BIBLIOGRAPHY

P. Anselmann et al, Gallex collaboration, Physics letters (1992) B 285, 376
R. Davis, "Inside the Sun", ed. by G. Berthomieu and M. Cribier (1989), p 171.
H. E. DeWitt, H. C. Graboske and M. C. Cooper, Astrophys. J. (1973), 439.
H. C. Graboske, H. E. DeWitt, A.S. Grossman and M. S. Cooper, Ap. J. 181 (1973).
H. E. Mitler (1977), Astrophys. J., 212, 513.
V. N. Gavrin, SAGE collaboration, (1992), Dallas symposium
K. S. Hirata et al, Phys. Rev. Lett. 66 (1991) 9.
Ch. Lagrange, (1991), Journal Canad. de Physique, 69, 833
Salpeter, E. E. (1954), Australian J. Phys., 7, 353.
Salpeter, E. E. and Van Horn, H. M. (1969), Astrophys. J., 155, 183.
S.Turck-Chièze and I. Lopes, (1992), to appear in Astrophys. J.
S.Turck-Chièze, W. Dappen, E. Fossat, J. Provost, E. Schatzman, D. Vignaud (1992), submitted to Physics Report.

THE HEAVIER THAN Fe ELEMENTS

The synthesis of the nuclides heavier than iron: Where do we stand?

M. ARNOULD[1], K. TAKAHASHI[2]

[1]Institut d'Astronomie et d'Astrophysique, Université Libre de Bruxelles, B-1050 Bruxelles, Belgium

[2]Max-Planck-Institut für Astrophysik, W-8046 Garching bei München, Germany

1 INTRODUCTION

The synthesis of the stable nuclides heavier than iron is classically ascribed to three different stellar mechanisms referred to as the s-, r- and p-processes (Burbidge et al. 1957; Cameron 1957). The s-process is called for in order to explain those stable heavy nuclides located at the bottom of the valley of nuclear stability (s-nuclides), while the r- and p-processes have to account for those which are neutron-rich (r-nuclides) and neutron-deficient (p-nuclides), respectively.

The bulk solar-system abundance distributions of the s-, r-, and p-nuclides are represented in Fig. 1. They exhibit some distinct features: (i) well-developed s-nuclide abundance peaks at mass numbers $A = 138$ and 208, (ii) peaks in the r-nuclide distribution at $A = 130$ and 195, and (iii) p-nuclides that are 100-1000 times less abundant than the corresponding more neutron-rich isobars, while their distribution roughly parallels the s- and r-nuclei abundance curves (note, however, the very low abundances of the two odd-odd p-nuclides ^{138}La and ^{180}Ta, and the relatively large amounts of ^{92}Mo and ^{94}Mo). Not represented in Fig. 1 are those heavy stable nuclides that can be produced in significant proportions by both the s-and r- processes (referred to as sr-nuclides). Figure 1 can be used to separate their abundances into an s- and an r-component. Such a splitting, however, involves some uncertainty.

In addition, there is at present ample observational evidence that the solar mix of the s-, r- and p-nuclides is far from being universal. This is testified most vividly by a quite large suite of meteoritic isotopic anomalies that involve s-, r- or p-nuclides (see e.g. Begemann 1992, this Conference; Ott 1992, this Conference). At the galactic scale, differential effects are also observed in the distributions of the s- and r- nuclides versus metallicity (e.g. Lambert 1989). In addition, stellar evolution effects on the abundances of a variety of heavy elements are apparent in the surface composition of chemically peculiar stars (e.g. Smith 1989). The galactic cosmic rays also exhibit s- and r- relative abundances that differ from solar values (e.g. Ferrando 1992). At this point, it has to be emphasized that no information is available about the p-nuclide abundances outside the solar system.

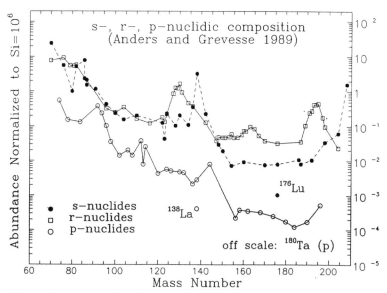

Fig. 1. *The solar system abundances of the s-, r-, and p-nuclides. The sr-species for which the estimated s- or r-process contribution exceeds 10% are not shown*

Soon after the studies of Burbidge et al. (1957) and Cameron (1957), Hubert Reeves clearly recognized the importance of a careful study of the nuclear physics and astrophysics aspects of the s-, r- and p-processes, and has contributed substantially to their understanding. He has also provided one of the most appealing scenarios for the origin of certain meteoritic isotopic anomalies, especially developing the notion that the solar system has been born in an OB association (Reeves, 1978).

Much further efforts have been devoted to the understanding of the stellar synthesis of the heavy nuclides. In particular, many dedicated experimental and theoretical works have tried to provide the nuclear physics input necessary for the study of the stellar alchemy of the heavy elements, while stellar modelling has attempted to unravel the adequate astrophysical sites for the operation of those processes.

It is fair to say that, despite the earnest efforts in the past, much remains to be done in order to gain a reliable picture of the s-, r- and p-processes, and of the observed abundances of the corresponding heavy nuclides in stars and in the solar system. It is beyond the scope of this paper to review all these problems in detail. Instead, we will limit ourselves to a brief overview of, and to some critical comments on the nuclear physics and astrophysical aspects of the three processes. Further details can be found in recent reviews by Käppeler et al. (1989), Arnould and Rayet (1990), Arnould (1991), Cowan et al. (1991) and Lambert (1992).

2 THE S-PROCESS

The s-process results from the production of neutrons and their captures on pre-existing ("seed") nuclei (from previous stellar generations or evolutionary phases) on timescales longer than most β-decay lifetimes. For some nuclides, neutron captures can compete with β-decays, leading to local "branches" along the main path. Even so, the s-process always flows very close to the bottom of the valley of nuclear stability, and "hits" on its way s- or sr- nuclides, while the r- and p-nuclei stay out of its reach. Typical s-process flow paths are obtained for neutron densities n_n in the approximate $10^7 - 10^9$ cm^{-3} range and temperatures in excess of 10^8 K.

There is now ample observational evidence that the s-process is presently at work in a variety of chemically peculiar Red Giants (e.g. Smith 1989). Special objects like FG Sagittae (Langer et al. 1974), or SN1987A (Mazzali et al. 1992, and references therein) may also exhibit enrichments of s-nuclides. Clearly, these observations help identifying the possible astrophysical sites of the s-process. In this respect, a key information is also provided by the time evolution of the galactic s-process content (e.g. Mathews et al. 1992, Baraffe and Takahashi 1992).

2.1 Basic nuclear data

A question of primary importance for the study of the s-process concerns the possibility of release of neutrons in stellar plasmas. Our present knowledge of the rates of potentially important neutron-producing reactions along with complementary astrophysical arguments indicate that ^{13}C$(\alpha,n)^{16}$O and ^{22}Ne$(\alpha,n)^{25}$Mg are likely to be the two most important neutron sources in stars.

Much experimental and theoretical effort has been devoted recently to those reactions, as discussed by Drotleff et al. (1992), as well as to the competing ^{22}Ne$(\alpha,\gamma)^{26}$Mg channel (Wolke et al. 1989). The derived rates differ in several respects from the ones recommended by Caughlan and Fowler (1988; CF88).

Another key nuclear physics input for s-process studies concerns neutron capture rates at astrophysical temperatures on nuclides in the whole $12 \leq A \leq 209$ mass range. The present state of our knowledge in the field is reviewed by Käppeler et al. (1989) and Beer et al. (1992).

While more precise measurements of some neutron capture cross sections on stable nuclides still appear desirable, much further effort will certainly be devoted to experiments on radioactive isotopes involved in "branched" s-process models (Käppeler 1992). Even so, theoretical estimates will remain for quite a while the vital source of information for the vast majority of the neutron captures on unstable nuclei. Model predictions are also required in order to evaluate the contribution of thermally populated excited states to the net neutron capture rates (e.g. Arnould 1972, Thielemann et al. 1986 for a Hauser-Feshbach approach to this problem).

Beta-decay rates are the indispensable complement to neutron capture rates in order to model the s-process, and in particular to describe the various branchings that can develop along its path. The β-decay half-lives under s-process conditions have been evaluated by Takahashi and Yokoi (1987).

It has to be emphasized that the proper evaluation of the stellar β-decay lifetimes and, to some extent, of the neutron capture rates is intimately related to the question of the thermal population of excited nuclear states in stellar plasmas. This problem is especially acute when dealing with long-lived isomeric states (e.g. Käppeler et al. 1989, for a review).

2.2 Astrophysical s-process models

The s-process models may be broadly classified into "classical" and "realistic" ones. In addition, various types of parametrized models trying to mimic more or less closely the realistic approaches have also been proposed.

The classical model represents a purely phenomenological approach to the s-process that takes detailed account of the necessary nuclear physics input, but avoids any reference to specific astrophysical environments (e.g. Käppeler et al. 1989). In addition to its simplicity which makes it amenable to an analytical treatment, this model owes its success to its ability to reproduce quite satisfactorily the solar system s-nuclide abundance distribution (Fig. 1). As reviewed by e.g. Käppeler et al. (1989), this fit is obtained under the reasonable assumption that different fractions of ^{56}Fe seeds are exposed to different neutron doses. In fact, at least two distinct s-process components, referred to as the "weak" and "main" components, are called for in order to account for the solar system s-nuclides in the $A \lesssim 90$ and $90 \lesssim A \leq 204$ mass ranges, respectively. In addition, a "strong" component appears to be required in order to fit the $204 < A \leq 209$ abundances. Another important conclusion of the many studies devoted to the classical s-process is that the main component is able to provide a good fit to the solar system abundances only if a distribution of neutron exposures $\tau = \int n_n v_T dt$ (where v_T is the neutron thermal velocity) is considered. In general, an exponential distribution of exposures is selected, the involved free parameters being frequently re-adjusted in the light of new available nuclear data (e.g. Käppeler et al. 1989, Beer et al. 1992).

Most analyses of *stellar* s-element abundances have also been performed in the framework of the classical model. Individual stars exhibit abundance distributions that require parameters values that span a range including those of relevance to the solar system.

2.3 The s-process in "real" stars

The classical s-process model is generally considered to provide some kind of "effective" s-process conditions that have to be obtained either in a single star (in order to account for its surface abundance), or in a collection of stars of possibly different masses and metallicities (in order to account for the solar system abundances through a model for the chemical evolution of the Galaxy). It remains to be seen if detailed stellar models can provide conditions compatible with the requirements of the classical model, or, much more modestly, stellar sites where some s-processing has a chance to develop.

Various such studies have been conducted recently. In what follows, we will limit ourselves to some comments about the s-process calculations in massive stars and in low- or intermediate-mass stars.

2.4 The s-process in massive stars

Several studies of the s-process during central He burning in massive stars of different masses and metallicities have recently been made (e.g. Langer et al. 1989, Raiteri et al. 1991b, Baraffe et al. 1992). They demonstrate that such conditions provide a plausible site for the solar system weak s-process component identified by the classical model. They also provide the necessary data for modelling the evolution of the abundances of the $A \lesssim 90$ s-nuclides in the galactic disk or halo (Raiteri et al. 1992; Baraffe and Takahashi 1992). We also note that the impact of the newly determined rates of α-captures on ^{13}C or ^{22}Ne upon the s-process during central He burning has been studied on grounds of detailed stellar models (Meynet and Arnould 1992). Although the calculation predicts some strenghtening of the neutron capture nucleosynthesis in such conditions, they certainly do not drastically affect the whole picture emerging from the many calculations using previous (CF88) rates. It is also shown that the neutron exposure associated with core He burning cannot be predicted with an accuracy that is better than a factor of about two. This is primarily due to the uncertainties remaining in the $^{22}Ne(\alpha,n)^{25}Mg$ rate. Without nuclear physics progress, it is thus quite premature to discuss in all details the subtleties of the massive star core He-burning neutron capture nucleosynthesis in relation with the solar-system s-process weak component.

The neutron capture nucleosynthesis during central carbon burning and shell helium burning has been calculated along with the structural evolution of massive stars (Arcoragi et al. 1991), while models of Nomoto and Hashimoto (1988) have inspired an evaluation of the possible s-process in shell carbon burning (Raiteri et al. 1991a). It is concluded that the s-process that can develop in such sites is so limited that it does not affect in any significant way the conclusions based on central He burning alone. This is still more so when the nuclear uncertainties mentioned above are duly taken into account. Namely, the additional neutron processing in the shell He, core C and shell C burning phases are well masked by the uncertainty in the neutron exposure accompanying core He burning.

Finally, we note that the suspected Ba (and possibly also Sr) overabundances claimed to exist in SN1987A (e.g. Mazzali et al. 1992, and references therein) have been studied by Prantzos et al. (1988). Their calculations rely on the operation of ^{22}Ne$(\alpha,n)^{25}$Mg during central He burning in a specific model for the SN1987A progenitor, and predict some Ba production that could account for the low overabundances derived from observation by Williams (1988) and Mazzali et al. (1992), but not for the higher values claimed by Höflich (1988). If these large Ba overabundances are real and not an artifact of the photospheric models, then another neutron capture process had to be at work in the SN1987A progenitor (see Sect. 3). This could also be true for other massive stars.

2.5 The s-process in low- and intermediate-mass stars

More than a decade ago, computations by Iben (e.g. Iben and Renzini 1983 for a review) have set up quite an appealing picture: thermal pulses in intermediate-mass AGB stars can generate neutrons through ^{22}Ne$(\alpha,n)^{25}$Mg, and synthesize s-nuclides in amounts that compare rather satisfactorily to the solar system s-process main component. In addition, that material can be transported to the stellar surface by the third dredge-up, this providing a natural explanation for the chemical peculiarities observed in certain classes of Red Giants. Through their winds (that were neglected in Iben's models), those objects could also be major agents for the galactic s-process enrichment.

Pessimism about this picture has grown, however, after the realization that it is far from obvious to observe counterparts of those theoretical intermediate-mass objects, or to identify stellar spectra with the ^{25}Mg excess expected from the operation of ^{22}Ne$(\alpha,n)^{25}$Mg. Further stellar and s-process modelling brought some additional touch of pessimism about the ability of ^{22}Ne$(\alpha,n)^{25}$Mg in intermediate-mass stars to account for stellar chemical peculiarities and for the solar system main s-process component.

The ^{13}C$(\alpha,n)^{16}$O neutron source in low-mass AGB stars has recently become a viable contender through a series of modelling experiments (e.g. Sackmann and Boothroyd 1991). However, it is clear that the efficiency of that source as well as the occurence of the dredge-up depend on an intricate physics, and that the predictions in that field remain a matter of "warm" debate and active research.

To make a long and complicated story short, let us just stress that the observed enrichment in s-elements at the surface of chemically peculiar Red Giants remains a puzzle. In contrast to what is sometimes claimed in the literature, it is impossible at this time to specify s-process yields in terms of the mass and metallicity of low-mass stars.

3 THE R-PROCESS

The global picture that emerges from the many studies devoted to the r-process can be summarized as follows: seed nuclei with abundances determined by the establishment of a nuclear statistical equilibrium (NSE) are subjected at the time of NSE freeze-out, but when temperatures still exceed $\approx 10^9$ K, to a very intense neutron flux, typical neutron densities lying in the $10^{20} \lesssim n_n \lesssim 10^{30}$ cm^{-3} range. Under such conditions, and in contrast to the situation encountered in the s-process, β-decays near the line of stability are far too slow to compete with neutron captures, so that typical r-process flow paths are shifted to very neutron-rich nuclei far from stability, and in particular to the neutron-magic nuclides ^{80}Zn and ^{130}Cd. In this scenario, the efficiency of the r-process depends on the neutron-to-seed abundance ratio that results from the composition at NSE, and from its alteration by the subsequent nuclear reactions. Various approximations of that general picture have been worked out in more or less great detail.

As the r-process involves very neutron-rich nuclides, it raises many difficult nuclear physics questions. Another (the most?) tantalizing problem concerns the site(s) where the required high density of neutrons can be obtained. Also enough, and just enough, neutrons must be available for synthesizing the actinides. In contrast to the situation encountered for the s-process, stellar spectroscopy does not really help identifying such sites by providing some direct evidence of the present-day operation of the r-process. A possible hint may come from the claim that the lines of Eu and Gd (elements normally blamed on the r-process) have been recently strenghtening in FG Sagittae (Wallerstein 1990). On the other hand, it cannot be totally ruled out that the suspected Ba overabundance in SN1987A is due to the r-process. A more useful information concerning the operating sites of the r-process is in fact provided by the time evolution of the galactic content of the r-nuclides (in practice Eu) (e.g. Mathews et al. 1992, and references therein).

3.1 Basic nuclear data

The NSE equations establishing the seed abundances can be solved accurately for a given set of temperature T, density ρ and neutron-to-proton ratio (or electron concentration Y_e) provided that the nuclear binding energies and the nuclear partition functions are precisely known. Mass differences and nuclear partition functions are also indispensable to the modelling of the subsequent neutron capture episode.

For most of the nuclei involved in the r-process, those basic quantities are not known experimentally, so that theoretical estimates are required. Being close to the basics, a recent mass formula that is referred to as the "extended Thomas-Fermi plus Strutinsky integral" (ETFSI) method (e.g. Pearson et al. 1991) leaves us with some hope for reliable mass extrapolations with a relatively small number of adjustable parameters to unknown regions of neutron-rich nuclei. As for the partition functions, the "back-shifted Fermi-gas model" of nuclear level density may be used for extrapolation once the parameter values are carefully chosen (e.g. Arnould and Tondeur 1981). Another key nuclear input data for r-process calculations concerns β-decay half-lives of nuclei along the flow paths. They not only determine the abundance pattern, but also provide a good measure of the dynamical timescale that fixes the duration of the nucleosynthesis. Again, existing experimental data are hardly sufficient, and one has to rely on models. State-of-the-art β-decay half-life calculations are based on Gamow-Teller β-strength functions obtained from a nuclear shell model within the so-called "quasi-particle random-phase approximation" (Möller and Randrup 1990, Staudt et al. 1990).

The evaluations of neutron capture cross sections on very neutron-rich nuclei are the necessary complement to the β-decay rate calculations, particularly when one deals with relatively low n_n values. So far, they have been done in the framework of a Hauser-Feshbach model (e.g. Cowan et al. 1991; Howard et al. 1992).

In addition, (i) β-delayed neutron emissions are known to be important in the re-shaping of the abundance distributions at the termination of the neutron irradiation phase, and (ii) in the regions where the r-process terminates, mechanisms like β-delayed or neutron-induced fissions may be at work, and may have to be considered in particular to predict production ratios for the actinides, and especially for the nucleo-cosmochronometers (see Sect. 3.4). A review of the nuclear physics and astrophysics aspects of these various processes can be found in e.g. Cowan et al. (1991). Let us simply note that the fission barriers for very neutron-rich actinides have been evaluated recently in the framework of the ETFSI model (Tondeur et al. 1992).

In closing the very brief overview presented in this section, it has to be emphasized that the reliability of the predictions of the various nuclear data necessary for the r-process modelling is still quite poor. It is beyond of the scope of this contribution to present a detailed discussion of the involved uncertainties. The reader is referred to e.g. Takahashi (1992) and Takahashi and Hillebrandt (1992) for further information and comments on these questions.

3.2 Astrophysical r-process models

As its s-process counterpart, the "classical" r-process model proposed by Seeger et al. (1965) is a purely phenomenological description based on several drastic approximations of the general scenario sketched in Sect. 3. In particular, it considers in general a very schematic seed distribution (often ^{56}Fe only), and constant values for T and n_n over a certain time interval τ, during which a (n,γ)-(γ,n) equilibrium can be established for each isotopic chain. After time τ, the neutron density is considered to fall abruptly to zero. Among many other simplifications, this model allows the isotopic abundances to be calculated by the mere application of the nuclear Saha equation, which just requires neutron separation energies and partition functions, and avoids the necessity of a detailed knowledge of neutron capture cross sections. This model has been widely used, in particular in order to evaluate the impact of new nuclear data on the r-process yield predictions (e.g. Kratz et al. 1992, Tondeur et al. 1992). A very difficult problem is to relate realistic astrophysical sites to the conditions identified by the classical model to be most appropriate for reproducing the solar-system r-process abundances. An apparent candidate for such sites is the type II supernova explosion of massive stars, for which some "parametrized" or "dynamical" r-process models have been developed. Other possibilities include carbon-rich He-burning layers in which $^{13}C(\alpha,n)$ can operate following an injection of protons, He/C-zones heated by the supernova shock, collisions of neutron stars and black holes, the (still putative) inhomogenous Big Bang, etc. (e.g. Cowan et al. 1991, for the list of proposed r-process scenarios). In the following, we just concentrate on the type II supernova model. With regard to the other possibilities mentioned above, we simply note that either they have not yet been studied in detail, or they concern expectedly rare events.

3.3 The r-process in "real" type II supernovae?

It has long been thought that appropriate r-process conditions could be found in the hot ($T \gtrsim 10^{10}$ K) and dense ($\rho \approx 10^{10} - 10^{11}$ g/cm^3) neutron-rich ("neutronized") material located behind the outgoing shock in a type II supernova event (e.g. Delano and Cameron 1971, Hillebrandt et al. 1976). A serious doubt about this scenario has, however, been cast by "modern" supernova calculations which suggest that the neutronized zone might well be left in the neutron star residue without being ejected. Even if this would not be the case, the model might suffer from another drawback that relates directly to the necessity of calling for a large enough range of Y_e values in order to reproduce the solar system r-process abundance curve. In practice, this translates into the requirement of an ejected mass of r-nuclides that is too large to be compatible with the galactic content of those species. This scenario could be resurrected if the r-processed layers characterized by different Y_e values were convectively mixed before the ejection, and if only a tiny fraction of their mass were ejected (by a jet, for example). A recent two-dimensional hydrodynamical simulation (without rotation and magnetic field) has shown that a strong convective mixing can develop in the layers at the surface of the proto-neutron star (Müller and Janka 1992, private communication). The possibility of a jet remains to be demonstrated by detailed magneto-hydrodynamical calculations of a rotating star.

An apparently more promising site for the r-process has been suggested by Woosley and Hoffman (1992) in connection with the delayed mechanism of type II super-novae. More specifically, it concerns the low density ($\rho \approx 10^5$ g/cm^3) and hot ($T \approx 6 \ 10^9$ K) region (termed "hot bubble") which forms and expands rapidly between the surface of the central neutron star and the overlying stellar mantle. This mate-rial is characterized by Y_e values in the $0.40 \lesssim Y_e \lesssim 0.48$ range (Janka 1992, private communication). Under such conditions, the NSE products are mostly α-particles and neutrons. As the temperature goes down, a tiny amount of heavy seeds can be synthesized by a chain of α-capture reactions ("α-process") starting from $3\alpha \rightarrow {}^{12}C$ and $\alpha + \alpha + n \rightarrow {}^9Be(\alpha, n)^{12}C$ (Woosley and Hoffman 1992). An appropriately large neutron-to-seed ratio could result, so that a successful r-process could develop without requiring the very low Y_e values that have to be called for in the high-T and high-ρ supernova model. This would help easing the galactic evolution problems mentioned above.

Using a schematic model for the evolution of the hot bubble, Meyer et al. (1992) and Howard et al. (1992) have demonstrated that the solar-system r-process abundance curve could be well reproduced with a certain distribution of Y_e values for different mass segments of the hot bubble.

On the other hand, Witti et al. (1992) and Takahashi et al. (1992) have studied the possible α-process and subsequent r-process by referring to numerical simulations to obtain the main features of the hot bubble, and in particular typical values of T, ρ and Y_e, and their evolution in time. It turns out that the model leads to such a strong α-process that the neutron-to-seed ratio is much too low for the r-processing to proceed beyond $A \approx 90$. Considering that this is based on just one background model, they have explored deviations from the model that are necessary for at least some sort of r-process to occur. They have shown that this can be the case when the ρ values are reduced by at most a constant factor of ≈ 10 if T, Y_e and time evolution of the background model are not altered. At this time, it is impossible to verify if these requirements can be met in type II supernovae.

Clearly, an urgent task is to improve the supernova modelling, or at least to test different models in due consideration of their uncertainties stemming from the complicated physics involved. A study of the sensitivity of the r-process predictions to the input nuclear physics (such as the rates of some key reactions involved in the α-process, masses, β-decay rates, or fission barriers) is also called for (e.g. Howard et al. 1992, Tondeur et al. 1992).

3.4 The age of the r-nuclides

Through the use of the abundances of the long-lived radionuclides ${}^{187}Re$, ${}^{232}Th$, ${}^{235}U$ and ${}^{238}U$ measured in meteorites, nucleo-cosmochronology aims at determining the age of the r-nuclides, a lower bound to the age of the Galaxy and of the Universe. In addition, some hope has grown recently of developing a novel nucleo-cosmochronology which does not rely on the solar system composition, but instead on the abundance of Th in stars with different ages, and hence directly on the variations of the Th content with time (e.g. François, this Conference).

Very much has been written on age determinations based on nucleo-cosmochronological techniques. Let us simply recall that, in order to establish a good chronometry based on radioanuclides, one needs to provide firstly a good set of (isotopic) abundances and nucleosynthesis yields, in addition to the radioactive half-lives. Another issue concerns the necessity, and then the possibility, of using detailed models for the chemical evolution of the Galaxy in order to gain a reliable chronological information. From a critical review of the status of those various requirements, Arnould and Takahashi (1990) concluded that the current uncertainties in the solar system abundances of U and Th coupled with the difficulties of evaluating their r-process yields made it impossible to obtain reliable ages for the r-nuclides from a galactic chemical evolution model satisfying various basic astronomical constraints. The ages that can be determined mathematically through the widely used exponential model are totally unreliable physically. They also stressed that the ^{187}Re-^{187}Os pair could not be regarded yet as a good chronometer, but that the prospect for improving the situation was not worse, and might in fact even be better, than that for establishing the ^{232}Th-^{235}U-^{238}U chronometry. In addition, the various uncertainties and intricacies involved in the use of the Th stellar data prompted them to state that any age determination derived on such grounds was not very reliable yet. It is our opinion that those conclusions remain fully valid today.

4 THE P-PROCESS

It is quite clear the the neutron-deficient p-nuclides cannot be made by neutron capture processes. In contrast, it seems natural to think of the transformation of pre-existing seed nuclei of the s- or r-type by the addition of protons (radiative proton captures), or by the removal of neutrons (neutron photodisintegrations).

The rates of (γ,n) photodisintegrations increase very rapidly with increased temperatures and decreased neutron separation energies. In fact, it appears that temperatures $T \gtrsim 10^9$ K are required in order for such transformations to have time to operate in realistic stellar situations. On the other hand, (p,γ) reactions are much less dependent on temperature and binding energy, but their rates decrease rapidly with increased Coulomb barrier heights. More specifically, those rates are decreased by factors $10^6 - 10^9$ at temperatures of a few 10^9 K when going from Fe to Bi. As a result, proton radiative captures can contribute at best to the production of only the lightest p-nuclides (e.g. ^{74}Se to ^{98}Ru), and in sufficiently proton-rich environments. Such considerations have been the main guidelines in the search for stellar environments where the p-nuclides could be synthesized.

The explosion of the H-rich envelopes of type II supernovae has long been held responsible for the synthesis of the p-nuclides, as originally suggested by Burbidge et al. (1957). It is quite clear today that this model is astrophysically implausible: the required explosion conditions ($\rho_{max} \approx 10^4$ gcm^{-3}, $T_{max} \gtrsim 2 \cdot 10^9$ K) indeed appear impossible to be attained in the considered supernova layers. In view of this difficulty, various other sites have been proposed, and especially the deep O/Ne layers of massive stars, either in their pre-supernova evolution or during their supernova explosion.

More recently, the development of the p-process in a very thin C-rich layer of massive white dwarfs exploding as type Ia supernovae has also been explored.

4.1 Basic nuclear data

As discussed by Rayet et al. (1990), the modelling of the p-process in the C- or O/Ne-rich layers of supernovae requires the setting up of an extended reaction network involving the stable to more or less neutron-deficient isotopes of the elements ranging from C to Bi. Those species are connected by numerous reactions with neutrons, protons and α-particles, as well as photodisintegrations. In hydrostatic (pre-supernova) conditions, β-decays also have to be taken into account.

Above silicon, the experimental information about the necessary capture rates is extremely scarce. They are commonly evaluated with the aid of a statistical Hauser-Feshbach model. The photodisintegration rates are derived from the application of the reciprocity theorem, which requires the knowledge of binding energies and partition functions (see Sect. 3.1). In contrast to the situation encountered in the r-process, most binding energies are already known experimentally. If not, model predictions generally agree quite satisfactorily. The problem raised by the lack of laboratory masses is thus much less acute than in the r-process case. This is also true for β-decay rates, as most of them have already been measured (Note that (free) e^--capture rates are too slow to be of practical significance in the astrophysical scenarios of relevance to the p-process).

4.2 The astrophysical p-process models

The development of the p-process in the deep O/Ne layers of massive stars or at the surface of exploding white dwarfs has most often been explored in the framework of parametrized models. More recently, "realistic" models have also been constructed. The first calculation demonstrating that the p-process can develop in the O/Ne-rich zone of massive stars has been performed for conditions appropriate to the pre-supernova stage, use being made of a detailed network of nuclear reactions (Arnould 1976). This work has been extended to a schematic explosive O-burning model based on parametrized temperature and density conditions by Woosley and Howard (1978) and, more recently, by Rayet et al. (1990).

This explosive p-process scenario has been further developed by the consideration of detailed type II supernova models for stars in the $M_{ZAMS} = 15 - 25M_\odot$ range (Arnould et al. 1992), as well as for a specific model of SN1987A (Prantzos et al. 1990). These calculations predict that the p-process yields from the different model stars are, on the whole, quite similar, and that most of the p-nuclides are synthesized in proportions close to solar. However, variations in the production of individual species are noticeable, and some p-nuclides are definitely underproduced in any of the type II supernova models considered thus far. This concerns in particular the Mo and Ru p-isotopes. This is also true for the odd-odd neutron-deficient nuclide ^{138}La. In contrast, significant amounts of ^{180}Ta are found to emerge from the considered explosions. As discussed in detail by Prantzos et al. (1990), this results from a subtle balance between the destruction by (γ,n) reactions of the relatively abundant ^{181}Ta seed and that of the relatively easy to photodisintegrate ^{180}Ta. This balance allows the latter nuclide to be produced in layers which do not experience temperatures much in excess of $\approx 10^9$ K during the supernova explosion. The ability of the p-process to synthesize ^{180}Ta puts constraints on the other models (reviewed by e.g. Németh and Käppeler 1992) that have been proposed in order to explain the amount of the rarest "stable" nuclide in the solar system.

The possibility of development of the p-process in a very thin C-rich external zone of massive white dwarfs exploding as type Ia supernovae has also been explored in the framework of a parametrized model (Howard et al. 1991), or of more realistic models (Howard and Meyer 1992). Contrary to some initial hope, this scenario does not appear able to cure some of the problems faced by the type II supernova models, and in particular the underproduction of the Mo and Ru p-isotopes.

Finally, very massive stars ending their lives as pair-creation supernovae have been proposed as a new site for the p-process. Detailed models show that these objects can produce the p-nuclides both (quasi-)hydrostatically and explosively, the non-explosive contribution representing in fact a significant fraction of the total yields (Rayet et al. 1992).

The p-process yield predictions have been confronted not only with the solar system bulk composition, but also with meteoritic isotopic anomalies that are attributed to the p-process (e.g. Rayet and Arnould 1992, and references therein). Those comparisons are blurred to some extent by the nuclear physics and astrophysics uncertainties that affect the abundances computed in all the scenarios. Let us note in particular that the calculated p-process yields are very sensitive to the adopted seed distribution. This problem is particularly acute in the type Ia supernova scenario. A first attempt to answer the question of the seed abundances in such a scenario has been conducted by José et al. (1992) on grounds of a preliminary model. As another example of an astrophysical source of uncertainty, let us cite the pair-creation supernova case, where the absolute and relative quantities of ejected p-nuclides depend more or less drastically on the mass of the possible remnant, which is difficult to predict reliably at the present stage.

5 CONCLUSION

Much dedicated experimental and theoretical effort has been devoted in recent years to the understanding of the nuclear physics and astrophysics aspects of the s-, r- and p-processes of nucleosynthesis. Much has been written about all that. Very schematically, one can try summarizing the situation as follows:

(1) Even if the s-process still poses some difficult nuclear physics problems, the most vexing questions appear to be of astrophysical nature. While the neutrons produced during central He burning in massive stars are likely to account for the weak component of the solar-system s-process abundance distribution, no reliable model is able to explain the corresponding main component, or the s-nuclide abundances observed at the surface of chemically peculiar Red Giants;

(2) the r-process raises very severe nuclear physics and astrophysics problems. The former ones have to do with the bulk properties of very neutron-rich nuclei, as well as with their neutron capture or β-decay rates. The measurement of all those quantities for a significant fraction of the nuclei of interest is of course not conceivable in any foreseeable future. In contrast, direct or indirect experimental information for a carefully selected set of unstable nuclides would provide very valuable checks of the validity of the adopted models. On the astrophysical side, type II supernovae, and more specifically their associated hot bubbles, remain to be modelled in detail in order to determine if some at least of them can indeed be the prividledged sites of the r-process, as suspected today;

(3) The nuclear physics and astrophysics questions that have to be dealt with in the description of the p-process are probably somewhat less acute than in the r-process case. The properties of the neutron-deficient nuclides are in general better known than those of the very neutron-rich ones. On the other hand, the modelling of the supernova layers where the p-process is expected to develop is put on safer grounds than the description of the r-process zone that is located closer to the forming neutron star. Even so, many puzzles remain to be solved, and in particular the source of the Mo and Ru p-isotopes, or the exact contribution of type Ia supernovae to the solar system abundances of the p-nuclides.

There is certainly much more to come about all that. *"Patience dans l'Azur"*! (Reeves 1981).

Acknowledgements. This work has been supported in part by the Science Program SC1-0065 of the European Communities. M.A. is Chercheur Qualifié FNRS (Belgium).

References:

Anders E., Grevesse N., 1989, Geochim. Cosmochim. Acta 53, 197

Arcoragi J.-P., Langer N., Arnould M. 1991, A&A 249, 134

Arnould M. 1972, A&A 19, 92

Arnould M. 1976, A&A 46, 117

Arnould M. 1991, in Evolution of Stars: The Photospheric Abundance Connection, eds. G. Michaud and A. Tutukov (Dordrecht: Kluwer Academic Publishers), p. 287

Arnould M., Rayet M. 1990, Ann. Phys. Fr. 15, 183

Arnould M., Takahashi K. 1990, in Astrophysical Ages and Dating Methods, eds. E. Vangioni-Flam, M. Cassé, J. Audouze and J. Tran Thanh Van (Gif-sur- Yvette: Editions Frontières), p. 325

Arnould M., Tondeur F. 1981, CERN Report 80-09, p. 229

Arnould M., Rayet M., Hashimoto M. 1992, in Unstable Nuclei in Astrophysics, eds. S. Kubono and T. Kajino (Singapore: World Scientific), p. 23

Baraffe I., Takahashi K. 1992, A&A, submitted

Baraffe I., El Eid M.F., Prantzos N. 1992, A&A 258, 357

Beer H., Voss F., Winters R.R. 1992, ApJS 80, 403

Burbidge E.M., Burbidge G.R., Fowler W.A., Hoyle F. 1957, Rev. Mod. Phys. 29, 547

Cameron A.G.W. 1957, Chalk River Report CRL-41, unpublished

Caughlan G.R., Fowler W.A. 1988, At. Data Nucl. Data Tables 40, 283

Cowan J. J., Thielemann F.-K., Truran J. W. 1991, Phys. Rep. 208, 267

Delano M.D., Cameron A.G.W. 1971, Ap&SS 10, 203

Drotleff H.W. et al. 1992, in Proc. 2nd Intern. Symp. on Nuclear Astrophysics "Nuclei in the Cosmos", Karlsruhe, to appear

Ferrando P. 1992, in Proc. 2nd Intern. Symp. on Nuclear Astrophysics "Nuclei in the Cosmos", loc. cit.

Hillebrandt W., Takahashi K., Kodama T. 1976, A&A 52, 63

Höflich P. 1988, Proc. Astron. Soc. Aust. 7, 434

Howard W.M., Meyer B.S. 1992, in Proc. 2nd Intern. Symp. on Nuclear Astrophysics "Nuclei in the Cosmos", loc. cit.

Howard W.M., Meyer B.S., Woosley S.E. 1991, ApJL 73, L5

Howard W.M., Goriely S., Rayet M., Arnould M. 1992, in Proc. 2nd Intern. Symp. "Nuclei in the Cosmos", loc. cit.

Iben I. Jr., Renzini A. 1983, Ann. Rev. Astron. Astrophys. 21, 271

José J., Rayet M., Arnould M., Hernanz M. 1992, in Proc. 2nd Intern. Symp. on Nuclear Astrophysics "Nuclei in the Cosmos", loc. cit.

Käppeler F. 1992, in Radioactive Nuclear Beams 1991, ed. Th. Delbar (Bristol: Adam Hilger), p. 305

Käppeler F., Beer H., Wisshak, K. 1989, Rep. Prog. Phys. 52, 945

Kratz K.-L et al. 1992, in Proc. 6th Intern. Conf. on Nuclei far from Stability, Bernkastel-Kues, to appear

Lambert D.L. 1989, in Cosmic Abundances of Matter, AIP Conference Proceedings 183, ed. C.J. Waddington (New York: AIP), p. 168

Lambert D.L. 1992, Astron. Astrophys. Rev. 3, 201

Langer G.E., Kraft R.P., Anderson K.S. 1974, ApJ 189, 509

Langer N., Arcoragi J.-P., Arnould M. 1989, A&A 210, 187

Mathews G.J., Bazan G., Cowan J.J. 1992, ApJ 391, 719

Mazzali P.A., Lucy L.B., Butler K. 1992, A&A 258, 399

Meyer B.S. et al. 1992, ApJ, in press

Meynet G., Arnould M. 1992, in Proc. 2nd Intern. Symp. on Nuclear Astrophysics "Nuclei in the Cosmos", loc. cit.

Möller P., Randrup J. 1990, Nucl. Phys. A514, 1

Németh Zs., Käppeler F. 1992, ApJ 392, 277

Nomoto K., Hashimoto M. 1988, Phys. Rep. 163, 13

Pearson J.M. et al. 1991, Nucl. Phys. A528, 1

Prantzos N., Arnould M., Cassé M. 1988, ApJ 331, L15

Prantzos N., Hashimoto M., Rayet M., Arnould M. 1990, A&A 238, 455

Raiteri C.M., Gallino R., Busso M. 1992, ApJ 387, 263

Raiteri C.M., Busso M., Gallino R., Picchio G. 1991a, ApJ 371, 665

Raiteri C.M., Busso M., Gallino R., Pulone L. 1991b, ApJ 367, 228

Rayet M, Arnould M. 1992, in Radioactive Nuclear Beams 1991, loc. cit., p. 347

Rayet M., El Eid M, Arnould M. 1992, in Proc. 2nd Intern. Symp. on Nuclear Astrophysics "Nuclei in the Cosmos", loc. cit.

Rayet M., Prantzos N., Arnould M. 1990, A&A 227, 271

Reeves H. 1978, in Protostars and Planets, ed. T. Gehrels (Tucson: University of Arizona Press), p. 399

Reeves H. 1981, Patience dans l'Azur (Paris: Editions du Seuil, Collection "Science Ouverte")

Sackmann I.-J., Brothroyd A. I. 1991, in Evolution of Stars: The Photospheric Abundance Connection, loc. cit., p. 275

Seeger P.A., Fowler W.A., Clayton D.D. 1965, ApJS 11, 121

Smith, V.V. 1989, in Cosmic Abundances of Matter, loc. cit., p. 200

Staudt A., Bender E., Muto K., Klapdor-Kleingrothaus H. V. 1990, Atom. Data Nucl. Data Tables 44, 79

Takahashi K. 1992, Comm. Astrophys., in press

Takahashi K., Hillebrandt W. 1992, J. Phys. G, to appear

Takahashi K., Yokoi K. 1987, At. Data Nucl. Data Tables, 36, 375

Takahashi K., Janka H.-Th., Witti J., Hillebrandt W. 1992, in Proc. 6th Intern. Conf. on Nuclei far from Stability, loc. cit.

Thielemann F.-K., Arnould M., Truran J.W. 1986, in Advances in Nuclear Astrophysics, eds. E. Vangioni-Flam, J. Audouze, M. Cassé, J.P. Chièze and J. Tran Thanh Van (Gif-sur-Yvette: Editions Frontières), p. 525

Tondeur F. et al. 1992, in Proc. 6th Intern. Conf. on Nuclei far from Stability, loc. cit.

Wallerstein G. 1990, ApJS 74, 755

Williams R.E. 1988, in Atmospheric Diagnostics of Stellar Evolution: Chemical Peculiarities, Mass Loss and Explosion, Lecture Notes in Physics Vol. 305, ed. K. Nomoto (Berlin: Springer-Verlag), p. 274

Witti J., Janka H.-Th, Takahashi K., Hillebrandt W. 1992, in Proc. Intern. Symposium "Nuclei in the Cosmos", loc. cit.

Wolke K. et al. 1989, Z. Phys. A334, 491

Woosley S.E., Hoffman R.D. 1992, ApJ, in press

Woosley S.E., Howard W.M. 1978, ApJS 36, 285

Infrared colors of S stars and the binary/Tc connection

A. JORISSEN

European Southern Observatory, Garching bei München, Germany

1 INTRODUCTION

The cool peculiar giants of type S were traditionally considered as a transition stage in the M – S – C evolutionary sequence on the asymptotic giant branch (AGB; see e.g. Iben & Renzini 1983). However, in recent years, it has been suggested that S stars lacking Tc (an element with no stable isotopes originally observed by Merrill in 1952 in the spectra of *some* S stars) are linked instead to the barium stars, which are G and K giants with chemical peculiarities similar to those exhibited by S stars. Like barium stars, Tc-deficient S stars appear to belong systematically to binary systems (McClure et al. 1980, McClure 1983, Jorissen & Mayor 1988, McClure & Woodsworth 1990, Brown et al. 1990). The binary nature of Tc-deficient S stars seems now beyond doubt since it relies on (i) periodic radial-velocity variations (Jorissen & Mayor 1988, 1992, Brown et al. 1990), (ii) direct observation of the ultraviolet light of the white dwarf companion (Smith & Lambert 1987, Johnson et al. 1990, 1992), (iii) observations of spectral lines of high excitation (like HeI λ 10830) that are lacking in normal, single giants (Brown et al. 1990), and (iv) infrared colors, which do not reveal any circumstellar dust contrarily to the situation prevailing for the Tc-rich S stars (see also Johnson 1992). This contribution analyzes the infrared colors of S stars in the framework of the Tc-rich/Tc-deficient dichotomy. More details will be given in a subsequent paper (Jorissen et al. 1992).

2 INFRARED COLORS OF S STARS

Figure 1 presents the $(K - [12], K - [25])$ color – color diagram for S stars with a known Tc content, taken from Little et al. (1987), Smith & Lambert (1988) or from new observations (Jorissen et al. 1992). The $K - [12]$ and $K - [25]$ color indices were computed from the (non-color corrected) fluxes at 12 and 25 μm provided by the 2nd edition of the IRAS *Point Source Catalog* (IRAS Science Team 1988) and from the K mag in the *Two-Micron Sky Survey* (Neugebauer & Leighton 1969), or if not available, from the magnitudes provided by Catchpole et al. (1979), Chen et al. (1988) or Noguchi et al. (1991). The absolute calibration of Beckwith et al. (1976) has been used to convert K magnitudes into fluxes (620 Jy correspond to $K=0$). The identification of S stars in the IRAS *Point Source Catalog* was performed on the basis of the match between the IRAS positions and the ones listed by Stephenson (1976).

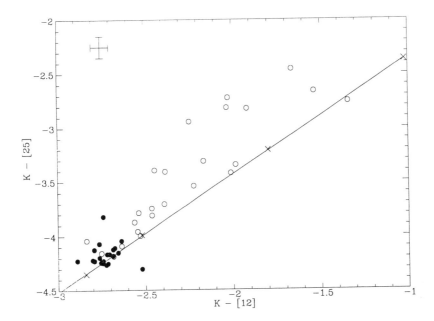

Figure 1. The $(K - [12], K - [25])$ color–color diagram for Tc-deficient (black dots) and Tc-rich S stars (open dots). The index $K - [i]$ is defined as $K - 2.5 \log(620/F(i))$, where $F(i)$ is the (non color-corrected) flux in filter i from the second edition of the IRAS *Point Source Catalog*. Only stars with good quality fluxes at 12 and 25 μm were plotted. The solid line represents black body colors, with crosses corresponding to temperatures of 4000, 3000, 2000 and 1500 K. The error box corresponds to uncertainties of 5 and 10% on the 12 and 25 μm fluxes, respectively.

Figure 1 reveals that Tc-deficient S stars are concentrated in the lower left corner of the color – color diagram, whereas Tc-rich S stars exhibit a much larger scatter. The colors of Tc-deficient S stars are in fact compatible with their photospheric temperature. On the contrary, the much larger scatter exhibited by Tc-rich S stars may be due to lower effective temperatures for these stars, to a flux deficiency in the K band or to excesses in both the 12 and 25 μm bands. A flux deficiency in the K channel (due for example to molecular bands of H_2O in the coolest stars; see e.g. Frogel & Elias 1988) would move the stars along a line of slope 1, close to the black body line represented in Fig. 1. The fact that several Tc-rich S stars are located far away from this line indicates instead that 12 and 25 μm excesses are present. Figure 2, presenting the IRAS ([12] - [25], [12] - [60]) color – color diagram, also suggests that Tc-rich S stars are surrounded by circumstellar dust emitting in these IR bands (van der Veen & Habing 1988). On the contrary, Tc-deficient S stars have normal 25 and 60 μm fluxes given their effective temperatures (note that only an upper limit to the 60 μm flux is available for many Tc-deficient S stars, confirming that they are fainter than their Tc-rich counterparts in the IR).

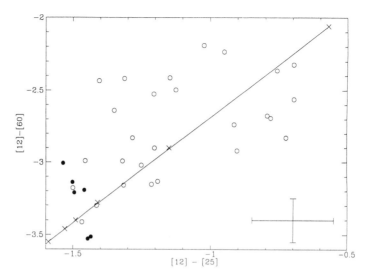

Figure 2. The ([12] - [25], [12] - [60]) color–color diagram for Tc-deficient (black dots) and Tc-rich S stars (open dots). The index $[i] - [j]$ is defined as $-2.5 \log(F(i)/F(j))$, where $F(i)$ and $F(j)$ are the (non color-corrected) fluxes in filters i and j from the second edition of the IRAS *Point Source Catalog*. Only stars with good quality fluxes at 12, 25 and 60 μm were plotted. The solid line corresponds to black body colors; crosses refer to temperatures of 10000 K, 4000 K, 3000 K, 2000 K, 1000 K and 500 K (from IRAS *Explanatory Supplement*, Table VI.C.6). The error box corresponds to uncertainties of 5% on the 12 μm flux and of 10% on the 25 and 60 μm fluxes.

The difference in the infrared colors of Tc-rich and Tc-deficient S stars is also well correlated to their pulsational properties, since Tc-deficient S stars show low-amplitude ($\Delta V < 1$) irregular or semi-regular light variations whereas Tc-rich S stars with large IR excesses are Mira variables (Little et al. 1987, Jorissen et al. 1992).

3 DISCUSSION

The absence of infrared excesses and large light variations for Tc-deficient S stars, contrarily to the situation prevailing for Tc-rich S stars, suggest that the former are less evolved than the latter. This hypothesis is well in line with the suggestion by Jorissen & Mayor (1992) that Tc-deficient S stars populate the upper part of the *first giant branch* instead of the AGB where Tc-rich S stars are located. Frogel & Elias (1988) have indeed shown that in mildly metal-deficient globular clusters, *only* long-period variables with infrared excesses are found *above* the He-core flash luminosity. Moreover, the clear segregation between Tc-deficient and Tc-rich S stars as revealed by their infrared colors provides interesting constraints on the dredge-up history. Busso et al. (1992) suggested that an AGB star could in principle oscillate between Tc-rich and Tc-deficient states, depending on the interpulse duration and frequency of third dredge-ups. In a model case corresponding to an interpulse duration of 10^5 y with dredge-ups every 9 pulses, the star spends about 20% of its AGB lifetime in a Tc-deficient state. That hypothesis is not supported by Fig. 1, since one would then expect to find *some* Tc-deficient S stars among evolved AGB stars exhibiting an IR excess (i.e. for the particular model of Busso et al., 4 to 5 Tc-deficient S stars would be expected among the 18 Tc-rich S stars observed).

4 REFERENCES

Beckwith S., Evans N.J., Becklin E.E., Neugebauer G., 1976, ApJ 208, 390

Brown J.A., Smith V.V., Lambert D.L., Dutchover E., Hinkle K.H., Johnson H.R., 1990, AJ 99, 1930

Busso M., Gallino R., Lambert D.L., Raiteri C.M., Smith V.V., 1992, ApJ, submitted

Catchpole R.M., Robertson B.S.C., Lloyd Evans T.H.H., Feast M.W., Glass I.S., Carter B.S., 1979, SAAO Circ. 1, 61

Chen P.S., Gao H., Chen Y.K., Dong H.W., 1988, A&AS 72, 239

Frogel J.A., Elias J.H., 1988, ApJ 324, 823

Iben I.Jr., Renzini A., 1983, ARA&A 21, 271

IRAS Science Team, 1988, IRAS Catalogues and Atlases. Beichman C.A., Neugebauer G., Habing H.J., Clegg P.E., Chester T.J. (eds.), NASA RP-1190

Johnson H.R., 1992. In: Evolutionary processes in interacting binary stars (IAU Symp. 151), eds. Y. Kondo, R. Polidan, R. Sistero, Dordrecht, Kluwer, in press

Johnson H.R., Ake T.B., Ameen M.M., Brown J.A., 1990, in: Miras to Planetary Nebulae: Which Path for Stellar Evolution?, eds. M.O. Mennessier & A. Omont, Editions Frontières, Gif-sur-Yvette, France, p. 332

Johnson H.R., Ake T.B., Ameen M.M., 1992, ApJ, in press

Jorissen A., Mayor M., 1988, A&A 198, 187

Jorissen A., Mayor M., 1992, A&A, in press

Jorissen, A., Frayer, D.T., Johnson, H.R., Mayor, M., Smith, V.V., 1992, in preparation

Little S.J., Little-Marenin I.R., Hagen-Bauer W., 1987, AJ 94, 981

McClure R.D., 1983, ApJ 268, 264

McClure R.D., Fletcher J.M., Nemec J.M., 1980, ApJ 238, L35

McClure R.D., Woodsworth A.W., 1990, ApJ 352, 709

Merrill P.W., 1952, ApJ 116, 21

Neugebauer G., Leighton R.B., 1969, *Two-Micron Sky Survey*, NASA SP-3047

Noguchi K., Sun J., Wang G., 1991, PASJ 43, 311

Smith V.V., Lambert D.L., 1987, AJ 94, 977

Smith V.V., Lambert D.L., 1988, ApJ 333, 219

Stephenson C.B., 1976, Publ. Warner & Swasey Observ. 2, 21

van der Veen W.E.C.J., Habing H.J., 1988, A&A 194, 125

MS and S Stars with and without Tc

M. BUSSO(1), R. GALLINO(2), D.L. LAMBERT(3), C.M. RAITERI(1), V.V. SMITH(3)

(¹) Osservatorio Astronomico di Torino

(²) Istituto di Fisica generale - Università di Torino

(³) University of Texas at Austin and Mc Donald Observatory

1. CONSTRAINTS ON s-PROCESSING IN MS AND S STARS

MS and S stars are members of a group of red giants (which includes also C–stars of N–type and Ba–stars) in which heavy (A > 80) s–process nuclei and carbon show correlated enhancements due to mixing of freshly synthesized elements from the interior (see e.g Lambert, 1989). They are known to be in the thermally pulsing asymptotic giant branch (TP-AGB) evolutionary phase; no other phase has been observed to produce s–elements with A > 80 and it is then natural to assume that they should allow a close analysis of how the abundances of these nuclei in the solar system are built. Several arguments were presented showing that common AGB stars have relatively low masses (M < 2 $M_\odot$: Feast, 1989) and low luminosities (L< 10^5 $L_\odot$: Blanco et al., 1980). This is confirmed by their normal abundance of heavy Mg isotopes (Clegg et al. 1979), that would instead be overabundant in more massive objects, where the He–burning shells have temperatures high enough (T > 3 10^8 K) to efficiently activate the neutron source $^{22}Ne(\alpha,n)^{25}Mg$.

Constraints from nuclear physics and meteoritic sciences have confirmed that the physical conditions found in models of low mass TP–AGB stars (i.e. of MS and S stars) are the same required to explain the solar system *main s*–component (i.e. that including s–nuclei heavier than A ≃ 80) and isotopic anomalies in interstellar SiC grains extracted from carbonaceous meteorites (Käppeler et al., 1990; Gallino et al., 1990). In particular, moderate temperature values (T ≤ 2 10^8 K) should dominate in the production site, in order to avoid the overproduction of critical nuclei like ^{80}Kr (Ott et al., 1988; Lewis et al., 1990). This indication has to be reconciled with the existence of the s–process thermometers (e.g. ^{152}Gd and ^{154}Er), and of other temperature-sensitive isotopes like ^{148}Sm and ^{150}Sm, which require that at least a small contribution to the neutron exposure is released at T> 2.5 10^8 K. The result is that the *main* component of the s–process has to be produced in two steps, at different temperatures, the bulk of the neutron flux being released at T < 2 10^8 K. All these requirements are met in the convective He–shells of low mass TP–AGB stars (M < 3 $M_\odot$) where, according to observations, neutrons cannot be produced by the ^{22}Ne$(\alpha,n)^{25}$Mg reaction. In such an environment, the alternative neutron source ^{13}C$(\alpha,n)^{16}$O has to be at work. Its possible occurrence is however a matter of discussion, mainly because it requires a small amount of protons to be mixed from the convective envelope into the ^{12}C–rich region, to produce ^{13}C, a mechanism whose modelling requires improvements in the current treatment of convective and semiconvective mixing (Hollowell and Iben, 1988; Boothroyd and Sackmann, 1988; Iben, 1983).

In order to establish whether independent constraints on the ^{13}C source can be derived from observations, in this paper we try to fit the C and *s*–abundances of MS and S stars using nucleosynthesis models for low mass TP-AGB stars. In the absence of other indications, the amount of ^{13}C burnt is considered as a free parameter. The observational material is mainly taken from Smith and Lambert (1990).

2. FITTING THE C- AND s-ELEMENT OBSERVATIONS

The essential features of our models are the following: (i) each case is made up of 40 subsequent pulses, in which the nucleosynthesis is computed using a network of α, p– and n–captures including 450 isotopes; cross sections are from Beer et al. (1992); (ii) standard AGB models are adopted, in which the main parameters (pulse mass, overlapping, interpulse time) are univoquely determined by the core mass M_H; (iv) dredge-up is assumed, with a parametrized repetition in time; the mixed mass is varied between 5 10^{-4} and 10^{-3} $M_\odot$ as in models by Lattanzio (1989) and by Chieffi et al.,(1992); (v) an initial mass of 1.5 $M_\odot$ is adopted, with metallicities in the range Z = 0.01 − 0.02; (vi) the envelope enrichment derives by the interplay of mixing from the intershell zones and mass loss, with various mass loss criteria. A similar, though somewhat simplified, approach together with a critical analysis of the observational material is presented in Busso et al. (1992), where the dependence of results on model parameters is also discussed. Here we shall summarize the main findings.

Figure 1 and 2 show some model envelope composition sequences compared to the loci of stellar observations. Both cases correspond to a final mass of 10^{-2} $M_\odot$ mixed to the envelope. In Figure 1 mixing is assumed to start at various pulse numbers, using the same thermal pulse sequence, reaching a neutron exposure $\tau_0 \sim 0.3$ mb^{-1}; in Figure 2 instead, the internal material is mixed to the surface only after an asymptotic distribution is reached in the shell, and using cases with different final exposures. The figures show how the details of the dredge–up occurrence control the results: stars with different Nd/Zr ratios can be interpreted either by a different neutron exposure or by a different mixing scheme. This implies that *s*–process nuclei alone are not sufficient to derive the effective τ_0 value experienced by a particular star.

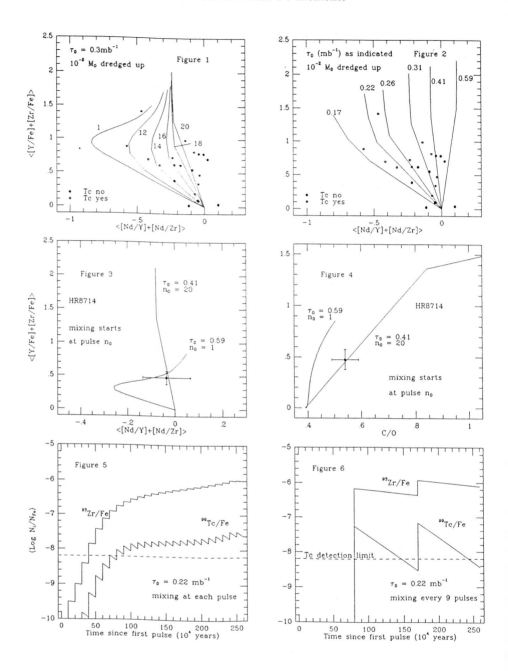

In order to disentangle the effects of nucleosynthesis from those of mixing, one can rely on the independent constraint given by the abundance of ^{12}C, a nucleus which is produced at practically the same efficiency in each pulse ($X_C \simeq 0.23$) and whose envelope enrichment depends on dredge–up but not on τ_0. Figures 3 and 4 show this for a particular star (HR8714). It has s–abundances (Figure 3) that can be fitted in more than one way, differently combining dredge-up and neutron exposure. However, when also the C–enrichment is monitored, only one sequence remains acceptable, thus providing a unique solution (Figure 4). In this way we find that S stars are characterized by τ_0 values in the range $0.2 - 0.4$ mb^{-1}. With the present observational uncertainties ($\simeq 0.2$ dex), however, this conclusion does not inform much on the ^{13}C production and burning: provided that a certain neutron exposure is attained and that mixing to the surface occurs at a certain rate (to match the s/C ratio), a fit to observations is possible independently on the other parameters (^{13}C ingestion rate, shell mass, mass loss rate).

Important constraints on the evolutionary status of the studied stars and on the frequency of mixing episodes can come from the observations of the long–lived unstable nucleus ^{99}Tc ($t_{1/2} = 2.~10^5$ y). Its decay in the envelope would deplete it if the dredge–up does not repeat frequently enough (see Figures 5 and 6). Hence the observational evidence that single S stars have Tc in their spectra and that non-Tc S stars are evolved representatives of mass-tranfer binaries suggests that dredge–up, once started, is a frequent phenomenon.

Finally, following carefully the temperature evolution in the region where ^{13}C is ingested (from Chieffi et al. 1992) we have evaluated the extra luminosity L produced by the $^{13}C(\alpha,n)^{16}O$ reaction and by neutron captures with respect to the normal 3α luminosity. With our shell bottom temperature at ^{13}C ingestion ($T_8 \geq 1.55$) and using standard ingestion rates ($\simeq 3$–$5~10^{-5}$ M$_\odot$/y), $L_{3\alpha}$ always prevails, at least if the amount of ^{13}C ingested does not exceed some 10^{-6} M$_\odot$. Different results have been obtained by Bazan and Lattanzio (1992), and are probably due to a too early ingestion of ^{13}C; indeed, the pulse temperature is slowly increasing in time while the ratio $\lambda(3\alpha)/\lambda(^{13}C(\alpha,n)^{16}O)$ roughly behaves as T^{20}.

REFERENCES
Bazan, G. and Lattanzio, J.C.: 1992 (preprint)
Beer, H., Voss, F., and Winters, F.F. 1992, *Ap. J.* (in press).
Blanco, V.M., McCarty, M.F. and Blanco, B.M.: 1980, *Ap. J.* **242**, 938.
Boothroyd, A. I., and Sackmann, I.-J.: 1988 *Ap. J* **328**, 653.
Busso, M., Gallino, R., Lambert, D.L., Raiteri, C.M. and Smith, V.V.: 1992, Ap.
    J. (in press).
Chieffi, A., Limongi, M. and Straniero, O.: 1992, (private communication).
Clegg, R., Lambert, D.L., and Bell, R.A.: 1979, *Ap. J.*, **234**, 188.

Feast, M.W.: 1989, in *Evolution of Peculiar Red Giant Stars*, IAU Coll. n. 106, ed. H. R. Johnson, B. Zuckerman, (Cambridge: Cambr. Univ. Press) p. 26.

Gallino, R., Busso, M., Picchio, G. and Raiteri, C.M.: 1990, *Nature* **348**, 298.

Hollowell, D.E. and Iben, I. Jr.: 1989, *Ap. J.* **340**, 966.

Iben, I.Jr.: 1983, *Ap. J. Lett.* **275**, L65.

Käppeler, F., Gallino, R., Busso, M., Picchio, G. and Raiteri, C.M.: 1990, *Ap. J.*, **354**, 630.

Lambert, 1989 in *Evolution of Peculiar Red Giant Stars*, ed. H.R. Johnson and B. Zuckerman, (Cambridge: Cambridge Univ. Press), p. 101.

Lattanzio, J.C.: 1989, *Ap. J.* **344**, L25

Lewis, R.S., Amari, S., and Anders, E.: 1990, *Nature* **348**, 293.

Ott, U., Begemann, F, Yang, J., and Epstein, S.: 1988, *Nature*, **332**, 700.

Smith V.V. and Lambert, D.L.: 1990, *Ap. J. Suppl.* **72**, 387.

On the Predictive Power of Nuclear Models

G. AUDI and C. BORCEA[1]

CSNSM, IN2P3-CNRS, Laboratoire René Bernas,
Bâtiment 108, F-91405 Orsay Campus, France

One of the most important nuclear physics ingredients that is determinant in nuclear astrophysics is certainly the mass of a nucleus. For example, in the rapid neutron capture process (r-process), Goriely and Arnould [92Gor] have listed already five entries for the nuclear masses. Either directly in the position of the neutron drip line, the neutron separation energies (instrumental for determining the neutron capture cross-sections) and the Q_β (for the β-decay rates), or indirectly in the fission rates or the β-delayed processes rates.

Unfortunately it is not an information easy to get, not to say impossible, in the regions of interest for astrophysics. One could try extrapolations starting from the experimentaly known masses. But it is a quite long way to the drip lines or at least to the regions of the chart where S_n, the neutron separation energy, is of the order of 1 or 1.5 MeV. Any tentative along these lines would certainly fail.

Experiment being powerless in this respect, one would like to turn to a good theory. But there is no complete theory of the nuclear matter and of the nucleus which is ready to give all the needed informations. The way out, up to now, has been to try to marry both of them. Based on sound ideas about the nucleus many theoreticians tend to formulate simple enough equations to be calculable in large regions of the chart, and the parameters of these formulae have to fit the known masses. It is a necessary condition, but alas not a sufficient one.

As a direct consequence, strong divergences often occur between theories as discussed recently twice: by a nuclear physicist [90Hau] and by astrophysicists [92Gor] who showed the dramatic differences in the results they derive for the r-process depending on the mass formula they start with: the mass mismatch between two formulae near the drip line could vary up to 20 MeV; the neutron drip line could move up to 16 mass units further away and the resulting r-process abundances could change by more than one order of magnitude.

---

[1] IN2P3-Visitor, on leave from the Institute for Atomic Physics, Bucharest.

One could distinguish two types among the existing models for nuclear masses: the ones that use local relations and the ones that start from a nuclear model calculation with different ingredients of phenomenology. While for the first type, one may expect that a gradual deterioration can appear when getting to far outer masses, for the second type, in principle, every physical premise is contained so that, even if fitted to a particular sample, one should expect not too large divergences. This should be especially true for the theories that use a relatively small number of parameters.

The aim of the present contribution is to examine thoroughly the reliability of the existing models through the predictions they yield for nuclear masses. We have examined the structures they predict for the Surface of Masses, their predictive power and their divergences.

1 STRUCTURE OF THE MASS SURFACE

The experimental Surface of Masses is characterized by the very smooth behaviour of its four sheets (one for each type of parity in N and Z) manoeuvering together in a 3-dimensional space (N, Z and Mass). This smoothness is interrupted in some places by "accidents" that almost always can be associated with a relatively violent change in one or more physical parameters (e.g. shell closures, shape transitions). Unfortunately, visualizing and handling these sheets in a tridimensional space together with the constraints concerning their relative distance is not simple. A way out is to look at the "derivatives" of the surface of masses, like separation energies S_n, S_{2n}, decay energies Q_α, Q_β or pairing energies Δ_{nn}, Δ_{np} [85Bos]. These derivatives, not only exhibit the same smoothness property as the surface itself, but still more important, they enhance the physical structures. The ability of a mass formula to reproduce the general smoothness of the Surface of Masses and the clearly superposed physical structures qualifies it as a candidate for the prediction of still unknown nuclei.

For example, fig. 1 shows that the predictions of Möller and Nix [88Mol] have some oscillatory trends not present in the actual mass surface. These oscillations may deteriorate the accuracy of a prediction on a local scale but do not *a priori* imply strong deviations when going towards the drip lines. At the opposite, another model [88Sat] does reproduce quite well the masses of nuclei close to the bottom of the valley of stability but exhibits too large oscillations beyond the known masses rendering its predictions questionable in the case where far extrapolations are needed, even though it is a "global" model. More generaly, the Wigner correction terms for N=Z which are very important for predicting masses of proton rich nuclei are considered only in a few mass predictions and often, when taken into account, they do not result in a structure with the same amplitude as that displayed by the experimental masses.

In these examples, deviations from the experimental surface and from the smoothness

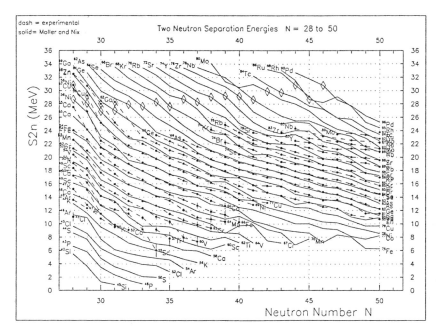

Figure 1 : Predictions of the theory of Möller and Nix [88Mol] (solid lines) are compared to the experimental masses (points, error bars and dashed lines) in an S_{2n} representation for N=28-50. Diamonds locate the so-called "Wigner" nuclei (N=Z). Oscillations in the theoretical curves indicate lack of smoothness of the formula. Constraints to limit the variation of some parameters from one nucleus to another could certainly improve the predicted Surface of Masses. Extension of the predictions up to the drip line is highly desirable here, as for all "global" models, in order to make "far" extrapolations possible.

property do not all have the same influence on the ability to predict masses very far from stability. This analysis, while making a first selection, has to be completed by a study of the predictive power and of the divergences among the models.

2 PREDICTIVE POWER OF THE MODELS

Since the 1986 atomic mass table [88Wap] some 175 new masses have been measured and this sample can be used to test the quality of predictions made by each model. To the 10 models already presented in the "1986-1987 Atomic Mass Predictions" of Haustein [88Hau] we added the ones of Duflo [92Duf] based on doublets and triplets and the ETFSI model (extended Thomas-Fermi plus Strutinsky integral) of Pearson et al. [92Pea], both adjusted also on the 1986 experimental masses, and also the "semi-empirical shell model formula" of Liran and Zeldes [76Lir].

To calculate the deviations of the models from the experimental values, we need

Table 1 : Compensated Linear Deviation (*cld*) of 13 mass models
calculated for 2 different sets of measured nuclear masses.

	MODEL	type	cld^a keV	nb	cld^b keV	nb	cpr
1	Pape and Antony	local	154	(85)	72	(3)	
2	Dussel,Caurier,Zuker	local	176	(1325)	135	(87)	
3	Möller and Nix	global	595	(1589)	632	(168)	1.06
4	Möller et al.	global	525	(1589)	593	(168)	1.13
5	Comay,Kelson,Zidon	local	149	(1628)	387	(174)	2.59
6	Satpathy and Nayak	global	298	(1589)	782	(171)	2.63
7	Tachibana et al.	global	321	(1653)	448	(175)	1.39
8	Spanier and Johansson	global	495	(884)	920	(63)	1.86
9	Jänecke and Masson	local	64	(1629)	287	(174)	4.52
10	Masson and Jänecke	local	132	(1578)	424	(168)	3.22
11	Duflo	global	288	(1381)	450	(130)	1.57
12	Liran and Zeldes	global	182	(1582)	755	(166)	4.14
13	Pearson et al.	global	543	(1489)	586	(146)	1.08

The figures in parentheses indicate the number of nuclei from the set of experimental
masses for which a given theory makes predictions.
a) calculated for the set of 1655 masses of the 1986 atomic mass table [88Wap]
b) calculated for the set of 175 new masses measured since 1986

to get a feeling of the "adherence" of the theoretical mass surface to the real one.
We made thus the choice of the **"compensated linear deviation"** (*cld*) [81Aud]
$cld = \frac{1}{n}\Sigma \max\{(|m_{th} - m_{exp}| - \sigma_{exp}), 0\}$, which in a way represents the volume comprised between the theoretical surface and the experimental one.

Table 1 presents the results of calculated deviations for the 13 theories and 2 sets of
experimental masses: the 1655 masses of the 1986 table [88Wap] and the set of the
175 new masses measured since 1986. There is a clear increase of the deviations for
the set of new masses. Last column gives the ratios between the deviations for the
two samples, which may be called "close predictive ratios" or **cpr**. The **cpr** for
the "global" models are systematically better than for the "local" ones.

3 DIVERGENCES OF THE MODELS

Now that the predictive power of the models have been evaluated, one may consider
to make an adequate statistical treatment of the predictions, provided the divergences
between models are within reasonable limits (see next paragraph) and that at least
[90Hau] six models give predictions for a given nuclear mass. A weighted average is
being considered. The uncertainty associated with each theory's prediction in that
average is taken as: $\sigma_i^{th} = cld_i \times cpr_i^2$. We observe that the dispersion among the
predictions from different models increase drastically when one approaches the drip
lines. However, for a relatively large region, this dispersion is in reasonable limits.

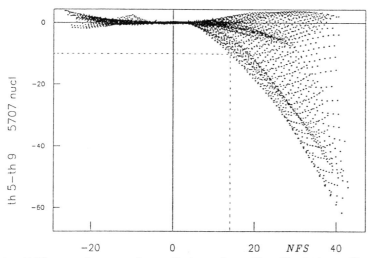

Figure 2 : Differences between the predictions of two "local" relations: Comay, Kelson and Zidon [88Com] and Jänecke and Masson [88Jan] plotted as a function of "neutrons from stability" for 5707 nuclei. The strong coherence around stability ($NFS = 0$) is due to the very good agreement of both predictions with the known masses. Divergences exceeding 10 MeV start at $NFS = +14$ on the neutron-rich side.

The averages of theoretical masses over the whole chart yielded results for 5018 nuclei of which 4227 had a dispersion below 1 MeV. These crude results will require naturally careful individual and global analysis.

Following Haustein [84Hau] we define a distance with respect to the stability line called "neutrons-from-stability": $NFS = N - Z - (0.4A^2)/(200 + A)$. A similar norm was proposed by Goriely and Arnould [92Gor] who defined a "neutron surplus", which is more precise but less convenient than Haustein's NFS. To get an overall idea about the zone where different predictions are still coherent, one may, similarly to [92Gor], compare any two theoretical predictions as a function of NFS, and also observe the deviations of a given model from the calculated "average" of predictions. Though very efficient, this method suffers however from the fact that the domains of predicted masses vary widely from one model to another. Many models do not venture to the vicinity of drip lines i.e. where we are interested to compare their predictions. This fact is in turn reflected by the size of the above set of averaged values. Anyway, as a general trend one can mention a strong divergence of "local" models among themselves and with respect to the "average" when departing from the measured masses (fig. 2). This tendency is also present but much attenuated, for those "global" models that make predictions in a large enough domain. Conclusion of the NFS analysis, confirmed by the statistical treatment above, is that going beyond $|NFS| = 20$ is hazardous for the time being. This number could define the frontier

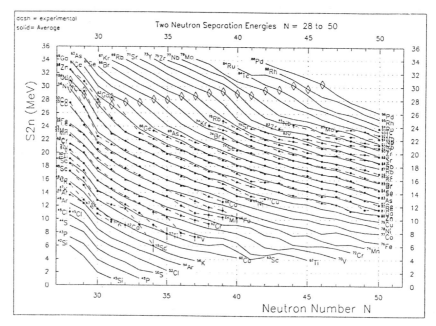

Figure 3 : Average values of theories, obtained as described in the text, are compared to the experimental masses (points, error bars and dashed lines) in an S_{2n} representation for N=28-50. The smooth character of the curves is preserved even for outer masses. Further smoothing of these averages may be considered in the production phase of our work. Extension of these predictions on the neutron-rich side is limited, due to the small number of models making predictions far enough from stability.

separating the medium-range extrapolations from the far extrapolations.

The averaged values obtained for the medium-range predictions can be compared with the experimental masses. An example is shown in fig. 3. The smooth character of the curves is preserved even for the region of the outer masses. Of course, the medium-range extrapolated values obtained in this way are far from being perfect, but they can be considered as the best estimate presently possible for these masses. Each theory contains some truth and certainly some approximations. Being constructed quite differently from each other, one may expect that the effects of these approximations do not coincide and would be random. If this is the case, then averaging would be licit and could be considered as the best way out until drastic improvements in theories is achieved.

Towards the neutron drip line, we have seen that the above procedure fails because of the important divergences among theories. This effect can be limited if in the far extrapolations, the "local" models are not used. To compensate the resulting impoverishment in models, it would be highly desirable to have all "global" models extend their predictions up to the neutron drip line.

4 CONCLUSIONS

We have compared the behaviour of the Surface of Masses, as predicted by 13 nuclear models, with the experimental surface. We have evaluated the predictive power of these models and analysed their divergences when predicting masses far from stability. The result of these analyses showed that, based on existing models, the medium range extrapolations are reliable while extrapolations toward drip lines (far extrapolations) are to be taken with much care untill further improvements in the theories make them congruent in these regions.

Presently, the methods for close and medium-range extrapolations are defined and tested. Serial production of such predictions is the next step, but that will require much care and time. Close extrapolations will be produced very soon to be included in the next table of experimental masses. Medium-range extrapolations may be evaluated and enter a special table, if strong interest for astrophysics is expressed. Special care will be taken to ensure the continuity property of the Surface of Masses at the border between the three regions: experimental masses, close extrapolations and medium-range extrapolations.

REFERENCES

[76Lir] S. Liran and N. Zeldes, 1975 Mass Predictions, ADNDT **17** (1976) 431.

[81Aud] G. Audi, PhD thesis, Université de Paris-Sud, Orsay (1981), unpublished.

[84Hau] P. E. Haustein, Proc. 7-th Int. Conf. on Atomic Masses and Fundam. Const. AMCO-7, Darmstadt 1984, O. Klepper ed., p.413.

[85Bos] K. Bos, G. Audi and A. H. Wapstra, Nucl. Phys. **A432** (1985) 140.

[88Com] E. Comay, I. Kelson and A. Zidon, in [88Hau] p.235.

[88Hau] P. E. Haustein, 1986-1987 Atomic Mass Predictions, ADNDT **39** (1988) 185.

[88Jan] J. Jänecke and P. J. Masson, in [88Hau] p.265.

[88Mol] P. Möller and J. R. Nix, in [88Hau] p.213.

[88Sat] L. Satpathy and R. C. Nayak, in [88Hau] p.241.

[88Wap] A. H. Wapstra, G. Audi and R. Hoekstra, in [88Hau] p.281.

[90Hau] P. E. Haustein, Exotic Nuclear Spectroscopy, W. C. McHarris editor, Plenum Press, New York, 1990.

[92Duf] J. Duflo, submitted to Nucl. Phys., March 1992.

[92Gor] S. Goriely and M. Arnould, submitted to Astron. Astrophys., February 1992.

[92Pea] M. Pearson and F. Tondeur, private communication and Y. Aboussir, J. M. Pearson, A. K. Dutta, F. Tondeur, preprint.

The ETFSI Nuclear Mass Formula

J.M. PEARSON[1], Y. ABOUSSIR[1], A.K. DUTTA[2],

F. TONDEUR[3]

[1] Laboratoire de Physique Nucléaire, Université de Montréal

[2] Laboratoire de Physique Nucléaire, Université de Montréal, and School of
Physics, Devi Ahilya University

[3] Institut d'Astrophysique, Université Libre de Bruxelles, and Institut Supérieur
Industriel de Bruxelles

ABSTRACT

In an attempt to develop an improved mass formula, and in particular to reduce the number
of free parameters, we have developed a high-speed approximation to the Hartree-Fock
method that uses the semi-classical extended Thomas-Fermi method with shell corrections
calculated by the Strutinsky-integral method. The 9 parameters of the underlying Skyrme
and δ-function pairing forces are fitted to all 1492 mass data for $A \geq 36$ with an rms error
of 0.730 MeV. We also calculate several fission barriers with the same force and find a
reasonable agreement with experiment. The present mass formula suggests that towards its
end-point the r-process path will pass through a region of significantly greater neutron
richness than predicted by the FRDM mass formula of Möller et al.

1 INTRODUCTION

The r-process depends crucially on the binding energies and fission barriers
among other properties, of nuclei so highly neutron (n)-rich that in most
cases there is no possibility of being able to measure them in the laboratory. It
thus becomes of the greatest importance to be able to make reliable
extrapolations of these properties away from the known region of nuclei,
where measurements *can* be performed, out towards the n-drip line.

Hitherto, the necessary extrapolation of binding energies and barriers was usually performed by means of a semi-empirical mass formula based on some form of the liquid-drop(let) model (DM) of the nucleus, with shell and pairing corrections added. This model has been developed to a high degree of refinement since the original proposals of Bethe and von Weizsäcker in 1935, mainly as a result of the efforts of workers at Los Alamos and Berkeley. In its latest form, the so-called finite-range droplet model (FRDM), 1540 mass-data points are fitted with an rms error of only 0.658 MeV [1, 2].

However, *any* mass formula will fit all the available data if it has enough free parameters, and while a precise fit to the data is a necessary condition for a reliable extrapolation, it is not sufficient. Rather, we can well imagine that two different mass formulas give comparable fits to the data but extrapolate differently. In such a case the one with the better theoretical foundation would be preferred; generally speaking, the theoretically superior of the two would be characterized by a smaller number of parameters. Now the FRDM has some 25 parameters, and astrophysicists might feel more comfortable if they could base their extrapolations on mass formulas that obtained a comparable fit to the data with a somewhat lower number of parameters. It is essentially this consideration that has motivated the present project.

One conspicuous theoretical weakness of all DM mass formulas, the rectification of which might be expected to lead to comparably high-quality fits with fewer parameters, is the lack of coherence between the macroscopic and microscopic parts, i.e., between the DM terms on the one hand, and the the shell-model and pairing corrections on the other hand. To be more specific, the calculation of the microscopic terms involves the use of a single-particle (s.p.) potential, and since this must be generated by the distribution of nucleons in the nucleus it constitutes a link between the macroscopic and microscopic parts. Now the actual way in which the s.p. potential is generated from the nucleon distribution is by folding some two-body force over the latter, but there is no unambiguous prescription in DM mass formulas for choosing this force. Furthermore, the density distribution itself is determined only in a very crude way in DM mass formulas (including the FRDM).

However, this difficulty is avoided when the binding energy is calculated by the Hartree-Fock (HF) method, simply because there is no separation of the energy into macroscopic and microscopic parts. Thus a mass formula based on the HF method (or rather on the HF-BCS method, since pairing always has to be taken into account) will be more theoretically secure than one based on the DM model. This method represents, in fact, the most fundamental approach to the mass formula that has any chance of reaching the required level of precision, even though it is much less rigorous than an approach based on the "real" nuclear forces.

The ideal procedure to be followed then would consist in taking some suitable form of effective interaction and fitting its parameters, along with those of the pairing force, to all the data on masses, fission barriers, radii, etc., exactly as with the DM mass formulas. Unfortunately, the large amount of computer time needed to calculate deformed nuclei in the HF-BCS method prohibits a systematic development of a mass formula on these lines.

Caught between the theoretical limitations of the DM mass formulas and the computational impracticality of the HF-BCS method, we have developed over the last few years the ETFSI approximation to the HF method. This approach is based on the extended Thomas-Fermi (ETF) approximation to the HF method for a Skyrme-type force, with shell corrections calculated by what we call the Strutinsky-integral (SI) method. Pairing is handled in the BCS approximation, with a δ-function pairing force.

The ETFSI-BCS method is essentially equivalent to the HF-BCS method in the sense that when the underlying force is fitted to the data the extrapolations out to the n-drip line are very close to those given by the HF-BCS method. Nevertheless, the ETFSI-BCS method is computationally so much more rapid than the HF-BCS method that it offers a practical approach to the ultimate problem of constructing a mass table, and the first version of such a mass table, ETFSI-1, has now been completed. It covers the range $36 \leq A \leq 300$, and goes out to both drip lines.

Detailed accounts of the ETFSI method have already been given in a series of four papers[3-6], referred to here as I-IV, respectively, and here we shall limit ourselves mainly to recalling the main features. But we shall also take this opportunity of comparing the ETFSI-1 table with the FRDM table, the full version of which became available to us only since our last publication.

2 THE ETFSI METHOD

2.1 ETF and the Skyrme Force

The basis of the method is the generalized Skyrme force

$$v_{ij} = t_0(1+x_0 P_\sigma)\delta(\, r_{ij})$$
$$+ t_1(1+x_1 P_\sigma)[p_{ij}{}^2\delta(\, r_{ij}) +\text{h. a.}]/(2\hbar^2)$$
$$+ t_2(1+x_2 P_\sigma)\, p_{ij}.\delta(\, r_{ij})\, p_{ij})/\hbar^2$$

$$+ t_3(1+x_3P_\sigma) [\rho_{qi}(\mathbf{r}_i)+\rho_{qj}(\mathbf{r}_j)]^\gamma \delta(\mathbf{r}_{ij})/6$$
$$+ (i/\hbar^2)W_0(\sigma_i+\sigma_j)\cdot \mathbf{p}_{ij}\times\delta(\mathbf{r}_{ij})\, \mathbf{p}_{ij} \qquad (1)$$

(see paper I for a discussion of the form we have adopted for the density dependence; the index q denotes n or p, according to whether the term in question relates to neutrons or protons, respectively). To this we add the constraints

$$x_2 = -(4+5x_1)/(5+4x_1) \qquad (2a)$$
$$t_2 = -t_1(5+4x_1)/3 \qquad (2b)$$

in order for the effective nucleon mass to be equal to the real mass, $M_q^* = M$, a choice which allows the density of s.p. states near the Fermi surface to be well reproduced without having to take particle-vibration coupling into account. Fitting the density of s.p. states near the Fermi surface helps to obtain correct masses, and has also been shown to be required for a good description of fission barriers.

For the ETF semi-classical approximation to the kinetic-energy and spin-current densities, τ_q and J_q, respectively, we use the full fourth-order (in powers of $\hbar$) expansions of Grammaticos and Voros[7, 8]. The energy density $\mathscr{E}(\mathbf{r})$, which gives the total energy as

$$E_{ETF}= \int \mathscr{E}(\mathbf{r})d^3\mathbf{r} \ , \qquad (3)$$

becomes a function of the nucleon densities ρ_q and their gradients, the full expressions being given in papers I and II.

For spherical nuclei the density distributions are parametrized according to the Fermi form,

$$\rho_q(\mathbf{r}) = \frac{\rho_{0q}}{1+\exp[(r-C_q)/a_q]} \qquad (4)$$

the total ETF energy being minimized with respect to the four parameters C_n, C_p, a_n, and a_p (the ρ_{0q} are fixed by normalization). As for deformed nuclei, we adopt the (c, h) parametrization of Brack et al[9], as described fully in paper II. This means that only axially symmetric deformations can be considered, but allowing triaxial deformations may not be of much consequence for total binding energies.

2. 2 Shell Corrections

It is inevitable with the ETF method that the energy will vary smoothly from one nucleus to another, since the shell effects are effectively lost in truncating the semi-classical expansions. Thus we are forced back to a microscopic-macroscopic approach, and are obliged to add shell corrections. However, compared to the DM mass formulas there is an important difference, since we can now determine quite unambiguously the s.p. fields appearing in the s.p. Schrödinger equation,

$$[(-\hbar^2/2M)\nabla^2 + U_q(\mathbf{r}) + \mathbf{W}_q(\mathbf{r}).\{-i\nabla\times\sigma\}] \, \phi(\mathbf{r}) = \epsilon\phi(\mathbf{r}) , \qquad (5)$$

the solutions to which are required for the calculation of the shell (and pairing) corrections: one simply folds the same Skyrme-type force involved in the ETF functional over the density distribution emerging from the macroscopic part of the calculation (see papers I and II). There is thus a high degree of coherence between the macroscopic and microscopic parts, the unifying factor being the Skyrme-type force that underlies both.

Neglecting pairing for ease of presentation, we note that the shell correct-ions, which determine the total energy according to

$$E \simeq E_{ETF} + \delta , \qquad (6)$$

can be written, according to the Strutinsky theorem, as

$$\delta = \sum_{q=n,p} \left({\sum_{\mu}}' \epsilon_{\mu, q} - \sum_{\mu} \tilde{\epsilon}_{\mu, q} \right) , \qquad (7)$$

where the first term on the right-hand side represents the sum over all occupied states (neutrons and protons) of the s. p. energies corresponding to Eq. (5), while the second term is a smoothed form of the first. To calculate this smoothed term we could in principle have used the conventional Strutinsky energy-averaging method, but it is well known that besides being cumbersome this method contains some ambiguities, mainly because of the continuum s. p. states, which means that they may be expected to become particularly troublesome towards the drip lines. These problems are avoided by a direct application of the Strutinsky theorem expressing the smoothed term in Eq. (7) as

$$\sum_{\mu, q} \tilde{\epsilon}_{\mu, q} = \sum_{q=n, p} \int d^3r \left(\frac{\hbar^2}{2M} \tilde{\tau}_q + \tilde{\rho}_q U_q + \tilde{\mathbf{J}}_q . \mathbf{W}_q \right) . \qquad (8)$$

Here τ_q, ρ_q and $\mathbf{J}_q$ are the smoothed densities obtained in the macroscopic part of the calculation, while U_q and $\mathbf{W}_q$ are the corresponding fields, appearing in Eq. (5).

This prescription for shell corrections is very simple to apply in our case and is quite unambiguous, even at the drip lines. Note, however, that it cannot be used with drop(let) models, mainly because the distribution of the DM energy between potential and kinetic terms is not specified, but also because the distributions ρ_q and J_q giving rise to the s. p. fields are not known. It is this prescription that we call the Strutinsky-integral (SI) method, to distinguish it from the familiar energy-averaging method, although in reality it does not deserve a special name, since it is nothing more than the Strutinsky theorem rendered explicit. Eq. (8) was apparently first written down by Chu et al[10], but it does not seem to have been used very much in the past. We are indebted to Byron Jennings for pointing out this equation to us, since it is absolutely essential to the feasibility of our project.

2. 3 Pairing
We handle this by doing BCS (blocking) with a δ-function force,

$$v_{ij} = V_\pi \, \delta(\, \mathbf{r}_{ij}) \, . \tag{9}$$

Although this increases the computer time as compared to the "constant-G" model, only a single parameter V_π, the same for all nuclei and for both neutron and proton pairing, is involved. Thus insofar as the data are well fitted, we can expect a more reliable extrapolation out to the exotic regions. However, as we shall see, our model might be a little too simple to reproduce the observed pairing effects with the required precision.

2. 4 Correction for Rotational Energy
For deformed nuclei and fission barriers we subtract from the total computed energy the spurious rotational energy

$$E_{rot} = \frac{\hbar^2}{2\,\mathcal{J}} <J^2>$$

$$\tag{10}$$

(see paper III for the determination of $\mathcal{J}$, the moment of inertia: it is not fitted to the mass data).

2. 5 Comparison with HF
For a given set of force parameters the ETFSI method overbinds nuclei by between 3 and 7 MeV, as compared to HF (this represents the difference

between the characteristic overbinding of the ETF method and the underbinding associated with the restricted form (4) of the density profile). Now while errors this large are inacceptable in modern mass formulas it is not inevitable that this will pose a problem in practice, since whatever method is being used the force parameters are always fitted to the data. Rather, the crucial question is: when the ETFSI method is extrapolated out to the n-drip line, will it agree with the HF extrapolation? In papers I and II we showed that the discrepancy between the two methods at the n-drip line is less than 1 MeV for the total energy and fission barriers, and less than 0. 5 MeV for neutron separation energies (S_n) and beta-decay energies (Q_β). We thus conclude that while the ETFSI method is very much faster than HF it gives essentially the same extrapolations from known nuclei out to the n-drip line, for a given *form* of force.

2. 6 Interpolation

Even though the total energy suffers from shell-model fluctuations, the ETFSI method expresses it entirely in terms of quantities that vary smoothly with respect to N, Z, and the deformation parameters c and h. This enables us to make extensive use of interpolation in constructing the mass table, with the complete deformed ETFSI calculation being performed only for a restricted number of key nuclei (see III). Without interpolation the amount of computer time required to calculate the more than 6000 nuclei that lie between the drip lines, only a very small fraction of which can be supposed *a priori* to be spherical, would have been prohibitive. We emphasize that this possibility of interpolation is entirely a consequence of the ETFSI method being based on the Strutinsky theorem, and is not available to the HF method itself. In fact, the overall computer time for constructing a mass table with the ETFSI method is some 2000 times shorter than with the HF method.

2. 7 Reiterative Procedure for Constructing Mass Table

Despite the comparative rapidity of the ETFSI method, the deformed-nucleus calculations are still too slow to include them explicitly in the fit of the force, since this requires that each nucleus be calculated many times as χ^2 is minimized. A reiterative approach to the problem was therefore adopted, with a preliminary force, SkSC1, first being fitted to the masses of just the spherical nuclei (see III). With this force we then calculated *all* nuclei, spherical and deformed, and defined the deformation energy according to

$$E_{def} = E_{sph} - E_{eq} , \qquad (11)$$

where E_{eq} and E_{sph} are the calculated energies in the deformed and spherical configurations, respectively. We then assume that the deformation energy will be insensitive to small changes in the parameters of the Skyrme force (but not the pairing force). Thus we can use the deformation energies calculated with force SkSC1 to renormalize all measured masses to their "equivalent spherical-configuration" values. Subsequent fits of the force are then made to both spherical and deformed nuclei, while always assuming a spherical configuration. In this way the computer time is kept within reasonable limits. The assumption that the deformation energy is the same for the new force must, of course, be confirmed *a posteriori*.

3 RESULTS

3.1 Masses

We impose a lower limit of $A = 36$ on the nuclei that we calculate, because the semi-classical ETF approximation and the interpolation procedures become unreliable for lighter nuclei. (Fortunately, there is little astrophysical need for extrapolation of masses in this region of the nuclear chart.) Following then the procedure outlined in Section 2.7, we fit our force, as defined in Eqns. (1), (2) and (9), to all the 1492 masses with $A \geq 36$ given in the 1988 data compilation[11]. The resulting parameter set, SkSC4, on which we base our mass table ETFSI-1, is

$$t_0 = -1789.42 \text{ MeV.fm}^3 \quad t_2 = -283.467 \text{ MeV.fm}^5 \quad t_3 = 12782.3 \text{ MeV.fm}^4$$
$$x_0 = 0.79 \qquad\qquad x_2 = -0.5 \qquad\qquad x_3 = 1.13871$$
$$W_0 = 124.877 \text{ MeV.fm}^5 \quad \gamma = 0.333333 \qquad V_{\pi\pi} = -220.0 \text{ MeV.fm}^3$$

The corresponding macroscopic parameters, relating to infinite and semi-infinite nuclear matter, are

$$a_v = -15.87 \text{ MeV} \qquad K = 234.7 \text{ MeV} \qquad J = 27.0 \text{ MeV} \qquad k_F = 1.335 \text{ fm}^{-1}$$
$$a_{sf} = 17.3 \text{ MeV} \qquad\qquad Q = 112.3 \text{ MeV}$$

(see III, for example, for the definitions of these quantities).

The rms error of the fit to the 1492 data points is 0.730 MeV. This is not quite as good as the FRDM fit (0.658 MeV for 1540 masses), but it has been achieved with only 9 parameters rather than some 25. Moreover, two of our parameters, x_2 and γ, are optimized only very roughly.

As far as absolute masses are concerned, there is little systematic difference between the fits to the data given by the two mass formulas, but this picture changes when we look at even-odd mass differences. In this respect the FRDM mass formula fits the data much better than does ETFSI-1, especially for large A values. The implications for the beta-decay energies $Q_{\beta-}$, which are of considerable relevance to the r-process, are shown in Fig. 1, where we plot the errors $\epsilon(Q_{\beta-}) \equiv Q_{\beta-}(\exp) - Q_{\beta-}(\text{calc})$. While the two formulas behave comparably for A = 154, we see that the FRDM formula is definitely superior for A = 234. For this reason, we believe that a calculation of the r-process based on ETFSI-1 would have to be interpreted with some caution.

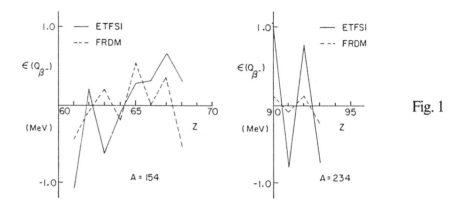

Fig. 1

We have shown in IV that the problem with force SkSC4 lies in a too strong *neutron* pairing, and that very good even-odd differences would be obtained if V_π were reduced to about -200 MeV for neutrons, while leaving its value for protons unchanged at -220 MeV. However, this would then result in our open-shell nuclei being badly underbound, since the pairing force not only produces even-odd fluctuations in the total energy, but also gives rise to a much smoother, though strongly shell-dependent, contribution.

One possible improvement would consist in using the weaker neutron pairing with the Lipkin-Nogami method[12] rather than with the BCS method. Another possibility, suggested by the fact that our problem becomes more serious with increasing A, is a density-dependent pairing force[13]. Finally, one can exploit the fact that the release of the constraint $M_q^* = M$ will have a considerable impact on the pairing properties. All three of these possible solutions to the problem are under active investigation, but it is quite likely that improved even-odd differences will be possible only at the price of a worse overall fit to the absolute masses.

In the meantime, it is instructive to compare the extrapolations that the two models make beyond the region of known nuclei, and in particular to compare their predictions for the r-process. Actually, there is no point to a complete r-process calculation with our model before the pairing problem is

resolved, but it is easy to show that there will be significant differences
between the two models. To a good approximation the r-process path under
canonical conditions is characterized by $S_n = 3.0$ MeV, but in order to avoid
the problems with our pairing we adopt the criterion of $S_{2n} = 6.0$ MeV. We
plot the corresponding curves in the N-Z plane and find that while they are
virtually identical for $Z < 90$, they diverge beyond uranium, as shown in Fig.
2, with the ETFSI model predicting that towards its end-point the r-process
path passes through a region of considerably greater neutron richness. This
disagreement between the two models is a consequence of the magic number
$N = 184$ being much stronger with ETFSI than with FRDM.

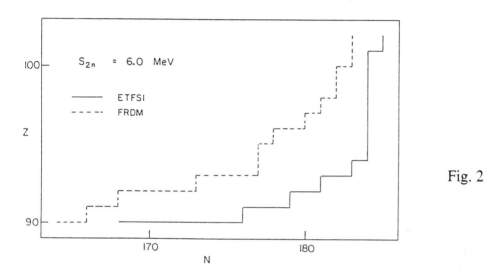

Fig. 2

3. 2 Fission barriers

Since the elucidation of the r-process requires the calculation of the fission
barriers of unknown nuclei it would be desirable to fit our forces not only to
masses but also to experimental barriers. However, this is quite impractical,
and we see here to what extent the force SkSC4, fitted exclusively to masses,
can reproduce experimental barriers. We calculate just the outer barriers of
18 nuclei between ^{228}Ra and ^{246}Cm, removing the condition of left-right
symmetry that was imposed for the mass calculations. A very good
agreement with experiment is obtained for the lightest of these nuclei, but as
we move towards heavier nuclei there is a tendency to underestimate the
barriers, the error in the worst case being 1.4 MeV. A preliminary
calculation shows that weakening the neutron pairing, as indicated by the
mass systematics, improves the barriers. (Further details are given in paper
IV.)

4 CONCLUDING REMARKS

We have constructed the first mass table to be based entirely on microscopic nuclear forces, and with 9 parameters have achieved a fit to the mass data that is almost as good as the one obtained by the most sophisticated mass formula based on the liquid-droplet model, the 25-parameter FRDM. This suggests that once we have an improved treatment of pairing in heavy nuclei our model will provide a more reliable extrapolation into the neutron-rich regions of the nuclear chart relevant to the r-process.

The most striking difference between the predictions of the two models (aside from the problem with pairing that our model has in its present form) lies in the stronger magic number that we have at $N = 184$, which will have significant implications for the end-point of the r-process path. This difference will presumably survive the rectification of our pairing problem.

We have shown that a force fitted to masses alone can reproduce fission barriers well, a point of some importance for calculations of the r-process.

REFERENCES

1. P. Möller, W. D. Myers, W. J. Swiatecki and J. Treiner, Atomic Data and Nuclear Data Tables **39** (1988) 225
2. P. Möller, J. R. Nix, W. D. Myers and W. J. Swiatecki, Atomic Data and Nuclear Data Tables, to be published (1992)
3. A. K. Dutta, J.-P. Arcoragi, J. M. Pearson, R. Behrman and F. Tondeur, Nucl. Phys. **A458** (1986) 77
4. F. Tondeur, A. K. Dutta, J. M. Pearson and R. Behrman, Nucl. Phys. **A470** (1987) 93
5. J. M. Pearson, Y. Aboussir, A. K. Dutta, R. C. Nayak, M. Farine and F. Tondeur, Nucl. Phys. **A528** (1991) 1
6. Y. Aboussir, J. M. Pearson, A. K. Dutta and F. Tondeur, Nucl. Phys., to be published (1992)
7. B. Grammaticos and A. Voros, Ann. of Phys. **123** (1979) 359
8. B. Grammaticos and A. Voros, Ann. of Phys. **129** (1980) 153
9. M. Brack, C. Guet and H.-B. Håkansson, Phys. Reports. **123** (1985) 275
10. Y. H. Chu, B. K. Jennings and M. Brack, Phys. Lett. **B68** (1977) 407
11. A. H. Wapstra, G. Audi and R. Hoekstra, Atomic Data and Nuclear Data Tables **39** (1988) 281
12. H. C. Pradhan, Y. Nogami and J. Law, Nucl. Phys. **A201**(1973) 357
13. V. E. Starodubsky and M. V. Zverev, Phys. Lett. **B276** (1992) 269

Isotopic R-Process Abundances and Nuclear Structure far from Stability

K.-L. KRATZ[1], J.-P. BITOUZET[2,3], P. MÖLLER[1], B. PFEIFFER[1], A. WÖHR[1], and F.-K. THIELEMANN[3]

[1]Institut für Kernchemie, Universität Mainz, Germany
[2]Ecole Polytechnique, Palaiseau, France
[3]Harvard-Smithsonian Center for Astrophysics, Cambridge, MA, USA

1 R-PROCESS SCENARIOS

Since the pioneering work of Burbidge et al. (1957), the rapid neutron-capture process has been associated with high neutron-density environments, where neutron captures are faster than β-decays even for nuclei up to 15–30 units away from stability. The r-abundance maxima - which are related to the magic neutron numbers - are then encountered for smaller mass numbers A than in the s-process. However, besides this basic understanding, the history of r-process research has been quite diverse in suggesting scenarios [for a recent review see, e.g., Cowan et al. (1991)]:

(I) Several investigators noticed that the overall shape of the solar r-process abundances ($N_{r,\odot}$) and the position of the main peaks may be reproduced by a superposition of several neutron densities. The suggested astrophysical sites were type II supernovae (SN), but until today a detailed understanding of the SN mechanism is still pending. (II) Alternatives were explosive He-burning environments, where the r-process acts on previously s-processed material, and a single neutron density is sufficient to transfer matter from each s-process peak into the next higher r-peak. (III) Independent of a particular site, Cameron et al. (1983) performed steady-flow calculations (but without an $(n,\gamma)\rightleftharpoons(\gamma,n)$ equilibrium) which qualitatively reproduced the main r-peaks with decreasing abundances for increasing A.

Only recently, first experimental information in the A$\simeq$80 and 130 regions [see, e.g. Kratz et al. (1988,1991)], together with improved accuracy in $N_{r,\odot}$ [Käppeler et al. (1989)] made it possible to even analyze isotopic abundance patterns in the first two r-peaks. The authors concluded that, indeed, a steady-flow *and* an $(n,\gamma)\rightleftharpoons(\gamma,n)$ equilibrium is required to reproduce specific features in these peaks. This 'waiting-point' concept implies approximate equality of progenitor abundance times β-decay rate ($\lambda_\beta = \ln 2/T_{1/2}$). In this case, apart from neutron density (n_n) and stellar temperature (T_9), the knowledge of nuclear masses (respectively neutron binding energies B_n) and β-decay half-lives ($T_{1/2}$) (as well as β-delayed neutron branching ratios P_n) alone would be sufficient to predict the whole set of $N_{r,\odot}$ as a function of A.

The aim of the present investigation is - without assuming a particular astrophysical site or model - to deduce stellar conditions which produce the r-abundance pattern and to check whether all of the above scenarios are still consistent with the present knowledge of the $N_{r,\odot}$ and the status of nuclear-structure theory and data very far from β-stability.

2   NUCLEAR-DATA SET

Since the vast majority of nuclei in or close to the r-process path is not accessible in terrestrial laboratories, a general understanding of their nuclear-structure properties may only be obtained through theoretical means. However, since a number of different quantities are needed in r-process calculations, in the past it was not possible to obtain them all from one source, thus raising the questions of consistency.

In the present paper, for the first time a *unified* macroscopic-microscopic approach [Möller et al. (1990,1992) has been applied, within which all properties can be studied in an internally consistent way. A microscopic (folded-Yukawa) single-particle model with extensions is combined with a macroscopic (finite range droplet) model which includes Coulomb redistribution effects and an improved formulation of the (Lipkin-Nogami) pairing model. As a first step, nuclear ground-state masses and shapes are calculated. Once these quantities are known, nuclear wave functions are derived for the appropriate shapes. Matrix elements giving β-decay rates and other quantities of interest may then be determined. $T_{1/2}$ and P_n-values are deduced from theoretical Gamow-Teller (GT) strength functions calculated within the QRPA [Möller and Randrup (1990)].

This *consistent* nuclear-data set is expected to yield more reliable predictions than earlier models. Nevertheless, being aware that also our new approach must have its deficiencies, we have further improved the data set by including into the GT strength-function calculations also first-forbidden transitions [Takahashi et al. (1973)], and by taking into account all recent experiments on Q_β, B_n, $T_{1/2}$ and P_n as well as known nuclear-structure properties, either model-inherently not contained (e.g. p-n residual interactions) or not properly described (e.g. the onset of deformation at A$\simeq$100) by our above - still too simplistic - approach. For a detailed discussion, see Kratz et al. (1992).

3   FITS TO SOLAR R-ABUNDANCES

First, we want to generalize our earlier fits to selected mass regions [see, e.g. Kratz et al. (1988,1991)] to the whole $N_{r,\odot}$ distribution by assuming a *global* steady flow. The result of such a calculation for freeze-out conditions, normalized to the A$\simeq$80 peak [$N_{r,\odot}(^{80}Se)\lambda_\beta(^{80}Zn)\simeq29.5$ s^{-1}], is displayed in Fig. 1. Our calculation does produce three r-abundance maxima; however, the A$\simeq$130 and 195 peaks are much too tall and are also shifted to higher A, which means that in these regions n_n is too low. Thus, we conclude that the $N_{r,\odot}$ distribution, with declining peak heights as a function of A, *cannot* be explained by a global steady flow with an $(n,\gamma)\rightleftharpoons(\gamma,n)$-equilibrium. This result rules out scenario (III) of Cameron et al. (1983) as a possible solution.

In a next step we have focussed on reproducing the details of individual mass regions, and finding the break points between steady-flow areas. The results of these *static* calculations can be summarized as follows: The r-process, indeed, has reached a steady-flow equilibrium which is, however, no longer global but only local in between the r-abundance peaks. It breaks down at the top of each peak, i.e. at the N=50, 82 and 126 closed shells. Similar to the $N_s\sigma$-curve for the s-process, we obtain a

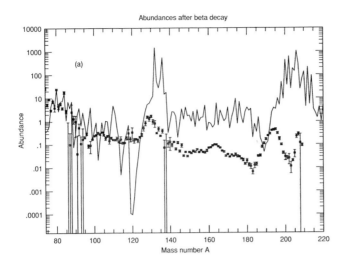

Fig. 1: *Global r-abundance curve, using the waiting-point and steady-flow approximation with $T_9=1.3$ and $n_n=10^{20}$ cm^{-3} and our unified nuclear-data set for B_n, $T_{1/2}$ and P_n, in comparison with the $N_{r,\odot}=(N_\odot-N_s)$ of Käppeler et al. (1989).*

three- (or four-) step r-process $N_r\lambda_\beta$-curve with different neutron number densities. The statistical weights of the first three $N_r\lambda_\beta$-components are 10:2.5:1. The 'best-fit' T_9-n_n conditions for the different mass ranges are shown in Fig. 2. There remains however, an uncertainty for the conditions which produce the $N_{r,\odot}$ beyond A≃200. Since their progenitors decay via β- and α-decay chains, there is no simple shape to fit, and our study is just exploratory. Nevertheless, our results rule out explosive He-burning (scenario II) as a possible solution and support the SN II origin of the r-process [scenario (I); see, e.g., Meyer et al. (1992)].

Finally, the question remains whether the $N_{r,\odot}$ curve can be the result of a super-position of three (or more) local steady flows, where each component dominates one peak. This is not possible, as already indicated by our *global* steady-flow calculations. A superposition will only make sense when the abundances for a specific T_9-n_n condition, appropriate to a region $A \leq A_{peak}$, will be set to zero for $A > A_{peak}$. Such a situation can occur in time-dependent calculations, where the r-peaks containing neutron-magic nuclei with the longest $T_{1/2}$ will act as 'bottle necks', over which only small amounts of matter will pass. When starting our time-dependent calculations with Z=26, the time scales for reaching the appropriate peaks are with 1.5 to 2.5 s comparable, and are consistent with the expected duration of the r-process in a SN. A typical result for a four-component time-dependent calculation is shown in Fig. 3. Clearly, the main features of the $N_{r,\odot}$ distribution are well reproduced, and in certain regions even *isotopic abundances* become meaningful. However, there remain several deficiencies which - apart from $A \leq 78$ where the $(N_\odot-N_s)$ residuals are not of

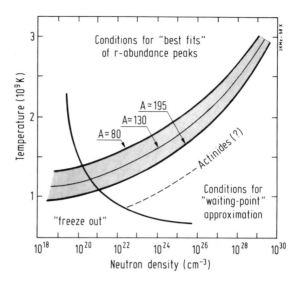

Fig. 2: T_9-n_n *conditions for the waiting-point and steady-flow approximation for different mass ranges of $N_{r,\odot}$ [Käppeler et al. (1989)]. Since r-process nuclei beyond Bi decay via β- and α-chains, no clear features emerge for a 'best fit'; hence, the dashed line indicates but a first guess for the fourth component.*

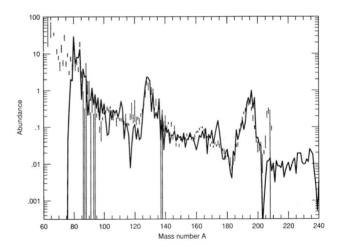

Fig. 3: *R-abundance distribution obtained from a superposition of four time-dependent calculations with the 'best-fit' T_9-n_n values for the $A \simeq 80$ peak and the $90 \leq A \leq 130$ and $135 \leq A \leq 195$ mass ranges, and a first guess for the $A > 195$ region.*

genuine r-origin - clearly indicate nuclear-structure effects very far from stability not accounted for properly in our present *unified* macroscopic-microscopic approach:

The solution to the $A \simeq 120$ abundance trough lies in the nuclear masses along the r-process path. In the Möller et al. (1992) mass model, the decrease of the B_n's occurs too slowly when approaching $N=82$. This implies that - in an extreme view - there would exist *not a single* $B_n \simeq 2$ MeV waiting-point isotope between $^{112}Zr_{72}$ and $^{125}Tc_{82}$. This model-deficiency presumably has its origin in the neglect of the p-n residual interaction which seems to manifest itself in an overestimation of the $N=82$ shell strength below $^{132}_{50}Sn_{82}$.

In the $103 \leq A \leq 115$ and $160 \leq A \leq 180$ mass regions, sinusoidal deviations show up. As is discussed in detail in Kratz et al. (1992), these deviations are due to model-deficiencies in the development of quadrupole deformation around neutron mid-shells $N=66$ and 104 which are clearly correlated with the other nuclear properties B_n, $T_{1/2}$ and P_n.

In summary, we can conclude that both the overall success in the reproduction of the r-abundance peaks, as well as the local failures discussed above, indicate nuclear-structure signatures of extremely neutron-rich isotopes, the vast majority of them not accessible in terrestrial but only in stellar laboratories. This offers fascinating perspectives for experimentalists and theoreticians in the entwined fields of nuclear and astro-physics.

This work was supported by grants from DFG (Kr 806/1), BMFT (06MZ106), the Fondation de l'Ecole Polytechnique and from NSF (AST 89-13799).

REFERENCES

Burbidge, E.M. et al. 1957, Rev. Mod. Phys. 29, 547

Cameron, A.G.W. et al. 1983, Ap. Space Sci. 91, 221

Cowan, J.J. et al. 1991, Phys. Rep. 208, 267

Käppeler, F. et al. 1989, Rep. Prog. Phys. 52, 945

Kratz, K.-L. et al. 1988, J. Phys. G14, 742

Kratz, K.-L. et al. 1991, Z. Phys. A340, 419

Kratz, K.-L. et al. 1992, Ap. J., in press

Meyer, B.S. et al. 1992, contrib. to this conf., and subm. to Ap. J.

Möller, P. et al. 1990, in *Nuclei in the Cosmos*, MPA/P4, 226

Möller, P. et al. 1992, subm. to At. Data Nucl. Data Tables

Möller, P., Randrup, J. 1990, Nucl. Phys. A541, 1

Takahashi, K. et al. 19973, At. Data Nucl. Data Tables 12, 101

The Freeze Out of the R-Process

B.S. MEYER

Clemson University, Clemson, SC 29634-1911, USA

1. INTRODUCTION

Inspection of the solar-system r-process abundance distribution versus mass number reveals that it is particularly smooth. Its smoothness is especially evident when compared with that of its rival process in the production of the heavy nuclei, the s-process. Because the s-process always operates in a regime in which neutron captures occur more slowly than beta decays, abundances build up at nuclei with small cross sections. This gives the distribution a jagged, odd-even aspect because even-mass nuclei typically have smaller cross sections than odd-mass nuclei.

By contrast, the r-process probably occurs in an environment in which neutron captures initially occur much more rapidly than beta decays. In this case, the neutron captures ((n,γ) reactions) and their reverse reactions–neutron disintegrations ((γ,n) reactions)–essentially come into equilibrium. This is the classical r-process phase. Here a strong jagged aspect is also found in the abundance distribution because of the odd-even effect in the nuclear partition functions and neutron separation energies that determine the abundance pattern in (n,γ)-(γ,n) equilibrium. The r-process is dynamical, however, because the supply of free neutrons is used up by the neutron captures and the temperature and density are both decreasing due to expansion of the material undergoing r-process nucleosynthesis. Both of these effects cause an increase in the timescales for the (n,γ) and (γ,n) reaction rates until they become comparable to the beta-decay and/or expansion timescales. At this point, (n,γ)-(γ,n) equilibrium can no longer be maintained. Non-equilibrium processing now occurs. Further neutron capture and expansion lead to an s-process-like phase of the r-process in which the neutron capture rates become less than the beta-decay rates. At some point the neutron abundance, density, and temperature are so low that the neutron capture and disintegrations freeze out (that is, become much slower than the expansion). Only beta decays change the abundances from this point on. We frequently term the nuclear processing occurring from the time the (n,γ) and (γ,n) reactions fall out of equilibrium to the time these reactions have frozen out the freeze out or freeze-out processing.

If the s-process-like phase of the freeze out were to last a long time, the r-process abundance distribution would have a jagged aspect similar to that of the solar-system s-process distribution. Because the solar-system r-process is in most places much smoother than the typical pattern in either the classical phase or in a long-lasting s-process-like phase, the smoothness must derive in large part from the passage of the r-process through the non-equilibrium, or freeze out, phase, but with with little time spent in the s-process-like phase.

Beta-delayed neutron emission may also smooth the r-process distribution (Kodama and Takahashi 1975). As the neutron-rich nuclei produced in the r-process beta decay back to the line of beta stability, the decays may leave daughter nuclei in excited nuclear states unstable to the emission of neutrons. The subsequent neutron emission diverts some of the beta-decay flow from atomic mass chain A to chain $A - 1$ (or even lower mass if multiple neutrons are emitted). If the abundance of nuclei in chain $A - 1$ is less than that in chain A, delayed neutron emission causes the abundance of chain $A - 1$ to grow at the expense of chain A. This can smooth out the abundance difference between chains $A - 1$ and A.

It is naive to think that beta-delayed neutron emission or freeze-out processing operate independently during the r-process. For example, the neutrons produced by delayed neutron emission during freeze-out processing must be captured back onto nuclei. Nevertheless, it is useful to attempt to determine which of the two–beta-delayed neutron emission or freeze-out processing–has the greater effect in smoothing the r-process abundance distribution. The answer to this question will allow us to determine to what extent we may neglect the freeze-out processing that occurs during the r-process. It will also provide us with a deeper understanding of the r-process in general and of the smoothness of the r-process abundance curve in particular. (For a review of the r-process, see Cowan et al. (1991).)

2. CALCULATIONS

In order to study the effects of freeze-out processing and beta-delayed neutron emission during the r-process, I used the computer code described in Meyer et al. (1992). This code follows the r-process in the high-entropy bubble in type II supernovae, which, as shown in Meyer et al. (1992), is a promising site for the r-process.

I made one run of the code with freeze-out processing and beta-delayed neutron emission included. The neutron excess per baryon parameter η for this run was 0.300 and all other parameters were as in Meyer et al. (1992). The result is shown as the lowest curve in figures 1 and 2. This individual calculation does not itself give a good representation of the solar-system r-process abundance distribution. As shown in Meyer *et al.* (1992), however, a series of runs at varying η all added together in a weighted sum give an abundance distribution that closely resembles the solar-system pattern.

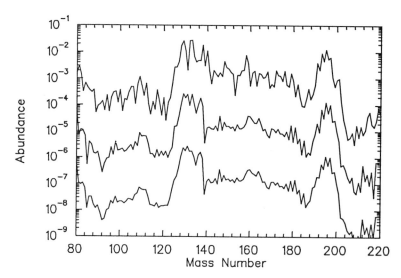

Figure 1

As discussed above, the classical phase of the *r*-process ends when the (n,γ) and (γ,n) reactions fall out of equilibrium. This of course does not happen globally at the same time, but for purposes of this paper, I arbitrarily define this to occur at a critical neutron density n_n of 10^{21} cm^{-3}. This is in accord with the critical n_n of found by Cameron *et al.* (1983). The abundance pattern at $n_n = 10^{21}$ cm^{-3} is the top curve in figures 1 and 2. The abundance curve at this point shows the strongly jagged aspect of the classical phase of the *r*-process. The difference between this curve and the lowest curve in each figure shows the smoothing caused by (n,γ) and (γ,n) reactions during freeze-out processing and by beta-delayed neutron emission.

I then ran two variations of the standard run. In the first I turned off beta-delayed neutron emission at $n_n = 10^{21}$ cm^{-3}. I did this by keeping the total beta-decay rate the same by setting the branching ratio to the channel with no neutrons emitted equal to one and the branching ratio to channels with neutrons emitted equal to zero. Since the beta decays now do not change the atomic mass A of a nucleus, the smoothing of the abundance curve is due solely to (n,γ) and (γ,n) reactions during freeze-out processing. The final result of this run is the middle curve of figure 1.

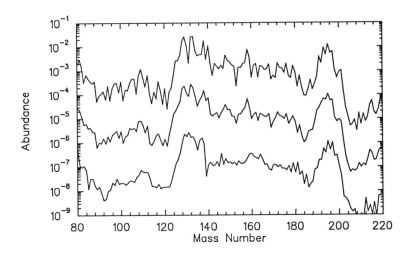

Figure 2

In the second variation of the standard run I turned off (n,γ) and (γ,n) reactions after $n_n = 10^{21}$ cm^{-3}. Smoothing in this case is due solely to beta-delayed neutron emission. The final result of this run is the middle curve of figure 2.

3. CONCLUSIONS

The bulk of the smoothing of the r-process curves clearly comes from the (n,γ) and (γ,n) reactions during freeze out, as may be seen from figure 1. Overall the middle and bottom curves of figure 1 are quite similar. The middle and bottom curves of figure 2 are not so similar, although beta-delayed neutron emission does provide a good deal of smoothing. A conclusion is that (n,γ)-(γ,n) equilibrium calculations of the r-process followed by smoothing simply by beta-delayed neutron emission do not give a complete picture of the r-process. These conclusions are of course made in the context of the high-entropy bubble r-process, but we should not expect the results from other sites to be too different because the r-process in other sites is quite similar to that in the high-entropy bubble.

(n,γ) and (γ,n) reactions during freeze out are themselves not the complete picture, however. Subtle differences between the middle and bottom curves in figure 1 show that delayed neutron emission does play a role in shaping the r-process abundance curve. For example, in the mass range $A = 100 - 120$ and on the low mass side of the $A = 195$ peak, the complete calculation results are smoother than the results with beta-delayed neutron emission not operating. Delayed neutron emission not only smoothes the r-process curve by shifting nuclei from one mass chain to chains of lower mass, but also by supplying neutrons to be captured on other nuclei. The extra neutrons can thus allow for more smoothing by keeping the supply of neutrons available for a longer time. In sum, detailed conclusions about the r-process should be made from dynamical calculations that include non-equilibrium processing and beta-delayed neutron emission, although the role of beta-delayed neutron emission seems less important than that of neutron captures and disintegrations during freeze out.

REFERENCES

Cameron, A.G.W., Cowan, J.J., and Truran, J.W. 1983, *Astrophys. Space Sci.* **82**, 123.

Cowan, J.J., Thielemann, F.-K., and Truran, J.W. 1991, *Phys. Rept.* **208**, 267.

Kodama, T. and Takahashi, K. 1975, *Nucl. Phys.* **A239**, 489.

Meyer, B.S., Mathews, G.J., Howard, W.M., Woosley, S.E., and Hoffman, R.D. 1992, *Ap. J.*, in press.

Investigation of very neutron-rich Fe, Co and Ni isotopes encountered along the r-process path

S. CZAJKOWSKI[1], M. BERNAS[1], J.L. SIDA[1], P. ARMBRUSTER[2], H. FAUST[2],
Ph. DESSAGNE[3], Ch. MIEHE[3], Ch. PUJOL[3], G. AUDI[4], J.K.P. LEE[5],
Ch. KOZHUHAROV[6], H. GEISSEL[6], G. MÜNZENBERG[6], and the FRS GROUP[6]

[1] Institut de Physique Nucléaire d'Orsay, 91406 Orsay Cedex, France
[2] Institut Laue Langevin, 38042 Grenoble Cedex, France
[3] Centre de Recherches Nucléaires, 67037 Strasbourg Cedex, France
[4] CSNSM, Laboratoire René Bernas, 91405 Orsay Campus, France
[5] Foster Radiation Laboratory, MacGill University, Montreal, Canada
[6] Gesellschaft für Schwerionenforschung, W-6100 Darmstadt, Germany

Abstract: Very neutron-rich nuclei beyond Fe must be investigated in order to understand the r-process. New Ni, Co and Fe isotopes, with ten to twelve neutrons more than the heaviest stable isotopes, were discovered in the thermal fission of ^{235}U and ^{239}Pu, and their half-lives were deduced from the analysis of time delayed coincidences between the fragment implantation and the detection of consecutive β-particles in the same Si-pin diode detector. In a pilot experiment, high energy projectile fragmentation was expoited to produce neutron rich species. The fragments of interest were separated and energy bunched with the FRS. After being slowed down, they were selectively implanted in a similar detecting system. The β decay half-life were obtained as in the previous case.

1. INTRODUCTION

In the solar system, half of the bulk of nuclei heavier than Fe have been synthesized by the r-process. The three narrow r-peaks, still observed today in the abundancies curve are finger prints of this stage of nucleosynthesis. They result from the neutron shell closures which generate sudden decreases in the last neutron binding energies thus causing large reduction of neutron capture cross-sections. In the very large neutron flux produced by the Super-Nova creating our solar system, at high temperature, after a chain of neutron captures, an equilibrium was established between (n,γ) and (γ,n) cross-sections for the boundary isotopes which then could only undergo β decay, a slow process compared to (n,γ). In this steady flow model, the A abondancies under the peaks are proportionnal to the β-half-life of the isotopes encountered along the neutron shell closure [1].

Knowing the properties of those nuclei and of their neighbours- binding energies, half-lives, neutron emission probabilities and β-γ decay scheme - r-process astrophysical conditions such as temperature and neutron flux can be retraced.

Concerning the A=80 peak, two isotopes are of major importance; the "waiting point" nucleus, ^{80}Zn and the near-by isotone ^{79}Cu. These have already been studied, using on-line mass separators [2] and [3] respectively. Up-to-now, these techniques failed in the Fe, Co and Ni-region, due to the low extraction efficiency from available ion sources for refractory elements.

In this work, in-flight separation techniques have been used to unambiguously identify new neutron-rich isotopes of the elements Fe, Co and Ni and to measure their β decay halflife.

2. THERMAL FISSION

Very asymmetric thermal fission $^{235}U(n_{th}, f)$ has been investigated at the neutron high flux reactor of the ILL. Mass yields and isotopic yields of the light fragments separated by the Lohengrin spectrometer and identified with a $\Delta E - E$ ionization chamber have been measured [4]. The lightest fragments ever seen in binary fission were observed, in spite of extremely low production yields. For Co and Fe measurements a ^{239}Pu target was used, because the yields were found larger by a factor 8 as compared to U. Even with this enhancement, a production rate as low as 10^{-9}/fission was measured for ^{68}Fe [5]. Being also the most neutron rich isotopes known today, they are close to the path followed by the r-process and thus the determination of their halflives is of great interest.

A row of 8 pin diodes (10x10x0.5 mm) was set in the chamber at the end of ion trajectories to detect ions and subsequent β particles. The half-life of an identified fragment was deduced from the analysis of time delayed coincidences between the ion implantation and the detection of the following β's. Maximum likelihood procedures have been used [6] because of the small number of fragments, of a β detection efficiency limited to 0.5 and of a non-negligeable β background rate (0.2/s/det).

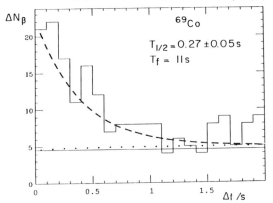

Figure 1 ^{69}Co-β time correlations histogram (all β's)

The full horizontal line is the constant background contribution and the dotted line includes the daughter decay. the dashed line is calculated with the half-life deduced from maximum likelihood analysis.

In spite of large error bars, the values obtained (see the table) are of interest given the scatter of theoretical expectations.

3. PROJECTILE FRAGMENTATION

Projectile fragmentation has been used for more than 10 years to produce isotopes far from stability [9]. The advantages of high energy beam are twofold: 1) Fragments being forward emitted, they are analysed with a good efficiency within a solid angle of 0.7 msr. 2) They are fully ionized, that is concentrated at a single atomic charge state.

The 500 MeV/u ^{86}Kr beam of 5 10^7 ions/pulse, delivered by the Unilac-SIS acceleration system(GSI) was directed on a 2 gr/cm^2 Be target. The objective was the determination of the ^{65}Fe half-life; this nuclei is the lightest isotope of unknown half-life and thus, the easiest to investigate. The fragments were separated with the new Fragment Recoil Separator (FRS), a momentum loss achromat [10]. It can be tuned to bunch the energies of fragments under study, although the momentum acceptance of the system is 2%, i.e. 1.3 GeV in energy for the ^{65}Fe. This dispersion was totally compensated for (see fig.2) and the choosen fragments emerged with a well defined range, different from the range of contaminants except for ^{66}Co. After having slowed down in thick degradors (12 gr/cm^2 of Al), the ^{66}Co and ^{65}Fe ions could be selectively implanted in one of the two rows of 10 Si-pin diodes of 0.1 gr/cm^2 thickness (see fig.2) where they decayed. The β's emitted were detected in the same Si detectors using a double preamplification, since energy signals due to the fragments are 10^5 larger than the ones left by β's. The analysis of the β-decay of the ^{66}Co contaminant, the half-life of which is known and short, has provided an excellent calibration for β detection efficiency and background conditions. The logarithmic time scale (fig.3) enable to picture both components of the curve [11]. With this experimental set-up many more unknown nuclei can be investigated in the future.

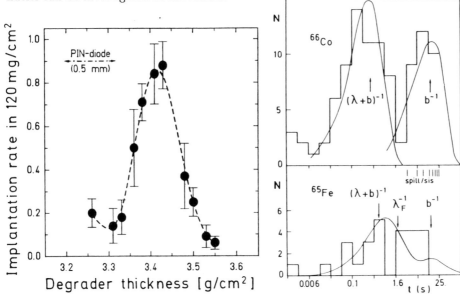

Figure 2 Range distribution of ^{66}Co **Figure 3** Time correlations for ^{66}Co and ^{65}Fe

Measured isotopes, are indicated in fig.4, with two tentative r-process paths based on different masses -and therefore half-life- evaluations . Since our values are substancially different from those, it can be worthwhile to update the r-process path calculations.

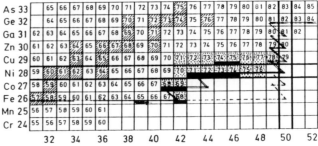

Figure 4 Isotopes investigated in this work-thick black mark. Dotted square are isotopes observed in fission around A=80. The r-process paths are indicated with a full line [12] and a dashed line [13]

	$T_{1/2}(s)$ this work	$T_{1/2}(s)$ revised GT[7]	$T_{1/2}(s)$ microscopic[8]
^{71}Ni	1.86 ± 0.35	2.18	5.44
^{72}Ni	2.06 ± 0.30	2.48	1.56
^{73}Ni	0.90 ± 0.15	0.69	0.81
^{74}Ni	1.10 ± 0.50	0.57	0.46
^{69}Co	0.27 ± 0.05	0.73	0.071
^{68}Co	0.18 ± 0.10	0.80	0.29
^{68}Fe	0.10 ± 0.06	0.37	0.16
^{65}Fe	0.40 ± 0.20	2.22	0.33

Table

4. CONCLUSION

Most of the decay properties of isotopes needed for the r-process calculations remain still unknown. The predictive power of the models decreases rapidly the further from stability we go. The β-decay half-lives measurements are of importance, especially for isotopes near neutron shell closures, since they support extrapolations to the values in question.

Improvements of the existing facilities, either with thermal fission [15] or using projectile fragmentation, would allow these studies to extend closer to the N=50 line, towards the doubly-magic isotope, ^{78}Ni.

REFERENCES

[1] J.J. Cowan et al *Phys. Rep.* **208** 267, 1991
[2] E. Lund et al., *Proc. of AMCO-7, THD Schrift. Wiss. und Techn.* **26**, 102, 1984 and R.L. Gill et al., *Phys. Rev. Lett.* **56**, 1874, 1986
[3] K.L. Kratz et al., *Z. Phys. A* **1**, 184, 1992
[4] J.L. Sida et al., *Nucl. Phys. A* **502**, 233c, 1989
[5] M. Bernas et al *Phys. Rev. Lett.* **67** 3661, 1991
[6] M. Bernas et al *Zeit. fur Phys. A* **337** 41, 1990
[7] T. Tachibana et al *Report of Sciences and Engin. Res. Lab. Wasada University* **88-4**, 1988
[8] A. Staudt et al *At. Data Nucl. Data tables* **44** 79, 1990
[9] D. Guillemaud-Müller *Contribution to this conference*
[10] K.H. Schmidt et al. *Nucl. Inst. and Meth. A* **260**, 287, 1988 and ref. therein and H. Geissel et al *GSI Preprint-91-46* and to be published in *Zeit.f.Phys.A 1992*
[11] K.H. Schmidt *Nucl. Phys. A* **318** 253, 1979
[12] A.G.W. Cameron *Space Sci. Rev.* **15** 21, 1973
[13] K.L. Kratz et al. *Journal of Phys.G* **24**, S331, 1988
[14] K.L. Kratz et al., *Contribution to this conference*
[15] H. Faust et al. 1992 *ILL Report 92FA 01T*

Approximating the r-Process on Earth with Thermonuclear Explosions

S.A. BECKER

Los Alamos National Laboratory

ABSTRACT

The astrophysical r–process can be approximately simulated in certain types of thermonuclear explosions. Between 1952 and 1969 twenty–three nuclear tests were fielded by the United States which had as one of their objectives the production of heavy transuranic elements. Of these tests, fifteen were at least partially successful. Some of these shots were conducted under the project Plowshare Peaceful Nuclear Explosion Program as scientific research experiments. A review of the program, target nuclei used, and heavy element yields achieved, will be presented as well as discussion of plans for a new experiment in a future nuclear test.

1 OVERVIEW

Nuclear explosions have been used in a number of scientific investigations and applications.[1] One such study was the Heavy Element Program which had as its objectives the production of heavy transuranic elements and the investigation of nuclear properties of very neutron–rich isotopes.[2,3] For such an experiment an intense flux of neutrons is produced primarily by the $D(T,n)^4He$ reaction. These neutrons which begin with a 14.1 Mev energy are thermalized and are then used to induce multiple captures in a target such as ^{238}U. The intense flux of thermal neutrons is only present for ten to twenty nanoseconds which means that the target nuclei undergo multiple neutron captures at constant Z because there is no time for beta decay. The rapid multiple captures of neutrons by a target in a thermonuclear explosion has been given the name "prompt capture."[4] Once the "prompt capture" process is over, the neutron–rich nuclei transform into longer lived nuclei primarily through a series of beta decays back to the line of beta stability. The term "decay back" is used to refer to the series of beta decays to the line of beta stability experienced by a neutron–rich nucleus.[5]

Table 1 compares the environments for four different neutron capture processes. For neutron capture in the High Flux Isotopic Reactor (HFIR) at the Oak Ridge National Laboratory and during the s–process, the time scale for beta decay is much less than the time scale for successive neutron captures which results in production of nuclei close to the line of beta stability. In contrast, both the r–process and the "prompt capture" process undergo neutron capture so rapidly that multiple captures can occur before the onset of beta decay, thus producing nuclei far from the line of beta stability. The r–process differs from the "prompt capture" process in that it takes place on a longer time scale and at hotter temperatures. The net effect of these differences is that the neutron capture cross sections are smaller for the r–process than for the "prompt capture" process and the longer time scale allows some beta decays to occur during the r–process at the same time that neutrons captures are occurring. The similarities between the r–process and "prompt capture" process are, however, such that the data from a number of nuclear tests have been used to calibrate the nuclear physics used in r–process nucleosynthesis codes.[6]

TOTAL

NEUTRON EXPOSURE ENVIRONMENTS

Type of Exposure	Flux $(n\bar{v})$	Duration (δt)	Fluence $(n\bar{v}\delta t)$	Temperature
HFIR	$5 \times 10^{15}/cm^2$–s	≈ 0.5 yr	$\approx 10^{23}/cm^2$	2.5×10^{-5} keV
s–process	$\approx 10^{16}/cm^2$–s	$\approx 10^3$ yr	$\approx 10^{26}/cm^2$	10 to 30 keV
r–process	$> 10^{27}/cm^2$–s	1 to 100s	$> 10^{27}/cm^2$	100 keV
"prompt capture"	$> 10^{32}/cm^2$–s	$< 10^{-7}$s	$\approx 10^{25}/cm^2$	10 to 20 keV

2 HISTORICAL REVIEW

The fact the heavy transuranic isotopes could be produced in a thermonuclear device was discovered serendipitously through the analysis of the Mike test (fired 31/10/52) debris. All told 15 new isotopes[4] and 2 new elements[7], einsteinium and fermium, were discovered. A maximum of 17 neutron captures happened on some of the ^{238}U target nuclei which became ^{255}U and later underwent 8 beta decays to become ^{255}Fm. The estimated thermal neutron flux on the target was 2 to 3 moles of neutrons/cm^2. The discoveries from the Mike test lead to the development of the Heavy Element Program as a result of which 23 nuclear tests were conducted which had as one of their objectives the production of heavy transuranic elements. Of these tests 15 were at least partially successful, but only the milestone events will be discussed here. Some of these shots were performed under the project Plowshare Peaceful Nuclear Explosion Program. The results of the Mike test were later approximately duplicated at a much lower device yield in the Anacostia test (fired 27/11/62).[2,8] New discoveries were made with the Par (fired 9/10/64) and Barbel (16/10/64) tests which achieved an estimated thermal neutron flux of 11 moles of neutrons/cm^2. The very neutron–rich isotope ^{250}Cm was discovered and 19 neutron captures were achieved by some target ^{238}U nuclei to produce ^{257}Fm.[2,9] Higher thermal neutron fluxes of approximately 18 moles of neutrons/cm^2 and 35 moles of neutrons/cm^2 were respectively achieved in the Cyclamen[10] (fired 5/5/66) and Hutch (fired 16/7/69) tests.[11] These higher neutron fluxes produced a greater yield of heavy transuranic isotopes, but disappointingly no new isotopes or elements were discovered. Figure 1 illustrates the heavy element production as a function of atomic number for the Cyclamen and Hutch devices.[11] The inability to produce new isotopes heavier than ^{257}Fm eventually lead to a loss of interest in the Heavy Element Program.

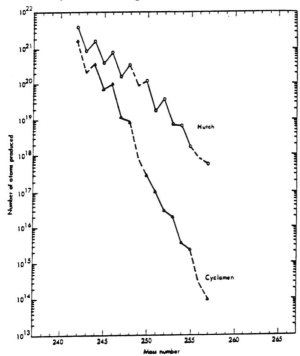

Fig. 1. Heavy element yields from the Hutch and Cyclamen experiments.

In the end, the Heavy Element Program was responsible for the discovery of 2 new elements and 16 new transuranic isotopes. The "prompt capture" process proved to be an efficient method of production for very neutron–rich isotopes like ^{250}Cm, ^{251}Cf, ^{254}Cf, ^{255}Es, and ^{257}Fm. Detailed studies were done to learn the nuclear properties of isotopes recovered from these tests. Sufficient quantities of ^{250}Cm and ^{257}Fm were extracted to use them as targets in accelerators to produce even heavier isotopes.

Besides ^{238}U, which produced the best results, target nuclei of ^{232}Th, ^{237}Np, ^{240}Pu, ^{242}Pu, and ^{243}Am were also fielded. Of these, positive results were obtained from ^{232}Th and ^{242}Pu where up to 13 neutron captures were observed. The other targets were destroyed by fission induced from the high energy neutrons.

Work in the Heavy Element Program came to an end with a number of questions remaining unanswered. One such question is, for example: "Is the saw tooth abundance pattern shown in Figure 1 produced as a result of a secondary capture chain on an odd Z nucleus resulting from the (n,p) reaction on the target nuclei, or is it the result of beta–delayed fission during the "decay back" to the line of beta stability?"[12]

3 FUTURE PROSPECTS

At present the only experimental test of theoretical predictions of beta–delayed fission and fission barrier heights for neutron–rich unstable nuclei come from the analysis of the products produced in heavy element nuclear tests.[12] As discussed in reference six, the data from nuclear tests for the uranium capture chain has been used to modify the input physics based on theoretical calculations for one r–process nucleosynthesis code. For example, nuclear test data have shown that some calculated theoretical fission barriers are too low for neutron–rich nuclei.

It now appears likely in an upcoming nuclear test that a heavy element experiment will be conducted using a ^{232}Th target. While ^{232}Th was fielded in the Hutch test, its performance was somewhat clouded by the presence of the more abundant ^{238}U target. In the upcoming nuclear test, ^{232}Th will be the only target which should provide a clear picture of the thorium capture chain and give additional experiment calibration for r–process nucleosynthesis codes. With a thorium target it may be possible to achieve more than 19 neutron captures. If this experiment is successful, an odd Z target like ^{231}Pa might be fielded in another future event.

REFERENCES

1.   Dorn, B. C. (1970) Ann. Rev. Nuc. Sci. 20, 79–103.
2.   Ingley, J. S. (1969) Nuc. Phys. A124, 130–144.
3.   Eberle, S. H. (1972) Kerntechnik 14, 65–71.
4.   Cowen, G. A. (1967) Los Alamos Scientific Laboratory Report LA–3738.
5.   Wene, C. O. and Johansson (1974) Phys. Scr. 10A, 157–162.
6.   Cowan, J. J. Thielemann, J. K., and Truran, J. W. (1991) Physics Reports 208, 267–394.
7.   Ghiorso, A. et al (1955) Phys. Rev. 99, 1048–1049.
8.   Hoff, R. W. and Dorn, D. W. (1964) Nuc. Sci. Eng. 18, 110–112.
9.   Bell, G. I. (1965) Phys. Rev. 139, B1207–B1216.
10.   Hoffmann, D. C. (1967) Arkiv For Fysik 36, 533–537.
11.   Hoff, R. W. and Hulet, E. K. (1970) in Engineering With Nuclear Explosives, Lawrence Radiation Laboratory Report CONF 700101 p1283–1294.
12.   Hoff, R. W. (1988) J. Phys. G. 14, S343–S356.

Neutrino Induced Nucleosynthesis of ^{182} Hf

S. RAMADURAI

Astrophysics Group,
Tata Institute of Fundamental Research,
Homi Bhabha Road, Navy Nagar, Colaba, Bombay 400 005, India

Abstract

The recording of neutrino signals from the $SN\,1987A$ in the Large Magellanic Cloud, has triggered the activity of calculating quantitatively, several neutrino induced physical processes, taking place in the astrophysical environments. Though cross-sections for the inelastic neutral current neutrino scattering off nuclei are small, because of the enormous flux of neutrinos produced during the collapse of the core of massive stars, these processes lead to substantial production of rare isotopes. The importance of neutrino induced nucleosynthesis in cosmochronometry is illustrated by calculating the abundance of ^{182}Hf by the neutrino excitation of ^{186}W followed by α emission.

1. INTRODUCTION

Neutrino induced nucleosynthesis was originally suggested as an alternative to p-process to account for those isotopes, which are bypassed by the conventional 's' and 'r' processes, namely slow and rapid neutron addition processes (Domogatskii, Eramzhyan and Nadyozhin 1977, 1978; Domogatskii and Nadyozhin 1977, 1978). Though the efficiency of this process in producing those p-process isotopes was questioned (Woosley 1977), the possibility of neutrino induced nucleosynthesis of light elements and other isotopes, was left open (Domogatskii et al. 1978; Domogatskii and Nadyozhin 1980a,b; Domogatskii and Imshennik 1982; Woosley 1977). However, the recent detection of neutrino signals from SN1987A (Bionta et al. 1987; Hirata et al. 1987) has once again triggered the interest in neutrino induced nucleosynthesis (Epstein, Colgate and Haxton 1988; Haxton 1991; Nadyozhin preprint 1991; Woosley and Haxton 1988; Woosley et al. 1990). The detailed models of the supernova explosion of SN1987A indicate the release of about $10^{53} ergs$ of energy in neutrinos of all types. While several papers have appeared highlighting the importance of nucleosynthesis by neutrino excitation of nuclei, none of them have specifically addressed the question of synthesis of an extinct radioactive nuclide, discovered recently-^{182}Hf. It should be noted that this nuclide is not bypassed by the neutron addition processes. This talk is aimed to highlight the importance of neutrino induced synthesis by one of the weakest mode of decay of excited nucleus, namely α decay. It is shown that under realistic model conditions almost the entire anomaly can be synthesized by neutrino excitation of ^{186}W followed by α decay.

2. NUCLEAR INTERACTIONS

The neutrino interactions of interest are the neutral current inelastic scattering of neutrinos off nuclei with mass number A and atomic number Z like

$$\nu + (A, Z) \rightarrow (A, Z)^* + \nu'$$

where ν represents any one of the three types of neutrinos or their anti-particles. The excited nucleon can decay depending on the branching ratios into any one of the following three channels

$$(A, Z)^* \rightarrow (A_i - 1, Z_i) + n$$

$$\rightarrow (A_i - 1, Z_i - 1) + p$$

$$\rightarrow (A_i - 4, Z_i - 2) + \alpha.$$

In order to make an estimate of the abundances produced by ν induced nucleosynthesis, the neutrino excitation cross-sections for excitation of target nuclei of mass number A and charge Z, have to be obtained. This has to be carefully determined, taking care to include the transitions to isobaric-analog as well as Gamow-Teller resonance states. Though, in general the calculation of cross-sections involve detailed shell model techniques (Haxton 1991), it is found for the purpose envisaged here, 'at T of about $10MeV$, all large differences between the nuclei have been smoothed out by the growing giant resonance response' and 'at temperatures characteristic of μ and τ neutrinos, the cross-sections per nucleon are roughly independent of the nuclear species'.

The second important quantity required are the branching ratios b_i into n, p, α or γ channels. Here the detailed Hauser-Feshbach studies are needed. Again, in the following, some simple estimate of probable value is used, the aim being to demonstrate the possibility of neutrino induced synthesis of the particular nuclide rather than the exact accordance with observations.

The number of the neutrino-nuclear interactions per unit volume per unit time, R_ν, is given by

$$R_\nu = n_{AZ} \frac{L_\nu(T)}{4\pi r^2 < E_\nu >} < \sigma_{\nu_{AZ}} > \qquad (1)$$

where n_{AZ} is the number density of target nuclei , L_ν is the total energy carried away by the neutrinos of given type per unit time, $< E_\nu >$, being the mean energy of individual neutrinos; r is the distance of the neutrino irradiated matter from the centre of the collapsing core. Here the neutrino excitation cross-section $< \sigma_{\nu_{AZ}} >$ has been averaged over the neutrino energy spectrum. The total energy emitted in terms of neutrinos of all flavours has been found to be of the order of the binding energy of a typical neutron star of about $3 \times 10^{53} ergs$ (Cooperstein 1988; Nadyozhin preprint 1976; Woosley, Wilson, and Mayle 1986). Most of these neutrinos are emitted during the cooling phase of the hot neutron star born at the final stage of the core collapse, which is of the order of $10-20$ seconds. The detailed supernova models constructed by the Lick group (Woosley 1988) as well as the Russian group (Nadyozhin preprint 1976) tend to show that the mean energy of the neutrinos $< E_\nu >$ will be about $10 MeV$. This is more representative of muon type and tauon type neutrinos, whereas the mean energy may be only about $5 MeV$ for the electron type neutrinos.

Having made a quick survey of the various parameters needed to calculate the abundances due to nucleosynthesis, we now proceed to calculate the production ratios of Hafnium isotopes.

3. NEUTRINO PROCESS PRODUCING ^{182}Hf

Since several authors have already done detailed studies of the ν induced nucle-osynthesis, in the following, the nucleosynthesis estimates are done specifically for ^{182}Hf, a newly discovered isotopic anomaly, live at the time of formation of the solar system and now extinct.

NTIMS measurements of W separated from metal of the Toluca iron meteorite show a deficit of $-2.4+1.4)\epsilon$ at mass 182, attributable to the presence of live $^{182}Hf(T_{1/2} = 9\ m.y.)$ in the solar system at the time of the meteorite's formation 4.5 $b.y.$ ago. This gives an upper limit to the $^{182}Hf/Hf$ ratio of about $3 \times (6 \pm 3) \times 10^{-5}$, 4.5 $b.y.$ ago (Harper, Volkening and Hermann preprint 1991). The production process which is advocated here is the neutrino excitation of ^{186}W, followed by α emission. The cross-section for this process can be estimated following the prescription of Woosley et al. (1990). However, as they have remarked, at temperatures characteristic of μ and τ neutrinos, the cross-section per nucleon (about $10^{-42} cm^2$) are roughly independent of the nuclear species and this is the value adopted here. Now the remaining prescription is about the branching ratio into the α channel. Here the interpolation methods involve detailed Hauser-Feshbach optical model calculations. Since the effort involved in such detailed studies proved very challenging, for the purposes of our calculation, a reasonable value of 10^{-3} is adopted, considering the branching ratios of several nuclei, computed using such detailed techniques by Woosley et al. (1990). With this we can now compute the abundance expected, in the ν irradiated neon shell of about $20 M_0$ star, at a distance of $10^9 cm$ from the centre as

$$\frac{\delta n_{AZ}}{n_{AZ}} = \frac{L_\nu}{4\pi r^2 < E_\nu >} < \sigma_{\nu_{AZ}} >$$

for the excitation of the parent nuclei. The ratio of the resultant nuclei of interest can be obtained by multiplying the result by the α branching ratio. In the case of Hafnium and Tungsten, the total Hf to the ^{186}W ratio happens to be 2 (Cameron 1982). Hence the ratio of $^{182}Hf/Hf$ is given as

$$N_{182_{Hf}}/N_{Hf} = \frac{\delta n_{186w}}{n_{186w}} \times b_\alpha \times 2n_{186w} \sim \frac{10^{53} \times 186 \times 10^{-42} \times 2 \times 10^{-3}}{4\pi \times 10^{18} \times 1.602 \times 10^{-5}} = 1.86 \times 10^{-4}$$

which is nearly the same as observed upper limit quoted above. Considering that we have made the most pessimistic estimate of the ratio, the abundance could indeed be much higher than computed here. This brings it almost in par with the estimate of another chronometer namely ^{26}Al, which also has been shown to be produced in reasonable amount by the neutrino processes (Domogadskii and Nadyozhin 1980a).

4. DISCUSSION AND CONCLUSION

Since the first demonstration of neutrino induced nucleosynthesis, several new developments have taken place like the refinement of the supernova progenitor evolution, studies of the explosion mechanism, incorporation of neutrino induced nuclear reaction network in the evolutionary code etc. As a result of this it has been shown (Woosley et al. 1990 and references cited therin; Ramadurai and Schramm 1991) that neutrino induced nucleosynthesis may be responsible for the observed abundances of ^7Li, ^{11}B, ^{19}F, ^{138}La, and ^{180}Ta. Though it is claimed that several other isotopes are synthesized by the process, as has been shown by Ramadurai and Schramm (1991), this claim is very marginal. This is especially true regarding the light isotopes, which have a lot of bearing on the nature of dark matter in a hot big bang model of the universe (see talks in this meeting by Reeves; Schramm).

It should be mentioned that the neutrino induced nucleosynthesis brings us to a situation, where it is absolutely essential to determine the details of the explosion mechanism and the subsequent evolution of the supernova debris. This is because of the possibility of the survival of the neutrino produced   isotopes critically depends on the nature of the envelope ejection. If the neon shell, in which the isotopes are made, is ejected before the nuclei had experienced photodissociation, then this indeed will be an ideal alternative to the ill-understood p-process. The case of ^{182}Hf illustrated here, demonstrates that even if the isotope is in the normal neutron addition path, significant contribution from neutrino induced processes can occur.

V. ACKNOWLEDGEMENTS

This work is supported by the INDO-US Project on Supernova Physics INT 87-15411. The author is indebted to David N Schramm for extensive discussions on all aspects of this problem. The financial assistance from the organizers of the Hubert Reeves' 60th Birthday Symposium on *'Origin and Evolution of Elements'* is gratefully acknowledged.

REFERENCES

Bionta, R.M., et al. 1987, *Phys. Rev. Letters,* **58**, 1494.

Cameron, A.G.W. 1982, *Essays in Nuclear Astrophysics* ed. C.A. Barnes, D.D. Clayton, and D.N. Schramm (Cambridge University Press) p. 23.

Cooperstein, J. 1988, *Phys. Rept.,* **163**, 95.

Domogadskii, G.V., Eramzhyan, R.A.. and Nadyozhin, D.K. 1977, in Neutrino 77 (Moscow: NAUKA), p. 115.

_____ 1978, *Ap. Space Sci.,* **58**, 273.

Domogadskii, G.V., and Nadyozhin, D.K. 1977, M.N.R.A.S., **178**, 33P.

_____ 1978, *Soviet Astr.,* **22**, 297.

_____ 1980a, *Soviet Astr. Letters,* **6**, 127.

_____ 1980b, *Ap. Space Sci.,* **70**, 33.

Domogatskii, G.V., and Imshennik, S.V. 1982, *Soviet Astr. Letters,* **8**, 190.

Epstein, R.I., Colgate, S.A., and Haxton, W.C. 1988, *Phys. Rev. Letters,* **61**, 2038.

Haxton, W.C. 1991, *Nucl. Physics,* **522**, 325c.

Hirata, K., et al. 1987, *Phys. Rev. Letters,* **58**, 1490.

Ramadurai, S., and Schramm, D.N. 1991, Abstract submitted to Annual Meeting of Astron. Soc. India, 1991.

Woosley, S.E. 1977, *Nature,* **269**, 42

_____ 1988, *Ap. J.,* **330**, 218.

Woosley, S.E., and Haxton, W.C. 1988, *Nature,,* **334**, 45.

Woosley, S.E., Hartmann, D.H., Hoffman, R.D., and Haxton, W.C. 1990, *Ap. J.,* **356**, 272.

Woosley, S.E., Wilson, J.P., and Mayle, R.M. 1986, *Ap. J.,* **302**, 19.

GALACTIC EVOLUTION AND COSMOCHRONOLOGY

The First Stars in the Galaxy

ROGER CAYREL

Observatoire de Paris
61, av. de l'Observatoire
75014 Paris,France

1 INTRODUCTION

The current model of chemical evolution of the Universe assumes that the first stars formed from primordial big-bang material, and subsequently from material gradually enriched in heavier elements, produced in stellar interiors by nuclear burning, and injected in the interstellar medium by violent mass-loss (supernovae explosions) or milder mass-loss (planetary nebulae, stellar winds, etc...). It is thence expected that stars exist with a full range of metallicity, from no heavy elements at all (this hypothetic class of stars has been called population III) and a large range of metallicity from almost no heavy elements to solar abundances and even more, as found in particular in the galactic bulge (Rich 1990). A first question which arises is if a complete sample of metallicity can be found in our own Galaxy, or if some pregalactic enrichment took place, cutting off the very low metallicity end of the distribution in the galactic sample.

This last idea was popular after Bond's survey (Bond 1981) and before COBE observations (Mather et al. 1990). Thanks to a new survey, which is today only 10 % complete (Beers *et al.* 1985,1992) we shall see that there is no reason to believe that it is necessary to search outside the galactic halo to reach the lowest metallicities or the lowest level of astration (this last word was greatly popularized by our friend Hubert Reeves).

Indeed, the observation of halo-star with metallicities less than one thousanth of the solar metallicity, and ages which are barely compatible with the age of the Universe, is giving an insight on events with are likely more "primordial" than quasars at a redshift of 4. At least it teaches us that star formation is a very old phenomenon in the Universe.

In section 2 we succintly describe the Beers et al. survey ,which is currently the major effort ever undertaken to obtain a significant sample of stars at extremely low levels of metallicity, and therefore mostly made of big-bang material.In section 3 we give an overview of the kinematical properties of these very metal-poor stars, properties which raise the important question of the origin of this stellar population, also present in most other extragalactic systems. In section 4 we discuss the problem of the age of the same stellar population and in section 5 we summarize this review.

2 BEERS ET AL. SURVEY

After Bond's survey, the reward for scrutinizing half a million spectra for finding very metal-poor stars turned out to be very slim. Only 3 stars more metal-poor than a factor of one thousand with respect to the sun (Bond 1981). Who was willing to do better? Beers, Preston and Shectman (1985,1992) have started in 1978 a deep survey (3.5 magnitudes deeper than Bond's survey) aimed at increasing the sample of very metal-poor stars by two order of magnitudes or so. Fig. 1 gives an idea of the volume of space probed by the BPS survey at the galactic scale.

The strategy was to take dedicated Schmidt plates of high latitude fields, with an objective-prism and a narrow band filter, limiting the sky background to 150 Å around the H and K line region. It is then possible to reach the magnitude 16 in the Johnson B-colour. The overwhelming majority of spectra show strong H and K lines. The selection criterion is the absence of these lines. This selection is done by visual inspection with a low power binocular. The objects found are only metal poor candidates. Further spectroscopic observations are needed to remove about one third of the objects which are not metal-poor halo stars, but very hot stars, degenerate stars, extragalactic objects, etc...These complementary observations are made with a slit-spectrograph, and a resolution of about 1 Å. Actually these medium-resolution observations are the bottleneck of the survey. At present, 8000 candidates have been found by inspection of the Schmidt plates, but only 1500 of those have gone to the medium-resolution phase. Out of these 1500 objects, 1000 turned out to be real metal-poor stars and from a comparison of the equivalent widths of the Balmer lines and the equivalent width of the K line a metallicity is derived.

For the first time a histogramme of metallicities in the range $10^{-4}Z_{\odot}$ to $10^{-2}Z_{\odot}$ has been obtained. We give it (Fig.2), limited to the unevolved stars, to avoid the biais with temperature, which occurs in a mixture of stars with a typical temperature of 6000 K, and much cooler (giant) stars, with temperatures between 4000 and 5000 K. It is clear from the histogramme, that there is no trend of weakening of the number of low-metallicity stars per δZ unit with Z, when Z tends towards zero. The opposite tendency is due to the fact that the completeness of the survey at the survey at $Z = 10^{-2}Z_{\odot}$ is not yet 100% . It is interesting to note that the so called "simple model" (closed system with constant initial mass function and instant recycling of the heavy elements produced) of galactic chemical enrichment predicts a flat histogramme. The concentration in heavy elements at time t being proportional to the number N of stars produced before time t, when the fraction of gas converted into stars is small in comparison of the of the total mass of gas. Z is proportional to $N(t)$, and in particular to the number n of low-mass unevolved stars, still there today. dn/dZ is thence constant at small Z in this simple model. It remains so if even if a constant fraction of the heavy elements is lost by the system, as shown by Hartwick(1976).

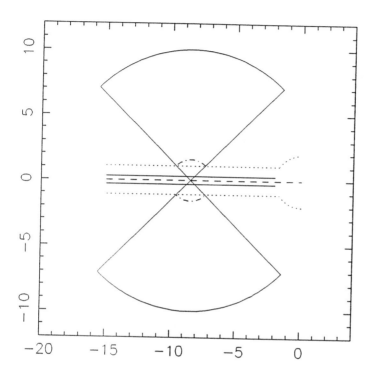

Fig. 1. This figure represents a cut by the galactic meridian plane containing the sun, of the double cone explored by the BPS survey. The scale is given in kpc, with the sun at 8.5 kpc from the galactic center. The galactic plane is represented by the dashed line, the thin disk by the full lines, and the thick disk and buldge by the dotted lines.The big cone is the volume explored for giant stars of zero absolute magnitude, and the small cone (dashed-dotted line) the volume explored for bright turn-off stars of absolute magnitude 4. Bond's survey reached for giants about the limit reached by BPS for turn-off stars. The volume of space probed for giants by the BPS survey is about 1500 kpc^3.

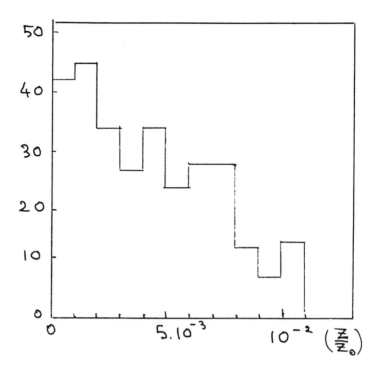

Fig.2 Histogramme of metallicities for turn-off stars from the
BPS survey (Beers et al. 1992). Note that there is no weakening
in the number of stars in the lower metallicity bins.

In conclusion, there is no cut-off in the number of stars at very low metallicity, *and the
galactic halo started from primordial gas, free from elements produced inside stars.*

3 KINEMATICS OF VERY METAL-POOR STARS

Already in the sixties, Eggen, Lyndell-Bell and Sandage (1962) have pointed out the
very peculiar kinematical properties of metal-poor stars. The sample of stars on
which they worked has been greatly increased during the 30 years in between, thanks
to surveys by Sandage and Fouts (1987), Norris (1986) and Carney et al. (1990).
A difficulty in analysing correlations between metallicity and kinematical properties
is the fact that, at the beginning, metal-poor stars were searched in high proper-
motions lists of objects, to increase the percentage of success. The elimination of the

biaises caused by this circumstance has been discussed by Norris (1986). Another new event is the identification by Gilmore and Reid (1983) of an intermediate population, known as the "thick disk", to be related with the metal-rich globular clusters (Zinn 1985). There is nothing better , to introduce the subject, than having a look at the diagramme $(V_{rot}, [Fe/H])$ giving the rotational velocity with respect to an inertial reference frame (not with respect to the local standard of rest participating to the galactic rotation) of stars binned by metallicity (fig.3).

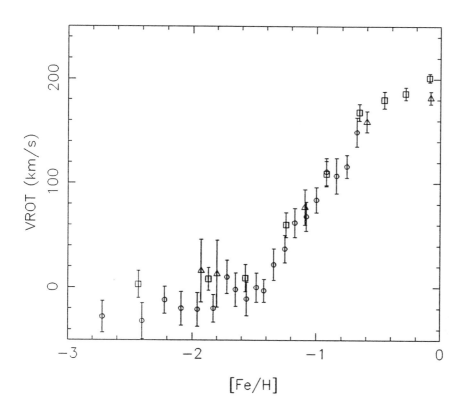

Fig. 3 Rotational velocities of high proper-motion stars, binned by metallicity intervals. The rotational velocity is with respect to an inertial rest-frame. The different symbols refer to three sources: circles are from Norris and Ryan 1989b,triangles from Sandage and Fouts 1987, and squares from Carney et al. 1990.

It is striking that, on the average, stars more metal poor than $[Fe/H] = -1.5$ do not participate at all to the galactic rotation. Because of the overlap in metallicity for individual stars, between thick disk stars and halo stars the diagramme of fig. 3 for individual stars is very noisy (fig. 4) and the clear trend shown by fig.3 with metallicity bins less visible. However, fig. 4 opens a very interesting interpretation of fig.3.

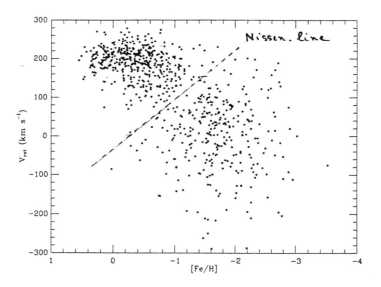

Fig. 4 Rotational velocities of individual stars versus metallicity, according to Laird et al. 1988. Notice the separation in two subsystems by the Nissen and Schuster line.

Nissen and Schuster (1991) have proposed that the physical separation between populations in the plane $(V_{rot}, [Fe/H])$ occurs along an oblique line, drawn in fig.4. Everything below the line would be the halo population, and everything above the line would be disk (thin+thick) population. The gradual transition in fig.3, from high rotation to zero rotation would then be understood, not as continuous change of population, but by the inclusion of different proportions of disk and halo population, with pure disk pop. at $[Fe/H] = 0.$ and pure halo pop. at $[Fe/H] = -2$. The thick disk, in fig. 4, appears as an excrescence of the thin disk, with rotational velocities between 100 and 180 km/s and metallicities between -.4 and -1.0 . It must be recalled that the sample of fig.4 is a sample of high proper motion stars and that the proportions of metal-poor stars versus metal-normal stars is considerably enhanced with respect to a volume limited sample in the solar neighbourhood. As discussed by Norris and Green (1989) a clear-cut separation of the thick disk from thin disk is not physically obvious. It is our interpretation that the thick disk reflects the state of the disk before a good circularisation of the gas motion was achieved.

The origin of the halo is one of the more fascinating problem of galactic physics. Apparently there is a triple gap between the halo and the disk properties. There is a kinematical gap, just explained above, a metallicity gap of 1.0 dex (thick disk $-.5$, halo -1.5) and an age gap that we shall see in the next section. Norris and Ryan (1989) have argued that the halo was produced by accretion of small subsystems, by dynamical friction. This concept is to be opposed to the Eggen, Lynden-Bell and Sandage model of the Galaxy being produced by the gravitational collapse of a single big cloud, with continuity between halo, thick disk and thin disk stages. The argument put forward by Norris and Ryan is the fact that metal-poor stars with velocities close to the escape velocity in the Galaxy, at the solar neighbourhood, have all retrograde orbits. This is very difficult to explain in the collapse scenario, where the initial cloud must have a prograde rotation. The accretion hypothesis is supported by dynamical friction computations, showing that prograde and retrograde systems react very differently in the accretion process.

In conclusion, the stars formed out of matter with a low astration level, do not partake the galactic rotation, and may very well have been formed elsewhere, and then accreted by our Galaxy. In that case, it is fair to say that the halo would be of extragalactic origin.

4 THE AGE OF THE FIRST STARS

two questions will be addressed in this section.
(i) What is the age ofthe oldest stars ?
(ii) Is there a spread in the epoch of formation of halo objects?

The first question is the same as finding the age of the globular clusters. At present it is not possible to date field halo stars , because their distance is not accurately known.The situation will evolve when the HIPPARCOS parallaxes will become available in 1995. Even with the globular clusters, for which a complete "isochrone" is observed in the colour-luminosity diagram, there are zero-point problems, due to an uncertain amount of absorption and reddening,lack of exact absolute magnitudes for horizontal-branch stars,etc... A thorough discussion of the problem can be found in "Astrophysical Ages and Dating Methods" a colloquium held in June 1989 in this same room. The stellar ages are also affected by a number of uncertainties in the theoretical models (mixing length to scale height ratio, opacities, etc...). Everything being considered, there is a general agreement that, taking into account the enhanced [O/Fe] in metal-poor stars, and using the proper value of $Y = 0.23$ for pop. II the age of the well studied globular clusters is 15 to 16 Gy, with a firm lower limit at 12.5 Gy and a shallow upper limit at 20 Gy (Rood 1990, Caputo and Quarta 1986, VandenBerg 1990, Buonnano et al. 1990).

Before comparing this range to the age of the Universe, let us see how much coeval are the globular clusters. This point has been a longstanding subject of debate, with some supporting ,as Sandage or VandenBerg, the concept that at least the globular clusters of the spheroid were practically coeval, whereas the Yale group led by Demarque insisted that the so called second parameter was nothing else than age, this implying of course a significant spread (at least 4 Gy) in the ages of the GCs. This problem has now come to some sort of consensus, with the statement that the vast majority of globular clusters are nearly coeval, but that there are cases of globular clusters younger that the mean by a few Gy. The undisputed example is the couple NGC 288 and NGC 362 (Green and Norris 1990, VandenBerg et al. 1990) for which an age difference of 2 to 3 Gy is agreed upon. Interestingly, some very distant globular clusters, as Pal 12, belong to the category of these "young" globular clusters.

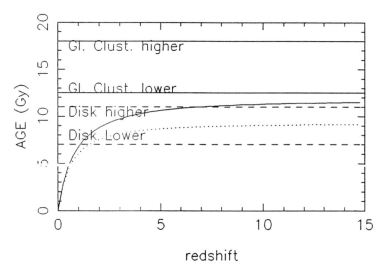

Fig. 5 Age versus redshift curve. The cosmological constant is zero, $H_0 = 70\ km/s/Mpc$. The full curve is for $\Omega_0 = 0.2$ and the dotted curve for $\Omega_0 = 1.0$. No globular cluster can be formed with these parameters.

Let us now compare the age of the most frequent, and "old" globular clusters with the age of the Universe.

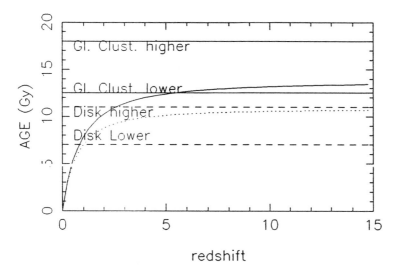

Fig. 6 Age versus redshift curve. Same parameters as in fig. 54,
except that now $H_0 = 60\ km/s/Mpc$.It is now possible to
form GCs at redshift 5 or larger if Ω_0 is small. $\Omega_0 = 1.0$ is not
allowed.

We shall limit ourselves here to the case of a null cosmological constant, Λ_0, on the ground that the attempts to connect Λ_0 to some "vacuum energy density" lead to unacceptable values for this constant. We shall parametrize the two other constants, H_0 and Ω_0 with usual notations. Avoiding sollicitating these numbers from theoretical arguments, it is fair to say that $H_0 = 70\ km/s/Mpc$, and $\Omega_0 = 0.2$ is a reasonable choice, based on purely observational data. Fig.5 shows that such values are uncompatible with the age of the old globular clusters. The $\Omega_0 = 1.0$ curve (value for an inflationary Universe is of course even more uncompatible. Fig. 5 gives the age-redshift relationship for $H_0 = 60\ km/s/Mpc$. It is now barely possible to have the GCs formed (at $z = 5$), accepting the lower limit of their age as their actual age, with $\Omega_0 = 0.2$. There is is still no solution for $Omega_0 = 1.0$ .We have also plotted in fig. 5 and 6 the strip corresponding to the age of the disk $(9 \pm 2Gy)$, based on the luminosity function of white dwarfs (Hernanz et al. 1990) discarding the unprobable extreme case of total separation of carbon and oxygen throughout the whole star. The substantial shift between the epoch of formation of the disk and of the GCs is important to understand the relationship between the two systems. This shift adds up with other evidences, gap in metallicity, gap in kinematical properties, to indicate that the halo is an independent body in the Galaxy.

5 CONCLUSION

The quest towards more and more primitive stars is still going on ,on a large scale, and, contrarily to what was thought in the last decade, is producing a large sample of very metal-poor stars below one thousandth solar metallicity. It is a reasonable expectation that ,when the ESO VLT will be in full operation at the turn of the century, stars with a level of astration below 10^{-5} solar or 10^{-7} absolute will be known.

It also now clear that the halo is a component of the galaxy which is almost there by accident, considerably more primitive than the disk from all points of view, chemically, kinematically, and also in age.

The comparison between the age of the oldest halo stars and of the expansion age of the Universe is very compelling for the values of H_0 and Ω_0 : $H_0 \leq 60 km/s/Mpc$, and $\Omega_0 = 1$. only if $H_0 \leq 50$.

REFERENCES
Beers T.C., Preston G.W., Shectman A. 1985 A.J. 90,2089.
Beers T.C., Preston G.W., Shectman A. 1992 A.J. 103,1987.
Bond H.E. 1981 Ap.J. 248,606.
Buonanno R., Cacciari C., Corsi C.E., Fusi,Pecci F. 1990 in *Astrophysical Ages and Dating Methods* eds. E. Vangioni-Flam et al. Editions Frontières p.261.
Caputo F., Quarta M.L. 1986, Mem.Soc.It. 57,437.
Carney B.W., Latham D.W., Laird J.R. 1990 A.J. 99,572.
Eggen O.J., Lynden-Bell D., Sandage A.R. 1962 Ap.J. 136,735.
Gilmore G., Reid I.N., 1983 MNRAS 202,1025.
Green E.M., Norris J.E. 1990 Ap.J. 353, L17.
Hartwick F.D.A., 1976 Ap.J. 209,418.
Hernanz M., Garcia-Berro E., Isern J., Mochkovitch R. 1990 in *Astrophysical Ages and Dating Methods* eds. E. Vangioni-Flam et al., Editions Frontières, p. 171.
Laird J.B., Carney B.W., Latham D.W. 1988 A.J. 95,1843.
Mather J.C. et al.1990 Ap.J. 354,L37.
Nissen P.E., Schuster W.J. 1991 A.A.251,457.
Norris J. 1986 Ap.J.Suppl. 61,667.
Norris J.E., Green E.M. 1989 Ap.J. 337,272.
Norris J.E., Ryan S.G. 1989a Ap.J. 336,L17

Norris J.E., Ryan S.G. 1989b Ap.J. 340,739.

Rich R.M. 1990 Ap.J. 362,604.

Rood R.T. in *Astr0physical Ages and Dating Methods* eds. Vangioni-Flam et al. 1990, Editions Frontières,p.313.

Sandage A.R., Fouts G. 1987 A.J. 92,74.

VandenBerg D.A. 1990 in *Astrophysical Ages and Dating Methods* eds. E. Vangioni-Flam et al., Editions Frontières, p. 241.

VandenBerg D.A., Bolte M., Stetson P.B. 1990 A.J. 100,445.

Zinn R.,1985 Ap.J. 293,424.

Chemical abundances in four metal-poor stars

P.MOLARO[1], F.CASTELLI[2], F.PRIMAS[3]

[1] Astronomical Observatory of Trieste
[2] C.N.R. - Gruppo Nazionale Astronomia, Unità di Ricerca di Trieste
[3] Department of Astronomy, University of Trieste

1 ABSTRACT

CASPEC spectra of four faint metal-deficient candidates obtained at the 3.6m ESO telescope at La Silla (Chile) are analyzed. The stars have been selected from the sample of Beers et al. (1985). From the fine analysis with an LTE model atmosphere we obtain a metallicity range $-3.7 \leq [Fe/H] \leq -3.4$ for the four objects. The α and the *iron-group* elements show the same trends of other halo stars. A plateau of [Al/Mg] at the lowest metallicity is found, which reveals a possible primary component for Al. A large scatter is present in the abundances of heavy elements Sr and Ba, which are undetected in CS 22968-14.

2 INTRODUCTION

The most metal-deficient stars are the first records of the chemical evolution of the Galaxy and knowledge of the elemental abundances in their atmospheres provides clues on the origin and evolution of the chemical elements. The search for metal-deficient stars has been carried on extensively in the last decades, but only a few objects have been found with very low metallicity. Before our analysis, only six extremely metal-poor stars with [Fe/H]<-3.5 were known (Beers et al., 1992). The metallicities of these stars are at the lowest levels ever observed in the universe and almost two orders of magnitude below the metallicity of the oldest globular clusters.

3 RESULTS

The observations consist of CASPEC spectra for 4 metal-poor candidates, selected fromthe survey of Beers et al. (1985), obtained at the 3.6m telescope at La Silla. The available spectral range is from 3800 to 5400 Å at a resolution of R $\approx$ 20000 and with a S/N between 20 to 50, varying with the spectral wavelength and with the position within individual orders of the echelle spectra. The data have been reduced using the package Echelle in MIDAS. The photospheric parameters (T_{eff}, $\log g$ and ξ) have been obtained following the procedure described in Molaro and Castelli (1990) and are summarized in Table 1. Model atmospheres have been computed with the version 9 of the ATLAS code (Kurucz, 1991), for a metallicity $[M/H] = -3.5$. Abundances and microturbolent velocities have been derived using the WIDTH code (Kurucz, 1991). Reduction and abundance analysis have been tested using our observations of the well studied CD-38° 245. Our measurements of EW and chemical abundances are in good agreement with those reported in the literature.

Table 1: Photospheric parameters of our sample

Star	T_{eff} K	$\log g$	ξ km·s^{-1}
CD -38°245	4700	1.4	2.2
CS 22891-209	4723	0.75	2.1
CS 22897-8	4915	1.5	2.1
CS 22948-66	5100	0.5	2.3
CS 22968-14	4750	0.5	2.0

The derived iron abundances of our sample of stars are: $[Fe/H] = -3.72\pm0.2$ for CS 22968-14, $[Fe/H] = -3.56\pm0.3$ for CS 22897-8, $[Fe/H] = -3.42\pm0.3$ for CS 22891-209 and finally $[Fe/H] = -3.36\pm0.4$ for CS 22948-66. The adopted $[Fe/H]_\odot$ is -4.37 from Anders and Grevesse (1989). The errors quoted are the dispersion of the abundances for the individual Fe I lines and reflect uncertainties in the EW measurements as well as in the gf values of the lines. Systematic errors in the stellar parameters may produce another 0.2 dex in the abundance error (see the discussion in Ryan et al., 1991).

Abundances for a number of other elements are derived. The abundances of the α-elements Mg, Si, Ca and Ti, show a nearly constant overabundance to iron, as it has been observed in higher metallicity stars. Calcium abundances are lower by ≈ 0.3 dex when derived only from the resonant transition at $\lambda = 4226.7$Å . This effect has been already noted by Ryan et al.(1991) and may result from non-LTE effects.

Deviations from the pattern shown by more metal abundant stars are observed for Al. Our determinations are shown in Fig.1 together with the available measurements for the metal poor stars ($[Fe/H]<-1.0$). The abundance pattern shows a distinct rise of $[Al/Fe]$ at metallicities lower than -3.5. When Al is compared to Mg this behaviour is seen in the form of a plateau around the value $[Al/Mg] = -1$ for $[Mg/H]$ lower than -2. This strongly suggests the presence of a primary component of Al as predicted by explosive nucleosynthesis in metal-weak supernovae (Woosley and Weaver, 1982).

The *iron group elements* maintain about the same ratio to iron which is observed at higher metallicities in the halo. The abundances of Cr, Ni, Sc, and Mn track very well iron, remaining constant along the entire metallicity range explored.

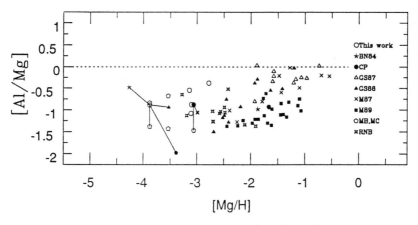

Figure 1: $[Al/Mg]$ vs. $[Mg/H]$

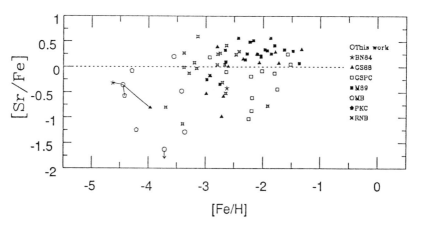

Figure 2: $[Sr/Fe]$ vs. $[Fe/H]$

The abundances of the *heavy elements* Sr and Ba show a large spread of nearly two orders of magnitude, which cannot be accounted for by observational errors (see Fig.2). It might be possible that we start to see chemical inhomogeneities in the gas from which these old stars have originated. Of particular interest is the case of CS 22968-14 which does not show any trace of the two elements with [Ba/H] and [Sr/H] upper limits at -1.4 dex and -1.6 dex respectively. In Fig.3 the spectral region of the

Sr II 4077.709 Å of CS 22891-209 and CS 22968-14 is shown for direct comparison. Other two dwarfs with the same property have been found by Ryan et al.(1991). These objects may form a particularly interesting group of ultra metal-deficient stars completely free of heavy elements. It has been suggested that they could be formed from gas that has suffered only one episode of chemical enrichment. Thus they should belong to the second stellar generation.

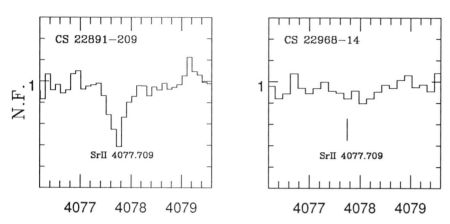

Figure 3: Spectra of CS 22891-209 and CS 22968-14 around Sr II 4077.709Å

4 REFERENCES

Anders,E. Grevesse,N. 1989, *Geochim. Cosmochim. Acta* **53**, 197.
Beers,T.C. Preston,G.W. Schectman,S.A. 1985: *As.J.* **90**, 2089.
Beers,T.C. Preston,G.W. Schectman,S.A. 1992: *As.J.* **103**, 1987.
Bessel,M.S. Norris,J. 1984, *Ap.J.* **285**, 622 (BN).
Carney,B.W. Peterson,R.C. 1981, *Ap.J.* **245**, 238 (CP).
Gilroy,K.K. Sneden,C. Pilachowski,C.A. Cowan,J.J. 1988, *Ap.J.* **327**, 298 (GSPC).
Gratton,R.G. Sneden,C. 1987, *Astron.Astroph.* **178**, 179 (GS87).
Gratton,R.G. Sneden,C. 1988, *Astron.Astroph.* **204**, 193 (GS88).
Kurucz,R.L. 1991: *private communication*
Magain,P. 1987, *Astron.Astroph.* **179**, 176 (M87).
Magain,P. 1989, *Astron.Astroph.* **209**, 211 (M89).
Molaro,P. Bonifacio,P. 1990, *Astron.Astroph.* **236**, L5 (MB).
Molaro,P. Castelli,F. 1990, *Astron.Astroph.* **228**, 426 (MC).
Peterson,R.C. Kurucz,R.L. Carney,B.W. 1990, *As.J.* **350**, 173 (PKC).
Ryan,S.G. Norris,J.E. Bessel,M.S. 1991, *As.J.* **102**, 303 (RNB).
Woosley,S.E. Weaver,T.A. 1982: in *Supernovae, A Survey of Current Research*, published by M.J.Rees and R.J.Stoneham (Reidel, Dordrecht), page 79.

Barium isotopes in metal-poor stars

P. MAGAIN[1,*] AND G. ZHAO[1,2]

[1]Institut d'Astrophysique, Université de Liège, Belgium
[2]Beijing Astronomical Observatory, Chinese Academy of Sciences
[*]Chercheur Qualifié au Fonds National de la Recherche Scientifique (Belgique)

1 INTRODUCTION

While most of the barium present in the solar system was produced through the s-process, it seems to be quite generally agreed now that the r-process contribution was much higher for the very metal-poor stars. This suggestion dates back to the work of Truran (1981) who suggested that the heavy elements abundance trends with metallicity can be better understood if these elements are products of r-process nucleosynthesis.

These ideas were confirmed by the works of Sneden and collaborators (Sneden and Parthasarathy 1983, Sneden and Pilachowski 1985, Gilroy et al. 1988) who showed that the heavy elements abundance patterns in very metal-poor giants are in good agreement with expectations from r-process nucleosynthesis and incompatible with the s-process.

On the other hand, as pointed out by Magain (1989), this hypothesis does not seem to agree with the observed variation of relative abundances with metallicity, which show a rise of, e.g., [Ba/Fe] at the lowest metallicities, followed by a more or less constant value for [Fe/H] > -2.

Given these problems, and the fact that past determinations of the r and s-process contributions were very indirect, it would be highly desirable if one could determine more directly the contribution of both processes to the nucleosynthesis of the heavy elements in metal-poor stars. In the present paper, we examine the feasibility of such a project in the case of barium and present some preliminary results.

2 BARIUM ISOTOPES AND HYPERFINE STRUCTURE

In solar system material, Ba is mainly represented by five stable isotopes, with mass numbers 134 to 138. While the even isotopes are primarily produced by the s-process, the contribution of the r-process to the odd isotopes is very significant (Cameron 1982, Anders and Grevesse 1989). In contrast with the spectral lines of the even isotopes, the lines of the odd isotopes of Ba are affected by hyperfine structure (HFS). The global effect of HFS on the Ba lines thus depends on the isotopic ratios, and, as a consequence, on the relative contributions of the r and s-processes to the total Ba abundance. This gives a means of estimating the relative contributions of these processes in stars of various metallicities.

However, due to the faintness of the very metal-poor stars and to the presence of other broadening mechanisms, it is very difficult to observe directly the HFS splitting of the lines. Fortunately, there is an alternative way of measuring the line broadening. Indeed, if the line is on the saturated part of the curve of growth, any additional broadening will produce a desaturation of the line, so that its equivalent width (EW) will increase. This is precisely the case for the 455.4 nm resonance line of BaII in metal-poor dwarfs with [Fe/H] ~ -2.

3 METHOD OF ANALYSIS AND FEASIBILITY

High resolution spectra were obtained with the CAT/CES at ESO for 4 dwarfs with [Fe/H] ~ -2. Effective temperatures $T_{\rm eff}$ and surface gravities $\log g$ were determined as in Zhao and Magain (1991). As the lines may be desaturated by microturbulence as well as by HFS, the precise determination of the microturbulence velocity $v_{\rm turb}$ is, of course, of crucial importance. It was determined from a set of CaI lines with precise oscillator strengths and damping constants. A value close to 1.5 km/s was obtained for all 4 stars. The barium abundance was deduced from two weak lines of BaII, at 585.3 and 649.6 nm. For the BaII oscillator strengths $\log gf$, we took means of the two recent determinations of Davidson et al. (1992) and Guet and Johnson (1991). The HFS data for the 455.4 nm line were taken from Rutten (1977). The damping constants were calculated from the Unsöld formula, multiplied by some empirical factor f_6 (here taken as 1.0). The barium abundance was then deduced from the 455.4 nm line for various isotopic mixtures, the proportion of Ba^{135} and Ba^{137} being changed until agreement with the weak lines was obtained.

For the case of HD 166913, the effect of uncertainties in various parameters was investigated in detail. The errors in the fraction of odd isotopes caused by errors in the input parameters are summarized in Table 1, which also indicates the required accuracy needed in any of these parameters if the isotopic ratio is to be determined with a 10 % accuracy.

It can be seen that, as expected, the microturbulence velocity is the most critical parameter. The effect of a 0.25 km/s uncertainty on the microturbulence velocity is illustrated in Fig. 1. It should be noted that such uncertainties cumulate, so that the quoted errors on all these parameters would result in a 25 % total uncertainty in the isotopic ratio for a single star.

It thus appears that, although at the limit of the present possibilities, this determination does not appear unfeasible if the analysis is carried out with great care.

Table 1. Error analysis

input parameter	input error	effect on isotopic ratio	required accuracy
T_{eff}	50 K	3 %	200 K
$\log g$	0.2	10 %	0.2
v_{turb}	0.2 km/s	20 %	0.1 km/s
$\log gf$	0.05	10 %	0.05
f_6	0.5	5 %	1.0
EW	0.3 pm	10 %	0.3 pm

4 PRELIMINARY RESULTS

The determination of the ratio of odd to even Ba isotopes in 4 metal-poor stars is illustrated in Fig. 2 and the results are summarized in Table 2.

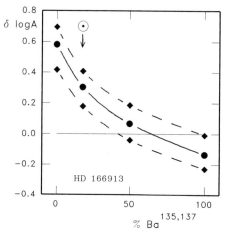
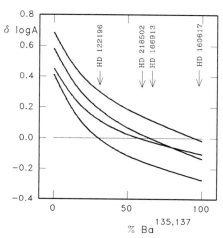

Figure 1. Difference in Ba line abundances as a function of isotopic mixture for HD 166913. Circles: $v_{\text{turb}} =$ 1.50 km/s. Lozenges: v_{turb} changed by 0.25 km/s

Figure 2. Determination of the isotopic mixture for the 4 stars considered

Table 2. Results

Star	T_{eff}	$\log g$	[Fe/H]	% Ba^{135,137}
HD 122196	5860	3.4	−2.0	30
HD 160617	5920	3.6	−1.9	90
HD 166913	6030	3.8	−1.7	65
HD 218502	6110	3.7	−1.9	60

On the average, the odd isotopes seem to account for more than one half of the total amount of barium in these stars. The scatter amounts to 25 %, which is compatible with the estimated uncertainty affecting a single determination.

As far as we can conclude from the present preliminary results, the r-process contribution is thus significantly larger than in solar system material, although s-process products might also be present in non-negligible amounts.

These results need of course to be refined and confirmed. Nevertheless, as this paper demonstrates, such a project is feasible provided all input data and every step of the analysis are carefully checked.

ACKNOWLEDGEMENTS

This work is based on observations collected at the European Southern Observatory (La Silla, Chile). We wish to thank Nicolas Grevesse for enlightening discussions. We also thank the Belgian National Fund for Scientific Research and the National Natural Scientific Foundation of China for financial support.

REFERENCES

Anders, E., Grevesse, N.: 1989, Geochimica et Cosmochimica Acta **53**, 197
Cameron, A.G.W.: 1982, ApSS **82**, 123
Davidson, M.D., Snoek, L.C., Volten, H., Dönszelmann, A.: 1992, A&A **255**, 457
Gilroy, K.K., Sneden, C., Pilachowski, C.A., Cowan, J.J.: 1988, ApJ **327**, 298
Guet, C., Johnson, W.R.: 1991, Phys. Rev. A **44**, 1531
Magain, P.: 1989, A&A **209**, 211
Rutten, R.J.: 1977, Solar Phys. **56**, 237
Sneden, C., Parthasarathy, M.: 1983, ApJ **267**, 757
Sneden, C., Pilachowski, C.: 1985, ApJ **288**, L55
Truran, J.W.: 1981, A&A **97**, 391
Zhao, G., Magain, P.: 1991, A&A **244**, 425

Evolution of light s-elements in the halo

I. BARAFFE[1,2] AND K. TAKAHASHI[2]

[1]Universitäts-Sternwarte Göttingen, W-3400 Göttingen, Germany

[2]Max-Planck-Institut für Astrophysik, W-8046 Garching bei München, Germany

1 INTRODUCTION The slow neutron-capture process in massive stars ($M \geq 10\,M_\odot$, and primarily during core helium burning) is known to produce elements ranging from Cu to Zr. Recent studies of such an "s"-process nucleosynthesis have shown that often the productivity of those "light" elements does not scale with the initial metallicity (Prantzos et al. 1990; Raiteri et al. 1991; Baraffe et al. 1992). With the theoretical s-yields at hand as a function of metallicity, one can now pursue a quantitative study of evolution of those elements to analyze observational data which have become available, albeit very uncertain, in a wide range of metallicity or in the $-3 \leq [Fe/H] \leq -1$ range (Magain 1989; Gilroy et al. 1988; Spite 1991). We briefly discuss some results based on a simple model of the chemical evolution of the galactic halo. The details will be presented elsewhere (Baraffe and Takahashi 1992).

2 INPUTS

2.1 The halo model We describe the chemical evolution of the halo by a "modified simple one-zone model", closely following the prescription by Pagel (1989 and references therein). In order to take into account the finite lifetimes of stars, however, we discard the instantaneous recycling approximation. For massive stars, the yields in metals and stellar lifetimes are taken from Baraffe and El Eid (1991); for low- and intermediate-mass stars, we adopt the values used by Yokoi et al. (1987). The star formation rate (SFR) is parametrized such that the time span of the chemical evolution of the halo is about 0.9 Gyr. A time-independent initial mass function (IMF) is then obtained from the present day mass function (Scalo, 1986). At those early epochs of the galactic history, the metallicity is mainly determined by the amount of oxygen coming from massive stars.

2.2 Yields of s-elements In this work, we aim at estimating just the contribution from the s-process in massive stars to the observed abundances of Cu, Sr, Y and Zr, and do not consider those from s-processing in low- and intermediate-mass stars that may become significant as the evolutionary time passes. The s-yields in massive stars are sensitive to the initial O/Fe ratio, particularly at low metallicities. We derive the time evolution of Fe abundance by utilizing the $[O/Fe]$ vs. $[Fe/H]$ relationship as inferred from observations (Cases 1 and 2 in Fig.1). The corresponding s-yields are given by Baraffe et al. (1992) for the initial metallicity of $10^{-8} \leq Z \leq 0.02$. The s-process in low-Z stars is more efficient in Case 2 than in Case 1. As mentioned in Baraffe et al. (1992), the efficiency is sensitive also to the $^{17}O(\alpha, \gamma)$ rate. As this rate is still very uncertain, the s-yields are calculated for two cases: Case A with the rate of Caughlan and Fowler (1988), and Case B with the rate set equal to nul. Case B favors the s-process at low Z. The productions of Sr, Y and Zr vary much from case to case and are not linear with respect to the initial Fe abundance, whereas a near linearity is found for Cu. A detailed discussion on s-process efficiency as a function of metallicity is found in Prantzos et al. (1990) and Baraffe et al. (1992).

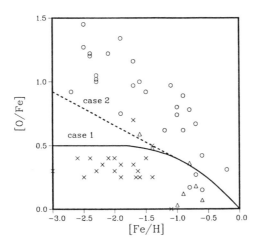

Fig. 1. *$[O/Fe]$ vs. $[Fe/H]$ relationship. The curves correspond to the two cases studied in this work. The observational data are from Barbuy (1988) ($\times$), Abia and Rebolo (1989) (O) and Spite and Spite (1991) (Δ).*

3 RESULTS   The model of chemical evolution is evolved until an iron abundance of $Fe/Fe_{\odot} = -1$ (or a highest metallicity observed in the halo $[Fe/H] \sim -1$; Gilmore and Wyse 1986) is reached. The predicted Cu and Sr evolutions are shown Fig.2 for Case 1A and 2B (cf. Sect. 2) and are compared with observations. As seen in Fig.2a, Cu is apparently underproduced by a factor of ~ 10. The Sr case is less obvious to interprete, because of the large scatter in the observational data. However, it does not seem likely that the s-process in massive stars can explain the observed abundance pattern at $[O/Fe] < -2.5$. The same holds for Y and Zr.

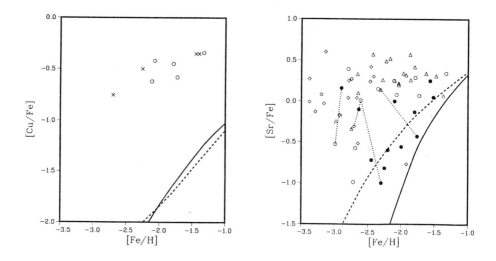

Fig. 2a and 2b. *Abundances of Cu (a) and Sr (b). The results are given for Case 1A (solid line) and Case 2B (dashed line: an optimal case; see text). The observational data are from Sneden and Crocker (1988) (×), Gratton and Sneden (1988) (O), Gilroy et al. (1988) (•), Magain (1989) (△), and Ryan et al. (1991) (◇). In b, the points linked by a dotted line are those reported for the same object but by different authors.*

The dashed line in Fig.2 represents an optimal case for the s-process production and corresponds to an upper limit for the slope of the s/Fe vs. Fe curve. Importantly, the variations of the s-yields with metallicity are found to be comparable for stars of different masses (Baraffe et al. 1992). The ratio s/Fe does increase with the stellar mass, but its gradient with respect to the inital Fe abundances does not change significantly with the mass. This implies that any change of chemical evolution model parameters (IMF and SFR and such) would only shift all the thoretical curves by nearly a constant amount but with their slopes virtually unchanged. Indeed, it is possible to find a set of parameters for the halo model to move the Cu curve close to the observed values. However, this results in an apparent overproduction of Sr and Y at $[Fe/H] > -2$ (cf. Fig.2b).

4 CONCLUSION   As it stands now, the s-process in massive stars does not likely explain Cu observed in halo stars. The situation is similar for Sr, Y and Zr. The observed behaviors of these elements with respect to Fe at $[Fe/H] \lesssim -2$ suggest that they may be better explained in terms of an operation of a primary process, an apparent candidate being the r-process (Truran, 1981). At $[Fe/H] > -2$, the observed Sr and Y could be the products of the s-process in massive stars, although the data are too poor so far to allow us to draw any difinite conclusions. This preliminary study dictates that the origin of those elements at early epochs of the Galaxy is far from being clarified, invoking further theoretical and observational studies. In particular, the scatter of the Sr data is much tantalizing. The acquisition of abundances in the $-1.5 \leq [Fe/H] \leq 0.02$ range is also called for.

REFERENCES

Abia C., Rebolo R., 1989, ApJ 347, 186

Baraffe I., El Eid M. F., 1991, A&A 245, 548

Baraffe I., El Eid M. F., Prantzos N., 1992, A&A 258, 357

Baraffe I., Takahashi K, 1992, in preparation

Barbuy B., 1988, A&A 191, 121

Caughlan G. R., Fowler W. A. 1988, Atom. Data Nucl. Data Tables 40, 283

Gilroy K. K., Sneden C., Pilachowski C. A., Cowan J., 1988, ApJ 327, 298

Gilmore G., Wyse R. F. G., 1986, Nature 322, 806

Gratton R. G., Sneden C., 1988, A&A 204, 193

Magain P., 1989, A&A 209, 211

Pagel B. E. J., 1989, in: Beckman, J.E. & Pagel, B.E.J. (eds.) Evolution Phenomena in Galaxies (Cambridge Univ. Press), p.301

Prantzos N., Hashimoto M., Nomoto K., 1990, A&A 234, 211

Raiteri C. M., Busso M., Gallino R., Picchio G., 1191, ApJ 371, 665

Ryan S. G., Norris J. E., Bessell M. S., 1991, AJ 102, 303

Scalo, J. H., 1986, Fund. Cosm. Phys. II, 1

Sneden C., Crocker D. A., 1988, ApJ 335, 406

Spite M., 1991, in: Barbuy B. (ed.) Proc IAU Symp. 149, The stellar populations of galaxies. Kluwer Dordrecht (in press)

Spite M., Spite F., 1991, A&A 252, 689

Truran J. W., 1981, A&A 97, 391

Yokoi K., Takahashi K., Arnould M., 1987, A&A 117, 65

Galactic Evolution of Cu and Zn

C.M. RAITERI[1], F. MATTEUCCI[2], M. BUSSO[1], R. GALLINO[3], R. GRATTON[4]

[1] Osservatorio Astronomico di Torino, [2] European Southern Observatory, [3] IFG dell'Universita' di Torino, [4] Osservatorio Astronomico di Padova

1 INTRODUCTION

Recent observational works have determined the abundance of Cu and Zn in stars of various metal content (Sneden & Crocker, 1988; Sneden et al., 1991). The resulting picture shows that the abundance of Zn scales with that of iron, while Cu is deficient in metal–poor stars, [Cu/Fe] increasing linearly as a function of [Fe/H], and [Cu/Fe] being ∼-1 at [Fe/H]=-3. In the framework of a simple model for the chemical evolution of the Galaxy, we would expect then a *primary* origin for Zn, while a more complex situation holds for Cu.

From the nucleosynthesis point of view, there is not a unique process that can account for the production of Cu and Zn, but several mechanisms are involved. Among the suggested sites of Cu production, Woosley (1986) individuated explosive Ne burning, and Woosley et al. (1990) found that a significant production of ^{63}Cu can be obtained from neutrino interactions during the explosion of a massive star. As for Zn, the results by Woosley (1986) indicated the α–rich freeze out of nuclear statistical equilibrium and/or explosive He burning. Another contribution to Cu and Zn production also comes from the s–process, in particular from the *weak*–component, that gives the dominant s–signature in the atomic mass zone 60<A<85. It is commonly believed that both the weak and the *main*–component (which accounts for the production of the heavy s–nuclei) are secondary mechanisms. Actually, the situation is more complex, since the final nature of the s–process is determined by many factors, among which the nature of the neutron source and the competition between iron seeds and neutron poisons (Raiteri et al. 1992).

2 STELLAR YIELDS

Table 1 shows the stellar yields (in $M_\odot$) of Cu and Zn from both the weak and the main s–component. In the first case a whole generation of solar metallicity massive stars is considered, from 10 to 50 $M_\odot$; neutron captures are followed through a nuclear network of ∼ 400 isotopes, whose cross sections are taken from the new compilation by Beer et al. (1992), that includes their temperature–dependence. The beta–decay rates of unstable isotopes are from Takahashi & Yokoi (1987). The *s*–processing is calculated during both core–He and shell–C burning, following the models by Nomoto & Hashimoto (1988). By performing calculations of a whole set of stellar models with different metal content we confirm that the weak component is to a good approximation a secondary process; hence the *s*–yields of Table 1 scale with Fe. As for the main *s*–component, that comes from thermally pulsing AGB stars of low mass (Käppeler et al. 1990), the contribution of the main component to the production of Cu and Zn is very low.

The yields from the nucleosynthesis in type I and II supernovae were derived from the papers by Thielemann et al. (1986; 1989), and are shown in Table 1 (Mod.A). They are supposed to be constant in time with respect to iron, that is of primary origin. Unfortunately, such yields are affected by large uncertainties due to both stellar models and nuclear physics.

3 MODELS OF CHEMICAL EVOLUTION

The model of chemical galactic enrichment that we used for analyzing the chemical evolution of the Galaxy is discussed in Matteucci & François (1989). The nucleosynthesis prescriptions quoted in the previous section were included using the matrices formalism of Talbot & Arnett (1973). Type Ia SNe are assumed to produce 0.6 $M_\odot$ of iron and traces of other heavy elements, while SNe of type Ib release 0.3 $M_\odot$ of iron. The amount of Fe produced in SNII is derived from a study of Arnett (1991), which reproduces the iron mass observed in the SN 1987A and predicts a maximum amount of iron of 0.3 $M_\odot$ from stars more massive than 25 $M_\odot$. Despite the number of assumptions, the model reproduces the majority of the observational constraints in the solar neighbourhood and in the whole disc.

4 RESULTS

Fig. 1 and 2 show the results of our calculations together with the observations by Sneden et al. (1991). In Model A we adopted "standard" prescriptions for the nucleosynthesis in supernovae (from Thielemann et al. 1986, 1989), as well as our results for the *s*–processing (see Table 1). The behaviour of [Cu/Fe] vs. [Fe/H] demonstrates that an important contribution is missing in the disc, for [Fe/H]>-1. As a matter of fact, the predicted solar abundance of Cu is too low with respect to the observed one (Anders & Grevesse, 1989). Also for Zn the theoretical [Zn/Fe] curve runs low in comparison with the observational points, and the Zn abundance at [Fe/H]=0 is too low. The *s*–yields are well determined, since they are constrained by the reproduction of the solar distribution of the *s*–isotopes. On the other hand, the results of the nucleosynthesis occurring in SNe are affected by large uncertainties, so that a series of models were run by changing the SN yields in order to find a best

TABLE 1

s-yields (M⊙)		
	Cu	Zn
s-weak (massive stars)	6.78(-5)	1.62(-4)
s-main (low mass stars)	2.8(-7)	3.6(-7)

SN-yields (M⊙)				
	Cu(SNII)	Cu(SNI)	Zn(SNII)	Zn(SNI)
Mod.A	3.31(-6)	1.05(-5)	3.06(-5)	3.24(-5)
Mod.C	7.00(-6)	5.00(-5)	7.00(-5)	3.24(-5)
Mod.P	5.00(-6)	2.61(-4)	1.40(-4)	6.30(-4)

Figure 1

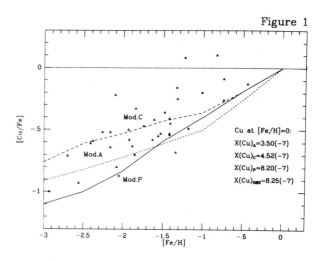

Figure 2

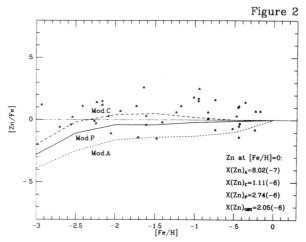

fit solution (Matteucci et al., 1992). We discuss here a couple of cases. In model C the contributions of both SNII and of SNI to Cu were enhanced. The result is a better reproduction of the [Cu/Fe] plot, but the derived solar Cu abundance is still low. Concerning Zn, the yield from type II SNe was roughly doubled; the [Zn/Fe] trend is satisfactory, but the solar abundance of Zn is underestimated by a factor of 2. The best fit model is model P, where the contribution from SNe of type I is dominant for both Cu and Zn, in agreement with the suggestion of Raiteri et al. (1992). The [Cu/Fe] behaviour is fairly good, and at the same time the solar value of Cu is completely accounted for. Also for Zn the predicted solar abundance and the trend of [Zn/Fe] with metallicity are satisfactorly reproduced.

5 CONCLUSIONS

In this paper we used a detailed model for the chemical evolution of the Galaxy in order to investigate the galactic enrichment of Cu and Zn. We found that the present models of stellar nucleosynthesis cannot give a satisfactory reproduction of both the solar abundance and the observed trends with metallicity of Cu and Zn. When taking account of the relative uncertainties in the production mechanisms and observational trends, a solution can rely in a more efficient production of these elements in supernovae. In particular, type I supernovae are expected to play the major role.

6 REFERENCES

Anders, E., Grevesse, N. 1989, Geochem. Cosmochem. Acta, 53, 197

Arnett, D.W. 1991, ApJ, 383, 295

Beer, H , Voß, F., Winters, F.F. 1992, ApJS, in press

Käppeler, F., Gallino, R., Busso, M., Picchio, G., Raiteri, C.M. 1990, ApJ, 354, 630

Matteucci, F., François, P. 1989, M.N.R.A.S., 239, 885

Matteucci, F., Raiteri, C.M., Busso, M., Gallino, R., Gratton, R. 1992, in preparation

Nomoto, K., Hashimoto, M. 1988, Phys. Rep., 163, 13

Raiteri, C.M., Gallino, R., Busso, M. 1992, ApJ, 387, 263 (RGB92)

Sneden, C., Crocker, D. 1988, ApJ, 335, 406

Sneden, C., Gratton, R.G., Crocker, D.A. 1991, A&A 246, 354

Takahashi, K., Yokoi, K. 1987, Atomic Data and Nuclear Data Tables, 36, 375

Talbot, R.J., Arnett, D.W. 1973, ApJ, 186, 69

Thielemann, F.-K., Hashimoto, M., Nomoto, K., Yokoi, K. 1989, in *Supernovae*, ed. S.E. Woosley, (New York: Springer Verlag)

Thielemann, F.-K., Nomoto, K., Yokoi, K. 1986, A&A, 158, 17

Woosley, S.E. 1986, in *Nucleosynthesis and Chemical Evolution*, B. Hauck, A. Maeder and G. Meynet (eds.), p.1

Woosley, S.E., Hartmann, D.H., Hoffman, R.D., Haxton, W.C. 1990, ApJ, 356, 272

[Th/Eu] ratio in halo stars

P. FRANÇOIS

DASGAL, Observatoire de Paris

1 INTRODUCTION

In July 1987, it was published a paper from Butcher concerning the abundance of thorium in metal poor G-dwarfs. Thorium has the peculiarity of being an unstable element with a half life of 14 Gyrs. He measured the abundance of thorium relative to Nd (which is a stable element) in a sample of 20 stars and found no significant variation with respect to the age of the stars as deduced from published stellar isochrones. Using a simple model of galactic chemical evolution (hereafter GCE)with a constant production ratio, he deduced an upper limit for the galactic disk age of 9.6 Gyrs. This determination raised a problem because the oldest star in his sample has an age of 19 Gyrs, according to stellar isochrones.

Later, Mathews and Schramm (1988) considered, in a GCE model, the detailed contribution of r-process and s-process to neodymium. They concluded that the real reduction in thorium is masked by a parallel reduction in the s-process component of neodymium. They derived an upper limit of $\simeq$ 18 Gyrs for the age of the Galaxy. Recently, Lawler et al. (1990) showed that the thorium line was blended with a CoI line. By using the model of Mathews and Schramm, they deduced an age ranging from 15 to 20 Gyrs.

Pagel (1989) suggested the ratio [Th/Eu] as an alternative cosmochronometer. A first determination of this ratio in a sample of 5 metal poor stars indicated a lower Th/Eu for increasing ages in agreement with the radio-active decay of Th.

In 1992, a careful analysis of the thorium blend in a sample of 18 mildly metal deficient stars led Morell and his collaborators to consider a contamination with unknown transitions that they mimic with fake FeI and FeII lines. They found a [Th/Fe] ratio increasing with decreasing metallicity in agreement with the belief that r-process are built in massive stars. This behaviour has already been found for europium which is another r-process element.

2 NEW DETERMINATION OF TH AND EU ABUNDANCE IN METAL POOR STARS

In this talk, we present new data concerning the [Th/Eu] ratio in a sample of 8 halo giants. Observations have been performed partly at the 1.4m CAT ESO telescope and partly at the 3.6m CFHT. The visual magnitude of the stars lies between 7 and 8.5 . The thorium line is at a wavelength ($\simeq \lambda 401$ nm) where the efficiency of the instrumentation and detector decreases significantly. Moreover, the transition is weak and blended requiring high S/N , high resolution spectra . Typical exposure time exceeds 1 hour and for the most difficult case, multiple exposures up to a total of 6 hours were needed. The analysis has been done assuming standard LTE analysis. All the measurements presented here have been made in stars already studied for other abundance measurements. Therefore, we took the atmospheric parameters from the literature. The line list of Morell et al. (1992) was adapted to dwarf stars. As we observed only giant stars, we had to adjust sligthly the oscillator strengths of fake FeII lines. The determination of the abundance has ben performed with spectrum synthesis methods.

3 RESULTS

Before going further, let us remind the observed trends of the element to iron ratios in metal poor stars as a function of their metallicity. There is now a "general agreement" among the observers concerning the behaviour of $[\alpha/\text{Fe}]$ vs $[\text{Fe/H}]$. The α-elements are found to be overabundant relative to iron by about 0.3-0.4 dex . The ratio flattens in the most deficient stars at a metallicity ranging from [Fe/H]=-1 to -1.5 dex according to the author. This behaviour has been interpreted (Matteucci and Greggio 1986) by the fact that α-elements are mostly produced by massive stars with short lifetime whereas Fe is mostly produced by SNIa with a longer progenitor lifetimes. It is interesting to compare this result with what is found for heavy elements like Ba and Eu. [Ba/Fe] is found to be solar in stars with a metallicity ranging from 0 to -2 .This lower value is still matter of debate, some authors preferring -1.5. Below this metallicity, [Ba/Fe] decreases following a secondary behaviour (i.e. a slope [Ba/Fe] vs [Fe/H] equal to 1). Europium has a quite different behaviour : the ratio [Eu/Fe] ratio becomes oversolar as the metallicity decreases to reach a mean value of about +0.5 dex at $[\text{Fe/H}] \simeq -1$. At [Fe/H] close to -2.5, the [Eu/Fe] ratio decreases very rapidly and becomes subsolar. Ba is mainly an s-element whereas Eu is a r-process element. It is interesting to note that for each type of element E there is a corresponding peculiar shape for the relation [E/Fe] versus [Fe/H]. In figure 1, we have plotted the [Eu/Fe] ratio determined for our sample of stars together with the data Gilroy et al. (1988). We confirm a sharp decrease of the [Eu/Fe] ratio in stars with metallicities

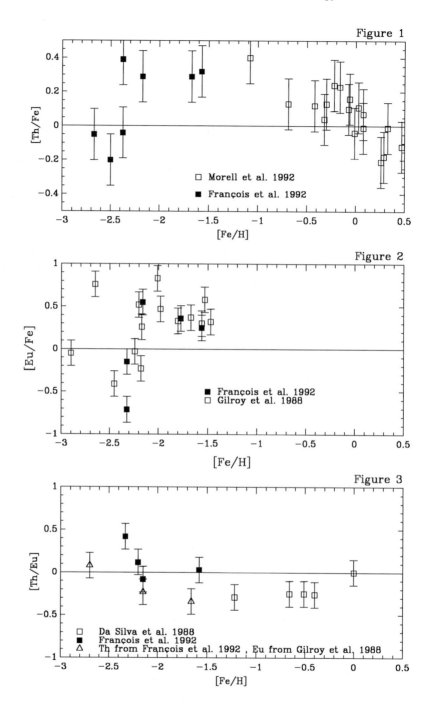

Figure 1

Figure 2

Figure 3

lower than [Fe/H] $\simeq$ -2.2. In Figure 2, we gathered the data for [Th/Fe] in our sample of giants and the data from Morell et al. (1992) for their sample of dwarf stars. The data of Morell et al. show an increase of the [Th/Fe] ratio when the metallicity decreases. For their most metal poor stars, they obtain a ratio of about +0.4 dex . For the most metal rich stars of our sample, we find a mean value in good agreement with theirs. In the range -2 to -2.5, we found the [Th/Fe] ratio to decrease sharply, a behaviour similar to what has been found already for europium, which is also an r-process element. Figure 3 shows our measured [Th/Eu] ratios as a function of [Fe/H]. We added the data of Da Silva et al. (1988). They observed dwarf stars with a metallicity ranging from solar to [Fe/H] $\simeq$ -1. They found a decreasing [Th/Eu] as [Fe/H] decreases, in agreement with the radio-active decay of thorium. We confirm that the [Th/Eu] becomes sub-solar in metal poor stars by a factor of $\simeq$ 2. However, in stars with metallicity lower than about [Fe/H]=-2.2 , the [Th/Eu] ratio seems to become oversolar. Mathews et al. (1992) showed that the [Eu/Fe] vs [Fe/H] relation can be understood if europium is mainly produced in massive stars in the range 8 to 12 $M_\odot$. A possible interpretation of the [Th/Eu] vs [Fe/H] relation could be that thorium is produced in stars more massive than the ones which produce europium. However, it is clear that before drawing firm conclusions, more data are needed and in particular new atomic data concerning the contaminating lines.

References :

Da Silva, L., De la Reza, R., Maghalaes, S. : 1989, in "Astrophysical ages and dating methods", eds. E. Vangioni-Flam et al. (Frontieres), p. 325

François, P., Spite, M., Spite, F. : 1992, AA, submitted

Gilroy, K.K., Sneden, C., Pilachowski, C.A., Cowan, J.J. : 1988, Apj, 327, 298

Lawler, J.E.L., Whaling, W., Grevesse, N. : 1990, Nature, 346, 635

Mathews, G.J., Schramm, D.N. : 1988, ApJ, 324, L67

Matteucci, F., Greggio, L. : 1986, AA, 154, 279

Morell, O., Kallander, D., Butcher, H.R. : 1992, Uppsala preprint 58

Pagel, B.E.J. :1989 in "Evolutionary Phenomena in galaxies ", eds J. Beckman, B.E.J. Pagel, Cambridge University Press, p. 201

Thorium Chronology and the Ages of Globular Clusters

B.E.J. PAGEL

NORDITA, Blegdamsvej 17, Dk-2100 Copenhagen Ø, Denmark

Abstract

Actinide abundance ratios in the Solar System only provide a lower limit to the age of the Galaxy, because of sensitivity to poorly known models of Galactic evolution. The Th/Nd criterion introduced by Butcher, when converted to a Th/Eu ratio, is capable in principle of acting as a check on ages of individual stars with much less sensitivity to details of Galactic evolution. Data sets now available are compared with two extremely different Galactic models and in either one they agree as well as can be expected with ages of the order of 15 Gyr deduced from HR diagrams of globular clusters and field stars of the Galactic halo, apart from a group of stars with [Fe/H]$\leq$-2.4 in which the Eu abundance appears to be anomalously low.

1 INTRODUCTION

It is a pleasure to join in paying tribute to Hubert Reeves on the occasion of his 60th birthday. I first met him at a conference in New York in 1963 (when he was still beardless) and have paid close attention since then to his outstanding work on spallation, protosolar D and ^{3}He, Big Bang nucleosynthesis and nuclear chronology, among other topics, and especially enjoyed his crisp, clear, concise and incisive way of explaining things.

In this talk I return to the subject of Galactic chemical evolution and nuclear cosmochronology (cf. Pagel 1989), updating some comments on the abundances of actinide nuclei in the Solar System and on the stellar thorium chronology introduced by Butcher.

2 CHRONOLOGY FROM ACTINIDES IN THE SOLAR SYSTEM

Following the pioneering work of Rutherford (1929), there have been numerous attempts to date the beginning of significant nucleosynthesis in the Galaxy by comparing meteoritic relative abundances of ^{238}U, ^{235}U, ^{232}Th and sometimes other nuclear "chronometers" with their mean lives and estimates of theoretical production rates (e.g. Fowler 1987; Cowan, Thielemann & Truran 1991:CTT); while Schramm & Wasserburg (1970), Tinsley (1977), Arnould, Takahashi & Yokoi (1984), Clayton (1988) and Pagel (1989), among others, have stressed the uncertainties arising from the effect of differing models of Galactic evolution. The issue is really whether nucleosynthesis in the disk of the Galaxy began in some fairly sudden event, which could be meaningfully dated, or whether (our region of) the Galaxy just "growed", as in inflow models (Larson 1976; Lynden-Bell 1975; Tinsley 1977, 1980; Clayton 1985, 1988) where the beginning of the process is essentially a non-event whose dating is just a consequence of the particular mathematical parameterization of the whole process that one happens to have chosen for the sake of simplicity. Both types of model can explain the "G-dwarf problem" (Pagel 1989, 1992). On top of this there are complications due to effects in stellar evolution such as the destruction of actinides by photo-fission in stellar interiors (Malaney, Mathews & Dearborn 1989); these are neglected in the present paper, apart from noting that allowance for photofission tends to increase the derived ages.

Assuming some simplifications such as instantaneous recycling, that the net star formation rate is ω times the mass of gas with $\omega = const.$, that inflowing gas is metal-free and that no gas is lost from the system, the basic equation governing the number of radio-active atoms of species i in a column perpendicular to the Galactic plane at time t is

$$N_i(t) = e^{-(\omega+\lambda_i)t}[N_i(0) + p_i \int_0^t \psi(t')e^{(\omega+\lambda_i)t'}dt'] \tag{1}$$

where λ_i is the decay constant, $N_i(0)$ the initial abundance, p_i the yield expressed as a number and $\psi(t)$ the net rate of star formation. (The product $\psi(t)e^{\omega t}$, or its generalisation to variable ω, is sometimes referred to as the "effective nucleosynthesis rate".) In the closed initial-spike or "prompt initial enrichment" model (Truran & Cameron 1971) used by Fowler (1987), the equation for $N_i(t)$ reduces to

$$N_i(\Delta) = e^{-\omega\Delta}p_i\psi_0[\frac{S}{1-S}\Delta e^{-\lambda_i\Delta} + \frac{1-e^{-\lambda_i\Delta}}{\lambda_i}] \tag{2}$$

where $\Delta = T - 4.6$ is the age of the Galaxy (in Gyr) at the time when the Solar System was formed and S is the proportion of a stable element, formed at the same rate as the r-process that gave rise to the actinides, resulting from the initial spike. For these parameters Fowler (1987) derived $S = 0.17 + 0.02$ and $\Delta = 5.4 + 1.5$ Gyr.

In considering inflow models, recent studies by Sommer-Larsen (1991) suggest taking a very simple case in which inflow occurs into an initially empty region at a rate $Ae^{-\omega t}$ with A constant. In this model the corresponding equation for an actinide abundance is

$$N_i(\Delta) = e^{-\omega\Delta} A\omega p_i \lambda_i^{-2} [\lambda_i \Delta - (1 - e^{-\lambda_i \Delta})].\tag{3}$$

The significant observational datum is

$$K_{ij}(\Delta) \equiv (N_i/N_j)(p_i/p_j)^{-1}\tag{4}$$

for each actinide pair, which is happily independent of A in the inflow model and of ω in both models. Recent data from Fowler (1987) and CTT give $K_{232,238} = 1.40\pm0.15$ and $K_{235,238} = 0.26 \pm 0.03$, where the errors are essentially inspired guesses. These ranges are compared to predictions from the two models, and from synthesis of all elements in a pure initial spike ($S = 1$), in Fig. 1, where it can be seen that all three models give reasonable concordance (shown by the vertical lines), but with ages for the Galaxy ranging from 6.5 Gyr to 21 Gyr (which could easily be extended to infinity), according to the model assumed. The one result that surprised me in this exercise was that the pure initial–spike model, with no nucleosynthesis in the course of Galactic evolution, and which is incompatible with stellar O/Fe ratios etc., but has been defended by Butcher (1988) in one of his wilder moments, actually fits the data within possible errors, and sets the absolute minimum Galactic age from radio-active chronology at the rather low value of 6.5 Gyr (cf. Butcher 1990). (Short-lived radio-activities in the Solar System could perhaps be explained as resulting from special events associated with its formation; cf. Reeves 1991.)

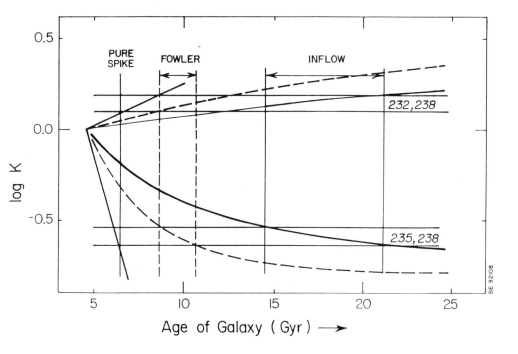

Figure 1: Observed K-ratios (estimated error limits shown by horizontal pairs of parallel lines) and their theoretical variation with the age of the Galaxy according to three models: a pure initial spike (full-drawn straight lines); Fowler's model (broken-line curves) and our inflow model (full-drawn curves).

3 BUTCHER'S THORIUM CHRONOLOGY

Butcher (1987) introduced the brilliant idea of checking ages of stars deduced from their position in the HR diagram by measuring their thorium abundances relative to a stable element. No knowledge of production ratios is needed and, because the formation of a given star is a well-defined event in the past, there is considerably less dependence on unknown Galactic evolutionary models than in the case of the Solar System data, as will be illustrated below. The problems lie (a) in the actual determination of relative Th abundances to the relatively good precision needed (still an outstanding problem); and (b) in Butcher's choice as a reference element of neodymium which has a substantial contribution from the s–process (Clayton 1987; Mathews & Schramm 1988; Butcher 1988). The latter difficulty can be overcome by using europium as the reference element (Pagel 1989; da Silva, de la Reza & Magalhães 1990).

The abundance of thorium in a stellar atmosphere is its abundance in the interstellar medium at the time t when the star was formed, as given by equation (1), modified by the factor $e^{-\lambda_i(T-t)}$ due to subsequent radioactive decay. The resulting expression for the thorium abundance in Fowler's model is

$$N_i(t,T) = p_i \psi_0 e^{-\omega t}[\frac{S}{1-S}\Delta e^{-\lambda_i T} + \frac{e^{-\lambda_i(T-t)} - e^{-\lambda_i T}}{\lambda_i}] \tag{5}$$

to be compared with the abundance of a stable element formed alongside

$$N_k(t) = p_k \psi_0 e^{-\omega t}[\frac{S}{1-S}\Delta + t], \tag{6}$$

whereas the corresponding expressions in our simple inflow model are

$$N_i(t,T) = p_i A\omega e^{-\omega t}\lambda_i^{-2}e^{-\lambda_i(T-t)}[\lambda_i t - (1 - e^{-\lambda_i t})] \tag{7}$$

and

$$N_k(t) = p_k A\omega e^{-\omega t}t^2/2. \tag{8}$$

These expressions can be compared with Th/Eu ratios observed in stars with age estimates from the HR diagram so as to see whether they are consistent or alternatively demand some correction to the stellar ages, e.g. bringing them down by a factor of about 2 as suggested by Butcher (1987). Since Butcher's paper first appeared, there have been two applications of major corrections to his Th/Nd ratios to allow for blends (Lawler, Whaling & Grevesse 1990: LWG; Morell, Källander & Butcher 1992: MKB), Eu/Nd/Th ratios have been measured in a few of Butcher's stars by da Silva, de la Reza & Magelhães (1990) and Eu/Th ratios have been measured in a number of halo stars by François (1991). The resulting data set, which should be an improvement on my previous crude attempts to estimate Th/Eu ratios indirectly (Pagel 1989), is presented in the accompanying Table and illustrated in Fig.2 along with predictions from the two models (Fowler's model shown by broken-line curves and the simple inflow model by thick full-drawn curves), assuming either (a) that the ages deduced from HR diagrams are correct, or (b) that they are overestimated by a factor of 2 (cf. Butcher 1987). Also shown for illustration are the limiting cases of a pure initial spike of production and of pure decay (which would apply if, for example, each star inherited just freshly produced elements from a nearby supernova; cf. the "Bing-Bang" model (Reeves 1972)).

Table 1

Star	Age (Gyr)	[Fe/H] MKB	[Th/Nd] LWG	[Th/Nd] MKB	[Th/Eu] (1)	[Th/Eu] (2)	[Th/Eu] (3)
Sun	4.6	0.0	0.00	0.00	0.00	0.00	0.00
HR 4523	8±3	-0.3	-0.14	-0.08	-0.26	-0.20	-0.27
HR 509 (τ Cet)	9±4	-0.5	0.08	0.19	-0.05	0.06	-0.23
HR 98 (β Hyi)	10±2	-0.3	-0.06	-0.03	-0.16	-0.13	-0.16
HR 3018	19±3	-1.0	0.12	0.22	-0.22	-0.12	-0.24
4 halo stars	15±3	>-2.4					-0.15
3 halo stars	15±3	<-2.4					≤0.4

(1) From [Th/Nd] by Lawler et al. and [Eu/Nd] by da Silva et al.
(2) From [Th/Nd] by Morell et al. and [Eu/Nd] by da Silva et al.
(3) From da Silva et al. or François

It can be seen that the predictions of the two Galactic evolution models are now not very different, especially for the case when ages are assumed to be overestimated. Such differences as there are are mainly in the opposite sense from those arising in the Solar System actinide chronology. Unfortunately all the effects are small in relation to the uncertainties inherent in stellar abundance analyses, relative to which the agreement between the different data sets in the last three columns of the Table is actually quite good. The data, such as they are, are quite consistent (as was also found by LWG) with even the largest ages deduced from HR diagrams and somewhat less consistent with the suggestion that they are overestimated, provided that there is no systematic offset error between the stars and the Sun. This result supports the arguments (though not the detailed models) of Clayton (1987) and Mathews & Schramm (1988). The data thus lend some modest independent support to the usual age determinations of globular clusters.

The one major anomaly is the large Th/Eu ratio found by François at extremely low metallicities, which sets in at about the same metallicity where Eu/Fe seems to go down (from a value somewhat above solar) with decreasing metallicity (Mathews

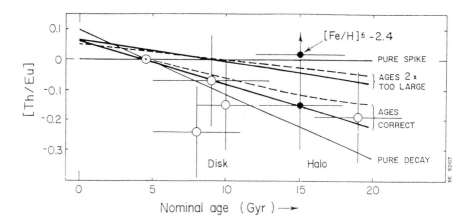

Figure 2: Logarithmic differential Th/Eu ratios plotted against stellar age derived from HR diagrams, assuming a straight mean of the entries in the last three columns of Table 1 and an error of ±0.15 dex in each case. Filled circles represent halo stars from François.

& Cowan 1990; François 1991). This phenomenon suggests that there has been a change in the relative yields of different r–process products at very early times and the possibility of such a thing happening inspires some caution in making deductions from radioactive chronology in general.

REFERENCES

Arnould, M., Takahashi, K. & Yokoi, K. 1984, *Astr. Astrophys.*, **137**, 51.
Butcher, H.R. 1987, *Nature*, London, **328**, 127.
Butcher, H.R. 1988, *ESO Messenger*, no. 51, p. 12.
Butcher, H.E. 1990, *in* **Astrophysical Ages and Dating Methods**,
    E. Vangioni-Flam, M. Cassé, J. Audouze & J.T.T. Van (eds),
    Paris: Ed. Frontières, p. 391.
Clayton, D.D. 1985, *in* **Nucleosynthesis: Challenges and New
    Developments**, W.D. Arnett & J.W. Truran (eds), Univ. Chicago Press,
    p. 65.
Clayton, D.D. 1987, *Nature*, London, **329**, 397.
Clayton, D.D. 1988, *Mon. Not. R. astr. Soc.*, **234**, 1.
Cowan, J.J., Thielemann, F.-K. & Truran, J.W. 1991, *Ann. Rev. Astr.
    Astrophys.*, **29**, 447.
da Silva, L., de la Reza, R. & de Magalhães, S.D. 1990, *in* **Astrophysical
    Ages and Dating Methods**, E. Vangioni-Flam, M. Cassé, J. Audouze
    & J.T.T. Van (eds), Paris: Ed. Frontières, p. 419.
Fowler, W.A. 1987, *Quart. J. R. astr. Soc.*, **28**, 87.
François, P. 1991, Poster Paper at IAU Gen. Assembly, Buenos Aires.
Larson, R.B. 1976, *Mon. Not. R. astr. Soc.*, **176**, 31.
Lawler, J.E., Whaling, W. & Grevesse, N. 1990, *Nature*, London, **346**, 635.
Lynden-Bell, D. 1975, *Vistas in Astr.*, **19**, 299.
Malaney, R.A., Mathews, G.J. & Dearborn, D. 1989, *Astrophys. J.*, **345**, 169.
Mathews, G.J. & Cowan, J.J. 1990, *Nature*, London, **345**, 491.
Mathews, G.J. & Schramm, D.N. 1988, *Astrophys. J. Lett.*, **324**, L67.
Morell, O., Källander, D. & Butcher, H.R. 1992, *Astr. Astrophys.*, **259**, 543.
Pagel, B.E.J. 1989, *in* **Evolutionary Phenomena in Galaxies**, J.E. Beckman
    & B.E.J. Pagel (eds), Cambridge Univ. Press, p. 201.
Pagel, B.E.J. 1992, *in* IAU Symposium 149: **The Stellar Populations of
    Galaxies**, B. Barbuy & A. Renzini (eds), Kluwer, p. 133.
Reeves, H. 1972, *Astr. Astrophys.*, **19**, 215.
Reeves, H. 1991, *Astr. Astrophys.*, **244**, 294.
Rutherford, E. 1929, *Nature*, London, **123**, 313.
Schramm, D.N. & Wasserburg, G.J. 1970, *Astrophys. J.*, **162**, 57.
Sommer-Larsen, J. 1991, *Mon. Not. R. astr. Soc.*, **249**, 368.
Tinsley, B.M. 1977, *Astrophys. J.*, **216**, 548.
Tinsley, B.M. 1980, *Fund. Cosmic Phys.*, **5**, 287.
Truran, J.W. & Cameron, A.G.W. 1971, *Astrophys. Space Sci.*, **14**, 179.

Problems with the Chemical Evolution of Irregular and Blue Compact Galaxies

J. LEQUEUX

Observatoire de Paris-Meudon, F-92195 Meudon CEDEX, et Laboratoire de Physique de l'Ecole Normale Supérieure, 24 Rue Lhomond, 75231 Paris CEDEX 05

1. INTRODUCTION

When the organizers of this conference asked me to speak about the evolution of galaxies, I thought naïvely that I would honor best Hubert Reeves by telling intelligent things of general nature. I soon realized that this would be impossible (at least for me) and that I should rather concentrate on simple, specific problems concerning the simplest of all galaxies: the irregular and blue compact galaxies. These galaxies are rather unevolved as they still contain a lot of gas and have low metallicities. Who can claim understanding more complex and/or more evolved galaxies if these simple systems are not undestood first? Still, as we will see, they are indeed not really understood!

I will describe the problems which arise when trying to understand the chemical evolution of irregular and blue compact galaxies. For many of these systems, we possess the following information of interest for this problem (unfortunately the complete information is no always available for a given galaxy):
- the abundance of several elements with respect to hydrogen in HII regions; I will discuss here only O/H and He/H. The abundances of several elements have also been measured in stars in the case of the Magellanic Clouds, but these abundances are somewhat controversial and it seems premature to use them in chemical evolution models (see Pagel 1992). Note that the abundance of oxygen is the only one that can be measured easily;
- the total mass of gas from 21-cm observations of atomic hydrogen; for such systems, the abundance of molecular hydrogen can be neglected contrary to the situation in spiral galaxies (Rubio et al. 1991, 1992).
- the total mass of the galaxy, from its rotation curve or velocity dispersion, generally measured from the 21-cm line.

2. CLOSED-BOX EVOLUTION AND PROBLEMS

The simplest possible model of evolution of galaxies is the closed-box model, also called the one-zone model. In this model, the galaxy has no exchange of matter with the external world and is assumed to have an homogeneous chemical composition, i.e. the material ejected by evolved stars is supposed to be instantaneously mixed with the rest of the galaxy. Moreover I assume that the instantaneous recycling approximation is valid, i.e. the stars which are active in the nucleosynthesis have lifetimes significantly smaller that the characteristic time for evolution of the galaxy. This approximation is valid for those stars which dominate the production of oxygen, which are massive and have short lifetimes. It is less valid for the evolution of helium. In these conditions, we have the famous equation which relates the abundance Z of heavy elements to the fraction μ of the mass of the galaxy in the form of interstellar matter ($\mu = M_{gas}/ M_{total}$):

$$Z = y \ln(1/\mu),$$

where y is the yield, i.e. the mass of heavy elements returned to the interstellar matter per unit net mass turned into stars in a stellar generation. For demonstration and a discussion, see e.g. Lequeux (1989). This equation can be applied to any element, e.g. oxygen: in this case it would write

$$Z_O = y_O \ln(1/\mu),$$

where Z_O is the abundance of oxygen (per mass) and y_O its yield.

Theoretical conventional estimates of the yield y in heavy elements from a "normal" initial mass function (IMF) of stars give $y \approx 0.010$. This value does not depend much on the upper mass limit provided it is $\geq 60 \ M_O$ or so, and on the exact slope of the IMF provided it is not very different from the Salpeter's one. Direct observations of the IMF in the Magellanic Clouds show that we have no reason to doubt these assumptions.

A number of studies have attempted at verifying this equation for irregular and blue compact galaxies. The first study is that of Lequeux et al. (1979); they find a rather good agreement between observations of a few galaxies and the predictions of the fundamental equation, but provided that the yield is $y \approx 2y_O \approx 0.003$ instead of the predicted 0.01. The following studies which concern a much larger number of galaxies, in particular that of Matteucci and Chiosi (1988), have shown that there is a large dispersion in the "observed" yields and that the value of 0.003 is rather an upper limit, making the discrepancy even worse.

Thus there is something wrong somewhere. But what?

Another problem concerns helium. The standard models of nucleosynthesis, based on a "normal" IMF, predict that the abundance Y of helium (by mass) increases at approximately the same rate as the sum Z of the abundances of all the heavier elements:

$$\Delta Y/\Delta Z \approx 1$$

while observations show larger values, roughly between 3 and 6 (Lequeux et al. 1979; Pagel et al. 1992)

Thus here also there is something wrong somewhere; what? Note that the calculated value of $\Delta Y/\Delta X$ is not very sensitive to details of the evolution of the galaxy, and the problem must lie elsewhere.

3. POSSIBLE EXPLANATIONS FOR THE DISCREPANCIES

If we turn back to the fundamental equation for closed-box evolution, we can foresee several possible explanations for the yield problem:

- the closed-box hypothesis is wrong: the galaxy can accrete or expel gas during its evolution

- the total mass M_{total} which enters into the equation, which should be the total mass of stars + gas + stellar remnants is wrong because of the presence of dark matter without relation to stars

- the nucleosynthesis calculations are wrong.

I will examine these possibilities in turn, discussing every time their implications on the helium problem.

3.1 Accretion of primeval gas by the galaxy

It is obvious that if the galaxy accretes gas without heavy elements, the gas it contains is diluted by pure H and He: this decreases the apparent yield. This idea has been quite popular and has been for example advocated by Matteucci and Chiosi (1983) to solve the yield problem: one will find in their paper a rather detailed treatment of this effect. However there is no direct evidence for gas infall on irregular galaxies, although marginal and controversial evidence exists in the case of our Galaxy (the High Velocity Clouds). Moreover, the dilution is by gas with primordial abundance of He and one cannot solve the helium problem in this way.

3.2 Selective ejection of gas with high metallicity (galactic wind)

Another way for decreasing the apparent yield is to get rid of part of the gas just enriched in heavy elements by massive stars, which will thus not be mixed with the pre-existing interstellar matter. This idea has been first proposed to my knowledge by Russell et al. (1988), and recently explored in details by Pilyugin (1992). It assumes that the energy liberated by the collective effect of stellar winds and supernova explosions in superassociations of young, massive stars is sufficient to eject out of the galaxy a large fracion of the stellar ejecta. This idea is physically sound and has some observational support: i) the existence of "chimneys" seen perpendicular to the galactic plane in edge-on galaxies suggests that hot gas heated by multiple supernova events - a superbubble- is ejected into the halo (Dettmar 1990); this gas might leave the galaxy in the weaker gravitational potential of irregular galaxies; ii) spiral and irregular galaxies seen more or less face-on exhibit superbubbles containing little gas which might be another manifestation of the same phenomenon (see e.g. Meaburn et al. 1987); and iii) the external parts of irregular galaxies (Hunter & Gallagher 1990) including the SMC (Le Coarer 1992; Okumura et al. 1992) also show loops and filaments which possibly reflect such ejections (but see Hunter & Gallagher 1992). Note that as the gas is ejected while young stars of intermediate and low mass did not have time to evolve it contains little new helium: thus oxygen is preferentially ejected with respect to helium and this model is able to solve the $\Delta Y/\Delta Z$ problem. This model predicts also other abundance differences with respect to the theoretical yields, for example a lower O/Fe ratio as Fe is believed to be produced by low-mass Type I supernovae. But this can barely be checked at present.

3.3 Dark matter not related to stars

As explained be earlier in this paper the total galactic mass M_{total} which should intervene in the fundamental evolution equation should not include dark matter unrelated to stellar evolution in the galaxy. However M_{total} is usually derived from the dynamics of the galaxy, thus contains any dark matter present. If there is a substantial fraction of unrelated dark matter, M_{total} is overestimated hence m is larger and the degree of evolution is less than it appears at first glance; the fundamental equation then requires a higher yield which can be in better agreement with the theoretical yield, as first suggested by Lequeux (1989). There are direct evidences for a substantial and sometimes very dominant quantity of dark matter in blue compact and dwarf irregular galaxies (I Zw18: Lequeux & Viallefond 1980; DDO 47, Sextans B: Comte et al. 1986). Carignan & Freeman (1988) and Carignan & Beaulieu (1988) have described the spectacular example of DDO 154. Recently, Kumai & Tosa (1992) have made an interesting study of the implications of dark matter on chemical abundances. They show that if one uses instead of the dynamical mass a total galactic mass derived from the luminosity using a reasonable mass/luminosity ratio, the results can be made consistent with the "theoretical" yield. Unfortunately the presence of dark matter cannot help solving the helium problem, as the relative yields of the different elements are not modified but only their "observed" absolute values.

3.4 Modified theoretical yields

As discussed earlier, the theoretical yield of 0.01 or so implies a "normal" IMF and "normal" nucleosynthesis receipes. The idea of a universal IMF, at least for those massive stars that contribute most to nucleosynthesis, is constantly gaining ground. But there has been recently an evolution in our ideas on the production of elements by stars. First it is not granted that the very massive stars end their lives by exploding as supernovae; it may well be that they collapse as a black hole instead, in which case everything which has not been expelled before as stellar wind is lost for chemical evolution. If for example every star with an initial mass higher than 20 $M_\odot$ ends as a black hole, the production of oxygen will be drastically decreased while that of helium will be only moderately affected. This offers a way of solving both the yield problem and the helium problem. Very recently, Maeder (1992) has explored in details the consequences of black hole formation and finds that for low metallicities of the stars ($Z = 0.001$) a corresponding limiting initial mass of 20 to 25 $M_\odot$ offers a satisfactory solution to both problems. A complication contained in his new models is that the yield is very much dependent on the initial metallicity of the stars, due to the effects of stellar winds. Stars at low metallicities have weak winds and most of the newly synthetized He and C has time to be converted into O; conversely, the winds are so strong at high metallicities that most of the newly synthetized He and C soon leave the star before being converted into O. It will be interesting to study anew galactic evolution using these new models: this has not yet been done, and the effects to be expected are probably quite large.

3.5 Another complication: self-pollution in HII regions

It is generally assumed that the abundances measured in HII regions are representative of those in the general interstellar medium, i.e. that the gas inside the HII regions has not yet be contaminated by the ejecta of the most massive stars of the ionizing star cluster. Substantial contamination will occur only 3 or 4 million years after the initial burst of star formation, when the most massive stars form Wolf-Rayet stars or explode as supernovae. There is a selection effect against its observation as the flux of ionizing photons decreases at the same time. But on the other hand the presence of Wolf-Rayet stars or red supergiants inside some HII regions (e.g. 30 Dor in the LMC) is well documented (supernovae explosions inside HII regions are difficult to detect). Thus it is well possible that a scatter in abundances will be observed in HII regions of the same galaxy even if the general interstellar medium is chemically homogeneus. This will in particular affect the He/O and the N/O abundance ratios and may account for the fact that Wolf-Rayet galaxies (i.e. those extragalactic HII regions exhibiting strong Wolf-Rayet emission features) seem to show more scatter in their He/O and N/O ratios than the "normal" extragalactic HII regions (see fig. 7 of Pagel et al. 1992). The effects of self-pollution of HII regions have very recently been explored by Pilyugin (1992). This is another cause of uncertainty when studying the chemical evolution of galaxies, as it yields scatter and a possible overestimate of the abundances.

4. CONCLUSIONS

There are several possibilities for explaining why the chemical evolution of irregular and blue compact galaxies does not seem to comply with the simple model of chemical evolution. In fact there are too many possibilities given the limited available data! Paradoxically, the initially much-favored idea of infall of primeval matter on these galaxies now seems more arbitrary than others and is not able to explain the high observed value of $\Delta Y/\Delta Z$. The presence of substantial dark matter is a quasi-certainty in several such galaxies.It can explain the low "observed" yield in heavy elements, while being unable to solve the helium enrichment problem. Other ideas I have detailed above are also appealing. These different hypotheses predict different behaviours for He, C, N, O and Fe and it will certainly be rewarding to compare observations (when available!) with predictions (if reliable!). We are only beginning to realize that the abundances measured in HII regions might be affected by self-pollution and may not reflect those in the general interstellar medium, and also that the theoretical yields can be very strongly affected by metallicity (through mass loss) and also by the possible formation of black holes as the end products of the evolution of the most massive stars. While there is certainly still work to do, the present situation is not encouraging enough to devote a life's activity to the chemical evolution of galaxies. My final remark will be: don't accept "definitive" conclusions on chemical evolution. This is an uncertain domain, full of pitfalls.

REFERENCES

Carignan C., Beaulieu S., 1988, ApJ 347, 760

Carignan C., Freeman K.C., 1988, ApJ 332, L33

Comte G., Lequeux J., Viallefond F., 1986, in *Star Forming Dwarf Galaxies*, ed.
        D. Kunth et al., p. 273

Dettmar R.-J., 1990, A&A 232, L15

Hunter D.A., Gallagher, J.S.III, 1990, ApJ 362, 480

Hunter D.A., Gallagher, J.S.III, 1992, ApJ 391, L9

Kumai Y., Tosa M., 1992, A&A 257, 511

Le Coarer E., 1992, in *New Aspects of Magellanic Cloud Research*, ed. B. Baschek, G.
        Klare and J. Lequeux, in preparation

Lequeux J., 1989, in *Evolution of Galaxies-Astronomical Observations*, ed. I
        Appenzeller, H.J. Habing & P. Léna, Springer-Verlag, Berlin, p. 147

Lequeux J., Peimbert M., Rayo J., Serrano A., Torres-Peimbert S., 1979, A&A 80,
        155

Lequeux J., Viallefond F., 1980, A&A 91, 262

Matteucci F., Chiosi C., 1983, A&A 123, 121

Meaburn J., Marston A.P., McGee R.X., Newton L.M., 1988, MNRAS 225, 591

Okumura K., Viallefond, F., Viton M., Rice W., 1992, in *New Aspects of
        Magellanic Cloud Research*, ed. B. Baschek, G. Klare and J. Lequeux, in
        preparation

Pagel B.E.J., 1992, in *New Aspects of Magellanic Cloud Research*, ed. B. Baschek,
        G. Klare and J. Lequeux, in preparation

Pagel B.E.G., Simonson E.A., Terlevich R.J., Edmunds M.G., 1992, MNRAS
        255, 325

Pilyugin L.S., 1992, A&A submitted

Rubio M., Garay G., Montani J., Thaddeus P., 1991, ApJ 368, 173

Rubio M., Lequeux J., Boulanger F., 1992, A&A in press

Russell S.C., Bessell M.S., Dopita M.A., 1988, in *Galactic and Extragalactic Star
        Formation*, ed. R.E. Pudritz & M. Fich, Kluwer, Dordrecht, p. 601

Origin of Heavy Elements in the Intra-cluster Medium and Galaxy Formation

M.HATTORI & N.TERASAWA

The Institute of Physical and Chemical Research (RIKEN),
Wako, Saitama 351-01, Japan

1.INTRODUCTION

The evidences which support the idea that the ICM is metal enriched by a wind from proto-galaxies become increasing (cg. Canizares et al. 1988; David et al. 1990;Tsuru 1992). Here we examine the possibility of early galactic wind and we constrain the condition of the star formation in the proto-galaxy. Our results predict that galaxies have an extended halo populated with large number of neutron stars. We show that the recent observational results of the distribution of γ-ray bursters suggest that the halo radius is very large if neutron stars distributed in the galactic halo are the sources of γ-ray bursters and are consistent with our theoretical prediction.

2.CONSTRAINT ON THE RADIUS OF THE PROTO-GALAXY

2.1 Star formation in proto-galaxy

Here we consider a burst of star formation in a proto-galaxy. The star formation period is assumed to be much shorter than a lifetime of $8M_\odot$ star ($\tau_{sn8} = 7 \times 10^7$yr) which is the longest lifetime among the progenitors of Type II supernova. IMF is assumed as having same form as the Salpeter IMF. An upper cutoff mass is fixed as $m_u = 40M_\odot$. We investigate two different lower cutoff masses. One is the normal disk-like case with lower cutoff mass $m_L = 0.1M_\odot$ and the other is the predominant massive star formation case with lower cutoff mass $m_L = 8M_\odot$. We control the star formation rate by introducing a star formation efficiency parameter, ϵ, which is defined as the mass fraction of the newly formed stars to the initial gas mass. In the following discussion we assume a uniform distribution of stars and gas when the burst of star formation occurs and we assume that the temperature of the interstellar medium is much less than $10^6 K$.

2.2 Critical Supernova rate

To realize the early galactic wind, it is neccesary to convert the kinetic energy of the supernova remnant into thermal energy of interstellar medium before the most of the energy in the remnant is lost by a radiative cooling. This would be occured when the supernova rate is high enough so that the supernova shells overlap before the each remnants cools significantly. By the comparison of a time scale of shell overlapping with a time scale of the cooling of the remnant (Shull & Silk 1979), the critical supernova rate, λ_{crit}, is calculated. Because the gas surrounding the supernova shell is the primodial gas, the critical supernova rate is calculated as

$$\lambda_{crit} = 9.7 \times 10^{-11} E_{51}^{-1.15} n_{ISM}^{1.7} \mathrm{SNepc^{-3}yr^{-1}}, \tag{1}$$

where E_{51} is the energy released by one supernova unit in 10^{51}erg, n_{ISM} is the density of interstellar medium in the proto-galaxy. The condition of the realization of the early galactic wind is expressed as

$$\lambda_{SNII} \geq \lambda_{crit}, \tag{2}$$

where λ_{SNII} is Type II supernova rate per unit volume in proto-galaxy. We have checked a validity of this equation by applying this to a star burst galaxy M82.

2.3 Results
From equation (2) we obtain the constraint equation for a radius of the proto-galaxy r,

$$\left(\frac{r}{10\mathrm{kpc}}\right)^{2.1} \geq \frac{100}{\chi/\chi_{0.1}} \frac{(1-\epsilon)^{1.7}}{\epsilon} E_{51}^{-1.15} \left(\frac{M_{primodial}}{10^{11}M_\odot}\right)^{0.7} \frac{\tau_{SNII}}{7 \times 10^7 \mathrm{yr}}, \tag{3}$$

where $\chi = \frac{\int_{m_L}^{m_u} \phi dm}{\int_{m_L}^{m_u} m\phi dm}$ and $\chi_{0.1} = \frac{\int_{0.1}^{40} \phi dm}{\int_{0.1}^{40} m\phi dm}$. The star formation efficiency in the nearby galaxies gives us a good reference of that in the proto-galaxy. The star formation efficiency in normal galaxies is regulated by the inonizing radiation from hot stars and is predicted to be as small as $\epsilon \sim 0.05$ whereas in the star burst galaxies this is as high as 0.5 (Larson 1987) where the Salpeter IMF is assumed. If the IMF contains a smaller proportion of low-mass stars, the observed efficiency is reduced. If the stars with mass less than $8M_\odot$ are not formed, the star formation efficiency in both environments is reduced by the factor of 0.1.

Results when $M_{primodial} = 10^{11}M_\odot$ are listed in Table I.

TableI

$m_L = 0.1M_\odot; \epsilon = 0.5 \Rightarrow r \geq 74\mathrm{kpc}; \epsilon = 0.05 \Rightarrow r \geq 380\mathrm{kpc}; r = 10\mathrm{kpc} \Rightarrow \epsilon \geq 0.94$

$m_L = 8M_\odot; \epsilon = 0.05 \Rightarrow r \geq 127\mathrm{kpc}; \epsilon = 0.005 \Rightarrow r \geq 400\mathrm{kpc}; r = 10\mathrm{kpc} \Rightarrow \epsilon \geq 0.6$

Our results show that even if the star formation efficiency in the proto-galaxy is as large as star burst galaxies the radius of the proto-galaxy must be more than 10 times larger than the present radius of the galaxy, ~ 10 kpc since a disk radius of the galaxy with luminous mass $10^{11}M_\odot$ is about 10kpc. If the radius of the proto-galaxy is as small as the present radius of the galaxy, a star formation in the proto-galaxy must be significantly more efficient than in a star burst galaxy. This follows from the observational requirement of ejection of metal from the proto-galaxy into the ICM. Here we ignore the effect of stellar winds from OB stars, because the kinetic luminosity of supernova explosions predominates over that of the stellar wind.

3. HALO NEUTRON STARS AND γ-RAY BURSTS

In the previous section the conditions of the star formation in the proto-galaxy have been deduced from the requirement of ejection of metal from the proto-galaxy into the ICM. We have shown that the radius of the proto-galaxy must be much larger than the present radius or the star formation in the proto-galaxy must be significantly more efficient than nearby star burst galaxies. If the star formation efficiency in the proto-galaxy is as low as that in star burst galaxies, our results predict that galaxies have an extended halo populated with large number of nuetron stars. If the proto-galactic star formation occurs isotropically, an isotropic distribution of the remnants neutron stars in galactic halo is expected. The angular distribution of γ-ray bursters observed by BATSE on Gamma Ray Observatory (GRO) satellite, on the other hand, is isotropic within statistical limits.

In order to derive required conditions for the distribution of neutron stars to be consistent with BATSE data, we assume a core-halo structure for the neutron star distribution.

$$\rho(R) = \rho(0) \left\{ 1 + \left(\frac{R}{R_c} \right)^2 \right\}^{-3\beta/2}, \tag{4}$$

where R denotes the distance to bursters from the Galactic Center and R_c is the core radius. The index β equals to 2/3 and the core radius equals to 13kpc when a neutron star distribution is similar to the dark halo of our Galaxy (Innanen 1973). In figure 1, we plotted $< \cos \theta >$ against $< V/V_{max} >$ for different values of R_c and r_m in the case of $\beta = 2/3$, where θ is the angle between the burst source and the Galactic Center, r_m is the observable maximum distance to the burst sources from the earth and the distance of the earth from the Galactic Center is taken to 8.5kpc. The current value of $< \cos \theta >$ is 0.008 ± 0.035 and that of $< V/V_{max} >$ is 0.33 ± 0.02 for 262 bursts (Meegan 1992a). A large core and maximum radius of the halo are required from these reaults if neutron stars distributed in the galactic halo are sources of γ-ray bursters. This result is consistent with our theoretical prediction. The distribution similar to the dark halo where $R_c = 16$kpc and $r_m = 80$kpc is consistent with the value of $< \cos \theta >$ and $< V/V_{max} >$ obtained by BATSE within 3σ deviation. We show an example of $\log N - \log S$ for $R_c = 16$kpc and $r_m = 80$kpc superposed with the $\log N - \log S$ for 210 BATSE burst events (Meegan et al. 1992b) in figure 2. This distribution explains observed $\log N - \log S$.

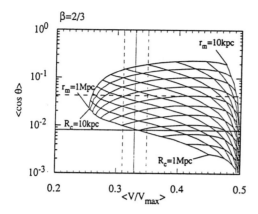

Figure 1.V/Vmax (Meegan,1992a)   **Figure 2.**Log$N$-Log$S$ (Meegan, et al. 1992b)

The total mass of the remnants neutron stars is costrained by the metalicity of the ICM. The mass ratio of the remnants neutron stars to iron ejected from the proto-galaxy is 20 (Tsujimoto & Nomoto 1990). Because the iron abundance in the ICM is arround $0.3Z_\odot$, the mass ratio of the remnants neutron stars to the ICM is 0.01. Recent observational works have shown that the ratio of the stellar mass to the ICM mass is $1 \sim 0.1$ (David, Forman & Jones 1990; Tsuru 1992). So the ratio of remnant neutron stars to the stellar mass is $0.01 \sim 0.1$. On the other hand, the ratio of the dark halo mass to the stellar mass in each galaxies is more than 3. Therefore, the remnant neutron stars can explain up to $0.3 - 3\%$ of the dark halo mass. However, the remnant neutron stars can not resolve the missing mass problem.

This work has been supported by the Special Researchers' Basic Science Program at RIKEN.

REFERENCES

Blaes, O. et al. 1990, *ApJ*, **363**, 612.
Canizares,C.R., Markert,T.H., & Donahue,M.E., 1988, in *Cooling Flows in Clusters and Galaxies*, p.63, ed. by Fabian,A.C., Kluwer, Dordrecht.
David,L.P., Forman,W., & Jones,C., 1990,*Astrophys.J.*,**359**,29.
Innanen,A., 1973, *Astrophys. Space Sci.*,**22**, 393.
Larson,R.B., 1987, in *Starbursts and Galaxy Evolution*, p.467, ed. byThuan,T.X., et al., Editions Frontieres.
Meegan, C. A. 1992a, private communication.
Meegan, C. A. et al. 1992b, *Nature*, **355**, 143.
Shull,J.M., & Silk,J., 1979,*Astrophys.J.*,**234**,427.
Tsujimoto,T., & Nomoto,K., 1990, private comunication.
Tsuru,T., 1992, Ph.D. University of Tokyo.

ISOTOPIC ANOMALIES

Isotope Abundance Anomalies and the Early Solar System: MuSiC vs. FUN

F. BEGEMANN

Max-Planck-Institut für Chemie (Otto-Hahn-Institut), Saarstraße 23, D-6500 Mainz, FRG.

INTRODUCTION

When meteorites, the earth and the moon came into being some 4.5 AE ago they presumably did so together with the sun. Since there is nothing exceptional about the sun as a star it can also be assumed that the sun formed, like other stars of similar mass, by the collaps of a dense (10-100 molecules/cm^3), cold (~ 10K) protostellar cloud. Although the masses of such protostellar clouds appear to be considerably larger than the mass of the stars formed from them it seems to be generally, almost invariably tacitly, assumed that no major element fractionation occurred upon the collaps. (For a divergent view cf. e.g. Manuel and Sabu 1975, 1988). To the extent that this assumption is justified the sun provides us with a snapshot of the kind of matter that was present in a particular region of space 4.5 AE ago.

Starting with Goldschmidt's (1937) abundance compilation of the elements it was the work of Suess (1947 a,b; 1956) in particular that demonstrated that certain nuclear properties of the nuclides are reflected in their abundance. This showed the way for all modern theories of the synthesis of elements heavier than the iron group elements. Present astrophysical wisdom has it that the abundances of the chemical elements in the solar system cannot be due to one single process of nucleosynthesis but that a number of different processes must be invoked. Depending on how specific one is with defining "different processes" between half a dozen and a dozen sources must have contributed their share to the protostellar cloud from which the solar system formed (Burbidge et al. 1957, Cameron 1957). As to the sites where the synthesis took place, hydrogen and most of the helium are supposed to owe their existence to the very first moments in the life of the universe after the Big Bang. The same is true, at least in part, for the next-heavier elements - Li, Be, and B - although another part of them originated from spallation reactions, possibly in a late-stage irradiation (Fowler et al. 1962, Gilmore 1992). All other elements, however, must have been synthesized in stars (Hoyle 1946, 1954; Burbidge et al. 1957, Cameron 1957). While the exact nature of some of the stars is still contentious it is clear that the matter synthesized within them must be ejected from the production site(s) into space in order to become available for the next round of astration - and in some cases for the formation of planets and planetesimals.

For practical reasons - because elemental and isotopic analyses can be performed in the laboratory with much better precision than is possible by remote observation of the sun - a pivotal question is whether there is available for analysis in the laboratory matter from somewhere in the solar system that matches the solar composition. A rare group of meteorites, the carbonaceous chondrites of type C1, do indeed seem to provide such a match. Of course, hydrogen, helium and other extremely volatile elements, i.e. more than 99% by number of the total inventory are missing from the meteorites but for more than two thirds of all elements is there a very good agreement between the abundances in C1 chondrites and the solar abundances which are normally taken to be equal to the abundances in the photosphere of the sun (Anders and Grevesse 1989).

Strictly speaking this observation, remarkable as it may be, does not help much for a determination with high precision of the solar abundances because, in order to rely on the meteoritic values it first must be established that they indeed do agree with the solar abundances - but then the problem has been solved anyhow. The importance rather lies in the fact that since their formation C1 chondrites have undergone but little thermometamorphic changes so that it may be feasible to look in great detail into the composition and the early history not only of the protosolar cloud as a whole but of at least some of its constituents.

Protostellar clouds consist of gas and dust. Astronomical observations suggest that the bulk of the non-volatile elements is present as dust which , for a nebula of solar composition, implies a dust/gas ratio by mass of ca. 10^{-3}. Astronomical observations suggest also that dust formation takes place already in the envelopes of stars that eject freshly synthesized matter into space. Since this matter will have had little chance to get mixed with that from other sources, this dust will carry the signature of the individual processes that produced the matter and that contributed to the overall composition of the presolar nebula. Judged from the obscuration of starlight, dust grains have typical sizes of about $0.1 \mu m$. Provided the primordial condensates survived their subsequent history in interstellar space and within the presolar nebula, the collaps of the nebula, the compaction of the carbonaceous chondrites and their thermal and aqueous history since compaction, it is stardust of this size that can be anticipated to consist of unadulterated matter from a single source of nucleosynthesis i.e. such stardust may have kept its "cosmic chemical memory" (Clayton 1977, 1992).

If it were just different chemical elements that owed their existence to different nucleosynthesis processes it would be impossible to identify specific reaction products because chemical reactions are very selective and the formation or not of specific molecules or compounds depends on many factors aside from the trivial one of availability of the elements in question. Fortunately, however, for many medium-heavy and heavy elements the constituent isotopes are produced in different processes also, and isotopes are (almost) non-selective in their chemical behavior. Hence, the fingerprint one searches for in the quest for stardust is deviations in the isotopic composition of elements that are of such a kind that they cannot have been produced by physical or chemical processes ongoing at the time when or since the solar system formed.

Murchison, Murray and Allende are the three meteorites that have yielded most of the presently existing data on anomalous isotope abundances. All three are carbonaceous chondrites, with Murchison and Murray belonging to group C2 and Allende to the more evolved group C3V. From Murchison and Murray it has been possible to separate diamond and silicon carbide grains of a size anticipated for stardust (Tang and Anders 1988) that are anomalous in the isotopic composition of their structural elements Si and C (Zinner et al. 1987) as well as in the trace elements they contain. For the noble gases Kr and Xe it has been known for quite some time that what has now been shown to be diamond contains an overabundance of r- and p-process isotopes (Lewis et al. 1975) while in what has now been identified as SiC the s-process isotopes are overabundant (Srinivasan and Anders 1978). From Murchison SiC (MuSiC) the list of isotopically anomalous elements has now been extended and includes the medium-heavy elements Sr, Ba, Nd, and Sm (Ott and Begemann 1990a,b; Zinner et al. 1991, Richter et al. 1992; Ott et al. 1992).

Each of these elements is of quite some interest from an astrophysical point of view. For Sr and Ba the heaviest isotopes ^{88}Sr and ^{138}Ba contain a "magic" number of neutrons (50 and 82, respectively). The associated small neutron capture cross sections of both nuclides make their abundance sensitive indicators of the effective neutron exposure, in the case of ^{88}Sr for the effects from both the weak and the main component of the s-process and for ^{138}Ba only the exposure from the main component. Nd and Sm, on the other hand, are located in the s-path of nucleosynthesis around the branching points at ^{147}Pm and ^{151}Sm so that their isotopic signature depends on neutron density (Ott et al. 1992).

In addition, however, the new MuSiC data make it possible to compare the anomalies with those reported for two Ca-Al-rich inclusions from Allende (Papanastassiou and Wasserburg 1978; Lugmair et al. 1978; McCulloch and Wasserburg 1978a,b) that so far have monopolized the isotope abundance anomalies in these elements. Since the first anomalies discovered there suggested both Fractionation and Unknown Nuclear effects they were dubbed FUN anomalies (Wasserburg et al. 1977).

MuSiC vs. FUN

The first thing to note about the MuSiC data is that the effects are huge. Depending on whether one prefers to describe the deviations from normal as deficits or as excesses, the anomalies amount to - 70% or + 300%, and even more (Fig. 1). This has the beneficial result that instrumental or natural mass-dependent fractionation effects are only important for unravelling second-order effects, a point that will be discussed in more detail below.

The four elements newly analyzed show the distinct signature of the (an) s-process. This is in keeping with the observation that Murchson SiC contains also s-Kr and s-Xe. If all these elements derived from the same star a

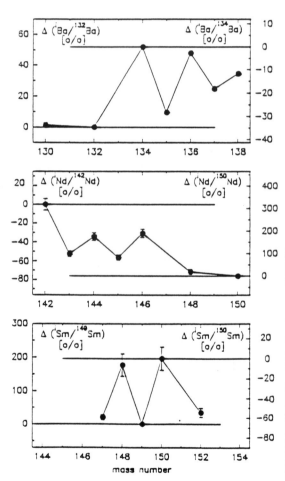

Fig. 1:
Isotopic composition of Ba, Nd, and Sm in SiC from Murchison meteorite. Plotted are the deviations of isotope abundance ratios (in %) from solar. Note, that the sign of the anomalies depends on the choice of index isotope. Data from Ott and Begemann (1990) and Ott et al. (1992).

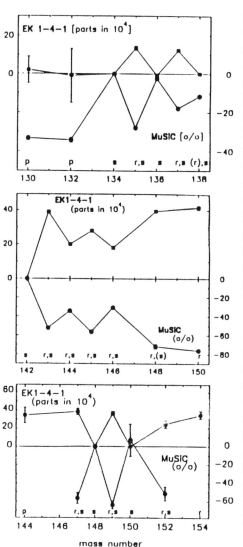

corollary would be that we are dealing with truly correlated anomalies in the sense that any information concerning details of the synthesis process derived from the anomalies in one element must be consistent with those derived from all other elements. This is not (yet) certain, however. Nor need it necessarily be true for the structural elements because it is conceivable that the s-process trace elements were implanted, possibly at different times as neutral atoms or as ions, into pre-existing interstellar dust grains that may or may not have been formed from matter ejected during an earlier phase in its life by the same star that produced the s-nuclides (Clayton 1981).

A comparison of the MuSiC abundance anomalies in Ba, Nd, and Sm with those reported for Allende inclusion EK1-4-1 shows them to be qualitatively complementary. Isotopes that are underabundant in MuSiC are overabundant in EK1-4-1, and vice versa (Fig. 2). This fact is not overly impressive, however, because it is to some extent an artefact of the normalisation procedure employed for the Allende data where it has always been assumed, explicitly or

Fig. 2: The isotope abundance anomalies in SiC from Murchison and Allende FUN inclusion EK1-4-1 (McCulloch and Wasserburg 1978a,b) are complementary: Isotopes overabundant in MuSiC are deficient in EK1-4-1 and vice versa. Data are normalized to ^{134}Ba (top panel), ^{142}Nd (middle) and ^{148}Sm (bottom), respectively. Note, that for Allende an internal normalisation was required in the data reduction that makes the anomalies at two isotopes per element vanish (^{134,138}Ba; ^{148,150}Sm) or equal (^{143,148}Nd).

implicitly, that the anomalies should be owing to an overabundance of r-process contributions. In retrospect, the complementarity of the effects with such that are unambiguously caused by an overabundance of s-nuclides vindicates this interpretation of the anomalies in the Allende inclusions(s).

Quantitativly, the (anti)-match of the anomalies pattern is not perfect for all elements. In the case of Ba the mismatch at ^{130}Ba and ^{132}Ba might be owing to the low abundance of the two isotopes of 0.1% each but the limits of error would have to be considerably enlarged in order to allow for agreement. But there is also disagreement at the not so rare ^{135,137}Ba that in Allende EK1-4-1 show overabundances of about equal size while in MuSiC the deficit at ^{135}Ba is twice that at ^{137}Ba.

In order to put these discrepancies into perspective, and to avoid their overinterpretation, it should be recalled that all mass spectrometric measurements of isotope abundance ratios must be corrected for mass-dependent fractionation effects that occur in the course of the analysis. This is done by assuming that in the unknown sample one ratio is known, by comparing the measured value of this ratio with the true one and then correcting all other measured ratios accordingly. Such a procedure not only removes from the competition for being anomalous in their relative abundance the two isotopes that are used for normalisation but it also influences in a qualitative and quantitative way the outcome of the data reduction. The extent to which the final result is affected depends on the size of the fractionation corrections as compared to the size of the anomalies one is dealing with. For MuSiC the corrections, normally of the order of a few tenth of a percent per atomic mass unit, are entirely negligible; as a matter of fact no corrections were applied to the data at all. In contradistinction the anomaly patterns derived for the Allende inclusions depend critically on the normalisation procedure. In the case of Ba from Allende EK1-4-1 McCulloch and Wasserburg (1978) found all abundance ratios to be anomalous, from about -40x10^{-4} for ^{130}Ba to +8x10^{-4} for ^{137}Ba, when they assumed the ^{135}Ba/^{138}Ba ratio in the sample to be the same as in terrestrial Ba. If, however, the ^{134}Ba/^{138}Ba ratio was used for the normalisation instead, the anomalies vanished for all isotopes but ^{135}Ba and ^{137}Ba (Fig. 2) and it is this pattern that was used in the above comparison with the MuSiC data. Obviously, there is quite some ambiguity in the results that cannot be resolved; indeed once the anomalies are as small as in the Allende inclusions it is not even possible to decide unambiguously the mass number of the isotope whose abundance is anomalous. Furthermore, it should be kept in mind that the uncertainties assigned to the anomalies do not, and perhaps cannot reasonably include any part that arises from uncertainties in the assumptions. In short, the above mismatch of the anomaly patterns in MuSiC and Allende inclusion EK1-4-1 may well be only an apparent one.

For Nd the agreement of the anomaly patterns in the two samples is excellent. After the afore said it is not clear, however, whether or not this is significant, in particular because the derivation of the Nd pattern in EK1-4-1 was even more involved than was the case for Ba. Here, none of the abundance ratios in the sample was found "suitable" for the removal of the instrumental fractionation effects. It was rather assumed that the r, s- ^{144}Nd in EK1-4-1 should include a contribution from the very same putative r-process whose effects one wants to deduce from the data in the first place. This contribution was somewhat arbitrarily assumed to amount to 20×10^{-4} of the measured abundance, the measured ^{144}Nd/^{142}Nd ratio was decreased accordingly and the so corrected ratio then used to remove the mass-dependent fractionation effects (McCulloch and Wasserburg 1978). While this procedure may be perfectly sound it is nevertheless true that to some degree the data reduction yielded what was put in as an assumption, namely that the relative contributions from the r-process to the abundance of mixed s,r-isotopes $^{143-146}$Nd and the pure r-isotopes ^{148,150}Nd are just about equal.

In the case of Sm the complementarity of the effects is also very good although in MuSiC the experimental uncertainties are still fairly large and the lightest (^{144}Sm) and heaviest (^{154}Sm) isotopes were compromised by interferences and could not yet be measured with sufficient accuracy (Ott et al. 1992).

Strontium, finally, is a particular case in that the abundance anomalies in MuSiC and Allende EK1-4-1 are not complementary at all (Fig. 3). In both samples only the abundance of p-only ^{84}Sr is anomalous, by -0.32% in EK1-4-1 (Papanastassiou and Wasserburg 1978) and by -64.2% in MuSiC (Richter et al. 1992). A possible

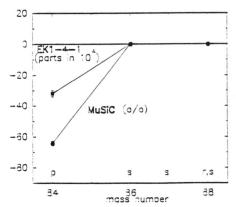

Fig. 3:

The complementarity in the anomalies pattern observed for Ba, Nd, and Sm from Murchison SiC and Allende inclusion EK1-4-1 fails for Sr that shows a deficit of p-only ^{84}Sr in both cases. No data are plotted for ^{87}Sr because of the presence or not of an unspecified radiogenic contribution from the decay of ^{87}Rb.

explanation would be that in the mass region around Sr the p- and r-processes are decoupled while for higher masses the effects of both processes always go together. On the other hand, it should be noted that the s-Sr in SiC from Murchison is peculiar in so far as the $^{88}Sr/^{86}Sr$ ratio is the same as in solar Sr to within $(3\pm8)^o/oo$ which leaves not much room for any contribution to the solar abundance of ^{88}Sr from the r-process.

CONCLUSIONS

The isotope abundance anomalies of medium-heavy elements in SiC separated from Murchison carbonaceous chondrite are quantitatively and qualitatively different from the anomalies in the same elements as they have been found in FUN inclusions from Allende carbonaceous chondrite. Quantitatively, the MuSiC anomalies are orders of magnitude larger. Qualitatively the data are different because the very size of the anomalies makes the identification of the mass number of the "anomalous isotope(s)" quite certain. There is none of the troublesome ambiguity any more that in the past has plagued the interpretation of very small effects and that has led to acrimonious and ludicrous exchanges.

For Ba, Nd, and Sm the surprisingly good (anti)-match of the anomalies pattern in MuSiC and Allende FUN inclusions appears too good as to be entirely serendipitous, the implication being that the s-Ba, Nd, Sm in SiC from Murchison are quantitatively complementary to the excess of r-Ba, Nd, Sm in Allende inclusions. That this should be so is not trivial, except if there were just one s-process and one r-process. The SiC from Murchison amounts to only a minute fraction (a few ppm) of the total meteorite and it contains only a tiny fraction of the s-nuclides in question. The bulk solar s-Ba, Nd, Sm will have been brought into the presolar cloud via different venues and may well have been synthesized in stars different from the one(s) that produced the MuSiC s-nuclides and carry a different isotopic signature. For Kr such evidence exists already because SiC from Murchison does not contain a unique type of s-Kr. Different SiC fractions rather yield Kr with differing s-$^{86}Kr/^{80}Kr$ ratios (Ott et al. 1988; Lewis et al. 1990) reflecting differing neutron densities and differing temperatures at the site of the synthesis. In the mass range of Kr the s-nuclides were apparently not produced in a single process nor in environments with narrowly fixed parameters (Gallino et al. 1990). Possibly, elements in the Kr-Sr mass range are particularly sensitive in this respect because both the "main" and the "weak" component of the s-process contribute to the bulk s-nuclides.

REFERENCES

Anders E. and N. Grevesse (1989) Abundances of the elements: Meteoritic and solar. Geochim. Cosmochim. Acta, **53**, 197-214.

Burbidge E.M., G.R. Burbidge, W.A. Fowler and F. Hoyle (1957) Synthesis of the elements in stars. Rev. Mod. Phys. **29**, 547-650.

Cameron A.G.W.(1957) Stellar evolution, nuclear astrophysics and nucleogenesis. Chalk River Report, AECL, CRL-41.

Clayton D.D. (1977) Solar system isotopic anomalies: Supernova neighbor or presolar carriers? Icarus **32**, 255-269.

Clayton D.D. (1981) Some key issues in isotopic anomalies: Astrophysical history and aggregation. Proc. Lunar Planet. Sci. Conf. **12B**, 1781-1802.

Clayton D.D. (1992) Meteorites and the origins of atomic nuclei. Meteoritics **27**, 5-17.

Fowler W.A., J.L. Greenstein and F. Hoyle (1962) Nucleosynthesis during the early history of the solar system. Geophys. J. Royal Astron. Soc. **6**, 148-220.

Gallino R., M. Busso, G. Picchio and C.M. Raiteri (1990) On the astrophysical interpretation of isotope anomalies in meteoritic SiC grains. Nature **348**, 298-302.

Gilmore G. (1992) Lithium given and taken away. Nature **358**, 108-109.

Goldschmidt V.M. (1937) Geochemische Verteilungsgesetze der Elemente IX. Skrifter Norske Videnskaps-Akad., Oslo I Mod. Natur. Kl. No. 4.

Hoyle F (1946) The synthesis of the elements from hydrogen. Mon. Not. Roy. Astron. Soc. **106**, 343-389.

Hoyle F. (1954) On nuclear reactions occurring in very hot stars, I. The synthesis of elements from carbon to nickel. Astrophys. J. Suppl.1, 121-146.

Lewis R.S., B. Srinivasan and E. Anders (1975) Host phase of a strange xenon component in Allende. Science **190**, 1251-1262.

Lewis R.S., S. Amari and E.Anders (1990) Meteoritic silicon carbide: Pristine material from carbon stars. Nature **348**, 293-298.

Lugmair G.W., K. Marti and N.B. Scheinin (1978) Incomplete mixing of products from r, p-, and s-process nucleosynthesis: Sm-Nd systematics in Allende inclusion EK1-04-1. Lunar Planet.Sci. **IX**, 672-674.

Manuel O.K. and D.D. Sabu (1975) Elemental and isotopic inhomogeneities in noble gases: The case for local synthesis of the chemical elements. Trans. Missouri Acad. Sci. **9**, 104-122.

Manuel O.K. and D.D. Sabu (1988) Isotopic anomalies in meteorites. Essays in Nuclear-, Geo- and Cosmochemistry, M.W. Rowe, editor, pp 1-42.

McCulloch M.T. and G.J. Wasserburg (1978a) Barium and neodymium isotopic anomalies in the Allende meteorite. Ap. J. **220**, L15-L19.

McCulloch M.T. and G.J. Wasserburg (1978b) More anomalies from the Allende meteorite: Samarium. Geophys. Res.Lett. **5**, 599-602.

Ott U., F. Begemann, J. Yang and S. Epstein (1988) s-process krypton of variable isotopic composition in the Murchison meteorite. Nature **332**, 700-702.

Ott U. and F. Begemann (1990a) Discovery of s-process barium in the Murchison meteorite. Ap. J. **353**, L57-L60.

Ott U. and F. Begemann (1990b) s-process material in Murchison: Sr and more on Ba. Lunar Planet. Sci. **XXI**, 920-921.

Ott U., S. Richter and F. Begemann (1992) s-process nucleosynthesis: the evidence from the meteorites. This volume.

Papanastassiou D.A. and G.J. Wasserburg (1978) Strontium isotopic anomalies in the Allende meteorite. Geophys. Res. Lett. **5**, 595-598.

Richter S., U. Ott and F. Begemann (1992) s-process isotope anomalies: Neodymium, samarium, and a bit more of strontium. Lunar Planet.Sci. **XXIII**, 1147-1148.

Srinivasan B. and E.Anders (1978) Noble gases in the Murchison meteorite: Possible relics of s-process nucleosynthesis. Science **201**, 51-56.

Suess H.E (1947a) Über kosmische Kernhäufigkeiten. I. Einige Häufigkeitsregeln und ihre Anwendung bei der Abschätzung der Häufigkeitswerte für die mittelschweren und schweren Elemente. Z. Naturforsch. **2a**, 311-321.

Suess H.E. (1947b) Über kosmische Kernhäufigkeiten. II. Einzelheiten in der Häufigkeitsverteilung der mittelschweren und schweren Kerne. **ibid. 2a**, 604-608.

Suess H.E. (1956) Abundances of the elements. Rev. Mod. Phys. **28**, 53-74.

Tang M. and E. Anders (1988) Isotopic anomalies of Ne, Xe, and C in meteorites. II. Interstellar diamond and SiC: Carriers of exotic noble gases. Geochim. Cosmochim. Acta **52**, 1235-1244.

Wasserburg G.J., T. Lee and D.A. Papanastassiou (1977) Correlated O and Mg isotopic anomalies in Allende inclusions: II. Magnesium. Geophys. Res.Lett. **4**, 299-302.

Zinner E., M. Tang and E. Anders (1987) Large isotopic anomalies of Si, C, N and noble gases in interstellar silicon carbide from the Murray meteorite. Nature **330**, 730-732.

Zinner E., S. Amari and R.S. Lewis (1991) s-process Ba, Nd and Sm in presolar SiC from the Murchison meteorite. Ap. J. **382**, L47-L50.

S-Process Nucleosynthesis: The Evidence from the Meteorites

U. OTT, S. RICHTER, AND F. BEGEMANN

Max-Planck-Institut für Chemie (Otto-Hahn-Institut), Saarstraße 23, D-6500 Mainz, F.R.G.

Judged by the isotopic composition of their constituent elements, the past few years have witnessed the isolation from primitive meteorites of clearly presolar diamond, graphite and silicon carbide. Among these, SiC carries trace elements that clearly show the signature of the s-process of nucleosynthesis resembling the so-called main component thought to occur during thermal pulses in AGB stars. In contrast to bulk solar system abundances, where contributions from different nucleosynthetic processes are always invariably mixed, excesses of nuclides produced in a single type of process allow better characterization of the isotopic abundance pattern produced in this process and, based on this, a better evaluation of the relevant parameters describing the process. We review data previously obtained on s-process Xe, Kr, Ba and Sr in such SiC grains and report new data for s-process Nd and Sm.

1 XENON, KRYPTON

Xenon was the first element in which excesses of isotopes produced in the s-process over those produced in the r- and p-processes, resp., were detected in acid-resistant fractions of primitive meteorites (Srinivasan and Anders, 1978) later identified as containing silicon carbide as the carrier phase of this xenon (Tang and Anders, 1988). S-process krypton is interesting since its composition varies as a function of the temperature at which SiC is combusted in order to release gases trapped within it (Ott et al., 1988) or with grain size (Lewis et al., 1990). These variations occur in the relative abundances of ^{86}Kr and ^{80}Kr and reflect the sensitivity of their abundances to the branchings at ^{85}Kr (depending on neutron density) and ^{79}Se (dependent, in addition, on temperature). The low ratio $(^{80}\text{Kr}/^{82}\text{Kr})_s$ indicates that s-process Kr in SiC was produced under conditions of higher neutron density/lower temperature than had been estimated. A good description of the relevant astrophysical conditions appears to be s-process nucleosynthesis during thermal pulses in AGB stars (Gallino et al., 1990).

2 BARIUM, STRONTIUM

The identification in meteoritic SiC of s-process barium (Ott and Begemann, 1990) has been a breakthrough in the field of isotopic anomalies in meteorites insofar as it tied isotopic anomalies found in noble gas elements to such in other elements. Because of the small neutron capture cross section of ^{138}Ba, $(^{138}Ba/^{136}Ba)_s$ is a sensitive measure of the effective neutron exposure, for which a value of 0.17 mb^{-1} (for kT = 30 keV) was derived, quite different from the 0.30 mb^{-1} used in order to reproduce the solar system elemental abundance curve (e.g. Käppeler et al., 1989). Later work by Zinner et al. (1991) showed the $(^{138}Ba/^{136}Ba)_s$ ratio to be slightly variable among SiC grain size fractions, indicating different neutron exposures for different grain populations.

The case of strontium is complex. Prombo et al. (1992) found variations between different SiC grain size fractions on the order of 5% in the ratio $(^{88}Sr/^{86}Sr)_s$. They note that the trend of this ratio with grain size is opposite to what would be expected from the AGB star model calculations and effects in the other isotopic systems. However, considering the complexity of the Sr system - $^{88}Sr/^{86}Sr$ is dependent both on neutron density and on neutron exposure - this is certainly not a strong argument against AGB star production.

3 NEODYMIUM AND SAMARIUM

Nd and Sm data reported for interstellar SiC first were obtained by ion microprobe (Zinner et al., 1991). A major obstacle for drawing conclusions about the composition of s-process Nd and s-process Sm from these results was due to the fact that isotopes with atomic masses 148 and 150 exist for both elements (r-only for Nd, s-only for Sm) and that both elements contributed about equally to the signal measured at these masses. We have shown that, using thermal ionization, it is possible to measure SiC-Nd virtually without introducing uncertainties from the correction for Sm interference (Richter et al., 1992). New data (labelled '5' and shown together with our old data in Fig. 1) extend the observed range, but the composition of s-process Nd derived from these data by extrapolating to r-only $^{150}Nd \equiv 0$ (i.e. $\delta(^{150}/^{144}Nd) \equiv -1000$) is $^{142}Nd/^{143}Nd/^{144}Nd/^{145}Nd/^{146}Nd/^{148}Nd/^{150}Nd$ = $2.13 \pm 0.08/0.293 \pm 0.006/\equiv 1/0.161 \pm 0.005/0.775 \pm 0.009/0.0281 \pm 0.0058/\equiv 0$, virtually identical with the composition we have reported previously (Richter et al., 1992). Noteworthy is the apparent finite s-process contribution to ^{148}Nd, which is considered r-only in the classical picture (cf. the small abundance predicted by e.g. Mathews and Käppeler (1984)).

In the case of Sm an interference at mass 154 prevented us from obtaining reliable data for this isotope, which also prevents us from using extrapolation to r-only

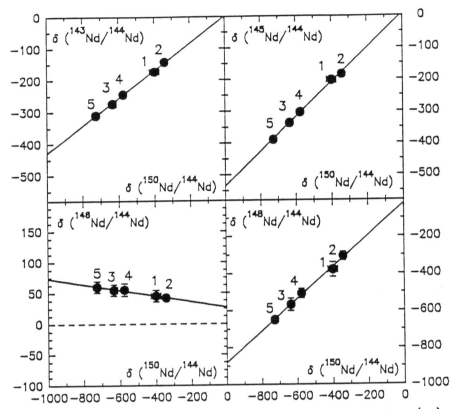

Fig. 1: Nd isotopic composition (deviation from normal in o/oo) measured in SiC from the Murchison meteorite

^{154}Sm $\equiv$ 0 in order to deduce the composition of s-process Sm. Data for the other isotopes are shown in Fig. 2. An s-process Sm composition calculated with the assumption that ^{150}Sm/^{149}Sm (not affected by branching) is equal to that predicted by the 'local approximation' is ^{147}Sm/^{148}Sm/^{149}Sm/^{150}Sm/^{152}Sm = 1.51±.14/4.83±.22/≡1/≡3.25/3.41±.26. ^{147}Sm/^{149}Sm (sensitive to the ^{147}Pm branching) is almost equal to the ratio expected from σN = const., indicating almost complete decay of ^{147}Pm into ^{147}Sm rather than neutron capture, which is quite different from the predictions of the classical model (Käppeler et al., 1989). There is agreement, on the other hand, in the case of ^{152}Sm/^{149}Sm (sensitive to the ^{151}Sm branching) where little decay of ^{151}Sm is indicated. For more far-reaching conclusions, however, the data have to be compared with realistic models which take into account time variability of neutron density and temperature such as, e.g. that of Gallino et al. (1990).

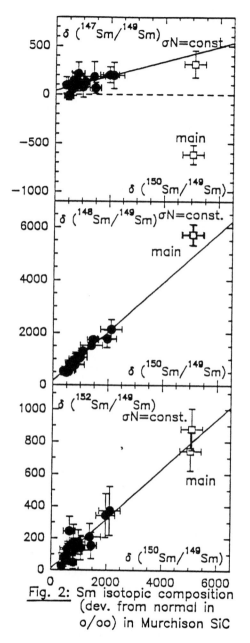

Fig. 2: Sm isotopic composition
(dev. from normal in
o/oo) in Murchison SiC

4 CONCLUSIONS

Analysis of trace elements contained in meteoritic presolar silicon carbide has just begun. Analyses of more elements certainly present and showing the signature of the s-process will undoubtedly be possible. The results should be able to shed more light on the details of the process.

REFERENCES

Gallino R., Busso M., Picchio G., and Raiteri C.M. (1990) Nature 348, 298-302.

Käppeler F., Beer H., and Wisshak K. (1989) Rep. Prog. Phys. 52, 945-1013.

Lewis R.S., Amari S. ,and Anders E. (1990) Nature 348, 293-298.

Mathews G.J. and Käppeler F. (1984) Ap. J.286, 810-821.

Ott U. and Begemann F. (1990) Ap. J. (Letters) 353, L57-L60.

Ott U., Begemann F., Yang J., and Epstein S. (1988) Nature332, 700-702.

Prombo C.A., Podosek F.A., Amari S., and Lewis R.S. (1992) Lunar Planet Sci. XXIII, 1111-1112.

Richter S., Ott U., and Begemann F. (1992) Lunar Planet. Sci. XXIII, 1147-1148.

Srinivasan B. and Anders E. (1978) Science 201, 51-56.

Tang M. and Anders E. (1988) Geochim. Cosmochim. Acta 52, 1235-1244.

Zinner E., Amari S., and Lewis R.S. (1991) Ap. J. (Letters) 382, L47-L50.

Correlated Isotopic Anomalies in Meteoritic SiC

R. GALLINO[1], M. BUSSO[2], C.M. RAITERI[2]

[1] Istituto di Fisica Generale dell'Universita' di Torino, Via P. Giuria 1, 10125 Torino, Italy [2] Osservatorio Astronomico di Torino, Italy

Patience dans les s
Hom(m)age to Hubert

1 INTRODUCTION

In a paper by Hubert Reeves (1966) on "Stellar Neutron Sources" and in a series of Lecture Notes on "Stellar Evolution and Nucleosynthesis" (Reeves 1968) the implications of neutron capture nucleosynthesis on the formation of the heavy elements were discussed. In particular thermonuclear reaction rates were first calculated in order to individuate the most important reactions for the release of neutrons.

For a long time the major role has been played by the $^{22}Ne(\alpha,n)^{25}Mg$ reaction, both considering massive stars (Lamb et al. 1977) and intermediate mass stars suffering thermal pulses on the AGB (3–8 $M_\odot$; Iben 1975; Truran & Iben 1977). However, its effectiveness in intermediate mass stars has been subsequently questioned on theoretical as well as observational grounds. At the same time, growing evidence has been gathered favoring low mass TP–AGB stars (1–3 $M_\odot$) as the best site for the production of the main s-component. These stars show correlated ^{12}C versus s-enhancements (Lambert 1989); the ^{22}Ne neutron source here is only marginally activated, while the alternative reaction $^{13}C(\alpha,n)^{16}O$ may efficiently operate (Gallino et al. 1988; Hollowell & Iben 1989; Gallino 1989; Käppeler et al. 1990). This source requires a primary production of ^{13}C, driven by the mixing of a small amount of protons from the envelope into the ^{12}C-rich zone. A ^{13}C-pocket of few 10^{-4} $M_\odot$ is then formed during the intershell period and ingested by the next convective instability. However, the efficiency of this mechanism is still a matter of debate, especially for galactic disc stars.

A completely independent confirmation of the low mass stars scenario comes from the analysis of isotopic anomalies carried by SiC microcrystals extracted from pristine meteorites (Lewis et al. 1990; Gallino et al. 1990). Carbon stars with initial metallicities slightly lower than solar are the best candidates for the formation of the bulge SiC grains in their extended mass–losing envelopes, accounting for most of the isotopic anomalies so far discovered. All heavy elements present in trace are affected by the s–process taking place during recurrent He–shell flashes.

2 INTERPRETATION OF SIC ANOMALIES

The analysis of isotopic anomalies in the TP–AGB framework must distinguish among: (i) light or intermediate elements below iron, that in the s–theory are commonly defined as "neutron poisons" and achieve low production (destruction) factors; (ii) s–only isotopes between Fe and Sr (^{70}Ge, ^{76}Se, ^{80}Kr, ^{82}Kr, ^{86}Sr and ^{87}Sr) that are mainly provided by the weak component through the activation in massive stars of the ^{22}Ne source during core–He or shell C–burning, and partly contributed by the main component; (iii) the heavier elements from Y up to Pb, belonging to the main component and reaching high production factors in TP–AGB stars. Moreover, one has to consider the modification in the envelope chemical composition induced by the I dredge–up episode, which occurred when the star first ascent the Red Giant Branch. Later on, the III dredge–up mixes with the envelope He–shell s–processed material and newly synthetized ^{12}C.

The measured enhancement in SiC grains of ^{14}N/^{15}N with respect to solar is essentially a direct consequence of the I dredge–up. The spread of nytrogen anomalies, not correlated with ^{12}C/^{13}C, finds its counterpart in the spectroscopic observations of M stars (Smith & Lambert 1990), pointing towards a multiple star interpretation for the formation of SiC grains. The observed variation of the ^{12}C/^{13}C ratio is a consequence of both the I and the III dredge–up. Indeed, the value of such ratio in the stellar envelope after the I dredge–up is $\sim 10 - -20$ as observed in M star (Lambert, 1989), and it is subsequently modified by progressive mixing of ^{12}C–rich material from the He shell. The excess of ^{22}Ne is a signature of the third dredge–up; at the beginning of He burning it is built by α–captures on ^{14}N and mostly survives further burning, because in low mas TP–AGB stars the maximum temperature reached by a thermal instability never exceeds T $\sim 3~10^8$ K. Hence, SiC grains are carriers of the well known Ne-E(H) component found in pristine meteorites, with no need of invoking the ^{22}Na–decay.

All noble gases have a non solar composition (Ott et al. 1988; Lewis et al. 1990; 1992), Kr and Xe showing a clear s–signature. While for xenon linear correlations in the three isotope plot indicate that the anomalous composition is a mixture of two components, one pure–s (the G–component), the other (the N–component) close to solar, the situation is more complex for krypton. A linear correlation of ^{84}Kr/^{82}Kr versus ^{83}Kr/^{82}Kr is found, but for ^{80}Kr/^{82}Kr, and especially for ^{86}Kr/^{82}Kr, a spread exists for grains of different sizes. This behavior can be understood in the framework of the s–theory as the consequence of branchings in the s–process path at ^{79}Se and ^{85}Kr and of different nucleosynthetic environments when SiC grains form. Noble gases are thought to be implanted as ions by hot stellar winds; on the other hand, other trace elements likely condense earlier together with SiC. Among them, Ba, Nd and Sm have been recently measured (Ott & Begemann 1990a; Zinner et al. 1991; Prombo et al. 1992a; 1992b; Richter et al. 1992). The interpretation of such anomalies strongly favours carbon stars (Gallino et al. 1992b). The amount of the measured mixture between N and G components represents typical envelope conditions. The non–solar s–anomaly is characteristic of mean neutron exposures $\tau_0=0.10-0.15$ mb^{-1}, which are half or less that responsible for the reproduction of the s–main component.

Together with C and N, also isotopic anomalies of Si (Zinner, et al. 1987, 1989; Tang et al. 1989, Stone et al. 1991), Ca and Ti (Amari et al. 1991; Ireland et al. 1991) have been discovered. For Si, the astrophysical interpretation is faced with difficulties. In fact, an observed slope of about 1 is found in the three isotope plot ^{29}Si/^{28}Si vs. ^{30}Si/^{28}Si, extending up to $\sim$200 permil, whereas a slope of about 0.5 results in typical s–process conditions, with a narrower spread of values. This fact is a consequence of the major role played by the ^{33}S(n,α)^{30}Si reaction (Wagemans et al. 1992). The solution can be obtained by considering a slightly non–solar N–component as it comes out from the present knowledge of the galactic chemical evolution of Si isotopes (Gallino et al. 1992a; see however Brown & Clayton (1992) for a different interpretation). The understanding of Ca and Ti anomalies involves the s–process too, despite the need of more precise neutron capture cross sections.

Finally, an unexpected solar–like isotopic composition has been detected for Sr (Ott & Begemann 1990b; Prombo et al. 1992a; Richter et al. 1992) whose explanation is still missing. Let us stress that the main s–component only partly accounts for the Sr solar abundance, and that the production of Sr isotopes is heavily affected by the difficult treatment of the branching at ^{85}Kr. Before analyzing this problem, improved measurements of Kr and Sr neutron capture cross sections are also highly desirable.

REFERENCES

Amari, S., Zinner, E. & Lewis, R. S. 1991, Meteoritics, 26, 314
Brown, L. E. & Clayton, D. D. 1992, ApJ, in press
Gallino R. 1989, in Evolution of Peculiar Red Giant Stars, ed. H. R. Johnson & B. Zuckerman (Cambridge Univ. Press: Cambridge), 176
Gallino, R., Busso, M., Picchio, G., Raiteri, C. M. & Renzini, A. 1988, ApJ, 334, L45
Gallino, R., Busso, M., Picchio, G. & Raiteri, C. M. 1990, Nature, 348, 298
Gallino, R., Raiteri, C. M. & Busso, M. 1992a, XVI General Assembly of the European Geophysical Society, Ann. Geoph. Suppl., 10, C465
Gallino, R., Raiteri, C. M. & Busso, M. 1992b, ApJ, submitted
Hollowell, D. E. & Iben, I. Jr. 1989, ApJ, 340, 966

Käppeler, F., Gallino, R., Busso, M., Picchio, G. & Raiteri, C. M. 1990, ApJ, 354,
       630
Iben, I. Jr. 1975, ApJ, 196, 525
Ireland, T. R., Zinner, E. & Amari, S. 1991, ApJ, 376, L53
Lamb, S. A., Howard, W. M., Truran, J. W. & Iben, I. Jr. 1977, ApJ, 217, 213
Lambert, D. L. 1989, in Evolution of Peculiar Red Giants, ed. H. R. Johnson & B.
       Zuckermann (Cambridge Univ. Press: Cambridge), 101
Lewis, R. S., Amari, S. & Anders, E. 1990, Nature, 348, 293
Lewis, R. S., Amari, S. & Anders, E. 1992, Geochim. Cosmochim. Acta, submitted
Ott, U. & Begemann, F. 1990a, ApJ, 353, L57
Ott, U. & Begemann, F. 1990b, Lun. Planet. Sci. XXI, 920
Ott, U., Begemann, F., Yang, J. & Epstein, S. 1988, Nature, 332, 700
Prombo, C. A., Podosek, F. A., Amari, S. & Lewis, R. S. 1992a, Lunar Planet. Sci.
       XXIII, 1111
Prombo, C. A., Podosek, F. A., Amari, S. & Lewis, R. S. 1992b, ApJ, submitted
Reeves, H. 1966, ApJ 146, 447
Reeves, H. 1968, Stellar Evolution and Nucleosynthesis, (New York: Gordon and
       Breach)
Richter, S., Ott, U. & Begemann, F. 1992, Lunar Planet. Sci. XXIII, 1147
Smith, V. V. & Lambert, D. L. 1990 ApJ Suppl. 72, 387
Stone, J., Hutcheon, I. D., Epstein, S. & Wasserburg, G. J. 1991, Earth Planet. Sci.
       Lett. 107, 570
Tang, M., Anders, E., Hoppe, P. & Zinner, E. 1989, Nature, 339, 351
Truran, J. W. & Iben, I. Jr. 1977, ApJ, 216, 797
Wagemans, C., D'Hondt, P. & Brissot, R. 1992, in Nuclei in the Cosmos (F. Käppeler,
       ed.), in press
Zinner, E., Amari, S. & Lewis, R. S. 1991, ApJ, 382, L47
Zinner, E., Tang, M. & Anders, E. 1987, Nature, 330, 730
Zinner, E., Tang, M. & Anders, E. 1989, Geochim. Cosmochim. Acta, 53, 3273

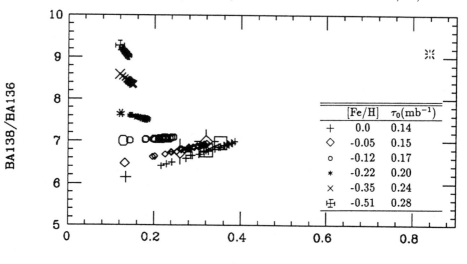

Study of Heavy Element Abundances in SiC Grains

W.M. HOWARD, M. ARNOULD

Institut d'Astronomie et d'Astrophysique
Universite Libre de Bruxelles
Brussels, Belgium

1 INTRODUCTION

Recent determinations of the isotopic composition of a series of heavy elements in small-sized (micron) SiC grains have provided nuclear astrophysics with a wealth of compositional clues as to the astrophysical origin of these grains. In particular, the isotopic compositions of Si, Ca, Ti, Sr, Kr, Xe, Ba, Nd, and Sm (Lewis, Amari and Anders, 1990; Amari, Lewis and Anders, 1992; for example) have been measured and exhibit significant deviations from their solar-system counterparts. In particular, the elements heavier than iron generally feature a depletion of the p-isotopes and an enrichment of the s-isotopes.

Although various astrophysical sources of these anomalous compositions have been proposed, e.g. Asymptotic Giant Branch (AGB) or Wolf-Rayet stars, *no consistent picture has emerged.* This is due in particular to the failure of existing AGB models to provide enough neutrons in order to account for the s-process enhancements observed at the surface of various types of chemically peculiar AGB stars. The amount of neutrons in calculations of s-process nucleosynthesis in such stars remains a *completely parametrized quantity.*

We take the point of view of studying a completely parametrized model in order to investigate in as a broad way as possible the neutron exposures or ranges of such exposures that could replicate the series of measured isotopic compositions without relying on any specific (and highly uncertain) astrophysical scenario. This study is in the same spirit in which one has studied the s-process, as well as the r- and p-processes with parametrized models in order to understand the bulk solar system composition. We find that we can reproduce the general trends of the isotopic data for the elements from krypton through samarium with a rather simple neutron exposure history. We also show how one can relax the classical assumption of a two component mixing curve. We interpret the linear correlations in the three-isotope plots as a mixture of components with slightly different neutron exposures.

2 THE STANDARD MODEL

The neutron captures are calculated by a method described earlier (Howard et al., 1986) where we use the latest compilation of neutron capture rates (Beer et al., 1992) and of the temperature and density dependent beta-decay rates. For the initial composition we take the solar composition (Anders and Grevesse, 1989). We define a standard model by the neutron exposure $\tau = 0.006 mb^{-1}$, a neutron exposure timescale of 5×10^6 s and a temperature $kT = 30$ keV, this implying a neutron density of $5 \times 10^9 cm^{-3}$. This model defines a standard set of abundances. The small neutron exposure ensures that the initial solar composition is only slightly modified, with the isotopes with the largest capture cross sections being modified the most. A series of computations have been performed with different τ values and initial compositions in order to analyze the sensitivity of the derived abundances to changes in these quantities.

It has to be emphasized that the material having experienced the considered low neutron exposure is not assumed to be mixed with a reservoir of solar composition before incorporated in grains. This is also in marked contrast with previously proposed models.

3 RESULTS

We find that the small neutron exposure characterizing the standard model replicates many of the features of the abundance pattern for the elements Kr, Sr, Xe, Ba, Nd, and Sm found in the KJ series of SiC grains in Murchison. Table 1 compares the solar-system values, the SiC values (Prombo et al., 1992) and our standard model results for Ba. The p-nuclei ^{130}Ba and ^{132}Ba are severely depleted because of their

Table 1: BARIUM

Ratio	Solar system	SiC value	Standard model
^{130}Ba/^{136}Ba	0.013	0.003	0.00076
^{132}Ba/^{136}Ba	0.013	0.003	0.0044
^{134}Ba/^{136}Ba	0.31	0.34	0.37
^{135}Ba/^{136}Ba	0.84	0.31	0.24
^{137}Ba/^{136}Ba	1.4	0.87	0.85
^{138}Ba/^{136}Ba	9.1	6.9	5.3

relatively large neutron capture cross sections, while ^{134}Ba and ^{136}Ba are enhanced by a factor of approximately two. The ^{135}Ba and ^{137}Ba isotopes are less enhanced because of their larger capture cross sections, and ^{138}Ba is only slightly modified because of its small capture cross section at the $N = 82$ closed neutron shell. We do not include the contributions from the ^{135}Cs and ^{137}Cs decays.

New results on neodymium and samarium compositions in SiC grains are presented at this conference (Ott, Richer and Begemann, 1992). Table 2 compares our standard model prediction for neodymium with these results. As in the case of Ba, the odd-mass isotopes of Nd are much more depleted than the corresponding even-mass ones. The standard model predicts a depletion of ^{143}Nd and ^{145}Nd, and a slight enrichment of ^{142}Nd, ^{144}Nd, ^{146}Nd and ^{148}Nd. In contrast to these even-mass isotopes, ^{150}Nd is seen to be depleted. This results from its rather large neutron capture cross section and, more importantly, from the fact that is does not lie on the standard model neutron capture path, in contrast to the other even-mass isotopes. As

Table 2: NEODYMIUM

Ratio	Solar System	SiC Value	Standard Model
^{142}Nd/^{144}Nd	1.14	1.71	1.12
^{143}Nd/^{144}Nd	0.508	0.38	0.35
^{145}Nd/^{144}Nd	0.349	0.24	0.29
^{146}Nd/^{144}Nd	0.721	0.76	0.76
^{148}Nd/^{144}Nd	0.242	0.12	0.27
^{150}Nd/^{144}Nd	0.237	0.095	0.08

seen in Table 2, these predictions are able to reproduce quite nicely the observed ^{143}Nd/^{144}Nd, ^{145}Nd/^{144}Nd and ^{146}Nd/^{144}Nd ratios. In contrast we cannot account for the near identical values of ^{148}Nd/^{144}Nd and ^{150}Nd/^{144}Nd, the first one being overestimated, while the second is underestimated for the reasons indicated above. Due to a s-process branching at ^{147}Nd, this situation depends quite sensitively on the adopted neutron density. Lowering the neutron density by a factor of two reduces the ^{148}Nd abundance and increases the ^{150}Nd abundance without substantially altering the abundances of the other Nd isotopes.

Furthermore, many of the linear trends found in the three-isotope plots consturcted from the esperimental data can be reproduced with a variation of neutron exposure about our standard model conditions.

For an astrophysical site, we suggest a star that has undergone a "last pulse" during its post-planetary nebula phase. It is believed that such a star has been observed (Langer, Kraft and Anderson, 1974) and is enriched in s-process elements. That star has presumably suffered significant mass loss prior to the s-processing, so that dilution effects are minimized.

4 CONCLUSIONS

We have presented a simple model calculation that replicates many of the isotopic abundance features that are found in SiC grains. This model is *quite different* from other parametrized models that also replicates many of the abundance patterns. The requirements of our model are: 1) s small but relatively rapid neutron exposure, 2) an initial solar-system like seed distribution of heavy elements, 3) a mixing of the products of such an exposure with a limited reservoir and finally, 4) an environment favorable for the formation of SiC grains. We speculate that a low- or intermediate-mass star experiencing a last pulse during its post-planetary nebular phase could provide an adequate astrophysical framework for our model, at least if those objects can indeed produce SiC grains. This possibility cannot be assertained or disproved by present observations.

A more complete description of the model presented here and of its predictions will be presented elsewhere (Arnould and Howard, 1992).

One of us (WMH) thanks the warm hospitality of the Institut d'Astronomie et d'Astrophysique of the Universite Libre de Bruxelles, and the financial support of the Fonds Bational Belge de la Recherche Scientifique (F.N.R.S.) during his sabbatical stay. He (WMH) is also grateful to the Institut Medical Edith Cavell, where an *unexpected visit* allowed some time for manuscript preparation. This work has been supported by the SCIENCE Programme SCI-0065 and the US Department of Energy through contract with the Lawrence Livermore National Laboratory. M. A. is Chercheur Qualifie F.N.R.S. (Belgium).

REFERENCES

Amari, S., Lewis, R. S. and Anders, E. (1992) Interstellar Grains in Meteorites. II. SiC and its nobel gases, preprint.

Anders, E. and Grevesse, N. (1989) Geochim. Cosmochem. Acta *53*, 197-214.

Arnould M. and Howard, W. M. (1992) Parametric Study of Heavy Element Isotopic Distributions in SiC Grains, submitted to Meteoritics.

Beer, H., Voss, F. and Winters, R. R. (1992) Ap. J. Suppl., *80*, 403.

Howard, W.M., Mathews, G. J., Takahashi, K. and Ward, R. A. (1986) Astrophys. J., *309*, 633-652.

Langer, G. E., Kraft, R. P. and Anderson, K. S. (1974) Astrophys. J., *189*, 509-521.

Lewis, R. S., Amari, S. and Anders, E. (1990) Nature, *348*, 293- 298.

Ott, U., Richter, S. and Begemann, F. (1992) S-process nucleosynthesis: the evidence from meteorites, this proceedings.

Prombo, C. A., Podosek, F. A., Amari, S. and Lewis, R. S. (1992) LPSC XXIII, to be published; also Astrophys. J., submitted.

Wolf-Rayet Stars as Generators of ^{26}Al

G. MEYNET[1], M. ARNOULD[2]

1) Geneva Observatory
2) Institut d'Astronomie et d'Astrophysique, Université Libre de Bruxelles

1 INTRODUCTION

It is widely recognized today that the radionuclide ^{26}Al ($t_{1/2} = 7.05\ 10^5$ yr) is of crucial importance in γ-ray astronomy and in cosmochemistry. On one hand, the 1.8 Mev γ-ray emission observed in the galactic disk is attributed to the decay of $\sim 3M_\odot$ of ^{26}Al that have been present in the interstellar medium (ISM) over the last $\sim 10^6$ yr (*e.g.* Schönfelder and Varendorff 1991 ; Clayton and Leising 1987 for reviews). On the other hand, there is now ample observational evidence that ^{26}Al has decayed in situ in various meteoritic inclusions, as well as in identified single grains of likely stellar origin (*e.g.* Wasserburg 1985 ; Zinner et al. 1991 ; Amari et al. 1992).

Various potential ^{26}Al production sites have been suggested and studied in more or less great details (*e.g.* Prantzos 1991 ; Forestini et al. 1991 ; Starrfield et al., this volume). In particular, Wolf-Rayet (WR) stars have been considered as possible ^{26}Al contaminating agents of some at least of the meteoritic material mentioned above (*e.g.* Arnould and Prantzos 1986). This is well in line with the "BING BANG" model promoted by Reeves (1978). The fact that certain WR stars are known to make dust episodically in their winds (Williams et al. 1992) may also be of interest in that respect. On the other hand, WR stars could be responsible for part at least of the present-day ISM $3M_\odot$ of ^{26}Al inferred from γ-ray observations. In fact, following Prantzos (1991), those objects could have contributed for up to about $0.5M_\odot$. This estimate relies on a modest nuclear physics update of previously computed WR models (Prantzos et al. 1986), and on a qualitative evaluation of stellar metallicity effects.

The aim of this paper is to revisit the ^{26}Al WR yields on grounds of detailed evolutionary models that incorporate extended nuclear reaction networks, as well as recent improvements in our knowledge of various basic ingredients, like mass loss rates, opacities, or nuclear reaction rates. Metallicity effects on the ^{26}Al yields are also studied quantitatively for the first time.

Section 2 presents the investigated stars, and the basic physical ingredients of the models. Our results are put in perspective with γ-ray astronomy and cosmochemistery in Sects. 3 and 4. Some conclusions are drawn in Sect. 5.

2 INVESTIGATED CASES AND BASIC MODEL INGREDIENTS

Three evolutionary sequences for a $60 M_\odot$ star are computed from the ZAMS to the end of the carbon burning phase with different physical ingredients (see below). Apart from the nuclear physics input, the stellar evolutionary models are computed as described by Schaller et al. (1992). In particular, the new radiative opacities of Rogers and Iglesias (1992) are used. On the other hand, the considered mass loss rates are such that Pop I stars with initial masses in excess of about $30 M_\odot$ end their lives with masses between 5 and $10 M_\odot$ (Maeder 1990). Finally, the nuclear reaction network of Schaller et al. (1992) is extended in order to include the Ne-Na and Mg-Al H-burning modes, as well as all the important neutron sources and sinks up to ^{27}Al that can operate during He-burning, the neutron poisoning by the heavier nuclides being approximated by an effective neutron sink (e.g. Jorissen and Arnould 1989). The rates of the reactions involved in the Ne-Na and Mg-Al chains are the same as those adopted by Forestini et al. (1991) at the exception of the reactions $^{25}Mg(p,\gamma)^{26}Al$, and $^{26}Mg(p,\gamma)^{27}Al$ for which the rates determined experimentally by Illiadis et al. (1990; IL90) are selected. The adopted rates of $^{13}C(\alpha,n)^{16}O$, $^{22}Ne(\alpha,n)^{25}Mg$ and $^{22}Ne(\alpha,\gamma)^{26}Mg$ that are important for the neutron budget during He-burning are those briefly discussed by Arnould and Takahashi (this volume), while the neutron capture rates are taken fron Beer et al. (1992).

3 ^{26}Al AND THE γ-RAY ASTRONOMY CONNECTION

The time integrated mass of ^{26}Al ejected by the wind of a WR star is given by

$$M_{26}^W = \int_0^t X_{26}^s(t')|\dot{M}(t')|dt'$$

where t is the total stellar lifetime, X_{26}^s the surface ^{26}Al mass fraction, and $\dot{M}$ the mass loss rate. This expression neglects the ^{26}Al β-decay in the calculation of the ^{26}Al wind content at a given time. This decay can be trivially taken into account. Table 1 gives M_{26}^W for our WR models and provides a comparison with previous estimates. From columns 1 and 5, it appears that our new stellar models approximately double the ejected mass of ^{26}Al that is predicted by previous calculations for a $Z = 0.02$ $60 M_\odot$ WR star. The contribution of WR stars to the present day galactic ^{26}Al could thus be larger than estimated previously.

From an inspection of columns 1 and 2 of Table 1, it appears that most of that change is due to our use of the IL90 rate instead of the one recommended by Caughlan and Fowler (1988, CF88) for $^{25}Mg(p,\gamma)^{26}Al$. This result may appear surprising, as the IL90 rate for that reaction is about 5 times lower than the CF88 value at temperatures $T < 8\ 10^7$ K. In fact, the lowering of the $^{25}Mg(p,\gamma)^{26}Al$ rate has the effect of producing ^{26}Al later in the evolution, so that ^{26}Al has less time to β-decay before being wind ejected.

An increase of ^{26}Al yields from WR stars is also expected to result from higher metallicities. Indeed, 1) the abundance of ^{25}Mg, progenitor of ^{26}Al, is proportional to the metallicity, and, 2) the mass loss rate increases. Walter and Maeder (1989) use simple arguments to derive a $(Z/Z_\odot)^2$ dependence of the mass of ejected ^{26}Al. Our two $Z = 0.02$ and $Z = 0.03$ $60M_\odot$ model stars (columns 2 and 3 of Table 1) indicate that the ejected mass varies like $(Z/Z_\odot)^{2.26}$. The possible contribution of WR stars to the present ^{26}Al content of the Galaxy is correspondingly larger than expected previously

Table 1

Values of M_{26}^W for $60M_\odot$ WR stars, as predicted by different models. IL90 and CF88 indicate that the rates of $^{25}Mg(p,\gamma)^{26}Al$, and $^{26}Mg(p,\gamma)^{27}Al$ are taken from Iliadis et al. (1990) and Caughlan and Fowler (1988).

present work Z = 0.02 (IL90)	present work Z = 0.02 (CF88)	present work Z = 0.03 (CF88)	Prantzos, 1990 Z=0.03 (IL90)	Walter & Maeder, 1989 Z = 0.02 (CF88)
$7.5\ 10^{-5}$ M$_\odot$	$4.0\ 10^{-5}$ M$_\odot$	$10.0\ 10^{-5}$ M$_\odot$	$4.7\ 10^{-5}$ M$_\odot$	$3.0\ 10^{-5}$ M$_\odot$

4. ^{26}Al AND THE METEORITIC CONNECTION

The $^{26}Al/^{27}Al$ ratios predicted by our models range between 2 and $5\ 10^{-2}$. These values are much higher than the "canonical" ratio of about $5\ 10^{-5}$ found in certain Ca-Al-rich meteoritic inclusions (Wasserburg 1985). They also clearly exceed the ratios ranging up to about 10^{-2} measured in most of the SiC grains from Murchison (Zinner et al. 1991 ; Amari et al. 1992). However, the family of so-called "X-grains" exhibits ratios in the approximate 0.1-1 range, with accompanying $^{12}C/^{13}C$ ratios varying from about 100 to 3000 (Amari et al. 1992). While our WR models can account for such C isotopic ratios, they fall somewhat short of explaining the corresponding exceptionally high $^{26}Al/^{27}Al$ ratios.

5. CONCLUSION

The main results of this work may be summarized as follows :

1) Our WR models predict amounts of ^{26}Al ejected into the ISM that are about two times larger than the quantities derived from previous models ;

2) The winds of WR stars are predicted to be characterized by $^{26}Al/^{27}Al$ ratios that are higher than in most of the SiC grains from Murchison. They are also larger than the ratios derived from AGB model stars (Forestini et al. 1991). Even so, the very high $^{26}Al/^{27}Al$ ratios measured in the series of X-grains cannot be fully explained by our WR models. In fact, the highest $^{26}Al/^{27}Al$ ratios could perhaps be accounted for by nova models only, as first predicted quantitatively by Arnould

et al. (1980) and Vangioni-Flam et al. (1980) for outbursts on CO white dwarfs. As shown more recently (Nofar et al. 1991 ; Starrfield et al. this volume), the same might be true for nova outbursts on $ONeMg$ white dwarfs. However, the C isotopic compostions emerging from novae are incompatible with those characterizing the X-grains. Our WR models certainly succeed in this respect. Obviously, much more work remains to be done before being able to assign unambiguously a stellar site to grains carrying very high $^{26}Al/^{27}Al$ ratios.

Acknowledgements. This work has been support in part by the SCIENCE Program SC1-0065. M.A. is Chercheur Qualifié F.N.R.S. (Belgium).

REFERENCES

Amari, S., Hoppe, P., Zinner, E., Lewis, R.S., 1992, Lunar Planetary Science XXIII

Arnould, M., Nørgaard, H., Thielemann, F.-K., Hillebrandt, W., 1980, ApJ 237, 931

Arnould, M., Prantzos, N., 1986, in *Nucleosynthesis and its implications on nuclear and particle physics*, eds. J. Audouze and N. Mathieu (Reidel, Dordrecht), p. 363

Beer, H., Voss, F., Winters, R.R., 1992, ApJS 80, 403

Caughlan, G.R., Fowler, W.A., 1988, Atomic Data Nuc. Data Tables 40, 283

Clayton, D.D., Leising, M.D., 1987, Phys. Rep. 144, 1

Forestini, M., Paulus, G., Arnould, M., 1991, A&A 252, 597

Iliadis, Ch., Schange, Th., Rolfs, C., Schröder, U., Somorjai, E., Trautvetter, H.P., Wolke, K., Endt, P.M., Kikstra, S.W., Champagne, A.E., Arnould, M., Paulus, G., 1990, Nucl. Phys. A512, 509

Jorissen, A., Arnould, M., 1989, A&A 221,161

Maeder A., 1990, A&AS 84, 139

Nofar, I., Shaviv, G., Starrfield, S., 1991, ApJ 369, 440

Prantzos, N., 1991, in *Gamma-ray line astrophysics*, eds. Ph. Durouchoux and N. Prantzos (AIP, New York), p. 129

Prantzos, N., Doom, C., Arnould, M., de Loore, C., 1986, ApJ 304, 695

Reeves, H., 1978, in *Protostars and Planets*, ed. T. Gehrels (University of Arizona Press, Tucson), p. 399

Rogers, F.J., Iglesias, C.A., 1992, ApJS 79, 507

Schaller G., Schaerer D., Meynet G., Maeder A., 1992, A&AS, in press

Schönfelder, V., Varendorff, M., 1991, in *Gamma-ray line astrophysics*, eds. Ph. Durouchoux and N. Prantzos (AIP, New York), p. 101

Vangioni-Flam, E., Audouze, J., Chièze, J.-P., 1980, A&A 82, 234

Walter R., Maeder A., 1989, A&A 218, 123

Wasserburg, G.J., 1985, in *Protostars and Planets II*, eds. D.C. Black and M.S. Matthews, (University of Arizona Press, Tucson) p. 703

Williams, P.M., van der Hucht, K.A., Bouchet, P., Spoelstra, T.A.Th., Eenens, P.R.J., Geballe, T.R., Kidger, M.R., Churchwell, E., 1992, MNRAS, to appear

Zinner, E., Amari, S., Anders, E., Lewis, R., 1991, Nature 349, 51

Closing Remarks

E. SALPETER

Cornell University, Ithaca, New York

"Birthday party" Symposia are always fun, but often suffer from having disconnected talks on many different subjects. The three musketeers of our organizing committee have done an unusually good job in organizing complimentary talks on related topics, so we got a real feeling for the remaining controversies amongst the experts, in addition to impressive results. I would not dare an attempt at conclusions for these closing remarks, but even that is all to the good: Having had such a successful symposium for Hubert's 60th birthday, we certainly want another one for the 70th and part of that will provide conclusions for this year's topics. Instead, I will give here an introduction to what might be one of the new topics.

Hubert mentioned that some topics on which there was progress in the eighties already were "in the air" around 1960. One topic was already "in the air" around 1950, but progress has been very slow: When Walter Baade delineated the two stellar populations, he already wondered whether the mix of stars produced (what we would call today the initial mass function, IMF) depended on environmental influences such as kinematics, but also on the metal abundance in the interstellar medium. There is now a possibility that an additional "stellar population III" makes up the extended dark matter halo which surrounds the population II halo which Baade already knew. There is also a possibility that spiral galaxies like ours have very extended, although tenuous, disks of (largely ionized, but partly neutral) gas which provide at least some of the socalled "Lyman-α clouds". Since there is likely to have been no star formation at all in these outermost disks, element abundances found via absorption lines there might give a rather clean picture of supernova debris which escaped from the population III halo.

The dark matter halo is known to extend to much larger radii than the visible stellar population I disk and population II sphere. As discussed earlier in this symposium, primordeal nucleosynthesis would allow these halos to be all baryonic out to radii of order 100 kpc (even more according to some estimates). The matter need not be baryonic, but if it is, it most likely is in the form of low-mass condensed objects (brown dwarfs, planets, asteroids and rocks) and/or neutron stars. The initial mass function for such a "stellar population III" must then have been extremely different from the standard IMF to have excluded all long-lived, hydrogen-burning visible stars (0.1 to $1.5M_\odot$, say). If one accepts this rather drastic requirement, it may also provide a location for most of the observed gamma ray bursts: These are thought to involve some form of energy release from neutron star surfaces and at least one of two interpretations of the data favors a very extended spherical halo for our Galaxy. One version of such models, which Ira Wasserman and I favor, releases the energy by asteroids accreting onto neutron stars. Such models would be possible if the non-standard IMF was brought about by a rather drastic form of "bimodal star formation", which makes stars of both very low and very large mass:

I feel that one will have to make rather drastic assumptions regarding kinematics and time-history, not just regarding the IMF, for a successful model of stellar population III. Although population II stars were formed earlier than population I, it would not work to merely form population III even earlier. As we heard in Cayrel's review talk: (a) the IMF must have been pretty standard at the beginning of the population II era, (b) there could have been very little metal contamination then from even earlier times and (c) there seems to have been a kinematic discontinuity between populations II and I (changing from little to much specific angular momentum). For quite different reasons I will argue below for population III stars in dense clusters with large velocity dispersion, which leads to the following scenario – just as one example of the complexity that will be required:

Imagine a proto-galaxy (plus proto-halo) made up entirely of gas clouds with the ensemble of clouds roughly an isothermal spheroid of radius ~100 kpc, velocity dispersion ~200 kms^{-1} plus a mean rotational velocity ~40 kms^{-1}. Consider first a small portion (say, ~ 1%) of the Maxwell-Boltzmann ensemble (with the small angular momentum displacement) of clouds which have small absolute values for the two non-radial components of their velocity (relative to the galaxy center). If their radial velocity component is inward (but otherwise arbitrary) this small fraction of the clouds will then fall towards the center in an almost straight line with angular momentum fairly small individually and an even smaller average value. The further kinematics depends on the sizes of the clouds, but it is hoped that violent relaxation could lead to the present-day population II kinematics, as suggested by Lynden-Bell, Eggen and Sandage a long time ago, and that these stars formed rapidly after cloud-cloud collisions. Imagine the following further dichotomy for the remaining 99% of the clouds. A small fraction (say, 10%), especially those at fairly small radial distances, undergo cloud-cloud collisions early, the gas is heated, cools radiatively, the orbits shrink and a fairly smooth rotating gas disk forms. As in Baade's days, one has to assume that star formation is slow and on-going in the disk to form population I (and intermediate population) eventually. The remaining 90% of the clouds do not collide rapidly and we assume that star formation inside an intact cloud is relatively fast and, partly because of the high internal velocity dispersion, has a peculiar IMF history. In the absence of collisional dissipation, this stellar population III extends out to the large initial radius (say, 100 kpc) but (in spite of the numerical label) the stars were formed after population II and before population I.

The high velocity dispersion in a cloud that will become a star cluster is both the excuse for a weird IMF but also makes escape from the cluster difficult. As one example of weirdness, assume that mostly massive stars form, and result in supernovae, until a large abundance of carbon and heavier elements has built up. Assume further that high metal abundance leads to a switch to mainly forming low mass objects, including asteroids. If the cluster mass is $\sim 10^7 M_\odot$ and $\sim 10^6$ massive stars are produced for a few times 10^8 years, most of the supernova blast waves dissipate and are contained inside the cluster, so that a large fraction of the metal-rich gas remains to make the low-mass stars later on. Nevertheless, those supernovae which occur very close to the cluster surface can have (half of) their blast wave exceed escape velocity, so that a few percent of the metal-rich gas escapes and fills that halo. Asteroid/neutron star collisions producing gamma-ray bursts is one observable consequence but the enriched halo gas is another, if the following conjecture on extended gaseous galaxy disks is true.

Spiral galaxies generally have neutral hydrogen disks with column densities of order $10^{21} cm^{-2}$ with radii only slightly larger than the stellar galactic disks (say, $\sim$20 kpc). Nevertheless, it is now likely that many spiral galaxies have extended disks of much lower column density but extending to radii of $\sim$100 kpc or more. There are three arguments for this conjecture. (a) There are now a few disk galaxies known where the neutral hydrogen column density falls off very slowly, from $10^{20} cm^{-2}$ to $10^{19} cm^{-2}$, say, over a radial distance $\sim$100 kpc; (b) theoretically, there should be a transition from neutral to ionized hydrogen at a column density of order $10^{19} cm^{-2}$, so that most really extended disks are not visible in HI emission but could still have appreciable (say, $\sim 10^{18} cm^{-2}$) HII column densities, which could explain (c) the few "Lyman-α clouds" which have already been found by the Hubble space telescope (HST), associated with ordinary, nearby galaxies but $\sim$100 kpc away. These extended disks are likely to have acted as "filter paper" to catch some of the nuclear debris from supernovae in the stellar population III halo. In ten years time there should be more HST data on Ly-α absorption, GRO data on gamma-ray bursts, MACHO data on brown dwarfs in the halo or their absence and possibly MACHO or LIGO data on neutron stars.

I finally want to thank the Institute for the gracious and low-key hospitality, the organizing committee for the fine program and, especially, Hubert Reeves for having inspired so many of the speakers as well as the committee.